Advances in Industrial Control

For further volumes:
www.springer.com/series/1412

Mohieddine Jelali

Control Performance Management in Industrial Automation

Assessment, Diagnosis and Improvement of Control Loop Performance

Springer

Dr.-Ing. Mohieddine Jelali
Faculty of Plant, Energy and Machine
Systems
Cologne University of Applied Sciences
Koeln, Germany

ISSN 1430-9491 ISSN 2193-1577 (electronic)
Advances in Industrial Control
ISBN 978-1-4471-4545-5 ISBN 978-1-4471-4546-2 (eBook)
DOI 10.1007/978-1-4471-4546-2
Springer London Heidelberg New York Dordrecht

Library of Congress Control Number: 2012951835

Printed on acid-free paper

Springer is part of Springer Science+Business Media (www.springer.com)

Für Doris, Yasmin und Dunja

Series Editors' Foreword

The series *Advances in Industrial Control* aims to report and encourage technology transfer in control engineering. The rapid development of control technology has an impact on all areas of the control discipline. New theory, new controllers, actuators, sensors, new industrial processes, computer methods, new applications, new philosophies..., new challenges. Much of this development work resides in industrial reports, feasibility study papers and the reports of advanced collaborative projects. The series offers an opportunity for researchers to present an extended exposition of such new work in all aspects of industrial control for wider and rapid dissemination.

Professor Thomas J. Harris's 1989 paper on the "Assessment of Closed-loop Performance" demonstrated how to benchmark the performance of commissioned controllers using industrial plant data without the need for intrusive control experiments on production plant. This paper appeared against a background of control system audits that showed a large proportion of control loops in many industries to be performing poorly and capable of much better performance. The causes of poor performance include poorly tuned controllers, controller stiction, and other problems. From 1989, and building on the foundation of Professor Harris's work, many new controller assessment techniques were devised and added to the published literature. Other academics and engineers began to look more precisely at the causes of poor control thus broadening the scope and analysis of the issues that control assessment techniques should address.

As Editors of the *Advances in Industrial Control* monograph series, we were keen to have a small corpus of influential volumes in the series on this growing and important field in industrial control. In consequence, the series contains the monographs:

- *Performance Assessment of Control Loops* by Biao Huang and Sirish L. Shah (ISBN 978-1-85233-639-4, 1999).
- *Process Control Performance Assessment: From Theory to Implementation* by Andrzej W. Ordys, Damien Uduehi and Michael A. Johnson (Eds.) (ISBN 978-1-84628-623-0, 2007).

- *Diagnosis of Process Nonlinearities and Valve Stiction: Data-Driven Approaches* by Md. Ali Ahammad Shoukat Choudhury, Sirish L. Shah and Nina F. Thornhill (ISBN 978-3-540-79223-9, 2008).
- *Detection and Diagnosis of Stiction in Control Loops: State of the Art and Advanced Methods* by Mohieddine Jelali and Biao Huang (Eds.) (ISBN 978-1-84882-774-5, 2010).

These influential monographs remain very relevant to the field of controller performance assessment studies and are still available.

Now, just over two decades on from Professor Harris's initial impetus there is an extensive body of controller assessment theory and methods in the archival published literature. Thus, it is an apposite time to have this work documented and described as a fully formed industrial science. We are therefore fortunate to welcome into the series Mohieddine Jelali's monograph, *Control Performance Management in Industrial Automation: Assessment, Diagnosis, and Improvement of Control-Loop Performance*. It is a comprehensive volume that covers the field in four parts: techniques for controller performance assessment, detection, and diagnosis of controller performance problems, techniques for controller improvement and closes with industrial applications and tools. Also useful is the reference section that archives the details of over 480 contributions to the extensive literature of the field.

Dr. Jelali's monograph shows that the science of controller performance assessment is well founded and thriving. However, it is interesting to reflect where future developments in the field might arise. One perennially vexing topic is that of establishing the relationship between investing in control performance management (and more generally, control systems per se) and economic benefit to the company enterprise. There is still a dearth of analysis and methods in this direction. In a similar vein, a salutary Chap. 15 in the monograph looks at industrial applications and the take-up of the ideas and methods presented. Dr. Jelali gives it as his opinion that only three commercial software products comprehensively cover the concepts and algorithms of control performance management found in the monograph, and goes on to say that of the list of ten or so packages described "only a small proportion of the methods found in this book can be found in these packages". Clearly, there is a wealth of ideas, methods, and algorithms collated in this monograph that are available for use by the industrial control and process engineer; the potential is there. In this way, Dr. Jelali's monograph meets one of the key objectives of the *Advances in Industrial Control* series; that of promoting new ideas and techniques that can help the industrial engineer enhance the performance of control systems technology.

Industrial Control Centre,
Glasgow, Scotland, UK

M.J. Grimble
M.A. Johnson

Preface

Aim of the Book

Control loops are the most important components in automation systems. Product quality, operation safety, material and energy consumption, and thus the financial performance, are directly or indirectly linked to the performance of control systems. To achieve, restore and sustain top performance of control loops is thus a vital interest for any company. Since process control systems are very complex, usually comprising different hierarchical levels, it is hopeless to maintain them on regular basis by plant personal. This is also the main reason why a large portion of industrial control loops has significant performance problems, as found out by audits carried out regularly since the 1990s. All these factors have contributed to the growth of the control performance monitoring (CPM) technology and applications in the last decade. CPM provides a framework for automatically and systematically assessing the performance of control loops, detecting and diagnosing root-causes of poor performance, as well as suggesting measures to improve control performance or avoid performance degradation. At this point, it is stressed that any CPM technique should be *non-invasive* to be accepted in industrial practice. That is, the analysis should always be carried out *based on only routine operating data with limited or no additional process knowledge and without the need for any experimentation with the plant, not even in closed loop.*

Since the key research by Harris (1989), the control community has developed numerous methods that focus on the performance assessment of control loops. At the centre of these approaches is the concept of minimum variance benchmarking and different modifications or extensions, which have attracted much attention. Meanwhile, there is a large number of techniques for basic and advanced performance assessment, detection and diagnosis of different sources of poor performance (bad controller tuning, process nonlinearities, oscillations, etc.). Usually, emphasis is placed on single techniques to detect special performance problems or plant faults.

This is the first monograph that deals with the *complete CPM technology*, from *controller assessment* (minimum-variance-control-based and advanced methods), over *detection and diagnosis of control loop problems* (process nonlinearities, oscillations, actuator faults), to the *improvement of control performance* (maintenance, re-design of loop components, automatic controller re-tuning). It provides a contribution towards the development and application of completely *self-contained and automatic* methodologies in the field. Moreover, within this work, many CPM tools have been developed that go far beyond available CPM packages. Industrial data from a large number of control loops in different industrial fields (building, chemicals, mining, mineral and metal processing) have been used to demonstrate the presented strategies and methods. Systematic procedures for automatic and continuous control performance monitoring, maintenance and optimisation have been recommended, combining different control performance metrics and assessment, diagnosis and improvement methods. The main objective is to sustain top control performance during the whole life cycle of the control system, despite different and changing operational conditions.

Contributions of the Book

In the light of the *control-system life-cycle management* introduced above, the first contribution of our work is to provide strategies and methods for establishing a new practice of *integrated control design and performance supervision* of technical processes, as illustrated in Fig. 1.

The focus of this book is only on the last three stages, i.e., performance monitoring, condition monitoring and diagnosis, and CPM-based controller re-tuning. The book is however NOT on control system design or implementation, which is the topic of many standard texts. To the knowledge of the author, all major aspects of CPM, from *controller assessment* (minimum-variance-control-based and advanced methods), over *detection and diagnosis of control loop problems* (process non-linearities, oscillations, actuator faults), to the *improvement of control performance* (maintenance, re-design of loop components, automatic controller re-tuning) are treated for the first time in this work from a common viewpoint.

More specifically, the contributions and messages of the present book are stated as follows:

1. *Review, Evaluation and Industrial Application of Available Methods and Systems.* A comprehensive and critical review of the current status in the complete CPM technology, including techniques for performance assessment, diagnosis and improvement. Standard methods and advanced new methods are presented as well. This provides an insight into the assumptions and fundamental limitations of each control performance assessment method, since each of the various techniques signifies specific information about the nature of the process. Most CPM algorithms presented in the book have been implemented and tested by the author in MATLAB/Simulink.
   An evaluation of publications during the 15 years after the key work by Harris (1989) shows some trends in the application of a number of control performance metrics and methods in different process industries. Also included is an

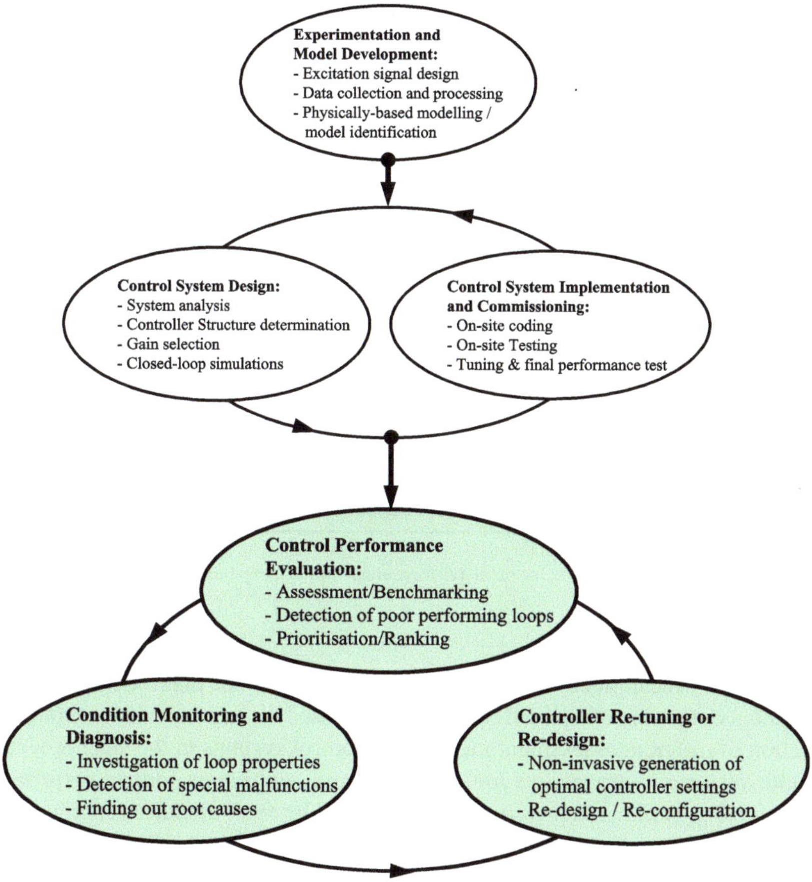

Fig. 1 Flow diagram of the process control design and supervision procedure

overview of CPM packages that have been developed and/or which are commercially available. Merits and drawbacks of the methods are also highlighted. Some control benchmarking techniques are compared in terms of parameters/data requirements and performance to draw guidelines for the choice of the most suitable technique or combination of techniques.
The majority of the methods presented are illustrated with real data from industrial control loops from different industrial fields, including chemicals, petrochemicals, pulp & paper plants, commercial buildings, power plants, mineral processing, mining and metal processing. Some information about these loops is given in Appendix C.

2. *Improved and New Methods for Control Performance Assessment and Diagnosis*. It is not sufficient and often dangerous to rely on a single index, or a single performance analysis method, by itself for performance monitoring and diagnosis, as each criterion has its merits and limitations. The best results are often

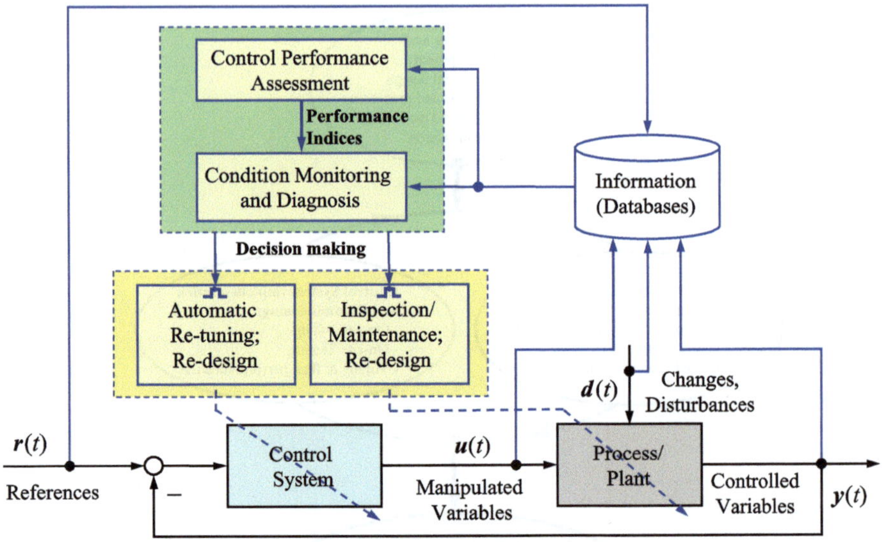

Fig. 2 Proposed framework for control-performance-monitoring-based optimisation of control loops

obtained by the collective application of several methods that reflect control performance measures from different aspects. Based on our experience from application of different assessment methods to control systems in the steel processing field, *systematic procedures for automatic and continuous control performance monitoring, maintenance and optimisation* will be recommended, combining different control performance metrics and assessment methods. Guidelines for how to select or determine the specific parameters, e.g., model orders and time delay, are worked out. Many improved and new CPM techniques developed by the author are presented. For instance, a new framework and method for detection and quantification of stiction in control valves is developed based on Hammerstein modelling and estimation using global search algorithms. Moreover, the method is extended for comprehensive oscillation diagnosis, i.e., discriminating between different causes generating the oscillation, even in the case of multiple faults.

3. *Introducing Anticipatory Control Maintenance Practices.* Bridging the fields "*Condition Monitoring and Diagnosis*", "*Control Performance Assessment and Monitoring*" and "*Automatic Controller Tuning and Adaptation*" is a core objective of this monograph; see Fig. 2. Loop condition monitoring and diagnosis is needed to investigate the properties of control loop components in terms of signal levels, noise levels, nonlinearities and equipment conditions. This particularly includes the detection of oscillations possibly generated by aggressive controller tuning, the presence of nonlinearities (e.g., static friction, dead-zone, hysteresis), or (internal and external) disturbances. Performance assessment is used to supervise the control loops during operation and ensure that they meet the performance specifications. Failure to meet the specifications should give an alert. It is

then decided to inspect/maintain a control loop component or to re-tune the controller. Methods and procedures for how to assist or partly automate this decision are presented.

4. *Automatic CPM-based Controller Re-tuning or Re-design*. When controller tuning is suggested, the control performance assessment results, i.e., indices, are used to generate new controller parameters, which can be downloaded to the controller on demand of the user or of a supervision mechanism. The main aim is to sustain top control performance despite different operational issues. For this purpose, new methods and procedures for CPM-based controller re-tuning are developed.
5. *Transfer the CPM Technology to Metal Processing Industry*. The comprehensive CPM review by Jelali (2006c) revealed a remarkable number of industrial applications to date, with a solid foundation in refining, petrochemical, chemical sectors and pulp & paper plants. However, only a few applications appeared in other industrial sectors. A substantial contribution of this book is thus to *transfer the CPM technology into a new industrial area, the metal processing*, where not much work has been done before. It is shown that the CPM algorithms can still perform well in *this more computationally demanding environment, where the speed is much faster and the time constants much smaller than in the traditional refining and chemicals applications*. Thus many special aspects have to be considered.

As stated by Shah et al. (2005), the challenges are primarily not related to whether the CPM technology itself is effective, but rather related to the human factors surrounding the use of CPM applications. The critical success factor is how an application integrates with existing work practices and maintenance procedures. A substantial part of this work *illustrates the monitoring methods in successful applications and tailored CPM tools integrated into maintenance procedures in rolling mills*, developed within research projects initiated or managed by the author.

Readership

The book is aimed primarily at researchers and engineers in process control and industrial practitioners in process automation, but it is also accessible to postgraduate students and final-year students in process-control engineering.

Most of the methods and algorithms presented in this book are the state-of-the-art and are demonstrated on industrial case studies from different industrial fields (building, chemicals and petrochemicals, mining, mineral and metal processing). The book will thus serve academic and industrial staff working in all these industries on control systems design, maintenance or optimisation.

Acknowledgements

The material in this book has been the outcome of a "private" project alongside my work at BFI in the period 2003–2008; only a part of the applications in Chap. 15

was carried out within the EU research project AUTOCHECK (Jelali et al. 2008). Not only the writing of this book, but also the thinking and implementing most of CPM routines took a big portion of my free time, at weekends and long nights.

I am very grateful to my three "women", *Doris*, *Yasmin* and *Dunja* for letting me use their time. Without their never-ending patience and love I probably would never have been able to finish the book.

It is a pleasure to be able to thank *Walter Ungerer* for motivating me to perform the project and for his continuous support. I thank my colleague *Andreas Wolff* for the initial idea to deal with the subject of CPM and for providing the basis for the common project AUTOCHECK.

The book has benefited from some studies and implementations done by students. In this context, I am particularly grateful to *Martina (Baart) Thormann* and to *Heinrich Ratjen* (who supervised the studies at the Cologne University of Applied Sciences).

During the years, I enjoyed immensely discussions with many international colleagues all around the world, e.g., *Biao Huang* (University of Alberta, Canada), *Alexander Horch* (ABB-Forschungszentrum Ladenburg, Germany), *Nina Thornhill* (Imperial College, London), *Claudio Scali* (University of Pisa, Italy), *Srinivas Karra* (Texas Tech University) and *Shoukat Choudhury* (Bangladesh University).

The industrial data (Appendix C) were friendly provided by: *Ashish Singhal* and *Timothy Salsbury* for loops BAS 1–8; *Alexander Horch* for loops CHEM 1–3 and PAP 1–10; *Peter He* and *Joe Qin* for loops CHEM 4–6; *Biao Huang* for loops CHEM 7–12; *Nina Thornhill* for loops CHEM 13–17 and 40–64; *Claudio Scali* for loops CHEM 18–28 and 32–39; *Shoukat Choudhury* and *Sirish Shah* for loops CHEM 29–31, PAP 11–13, POW 1–5 and MIN 1.

Last but not least, I would like to acknowledge the *anonymous reviewers* of the book (proposal) for their valuable comments and suggestions, as well as *Oliver Jackson*, *Charlotte Cross* (Springer) and *Donatas Akmanavičius* (VTeX) for their patience, editorial comments and detailed examination of the book.

Duisburg, Germany Mohieddine Jelali

Copyright Acknowledgements

Parts of the book appeared in the archived literature, and the authors gratefully acknowledge permissions to re-use material from the following papers.

Chapter 11 is an expanded and revised version of Jelali M., Estimation of valve stiction in control loops using separable least-squares and global search algorithms, *Journal of Process Control* 18:632–642, © 2008, Elsevier Ltd., with permission from Elsevier.

Chapter 13 is expanded and revised version of Jelali M., Automatisches Reglertuning basierend auf Methoden des Control Performance Monitoring, *at – Automatisierungstechnik* 55:10–19, © 2007, Oldenbourg Wissenschaftsverlag, with permission from Oldenbourg.

Chapter 14 is an expanded and revised version of Jelali M., An overview of control performance assessment technology and industrial applications, *Control Engineering Practice* 14:441–466, © 2006, Elsevier Ltd., with permission from Elsevier.

Parts of Chapter 15 are reprinted (with revision) from Jelali M., Performance assessment of control systems in rolling mills—Application to strip thickness and flatness control, *Journal of Process Control* 17:805–816, © 2007, Elsevier Ltd., with permission from Elsevier.

Contents

1 **Introduction** 1
1.1 Need for Control Performance Management (CPM) 2
1.1.1 Objectives and Importance of Control Assets 3
1.1.2 State of Industrial Process Control Performance 6
1.1.3 Root Causes of Control Performance Problems 8
1.2 Principle and Tasks of Control Performance Management 11
1.2.1 Control Performance Indices 12
1.2.2 Basic Procedure for Control Performance Management . . 14
1.2.3 Controller Performance Assessment Benchmarks 16
1.2.4 Challenges of Performance Monitoring Applications 17
1.3 Key Dates of the Development of CPM Technology and Literature Survey 18
1.4 Outline of the Book 22

Part I Evaluation of the Level of Control Performance

2 **Assessment Based on Minimum-Variance Principles** 29
2.1 System Descriptions and Basics 29
2.1.1 Note on Input–Output Models 31
2.2 Minimum-Variance Control (MVC) 32
2.3 Auto-Correlation Test for Minimum Variance 36
2.4 Minimum-Variance Index/Harris Index 38
2.4.1 Estimation from Time-Series Analysis 39
2.4.2 Estimation Algorithms 42
2.5 Assessment of Feedback/Feedforward Controls 49
2.6 Assessment of Set-Point Tracking and Cascade Control 54
2.6.1 Performance Assessment of Cascade Control Systems . . . 54
2.6.2 Assessment of Different Tuning Strategies 60
2.7 Summary and Conclusions 62

3 **User-Specified Benchmarking** 65
3.1 General Setting 65

3.2 IMC-Achievable Performance Assessment . . . 66
3.2.1 IMC Design . . . 67
3.2.2 IMC Benchmark . . . 70
3.3 Extended Horizon Approach . . . 72
3.4 Performance Index Based on Desired Pole Locations . . . 75
3.5 Historical or Reference Benchmarks . . . 76
3.6 Reference-Model/Relative Performance Index . . . 76
3.6.1 Selection of the Reference Model . . . 77
3.6.2 Disturbance Estimation . . . 77
3.6.3 Selection of Critical Index Value . . . 78
3.6.4 Index Properties . . . 78
3.7 Summary and Conclusions . . . 78

4 Advanced Control Performance Assessment . . . 81
4.1 Generalised Minimum-Variance Control (GMVC) Benchmarking . 81
4.1.1 GMV Control . . . 82
4.1.2 Selection of Weightings . . . 83
4.1.3 GMVC with Static Weightings . . . 84
4.1.4 Assessment Index and Procedure . . . 85
4.2 Linear-Quadratic Gaussian (LQG) Benchmarking . . . 86
4.2.1 Classical LQG Framework . . . 89
4.2.2 Polynomial Domain Approach . . . 89
4.2.3 LQG Framework as Special Case of MPC . . . 90
4.2.4 Subspace-Based LQG Design . . . 90
4.2.5 Generation of the LQG Performance Limit Curve . . . 91
4.2.6 LQG Assessment Using Routine Operating Data . . . 92
4.3 Model Predictive Control (MPC) Assessment . . . 96
4.3.1 Basic Principle and Properties . . . 97
4.3.2 Constrained Minimum-Variance Control . . . 101
4.3.3 Design-Case MPC Benchmarking . . . 101
4.3.4 Infinite-Horizon Model Predictive Control . . . 102
4.3.5 Assessing the Effect of Constraints . . . 110
4.3.6 Economic Performance Assessment of MPC . . . 110
4.4 Summary and Conclusions . . . 117

5 Deterministic Controller Assessment . . . 121
5.1 Performance Metrics . . . 121
5.2 Controller Assessment Based on Set-Point Response Data . . . 123
5.2.1 Normalised Criteria . . . 123
5.2.2 Assessment Methodology . . . 124
5.2.3 Determination of Time Delay from Step Response . . . 126
5.2.4 Application Examples . . . 129
5.3 Idle Index for Detecting Sluggish Control . . . 131
5.3.1 Characterisation of Sluggish Control . . . 132
5.3.2 Idle Index . . . 133

5.3.3 Practical Conditions and Parameter Selection 134
5.3.4 Illustrative Example 136
5.4 Assessment of Load Disturbance Rejection Performance 138
5.4.1 Methodology 138
5.4.2 Practical Conditions 140
5.4.3 Illustrative Example 141
5.5 Comparative Simulation Studies 143
5.6 Summary and Conclusions 144

6 Minimum-Variance Assessment of Multivariable Control Systems . 147
6.1 Interactor Matrix: Time-Delay Analogy 147
6.1.1 Definition and Special Forms 147
6.1.2 Recursive Determination of Unitary Interactor Matrices . . 149
6.1.3 Estimation from Closed-Loop Identification 153
6.2 Interactor-Matrix-Based Minimum-Variance Control Law 156
6.3 Assessment Based on the Interactor Matrix 157
6.4 Assessment Without Knowledge of the Interactor Matrix 162
6.4.1 Lower Bound of MIMO Performance Index 163
6.4.2 Upper Bound of MIMO Minimum Variance Benchmark . . 165
6.4.3 Recommended Procedure for the Assessment of MIMO Control Systems 165
6.5 Summary and Conclusions 169

7 Selection of Key Factors and Parameters in Assessment Algorithms . 171
7.1 Data Pre-processing 171
7.1.1 Selection of Sampling Interval 172
7.1.2 Selection of Data Length 173
7.1.3 Removing of Outliers, Detrending and Pre-filtering 175
7.1.4 Effect of Smoothing, Compression and Quantisation 176
7.2 Prediction Models and Identification Methods 178
7.2.1 Implication of the Use of Routine Operating Data 178
7.2.2 Role of the Estimated Model 179
7.2.3 AR(X)-Type Models 180
7.2.4 ARI(X)-Type Models 182
7.2.5 Laguerre Networks 182
7.2.6 Model-Free (Subspace) Identification 185
7.2.7 Estimation of Process Models from Routine Operating Data 186
7.3 Selection of Model Parameters 195
7.3.1 Time Delay Estimation 196
7.3.2 Model Order Selection 198
7.4 Comparative Study of Different Identification Techniques 200
7.4.1 AR vs. ARMA Modelling 202
7.4.2 Subspace Identification 203
7.4.3 Performance of the FCOR Algorithm 203
7.4.4 Use of Laguerre Networks 204

7.5 Model Estimation for Multivariable Processes 204
7.6 Summary and Conclusions 205

Part II Detection and Diagnosis of Control Performance Problems

8 Statistical Process Control 209
8.1 Univariate Statistical Process Control 209
8.2 Multivariate Statistical Process Control 212
8.2.1 Principal Component Analysis 212
8.2.2 Partial Least Squares 214
8.2.3 Advanced Methods for Multivariate Statistical Process Control 215
8.3 Use for Control Performance Assessment 216
8.4 Summary and Conclusions 217

9 Detection of Oscillating Control Loops 219
9.1 Root Causes of Poor Performance 219
9.2 Characterisation and Sources of Oscillations in Control Loops . . . 221
9.3 Detection of Peaks in the Power Spectrum 222
9.4 Regularity of "Large Enough" Integral of Absolute Error (IAE) . . 224
9.4.1 Load-Disturbance Detection 224
9.4.2 Basic Appraoch 225
9.4.3 Detection Procedure 226
9.4.4 Method Enhancement for Real-Time Oscillation Detection 226
9.5 Regularity of Upper and Lower Integral of Absolute Errors and Zero Crossings 228
9.5.1 Basic Methodology 228
9.5.2 Practical Conditions and Parameter Selection 229
9.6 Decay Ratio Approach of the Auto-covariance Function 230
9.6.1 Methodology 230
9.6.2 Practical Conditions and Parameter Selection 232
9.7 Regularity of Zero Crossings of the Auto-covariance Function . . . 233
9.8 Pitfalls of Multiple Oscillations—Need for Band-Pass Filtering . . 234
9.9 Detection of Intermittent Oscillations 237
9.10 Summary and Conclusions 237

10 Detection of Loop Nonlinearities 239
10.1 Methods Review 240
10.2 Bicoherence Technique 241
10.2.1 Non-Gaussianity Index 242
10.2.2 Nonlinearity Index 243
10.2.3 Procedure and Practical Conditions 244
10.2.4 Modified Indices 246
10.3 Surrogate Data Analysis 248
10.3.1 Generation of Surrogate Data 248
10.3.2 Discriminating Statistics—Nonlinear Predictability Index 250

10.3.3 Nonlinearity Detection Procedure 251
10.3.4 Spurious Nonlinearity—Pitfalls in the Surrogate Data . . . 252
10.3.5 Default Parameter Values and Practical Issues 255
10.4 Detection of Saturated Actuators 256
10.4.1 Saturation Test Based on Statistical Distribution 257
10.4.2 Saturation Index for Valve Monitoring 258
10.5 Comparative Studies 258
10.5.1 Unit-Wide Oscillation Caused by a Sensor Fault 259
10.5.2 Plant-Wide Oscillation Caused by a Valve Fault 261
10.6 Summary and Conclusions 263

11 Diagnosis of Stiction-Related Actuator Problems 265
11.1 Typical Valve-Controlled Loop 266
11.2 Effects Relating to Valve Nonlinearity 268
11.3 Stiction Analysis 269
11.3.1 Effect of Stiction in Control Loops 269
11.3.2 Physically Based Stiction Modelling 271
11.3.3 Data-Driven Stiction Modelling 272
11.3.4 Typical Trends of Variables and Input–Output Shape Analysis 274
11.4 Stiction Diagnosis Based on Shape Analysis of MV–OP Plots . . . 277
11.4.1 Stuck Indices 278
11.5 Cross-Correlation-Based Stiction Detection 281
11.6 Diagnosis Based on Curve Fitting 284
11.6.1 Sinusoidal Fitting 285
11.6.2 Triangular Fitting 286
11.6.3 Stiction Index and Detection Procedure 286
11.6.4 Similar Techniques 288
11.7 Nonlinearity Detection and PV–OP Pattern Analysis 289
11.7.1 Stiction Detection and Estimation Procedure 289
11.7.2 Practical Issues 292
11.8 Tests to Confirm Stiction 294
11.8.1 Controller Gain Change Test 295
11.8.2 Valve Travel or Bump Test 296
11.9 Stiction Diagnosis Procedure 297
11.10Summary and Conclusions 299

12 Complete Oscillation Diagnosis Based on Hammerstein Modelling 301
12.1 Features of the Proposed Framework 302
12.2 Identification Model Structure 303
12.3 Identification Algorithm 305
12.3.1 Linear Model Estimation 305
12.3.2 Nonlinear Model Estimation 308
12.4 Key Issues 311
12.4.1 Model Structure Selection 311

12.4.2 Determination of Initial Parameters and Incorporation of Constraints . . . 312
12.5 Application and Results . . . 313
12.5.1 Simulation Studies . . . 313
12.5.2 Industrial Case Studies . . . 314
12.6 Detection of Multiple Loop Faults . . . 320
12.6.1 Simulation Examples . . . 323
12.6.2 Industrial Examples . . . 324
12.7 Summary and Conclusions . . . 329

Part III Performance Improvement

13 Performance Monitoring and Improvement Strategies . . . 333
13.1 Performance Improvement Measures . . . 333
13.2 Loop Monitoring Paradigms . . . 335
13.2.1 Bottom-Up and Top-Down Approaches . . . 335
13.2.2 Loop Prioritisation and Ranking . . . 337
13.2.3 Relationship to Economical Benefits . . . 337
13.3 Comprehensive Procedure for Performance Monitoring . . . 338
13.4 Summary and Conclusions . . . 341

14 Controller Auto-Tuning Based on Control Performance Monitoring . 343
14.1 Basic Concepts of Controller Auto-Tuning and Adaptation . . . 345
14.2 Overview and Classification of CPM-Based Tuning Methods . . . 346
14.3 Optimisation-Based Assessment and Tuning . . . 347
14.3.1 Methods Based on Complete Knowledge of System Model . . . 347
14.3.2 Techniques Based on Routine and Set-Point Response Data 356
14.4 Iterative Controller Assessment and Tuning . . . 362
14.4.1 Techniques Based on Load Disturbance Changes . . . 362
14.4.2 Methods Based on Routine Data and Impulse Response Assessment . . . 366
14.5 Strategies for Variation of Controller Parameters . . . 380
14.5.1 Variation of Proportional Gain Alone and Fine Tuning of Integral Time . . . 380
14.5.2 Simultaneous Variation . . . 381
14.5.3 Successive Variation . . . 381
14.5.4 Constraints and Loop Stability . . . 382
14.6 Comparative Studies . . . 383
14.7 Summary and Conclusions . . . 385

Part IV Applications and Tools

15 Industrial CPM Technology and Applications . . . 389
15.1 Demands on Performance Monitoring Algorithms . . . 389
15.2 Review of Control Performance Monitoring Applications . . . 391
15.2.1 Analysis of Fields of Application . . . 393

15.2.2 Analysis of Type of Implemented Methods 393
15.3 Review of Control Performance Monitoring Systems 393
15.3.1 CPM Tools and Prototypes 393
15.3.2 Commercial Products 394
15.4 Summary and Conclusions 396
15.5 Summary of Industrial Case Studies 397

16 Performance Monitoring of Metal Processing Control Systems . . . 403
16.1 Introduction to the Metal Processing Technology 404
16.1.1 Steel Processing Route and Control Objectives 404
16.1.2 Control Objectives 406
16.1.3 Mill Automation 409
16.1.4 Overview of Metal Processing Control Systems 411
16.1.5 Technological Control Systems 414
16.2 Practical Aspects of Performance Assessment in Metal Processing 415
16.2.1 Online vs. Batch-Wise Evaluation 415
16.2.2 Oscillation Diagnosis 416
16.2.3 Time-Based vs. Length-Based Assessment 417
16.2.4 User-Specified Indices 418
16.3 Industrial Cases Studies and Developed Monitoring Tools 418
16.3.1 Gauge Control in Cold Tandem Mills 419
16.3.2 Flatness Control in Cold Tandem Mills 424
16.3.3 Temperature Control in Annealing Lines 427
16.4 Summary and Conclusions 439

Appendix A Basic Signal Processing and Statistics 441
A.1 Ergodicity 441
A.2 Expectation and Variance 441
A.3 Correlation and Covariance 442
A.4 Discrete Fourier Transform 443
A.5 Power Spectrum and Coherence Function 444

Appendix B Higher-Order Statistics 447
B.1 Moments and Cumulants 447
B.2 Polyspectra and Coherence Functions 449
B.3 Estimating the Bispectrum from Data 451
B.4 Skewness and Squared Bicoherence Functions 452

Appendix C Control Loops from Different Industries 457

References 461

Index 481

Abbreviations and Acronyms

AC	Alternating current
	Analyser control
ACF	Auto-correlation/covariance function
AE	Area index
AIC	Akaike's Information Criterion
ANOVA	Analysis of variance
ANSI	American National Standard Institution
AR	Auto-regressive
ARMA	Auto-regressive moving average
ARIMAX	Auto-regressive integrated moving average exogenous
ARMAX	Auto-regressive moving average exogenous
ARX	Auto-regressive exogenous
BAS	Building automation system
BJ	Box–Jenkins
BR	Backup roll
CC	Cohen–Coon
CCF	Cross-correlation function
CHEM	Chemicals
CHR	Chien–Hrones–Reswick
CPI	Control-performance index
CPM	Control-performance management
	Control-performance monitoring
CPU	Central processing unit
CVA	Canonical variate analysis
DC	Direct current
DCS	Distributed control system
DFT	Discrete Fourier transform
DMC	Dynamic matrix control
DR	Decay ratio
EPI = EHPI	Extended performance index = extended-horizon performance index
EWMA	Exponentially weighted moving average

EWMASE	Exponentially weighted moving average of squared error
FB(C)	Feedback (control)
FC	Flow control
FCOR	Filtering and correlation analysis
FF(C)	Feedforward (control)
FFT	Fast Fourier transform
FIR	Finite impulse response
FOPTD	First-order-plus-time-delay
GA	Genetic algorithm
GADS	Genetic algorithms and direct search
GC	Gauge control
GMV	Generalised minimum variance
GMVC	Generalised minimum variance control
GPC	Generalised predictive control
GUI	Graphical user interface
HIS	Historical data benchmark
HOS	Higher-order statistics
IAE	Integral of absolute error
IDFT	Inverse discrete Fourier transform
IFT	Iterative feedback tuning
IMC	Internal model control
IIR	Infinite impulse response
IPTD	Integrating plus time delay
IR	Impulse response
ISE	Integrated squared error
ITAE	Integral of time-multiplied absolute value of error
IV	Instrumental variable
LQ	Linear-quadratic
LQG	Linear-quadratic Gaussian
LQR	Linear-quadratic regulator
LS	Least square
LTI	Linear time-invariant
LTV	Linear time-variant
MA	Moving average
MAE	Mean absolute error
MC	Monte Carlo
MET	Metal
MIMO	Multiple-input multiple-output
MISO	Multiple-input single-output
MOESP	Multivariable output-error state space
MPC	Model predictive control
MSE	Mean squared error
MV	Manipulated variable Minimum variance
MVC	Minimum variance control

NCNP	New controller & new process model
NGI	Non-Gaussianity index
NLA	Nonlinearity analysis
NLI	Nonlinearity index
NPI	Non-predictability index
OBF	Orthonormal basis function
OCNP	Old controller & new process model
OCOP	Old controller & old process model
ODE	Ordinary differential equation
OE	Output error
OP	Controller output
P	Process
	Proportional
PAP	Pulp and paper
PC	Pressure control
PCA	Principal component analysis
PD	Principal domain
PDF	Probability density function
PEM	Prediction-error method
PI	Proportional-integral
PID	Proportional-integral-derivative
POW	Power
PS	Pattern search
PSD	Power spectral density
PV	Process variable
QF	Quantisation factor
RDI	Relative damping index
RHP	Right-half plane
RLS	Recursive least square
RPI	Relative performance index
RS	Restricted structure
SI	Stiction index
SISO	Single-input single-output
SOM	Self-organising map
SOPTD	Second-order-plus-time-delay
SP	Setpoint
SPC	Smith-predictor control
	Statistical process control
SVD	Singular value decomposition
TC	Temperature control
TCM	Tandem cold rolling mill
TDE	Time-delay estimation
TNLI	Total nonlinearity index
VP	Valve position
WR	Work roll

Chapter 1
Introduction

Control engineering deals with the theory, design and application of control systems. The primary objective of control systems is to maximise profits by transforming raw materials into products while satisfying criteria such as product-quality specifications, operational constraints, safety and environmental regulations (Seborg et al. 2004). The design, tuning and implementation of control strategies and controllers are undertaken within the first phase in the solution of control problems. When properly carried out, the result of this phase should be a well functioning and performing control system. However, after some time in operation, changes in the characteristics of the material/product being used, modifications of operation strategy and changes in the status of the plant equipment (aging, wear, fouling, component modifications, etc.) may lead to the degradation of control performance. Problems can arise even in well-designed control loops for a variety of reasons, ranging from a need for retuning due to the changes mentioned, to difficulties with the sensors, or actuator operation, which can occur in an unpredictable fashion.

Therefore, the second phase in the solution of control problems should be the supervision of the control loops and the early detection of performance deterioration. This task has traditionally been made by the plant personnel, i.e. maintenance and control staff. However, the rationalisation pressure in the process industries during the last decades has led to a drastic reduction of personnel. Moreover, the process industries are faced with ever-increasing demands on product quality, productivity and environmental regulations. These force companies to operate their plants at peak performance, hence the need for control systems with consistently high performance. Control systems are thus increasingly recognised as capital assets that should be maintained, monitored and revised routinely and automatically. These tasks are performed today within the framework of *control performance monitoring* (*CPM*), which has got considerable attention from both the academic and industrial communities in the last decade.

Evidence of this was shown at recent conferences, e.g. American Control Conference 2000, Chemical Process Control Conference 2001, European Control Conference 2001, IFAC World Congress 2002, Control 2004, ADCHEM 2006, where entire sessions or workshops were devoted to the topic of CPM. Also, a special is-

M. Jelali, *Control Performance Management in Industrial Automation*,
Advances in Industrial Control, DOI 10.1007/978-1-4471-4546-2_1,

sue of the International Journal of Adaptive Control and Signal Processing (2003: Vol. 17, Issues 7–9) has been devoted to this subject.

This young field of research is emerging towards new and more efficient maintenance and plant-asset management practices. This book tries to speed up the development by sharing our experiences in this exciting and useful area of automation and working out a new integrated framework for control performance monitoring, diagnosis and optimisation. The aim is to shift the control maintenance practice in the process industries from either scheduled or reactive to *anticipatory*, centred around *continuous assessment* and prediction of the performance degradation of control systems. Such a paradigm shift means a *control-system life-cycle management*, in which control systems are assessed and improved throughout their life cycle, starting from commissioning and continuing through the entire manufacturing process and usage phase. The performance of the controllers and other loop components can thus be improved continuously, ensuring products of consistently high quality. Of course, it may be too idealistic to hope that all control loops could be on a scheduled maintenance list, and perhaps this is not the best way to proceed in any case because of the random nature of faults and the minor importance of some loops. But there is no question that control system performance need supervision and maintenance that is *pro-actively* data-driven, not reactively complaint-driven.

1.1 Need for Control Performance Management (CPM)

A control system is an interconnection of components, i.e. sensor, process/plant, actuator and controller, forming a system configuration that has the general objective to influence the behaviour of the system in a desired way; see the block diagram in Fig. 1.1. The central component is the *process* whose output is to bc controlled. The *controller* seeks to maintain the measured process variable (PV) at a specified set point (SP) in spite of disturbances acting on the process. The *actuator* is the device that includes the final control element (a valve, damper, etc. and its associated equipment such as a positioner). This receives the controller output (OP) signal, reacts in appropriate fashion to impact the process, and consequently causes the PV to respond in the desired manner. The combination of process and actuator is usually called the *plant*, but the terms "process" and "plant" are often used interchangeably, since process and actuator are intimately connected.

Optimal process control can only be achieved when all aforementioned components are working properly. Hence, before tuning a loop, one must verify that each component is operating as specified and that the design is appropriate. Already for single control loops, it is clear that the task of getting and keeping all components in good health is not trivial. The fact that a plant in the process industry typically comprises hundreds to thousands control loops reveals the huge challenge of monitoring and ensuring peak performance of such complex control systems.

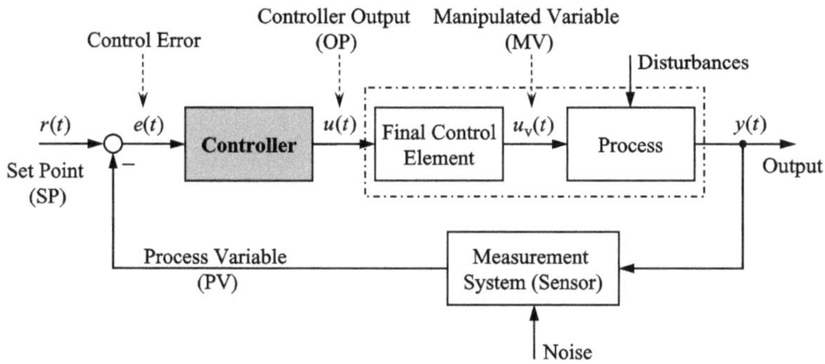

Fig. 1.1 Component block diagram of a closed-loop system

1.1.1 Objectives and Importance of Control Assets

Before discussing specific control performance monitoring methods, it is essential to recall that there is a rich variety of control problems with very diverse goals. The most important control problems to be distinguished are:

- *Steady-State Regulation.* This refers to the process of holding the system output y close to an operating, usually constant reference or set point r. The controller is usually called *regulator*. The key problems are load disturbances, measurement noise and process variations. In process control, set points are normally kept constant most of the time; changes are typically made only when material or production is altered. Rejection of load disturbances is thus the key issue in process control.
- *Set-Point Tracking (or Servo Control).* The controller is designed such that certain controlled variables are forced to follow prescribed *trajectories* or references as closely as possible. Tracking typically occurs in motion control and robotics.
- *Surge Attenuation.* For some applications, such as tank level controllers, averaging level control is the more appropriate strategy rather than to hold the tank level at a particular set point.

When tuning a control loop, compromises between robustness, i.e. sensitivity to changes in the plant parameters, and speed of response for good regulation or tracking must be taken. Tuning means the proper selection of the controller settings, e.g. the proportional gain K_c, the integral time T_I and the derivative time T_D for a PID controller. Also, the control effort is generally of main concern, as it is related to the final cost of the product and to the wear and life span of the actuators. It should be therefore kept at a minimum level.

The performance of a control system is usually specified by different criteria, which can be divided into the following categories (Fig. 1.2):

- *Deterministic Performance Criteria.* These are the traditional performance measures used in the case of deterministic disturbances, i.e. set-point changes or sudden load disturbances, such as the rise time, settling time, overshoot, offset from set point and integral error criteria; see Åström (1991).

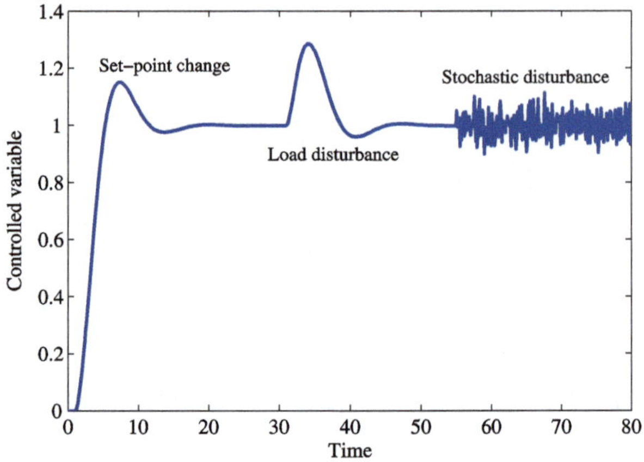

Fig. 1.2 Disturbances usually considered for control design

- *Stochastic Performance Criteria.* These typically include the variance, or equivalently the standard deviation, of the controlled variable or control error. Such criteria have direct relationship to process performance, product quality and energy or material consumption. In process control, steady-state regulation is the essential problem. Therefore, load-disturbance responses are more important than those to set points, as emphasised by Shinskey (1996). The most widespread (stochastic) criterion considered for performance assessment in process control is the *variance* (or, equivalently, the standard deviation), particularly for regulatory control:

$$\sigma_y^2 = \frac{1}{N-1}\sum_{k=1}^{N}\big(y(k)-\bar{y}\big)^2 \quad \text{with } \bar{y} = \frac{1}{N}\sum_{k=1}^{N} y(k). \tag{1.1}$$

In many applications, it is useful to combine stochastic and deterministic criteria. This is standard in the field of optimal and model-predictive control, where the control effort is penalised.

The widespread use of the variance as performance criterion is due to the fact that it typically represents the product-quality consistency. The reduction of variances of many quality variables not only implies improved product quality but also makes it possible to operate near the constraints to increase throughput, reduce energy consumption and save raw materials. This relationship is illustrated in Fig. 1.3.

By identifying, diagnosing and tuning key control loops, the variance can be reduced from $\sigma_{y,1}^2$ to $\sigma_{y,2}^2$, so that the set point can be moved closer to the plant boundary, i.e. from SP1 to SP2. For example, the rolling force in a rolling mill can be scheduled higher near the boundary limit when it is ensured that the strip flatness variance is sufficiently low. Increasing the rolling force directly implies higher throughput and thus higher profit.

It is well recognised that control assets are the foundation of plant performance. It is estimated that 75 % of physical assets in a plant are under process control (ARC Advisory Group). Control delivers the following benefits (Brisk 2004), affecting the financial performance (Fig. 1.4):

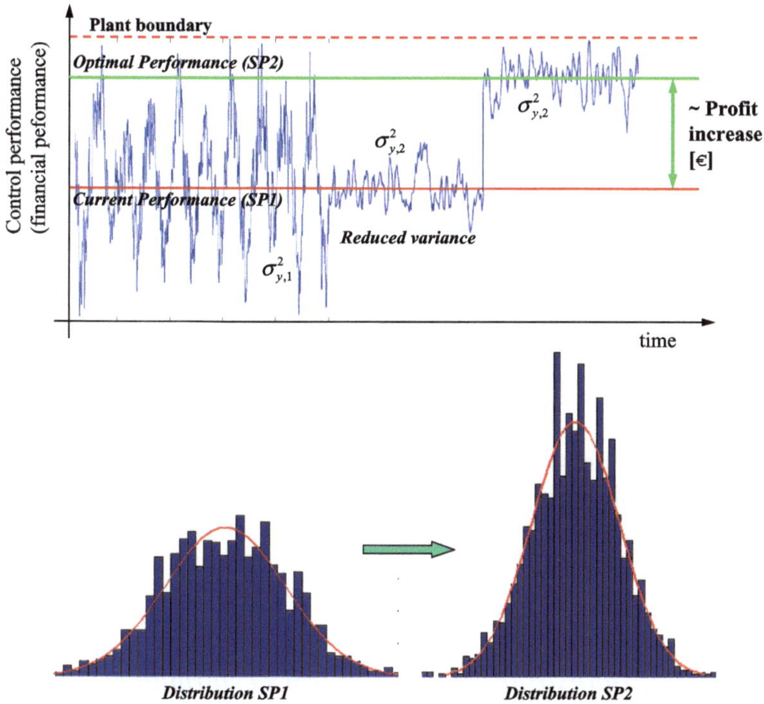

Fig. 1.3 Relationship between economic performance and variance reduction

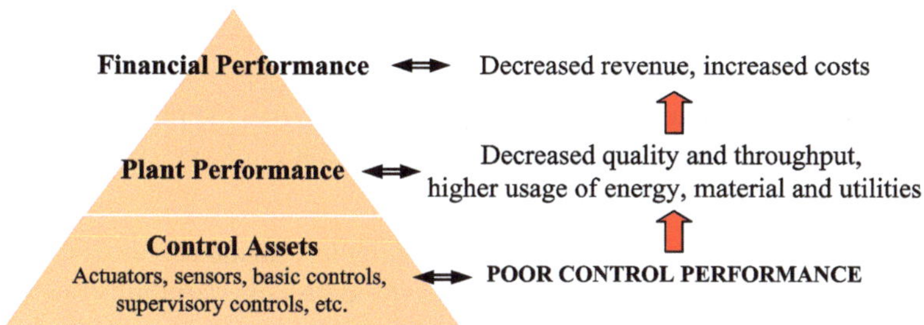

Fig. 1.4 Effect of poor control performance

- *Safer Operation and Reduced Environmental Impact*. Keeping process operation steady will always help reduce incidents, which may create hazardous conditions or undesirable emissions to the environment.
- *More Sustainable Manufacturing*. Better control can achieve more efficient raw material and energy usage per unit of product. Apart from the financial benefits, this yields waste reduction and better conserving of non-renewable resources.

- *Efficiency Gains*. These have been achieved from the very earliest days of process control application, and certainly for advanced control, from the 1970s onwards until today.
- *Quality Gains*. Maintaining consistent product quality is a key factor in ensuring and potentially growing a company's market share. From the early 1990s onwards, in an increasingly competitive and often global marketplace, control focusing on product quality became particularly important.
- *Agility Gains*. From the turn of the century and into the immediate future, new factors governing processing profitability include manufacturing flexibility, customer responsiveness and the related need to reduce working capital by processing to order, not to stock. This requires an agile processing capability, with responsive plant exploiting the full potential that well performing control can provide.

Poor control performance, therefore, leads to poor plant performance, and that in turn implies poor financial performance. This, again, underlines the need for some form of regular scheduled maintenance of control loops to ensure consistently high levels of performance.

1.1.2 State of Industrial Process Control Performance

Many surveys analysed the state of performance of control loops in different process industries. The main conclusion was that too often basic control principles are ignored, control algorithms are incorrectly chosen and tuned, while sensors and actuators are poorly selected or maintained. Consequently, the control performance of many loops can be significantly improved by proper loop re-tuning, controller redesign or equipment maintenance. In the following items, the results of some published audits are summarised to determine the "health" of control systems in the process industries (see Table 1.1):

- Bialkowski (1993). These audits of paper mills in Canada reveal that only 20 % of the control loops worked well and decreased process variability. The reasons why the performance was poor are bad tuning (30 %) and valve problems (30 %). The remaining 20 % of the controllers functioned poorly for a variety of reasons, such as sensor problems, bad choice of sampling rates and poor or non-existing anti-aliasing filters.
- Ender (1993). Similar observations are given in this study, where it is claimed that 30 % of installed process controllers operated in manual mode, 20 % of the loops use default parameters set by the controller manufacturer (so-called, "factory tuning"), and 30 % of the loops showed poor performance because of equipment problems in valves and sensors.
- Desborough and Miller (2002). This comprehensive audit of thousands of industrial basic control loops in the U.S. process industry led to the conclusion that,

Table 1.1 Control performance classification code (Paulonis and Cox 2003)

Class (colour)	Description
Best/ Excellent (dark green)	Loops are performing well and do not need attention. They are typically tracking the set point well, with very few or no significant deviations.
Good (light green)	Loops are performing adequately, but may have some component of performance that could be improved. Benefit to cost ratio for making improvements is, however, likely to be small.
Fair (orange/yellow)	Loops are not performing up to potential. Control is probably being maintained in a broad sense, but there is clear potential for performance improvement. It is recommended to improve these loops.
Poor (red)	Loops typically have a serious performance problem, e.g., high oscillations or large and frequent control errors. Investigation of these loops is imperative and promises substantial performance improvements.

despite high service factor (i.e. time in-auto-mode), only one third of the controllers were classified as acceptable performers, and the rest had significant improvement opportunity. 32 % of the controllers classified as "poor" or "fair" in this survey showed problems/faults in control valves.

- Paulonis and Cox (2003). This assessment study covered more than 9.000 PID controllers (for flow, pressure, level and temperature) in 40 (chemical) plants at 9 sites worldwide. 41 % of the control loops was found to belong to the "poor" and "fair" performance class, particularly due to hardware problems (valve/positioner/transducer).
- Ruel (2003). Different studies as well as observations confirm that typical performance distribution of North American control loops delineates as follows: 30 % of the loops has control valves in poor quality or in poor condition, 60 % poor controller tuning, 85 % poor loop design, 15 % controller in manual mode, 30 % not performing according to control objectives, 85 % performing better in automatic than manual mode.
- Torres et al. (2006). Control loop auditing on 700 control loops from 12 different Brazilian companies (petrochemical, pulp and paper, cement, chemical, steel and mining segments) from July 2004 to October 2005 showed in average that 14 % of loops showed excessive valve wear, 15 % of valves showed problems with stiction and hysteresis, 16 % of loops were in manual mode, 16 % of loops had severe tuning problems, 24 % of loop's controller outputs were saturated most of the time, and 41 % of loops oscillated due to tuning problems, coupling, disturbances and actuator problem.

Certainly, some problems may have been solved within revamping measures carried out in recent years. However, the performance has not improved much in this period, as reported in recent studies, although there has been great deal of academic work in the CPM area. In the experience of the author, a similar situation is found in the metal processing field. Figure 1.5 illustrates some time trends of process variable and controller output typically found in different process industries. It is difficult to distinguish between the sources of bad control performance of these loops from just looking at the measured trends, except the fact that all loops are more or less oscillating. For discriminating between the causes, appropriate performance indices

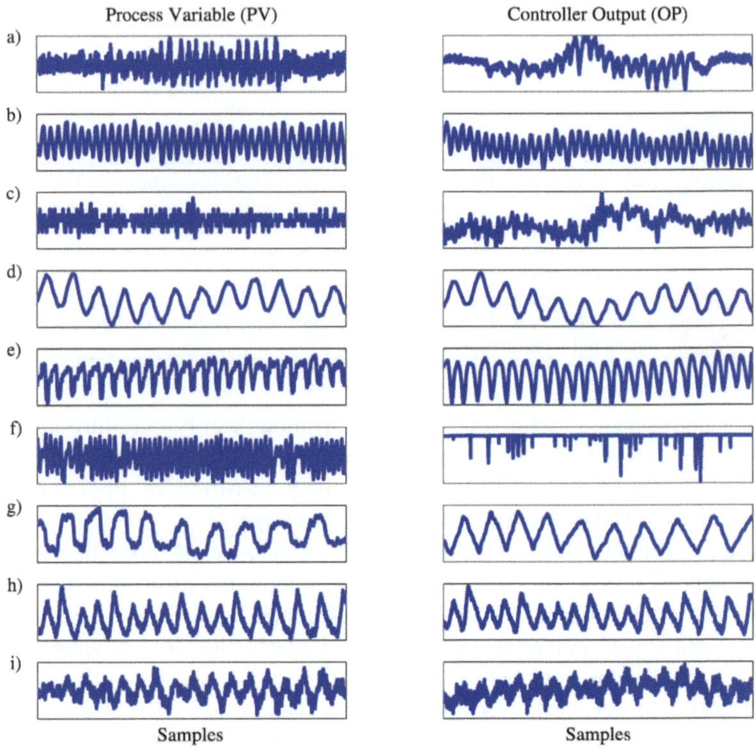

Fig. 1.5 Typical time trends (measured data) found in the process industries showing poor control performance due different causes: (**a**) intermittent oscillation; (**b**) stiction and tight tuning; (**c**) quantisation; (**d**) tuning problem; (**e**) sensor fault; (**f**) saturation; (**g**) stiction; (**h**) stiction; (**i**) external disturbances

and diagnosis techniques are needed. Most of these and similar data sets will be analysed later in this book.

1.1.3 Root Causes of Control Performance Problems

It is essential to understand the possible root-causes of control performance problems to be in the best position to suggest the proper measures for their solution. Generally, once commissioned and properly tuned, control assets deliver performance benefits quickly. The control performance then degrades over time; see Fig. 1.6. Various studies indicate that the half-life of good control loop performance is about six months (Bialkowski 1993; Ruel 2002). This can be caused by many effects, including reliability issues, operational issues, human factors and maintenance aspects. Some key issues are discussed below.

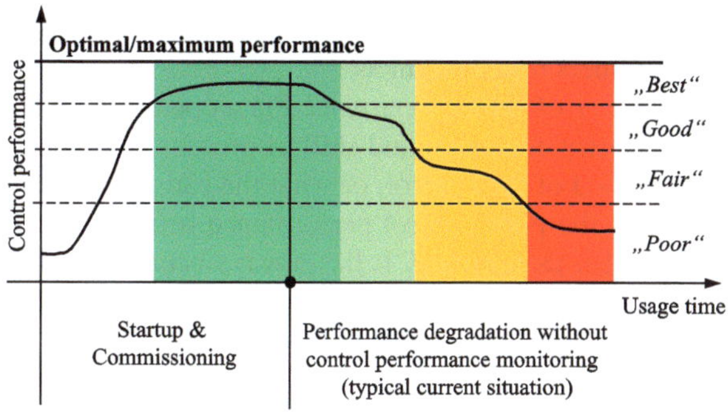

Fig. 1.6 Typical performance decay of industrial process control due to different factors

1.1.3.1 Inadequate Controller Tuning and Lack of Maintenance

This may be due to the fact that the controller has never been tuned or that it has been tuned based on a poor model, or even an inappropriate controller type has been used. More than 90 % of the controllers installed in automation systems are of the PID type, even in cases where other controllers are more appropriate. The most common cause of poor control performance is, however, that controllers are normally designed and tuned at the commissioning stage, but left unchanged after that for years or even decades, although the performance of many control loops decays over time owing to

- *Changes in the characteristics of the material/product being used*;
- *Modifications of operating points/ranges, strategies, or feed stocks*;
- *Variations in the status of the plant equipment, such as wear, increased friction and plant modifications*;
- *Failures in software or hardware*.

The controller settings are then no longer adequate, and the loop may become underdamped or too sluggish.

In the industrial practice, the main reasons quoted for lack of tuning and maintenance are:

- *Limited Time and Resources for Commissioning*. The commissioning engineers often tune the controllers until they "work", even poorly. They do not have enough time to undertake rigorous testing or optimise the control performance. Most controllers are tuned once they are installed and then never again.
- *Conservative Tuning*. Often, the controllers are conservatively tuned, i.e. for the "worst case", to retain stability when operating conditions change in nonlinear systems. This leads to sluggish controller behaviour.
- *Limited Maintenance Resources for a Huge Number of Control Loops*. There are few people responsible for maintenance of automation systems, and all are fully busy with keeping the control systems in operation, i.e. they have no or very little time for improving controllers. Moreover, typically a remarkable number of controllers have to be maintained by a very small number of control engineers, who survived decades of downsizing and outsourcing in the production factories.

Usually, nothing happens until the operators complain very loudly, or even final customers reject products.

- *Shortage of Skilled Personnel.* Plant operators and engineers often do not have the necessary education and skill of process control to be able to know what can be expected of the control, or what the causes are behind poor performance. Sometimes, the poor control performance becomes the norm, and production people accept it as normal ("It has always been like this"), or the operators switch to manual control, and in many cases the economic benefits drop to zero.

1.1.3.2 Equipment Malfunction or Poor Design

To achieve the control performance target, all elements of a control loop must be "healthy and work in harmony". Poor control performance may thus be the result of failing or malfunctioning sensors or actuators, e.g. due to excessive friction. More serious is the problem when a process or a control-loop component is not appropriately designed. The relation between process design and control can be succinctly summarised by the following quotation from a paper by Ziegler and Nichols (1943): "In the application of automatic controllers, it is important to realize that controller and process form a unit; credit or discredit for results obtained are attributable to one as much as the other. A poor controller is often able to perform acceptably on a process which is easily controlled. The finest controller made, when applied to a miserably designed process, may not deliver the desired performance. True, on badly designed processes, advanced controllers are able to eke out better results than older models, but on these processes there is a definite end-point which can be approached by the instrumentation and it falls short of perfection". Thus, the problems mentioned in this item cannot be overcome by re-tuning the controller. This underlines the importance of checking the control loop properties, e.g. signal levels, noise levels, nonlinearities and equipment conditions, before applying an automatic tuning procedure. It is within the framework of CPM to continuously check these properties and detect malfunctions of the loops based on routine operating data.

1.1.3.3 Inappropriate Control Structure

Inadequate input/output pairing, ignoring mutual interactions between the system variables, competing controllers, insufficient degrees of freedom, the presence of strong nonlinearities and the lack of time-delay compensation in the system are frequently found as sources for control-structure problems. If not properly addressed by means of feedforward control actions, external disturbances may also deteriorate the control performance. In the experience of the author, PID control is often used for systems with dominant time delays, instead of the more suitable time-delay compensators, i.e. Smith predictor, internal model control or even model predictive control. We also often find controllers operating with fixed settings for the whole operating range enforcing very conservative tunings. Implementing just gain-scheduling (as a special case adaptive control) in these situations would significantly improve the control performance.

1.1.3.4 Automation System (Platform) Constraints

Some of the problems for practical controller-performance analysis can be the required data access (with appropriate sampling time) and the high computing power needed for index calculation. Many plants in the process industries have control systems, which are between ten and fifteen years old, thus are not up to the task from a computing power perspective. However, the situation has changed in the last few years owing to major upgrading steps, so that the introduction of powerful CPM tools should now be possible more than ever.

1.2 Principle and Tasks of Control Performance Management

The main objective of control performance monitoring (CPM) is to provide online automated procedures that evaluate the performance of the control system and deliver information to plant personal for determining whether specified performance targets and response characteristics are being met by the controlled process variables; see Fig. 1.7. This should help detect and avoid performance deterioration owing to variations in the process and operation. Recommendations and/or actions are generated to inspect/maintain control loop components, e.g. sensors, actuators, or to re-tune the controller based on the calculated performance metrics within the assessment step.

The term *monitoring* means the action of watching out for changes in a statistic that reflects the control performance over time. The term *assessment* refers to the action of evaluating the considered statistic at a certain point in time. Note, however, that both terms are used somewhat interchangeably in the literature. Other synonyms used are loop *auditing*, control loop *management*, control performance *supervision* and control loop *benchmarking*.

Control performance management techniques ponder important process diagnostic questions, such as:

1. *Benchmark Selection*. What is the best achievable performance against which the performance of the installed controller should be assessed?
2. *Assessment*. Is the controller "healthy"? Is it doing its job satisfactory? Is the current control system achieving the best performance, i.e. the performance of benchmark?
3. *Diagnosis*. If not, why is it in "poor health"? How can one arrange for a performance benchmark to determine the improvement potential without disturbing the running system?
4. *Improvement*. What measures and steps would improve the performance of a problematic loop? Is it sufficient to re-tune the controller, or should some loop components be maintained, or even re-designed?

In this monograph, the word *control performance management* is introduced as an umbrella term for the sets of different activities and techniques of control performance assessment, monitoring, problem (e.g. oscillation and stiction) detection and diagnosis as well as performance improvement.

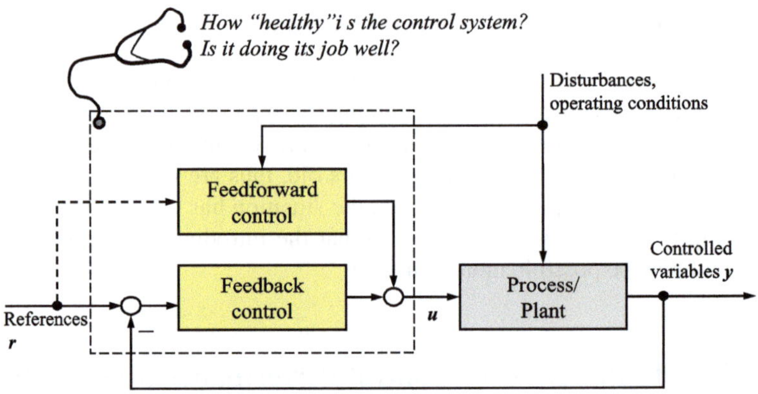

Fig. 1.7 Simplistic statement of control performance assessment problems

1.2.1 Control Performance Indices

The key features of performance metrics should ideally include (Hugo 1999; Xia et al. 2003)

- *Controller Orientation.* Metrics should be sensitive to detuning and process model mismatch or equipment problems and independent of disturbance or set-point spectrum, which can vary widely in a plant.
- *Easy Computation.* Metrics should be non-invasive, i.e. should not require plant tests, able to be automated and require minimum specification of process dynamics. All this implies the use of normal (closed-loop) operating data for the metric calculation. The capability of being automated is important since usually a large number of loops in a plant have to be assessed.
- *Objectivity and Accuracy.* The confidence interval of the metric should be provided or the accuracy can be tested by plant data. They should be non-arbitrary measures that compare the current quality of control to some universal standard (perfect control, minimum variance control, optimal control, best possible control, etc.).
- *Improvement Indication.* Ideally, the metrics should be realistic and achievable under the physical constraints and indicative of why the controller is performing poorly. It should also measure the improvement in profit (reduction in variance, pushing the process to constraints, etc.) due to the controller.

These guidelines serve as a reference when selecting a benchmarking criterion. Obviously, any benchmarking metric does not need to possess all these features. However, it should fulfil as many aforementioned properties as possible.

In the industrial practice, it is useful to select and present the performance figures in such a manner that different types of production personnel (executives and managers, production unit managers, production and automation engineers) have a presentation, which gives the results in a manner focused for their purposes; see Åkesson (2003) and Rakar et al. (2004) for more details.

Typically, the integrated squared error (ISE) and the output error variance have been commonly used as statistics for monitoring the effectiveness of a control strategy. Such statistics are, without any doubt, important with respect to rating the overall process performance. Historically, standard deviation monitoring of the controlled variable error from set point has often been carried out and found useful. From the perspective of controller performance monitoring, experience has shown that this information can be both limited and misleading. Set-point error standard deviation is a function of the magnitude of loop upsets. Changes in this statistical information can be functions of changing process conditions and may not necessarily reflect the performance of a controller. During periods of large plant upsets, higher standard deviations are to be expected despite the fact that controllers are responding as designed. During calm periods of operation, low standard deviations can be observed with poorly-designed controllers (Kozub 2002).

Moreover, there are always some random disturbances that are inherent to the process itself and cannot be normally compensated by the control system. This limits the control loop to a certain lower bound of variance representing the optimum (Shunta 1995). A further variance decrease below this minimum can only be achieved by changing plant equipment or instrumentation. This again underlines the need for considering performance metrics relative to the optimum to get an objective performance assessment of control systems.

Therefore, for the purpose of controller performance monitoring and diagnosis, relative measures, called *performance indices*, setting certain performance metrics in relation to what can be achieved by an optimal controller or a controller with desired properties are of key concern. This is the reason why performance monitoring approaches usually have certain performance benchmark with which the current performance of the loop is compared. The benchmark gives an indication of the inherent optimum that is set by the process design and equipment.

Moreover, with respect to the quality of feedback control, a variance or ISE statistic is essentially meaningless unless it can be compared to the controller with best/optimal performance. Therefore, within the CPM framework, the performance of a control system is always quantified by a relative metric, the *control performance index* (*CPI*), generally defined as

$$\eta = \frac{J_{\mathrm{des}}}{J_{\mathrm{act}}}, \tag{1.2}$$

where J_{des} is any ideal, optimal or desired/expected value for a given performance criterion (typically the variance), to be minimised. J_{act} is the actual value of the criterion, to be extracted from measured process data under the installed controller. In this context, two important cases have to be distinguished:

- *Perfect or Optimal Control as Benchmark*. The control performance index is a single scalar usually scaled to lie within [0, 1], where values close to 0 indicate poor performance, and values close to 1 mean better/tighter control. This indeed holds when perfect control is considered as benchmark. In this case, the cost function is split into two terms: the feedback-invariant, i.e. unpredictable, part J_{min} and the controller-dependent, i.e. predictable part J_c. Thus, CPI can be written as

$$\eta = \frac{J_{min}}{J_{min} + J_c}. \tag{1.3}$$

J_{min} results from fundamental limitations on performance in control systems, e.g. imposed by time delays, right-half plane (RHP) zeros, or constraints. It is obvious to see that for perfect/optimal control, i.e. $J_c = 0$: $\eta = 1$. The maximum possible percent improvement can be calculated by

$$I_p = (1 - \eta) \cdot 100\,\%. \tag{1.4}$$

- *User-specified Benchmark*. When a more realistic or less severe benchmark is specified, a performance index may take values higher than unity, thereby indicating that the current controller is doing better than required, i.e. $J_{act} < J_{des}$, in terms of the specified criterion.

It must be stressed that the performance indices are usually *relative quantities*, i.e. set into relation to a specified performance benchmark, as defined in Eq. 1.2, rather than using absolute performance criteria. This is a key feature and difference of the CPM technology when compared to the traditional performance evaluation practice. CPM can thus be regarded as the science of control performance indices, which are introduced to simply assess many phenomena in control loops. Examples are oscillation indices, nonlinearity indices, saturation indices and stiction indices. All these and other new indices will be presented throughout this book.

1.2.2 Basic Procedure for Control Performance Management

The assessment of the performance of a control system is a complex task, which should be generally performed by applying the following main stages (Jelali 2006c; Fig. 1.8):

1. *Determination of the Capability of the Running Control System*. This is concerned with the quantification of current performance (J_{act}). Measured routine-operating data are used and analysed to compute the performance figures of the current control system, e.g. the output variances.
2. *Selection or Design of a Benchmark for Performance Assessment*. This step specifies the benchmark (J_{des}), against which the current control performance will be evaluated. This may be the minimum variance, as an upper but not achievable performance bound, or any other user-specified criterion, which defines the desired or best-possible performance given the existing plant and control equipment. It is important to note that it is not always required to implement the benchmark at the plant.
3. *Assessment and Detection of Poor Performing Loops*. Based on calculations using measured data, the closeness of the current control performance to the selected benchmark is tested for. This results in the performance classification

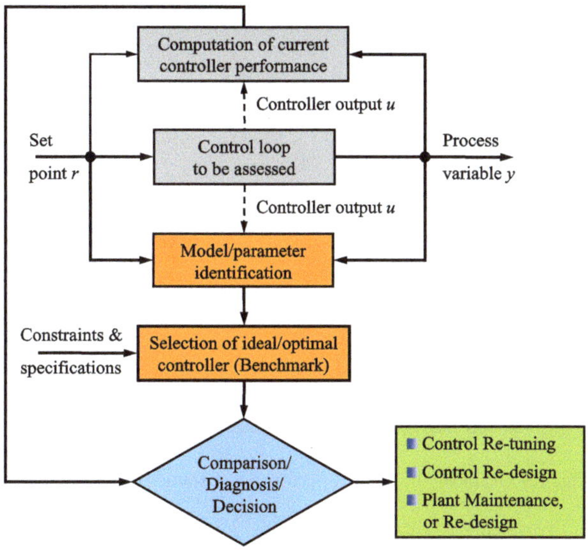

Fig. 1.8 Generic procedure for control performance management

best/good/fair/ poor of the control loop based on the performance index (η). Since most plants have at least hundreds of control loops located in different levels of hierarchy, it is important to first select the suitable monitoring paradigm (prioritisation/ranking approach, bottom-up/top-down strategy) to be followed. It is not necessary to further diagnose a process and controller when its performance is entirely satisfactory with respect to safety, product quality and plant profit. Only those control loops which are not adequately performing and offer potential benefit are considered in the subsequent diagnostic steps.

4. *Diagnosis of the Underlying Causes*. When the analysis indicates that the performance of a running controller deviates from good or desired performance, i.e. when the control loop performance is classified as "fair" or "poor", the reasons for this should be fixed in one of the problems/sources mentioned in Sect. 1.1.3. The diagnostic step is the most difficult task of CPM, where only a few approaches and studies are available. Until recently, the plant personnel must do this time consuming "detective job".
5. *Performance Improvement/Optimisation*. After isolating the causes of poor performance, corrective actions should be suggested to restore the health of the control system. In most cases, poor working controllers can be improved by retuning, i.e. adjusting their parameter settings. When the assessment procedure indicates that the desired control performance is not possible with the current process and control structure, more substantial modifications to improve the control system performance are required.

It can be easily deduced that different approaches can be selected for each stage of the procedure, and they have to be properly integrated to design a consistent overall strategy for performance monitoring. Complete and detailed CPM procedures will be developed and discussed throughout this book.

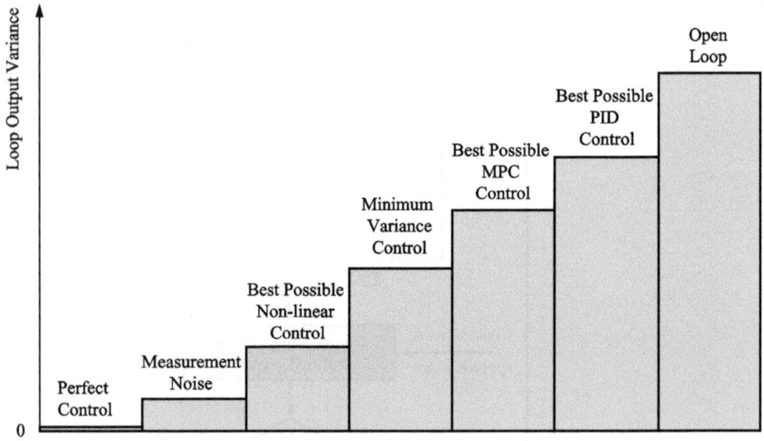

Fig. 1.9 Ranking of control performance standards in terms of achievable loop output variance

1.2.3 Controller Performance Assessment Benchmarks

Control performance assessment involves a comparison of the performance achieved by the current controller to the performance that could be attained by some standard or benchmark. Various benchmarks exist, which can be ranked based on the tightness of control quantified by the loop output variance, as shown in Fig. 1.9. Therefore, a key decision at the early stage of controller performance assessment is to select the most suitable benchmark for the application at hand. Some benchmarks are briefly introduced as follows (Hugo 1999):

- *Perfect Control.* While this may appear to be an unrealistic standard, it is in fact commonly invoked, at least implicitly. Assessing controllers based on output variance implicitly compares the performance to zero variance. Without any doubt, this is too high as standard, having in mind that the output variance largely depends on the set-point change and disturbance spectrum.
- *Best Possible Nonlinear Controller.* Two fundamental limitations to controller performance are the measurement error and the dead time: no controller can have tighter control than the random measurement error variance; and no controller can affect the process before the dead time has elapsed. The best possible nonlinear controller therefore represents a lower bound on what is achievable using a controller. However, nonlinear controllers are rare in industry due to both their complexity and the difficulty of obtaining a nonlinear model. For these reasons, there does not appear to be any performance index based on nonlinear controllers.
- *Minimum Variance Control.* For linear systems, a minimum variance controller results in the smallest possible closed-loop variance. Whereas the MVC itself may require specification of process and disturbance models, it is possible to assess the controller performance against MVC using *only* closed-loop data and an estimate of the process time delay. This makes MVC-based assessment so valuable to be adapted as a standard benchmark in CPM.

- *Best Possible MPC Controller/LQG Controller*. A more realistic index accounts for the simplified step-disturbance model of model predictive controllers (i.e. DMC), in that it determines what the closed-loop variance would be if an MPC were applied that had no process model error or move suppressions. This index explicitly addresses the fact that disturbance model in MPC may differ from the true disturbance.
- *Open-Loop*. Obviously, the variance of the open-loop process is a very perfunctory standard. Nonetheless, it is somewhat surprising that many control loops do not meet even this criterion. The study by Spencer and Elliot (1997/1998) has found that up to 80 % of controllers lead to an *increase* in variance over open-loop. The open-loop standard is however useful for determining whether *any* control should be applied—the usual benefits of introducing control have to be balanced against costs for measurement, control valve, installation and tuning.

Note that the ranking given above should be understood as flexible in the sense that under certain circumstances, e.g. the best possible MPC control can achieve the same performance as the MVC, or the best possible nonlinear control can yield a variance identical to measurement noise.

The selection of the performance standard particularly depends on the type and level of hierarchy of the controller to be assessed. For lower-level controllers, usually operated at regulatory mode, MVC-based benchmarks are the standard choice. When higher-level controllers are to be evaluated, LQG or MPC may be the right option, particularly, if the running controllers are of the MPC or even nonlinear type.

1.2.4 Challenges of Performance Monitoring Applications

As for the application of advanced control methods, there are obviously some serious problems, which prevent the widespread use of performance monitoring techniques in many process industries, even in the cases where the need is obvious and the potential benefit is known to be substantial:

- *Complexity of Process Control Systems*. There are a host of different types of industrial systems with different automation systems and hierarchies, which have to be optimised individually. Even in the same factory, there exist a large number of control loops to be monitored. Typical plants in the process industry usually contain some hundreds or even thousands control loops. The performance assessment of such complex process-control systems requires the use of systematic monitoring paradigms and strategies. It is impossible to evaluate each loop manually and individually, having in mind the sheer amount of data to be analysed. Also the number of measurement and control staff is usually limited, a fact making loop maintenance difficult.
- *Non-invasiveness*. Efficient CPM systems should be able to work only with routine operating (closed-loop) data without requiring any experimentation with the plant. This property is one key prerequisite for the acceptance of CPM in industrial practice. On the other hand, it makes the CPM task very challenging.

- *Diversity of Sources of Performance Degradation.* The causes of poor control performance are numerous and can be inherent to a particular control loop itself, to the plant design, to the interaction between plant components or loops, etc. For instance, an oscillation in a control loop may be due aggressive controller tuning, sensor or actuator faults (such as too excessive valve stiction), external disturbances, or combination of these problems.
- *Need for Specific Process Knowledge for Final Diagnosis.* The formal application of performance monitoring methods without understanding the physical fundamentals of the processes considered and without tailoring the methods to the existing automation structures may not lead to satisfactory results. Particularly for the diagnostic task, specific process knowledge is essential. Thus, a close co-operation between process engineers, control engineers and control technology consultants is necessary. Note that no kind or amount of "artificial intelligence" can replace real process knowledge/understanding.
- *Need for Specialised Software and Hardware.* The implementation and maintenance of monitoring systems requires specialised software and hardware resources, as well as high qualified and thus expensive staff. Of major concern is the data sampling rate. In theory, the data should be sampled as fast as the control interval. In practice, a sampling period of approximately $1/3$ of the process open-loop time constant is sufficient (Hugo 1999). This fast sampling presents a problem for many plant historians. Generally, these databases log data at a higher sample time, so that they are not adequate for performance assessment of fast loops. To overcome this limitation, specialised data collection programs have to be installed. However, the increasing availability of automatic data analysis tools greatly reduces the effort spent to profit from CPM technology.
- *Integration into the Maintenance Practice.* Perhaps the biggest challenge of introducing CPM systems is related to the human factors surrounding their use. The critical success factor is how a CPM application or tool integrates with existing work practices and maintenance procedures. A CPM system should work with high reliability to be accepted by the plant staff: too many false alerts or missed detections result in a reduced trust and use of the CPM system.

1.3 Key Dates of the Development of CPM Technology and Literature Survey

Control-performance monitoring and assessment (CPM) is an important technology to maintain the efficient operation of automation systems in production plants. This is achieved by indicating how far a control system is operating from its inherent optimum and what can be done to ensure that the gap between the optimum and the current performance is as small as possible over the longest possible period of operation. Usually, CPM is concerned with the assessment of the output variance due to unmeasured, stochastic disturbances, which are further assumed to be generated from a dynamic system driven by noise. For this reason, this class of CPM

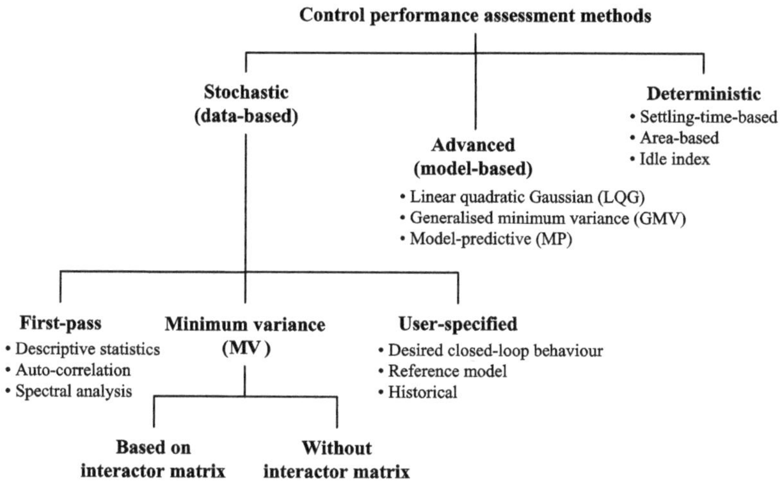

Fig. 1.10 Family tree of methods for evaluating the level of control performance

methods is referred to as *stochastic* performance monitoring. Whereas these methods provide an important aspect of the controller performance, they do not provide any information about the traditional performance, such as step changes in set-point or disturbance variables, settling time, decay ratio and stability margin of the control system. This class of CPM techniques is known as *deterministic* performance monitoring; see Fig. 1.10.

Control performance monitoring is a relatively young research field. Most theory and applications of CPM evolved during the last decade. The objective of this section to point out some of the moments in the evolution of the field, which we believe were particularly important; see Table 1.2.

Interest in theory and methods for the online analysis of control performance can be traced back to Åström (1979) and DeVries and Wu (1978). However, it was Shinskey (1990), Ender (1993) and Bialkowski (1993) who brought control performance problems in the process industry to a broader audience in the early 1990s. Since then, there was a widespread awareness that is beneficial to assess the performance of control loops. The breakthrough of the topic was due to Harris (1989), who demonstrated that the *minimum-variance benchmark* can be estimated from normal closed-loop operation data. The celebrated *Harris index*, which is based on comparison with performance obtained by minimum-variance control, was then born. The underlying principles originate from work by Åström (1979) and Box and Jenkins (1970), who established the theory of minimum-variance control (MVC), and DeVries and Wu (1978), who used these ideas for performance assessment. Desborough and Harris (1992) connected the Harris index to the squared correlation coefficient usually calculated in multiple regression analysis. The concept of MV benchmarking has then later been extended to feedback/feedforward loops; see Desborough and Harris (1993), Stanfelj et al. (1993) and Huang et al.

Table 1.2 A few key dates of the development of control performance assessment technology

1989	Harris: minimum-variance benchmark.
1993/1994	Ender/Bialkowski: audits of industrial control performance.
1995	Jofriet et al.: first control-performance assessment package.
1996	Harris et al.: assessment of MIMO control systems.
1999	• Thornhill et al.: plant-wide assessment, prediction horizon method. • Huang and Shah: FCOR algorithm, LQG benchmark. • Hägglund: idle index.
2001	• Ko and Edgar: MPC assessment. • Paulonis and Cox: Honeywell's study of large-scale CPM.
2002	Grimble: GMV benchmark; restricted-structure assessment.

(2000). The Harris concept has been applied in various process control applications over all process industry sectors and is still the standard for benchmarking control loops.

Extensions of the Harris index to unstable and non-minimum-phase systems were reported by Tyler and Morari (1995, 1996), who introduced statistical likelihood ratio tests. Lynch and Dumont (1996) used Laguerre networks to evaluate the performance index.

Soon, the drawbacks of minimum variance benchmarking have been recognised and several modified versions of the Harris index introduced. They include design specifications of the user, leading to more realistic performance indices, referred to as *user-specified benchmarks*. Work in this area was done by Kozub and Garcia (1993), Huang and Shah (1998) and Horch and Isaksson (1999). To the same category belong *historical data benchmarks* or *reference data set benchmarks* (Patwardhan et al. 1998; Huang et al. 1999; Gao et al. 2003).

The great majority, i.e. more than 90 %, of practical controllers are of PID type, and have order, structure and action constraints. Therefore, realistic performance indicators should be applied for their assessment, as first proposed by Eriksson and Isaksson (1994) and Ko and Edgar (1998). These approaches calculate a lower bound of the variance by restricting the controller type to PID only (optimal PID benchmarking) and allow for more general disturbance models. An explicit "one-shot" solution for the closed-loop output was derived by Ko and Edgar (2004) as a function of PID settings. Recent developments in this pragmatic direction have been worked out in Horton et al. (2003) and Huang (2003). Moreover, Grimble (2002b, 2003) provided a theoretical framework on the topic, referred to as *restricted-structure controller benchmarking*.

In 1995, the (likely) first CPM (expert) system called QCLiP (Queen's/QUNO Control Loop Performance Monitoring) making use of an MVC-based performance index and other analyses of closed-loop process data was reported by Jofriet et al. (1995); see also Harris et al. (1996b).

In the same year, Hägglund (1995) has shown that one of the main problems with control loop performance is the presence of oscillations in the loops. An *oscillation index* based on the magnitude of the integrated absolute error (IAE) between successive zero crossings of the control error and a procedure for detecting oscillations in control loops were introduced. Similar methods have been proposed by Forsman and Stattin (1999) and Mia and Seborg (1999). Since then, this topic has attracted much attention of research and application, starting with a series of papers by Thornhill and Hägglund (1997) and Thornhill et al. (2001, 2003b). Oscillation detection and diagnosis is still one of the very active areas in CPM today. A special, very rich topic is the diagnosis of valve stiction, where many methods have been developed, such as by Horch (1999, 2000), Kano et al. (2004), Choudhury et al. (2005a), Singhal and Salsbury (2005), Yamashita (2006) and He et al. (2007). Recent contributions are found by Choudhury et al. (2006a, 2006b) and Jelali (2008).

In the mid-1990s, academic research interest has shifted to the assessment of *MIMO* control systems using the minimum variance benchmark (Harris et al. 1996a; Huang et al. 1997a, 1997b; 1997c). In the performance assessment of MIMO feedback systems, the so-called interactor matrix plays an important role. The assumption that the interactor matrix be known turned out to be a major restriction on the generality of the method, since its estimation is quite involved and the accuracy problematic in general. Thus, it is highly desirable to surmount the problem. Approaches in this practical direction have been proposed by Ettaleb (1999), Ko and Edgar (2001b), McNabb and Qin (2003) and recently Huang et al. (2005a, 2005b) and Xia et al. (2006).

The other important research direction was aimed towards *plant-wide* (large-scale) control-loop-performance assessment. A significant advance in this topic was due to Thornhill et al. (1999), who showed that it is useful to provide default parameters for the performance index algorithm for various generic categories of refinery control loops. This work substantially lowered the barrier to large-scale implementation of performance-index-based monitoring. Especially, the extended horizon performance index (EHPI) method has been systematically developed further. Advances and new directions in this topic are well documented in Thornhill and Horch (2007); see also Thornhill et al. (2003c) and Xia and Howell (2003).

In 1999, the first textbook on "Performance Assessment of Control Loops" by Huang and Shah appeared. The book authors presented an efficient, stable filtering and correlation (*FCOR*) method to estimate the MV benchmark for SISO, MIMO and feedback/feedforward control systems. They also proposed the linear-quadratic Gaussian (*LQG*) regulator as an alternative to the MV benchmark to take into account the control effort in the performance assessment.

As a deterministic performance-assessment method, Swanda and Seborg (1997, 1999) proposed the *dimensionless settling time* of the closed-loop and the *dimensionless integral of absolute value* of control error as performance indices. Also in 1999, Hägglund presented a method to detect sluggish control loops by using the so-called *idle index* to detect conservatively tuned controllers when load disturbances occur.

As the ultimate multivariable controller in many process industries is model predictive control (MPC), active research is going on to assess the performance of

MPC, initiated by Patwardhan et al. (1998), Patwardhan (1999) and Zhang and Henson (1999). Recent work on the subject is found by Ko and Edgar (2001a), Shah et al. (2001), Schäfer and Çinar (2002), Gao et al. (2003) and Julien et al. (2004). Given the complexity of MPC that involves model errors, disturbance changes, optimal target settings, active constraint sets and controller tuning, the MPC performance monitoring is largely an unsolved problem.

Also, it is worthwhile to mention the *large-scale* performance assessment study by Desborough and Miller (2002), already published in 2001. They provided a very good documentation of the current status of industrial controller performance and suggested future directions of research and desired attributes of CPM systems.

A straightforward extension of the MV benchmark by considering control action penalisation leads to the more flexible approach of generalised MV (*GMV*) benchmarking suggested by Grimble in 2002. A multivariable version of the GMV control assessment was derived by Majecki and Grimble (2004b) using the concept of the interactor matrix of the generalized plant.

Recent trends in the field are dealing with plant-wide (large-scale) monitoring, approaches to automate the controller diagnosis, transforming the performance indices into economic values and integration of CPM into maintenance practices and asset management strategies (Grimble and Uduehi 2001; Ahsan et al. 2004; Farenzena and Trierweiler 2006; Xu et al. 2006a, 2006b). Also recent work is on extending the CPM techniques for time-varying systems (Huang 2002; Olaleye et al. 2004a, 2004b) and nonlinear systems (Majecki and Grimble 2004a, 2004b).

Several authors have published reviews of CPM theoretical issues and applications, such as the papers by Qin (1998), Harris et al. (1999), Harris and Seppala (2001) and Shah et al. (2001), Dittmar et al. (2003) and Thornhill et al. (2003a), and the book by Ordys et al. (2007). Jelali (2006c) provides an overview of the latest technical developments and industrial applications in the CPM field.

Two recently published books (Choudhury et al. 2008; Jelali and Huang 2010) deal with a special topic of CPM: the detection and diagnosis of stiction in control valves. It is well known that valve stiction in control loops causes oscillations in the form of periodic finite-amplitude instabilities, known as *limit cycles*. This phenomenon leads to increased variability in product quality, accelerated equipment wear, or even to control-system instability.

1.4 Outline of the Book

The overall structure and an overview of the main contents of the book are illustrated in Fig. 1.11. The book is divided into four parts comprising 16 chapters briefly described as follows.

The first step in the solution of any CPM task is to automatically identify problematic loops that present the best opportunities for improvement on a plant-wide scale. Part I is devoted to reviewing the state-of-the-art in performance assessment,

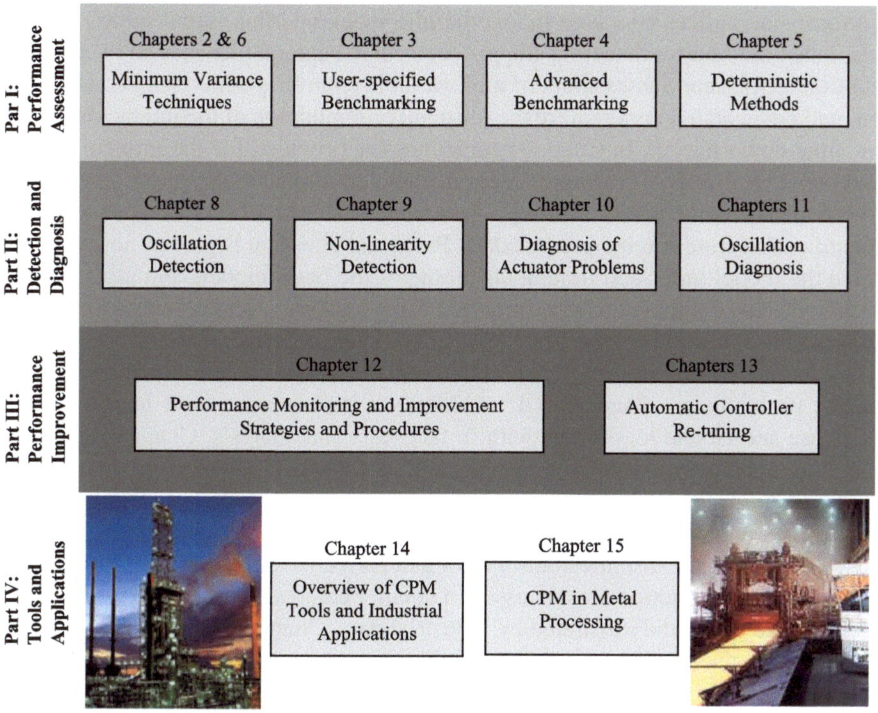

Fig. 1.11 Overview of the main contents of the book

including basic and advanced performance benchmarking methods, and is divided into six chapters. An in-depth presentation of the-state-of-the-art in control performance *assessment*, i.e. the evaluation of the level of performance of control loops, is provided in Chap. 2. The assumptions and fundamental limitations of the methods are described as well as their strengths and weaknesses. The review starts with giving some basic system descriptions. The main focus of the chapter is on presenting assessment methods based on minimum variance control (MVC) for single feedback control, combined feedback and feedforward control and cascade control loops. Since MVC benchmarking has drawbacks in practice, many alternative benchmarks have been proposed. User-specified assessment methods are treated in Chap. 3. They mainly consider user design preferences to evaluate control performance.

Chapter 4 is devoted to detailed discussion of advanced assessment techniques, including linear-quadratic Gaussian (LQG) benchmarking, generalised minimum variance (GMV) benchmarking and model predictive control (MPC) assessment. The notion of performance limit curve is introduced and used as the main vehicle for evaluating control loop performance by considering a combined performance objective with penalty on control moves. Chapter 5 contains three deterministic assessment techniques, namely: an assessment method based on set-point response data, a method for the load-disturbance assessment of PI controllers, combining some performance indices (the area index, the idle index and the output index) and the idle index method for detecting sluggish control. Chapter 6 provides an overview of control performance assessment of multivariable control systems, with an emphasis on methods which do not require the (difficult) computation of the interactor matrix, i.e. time-delay matrix. In Chap. 7, guidelines are provided for the implementation and parameterisation of the assessment methods described in the previous chapters. The focus is on the pre-processing of data, the selection of model types and their identification from routine process data. Particular attention is paid to how to determine the model order and time delay. Some of the basic models and identification techniques are compared, concerning assessment accuracy and computational load, to provide suggestions of the best suited approaches in practice.

Performance monitoring can be effective and return value only if the performance problems are fixed. Part II of the book is concerned with how to detect, diagnose and isolate loop faults with further tests and analysis. Chapter 8 contains a brief description of selected univariate and multivariate statistical process control (SPC) techniques. Methods and procedures for automatic and non-invasive detection of unwanted oscillations in control loops are treated in Chap. 9. Techniques based on integrated absolute control error and zero crossings and those based of the auto-covariance function are discussed in detail. Features and practical issues of the methods are given and illustrated by industrial data. Chapter 10 presents techniques for detecting the presence of process nonlinearities that may lead to limit cycles in the loop. This includes non-invasive data-based methods based on the analysis of bicoherence and surrogates of time series. The bicoherence and surrogates methods are demonstrated and compared with two industrial case studies, in which the main task is to find out the source of oscillations propagating through whole plants. Also discussed in this chapter is how to detect saturating controller outputs that may also lead to limit cycles. In Chap. 11, it is shown how actuator problems in control loops can be automatically detected and diagnosed. The main focus is on analysing and detecting static friction (stiction) in control valves, being the most common cause of oscillations found in the process industry. Non-invasive stiction detection methods based on cross-correlation, curve fitting and elliptic patterns of PV–OP plots are presented and discussed in detail. Tests for confirming the presence of stiction are then described. An oscillation diagnosis procedure that combines different techniques is proposed. In Chap. 12, we develop a novel technique for detection and quantification of valve stiction in control loops from normal closed-loop operating data based on two-stage identification of a Hammerstein model. This method ends up with a diagnosis algorithm that helps discriminate between loop problems induced by valve stiction, external disturbances, or aggressive controller.

Part III of this monograph provides remedies for control loop faults and performance improvement. In Chap. 13, some paradigms and strategies for monitoring the performance of complex process-control systems are introduced and discussed. A comprehensive procedure for performance monitoring is suggested that combines different methods described in the previous chapters. Chapter 14 presents a new contribution towards further development of the field of controller auto-tuning within the framework of CPM. The main objective is to propose novel methods for controller tuning based on control performance assessment results, i.e. performance criteria and indices. This includes optimisation-based assessment and tuning techniques as well as iterative assessment and tuning based on normal operating data, aiming to maximise the control performance index. New control performance indices and controller re-tuning techniques are presented. Illustrative examples demonstrate the applicability and efficiency of the proposed methods.

Part IV of the book is concerned with the CPM technology in industrial practice. The detailed current status in industrial CPM technology and applications is given in Chap. 15. Some trends in the application of a number of control-performance metrics and methods in different process industries are shown based on an evaluation of publications during the 20 years after the key research by Harris (1989). The chapter also includes an overview of CPM packages that have been developed and/or which are commercially available, to illustrate how some control performance indices and monitoring methods are already used in products. Chapter 16—one of the main contributions of the present book—is devoted to transfer the CPM technology into the challenging field of metal processing. First, an introduction to the metal processing technology and automation is provided. Successful application studies and tailored CPM tools from cold rolling area are described and discussed. It is finally shown how the developed CPM systems are integrated into the mill infrastructure (automation) and the maintenance practices of the customer.

Appendix A and Appendix B contain reviews of some definitions, relationships and properties of basic and higher-order statistics, respectively. These concepts are used throughout the book. Industrial control loops from which data are used within the book to illustrate the methods are described in Appendix C.

Part I
Evaluation of the Level of Control Performance

Chapter 2
Assessment Based on Minimum-Variance Principles

An important goal of quality improvement in manufacturing is the reduction of variability in product attributes. Producing more consistent output improves product performance and may reduce manufacturing costs. Therefore, the standard control performance assessment methods are based on the MV principle or modifications of it. The key point is that the MV benchmark (as a reference performance bound) can be estimated from routine operating data without additional experiments, provided that the system delay is known or can be estimated with sufficient accuracy.

This chapter provides an introduction to the theory of MV performance assessment. Some basic notations and concepts are given in Sect. 2.1. The derivation of minimum variance control is recalled in Sect. 2.2. In Sect. 2.3, the autocorrelation test to check minimum variance is considered. Section 2.4 presents the celebrated MVC-based performance index, known as the Harris index, and how to estimate it using different algorithms. In Sect. 2.5, the extension of MV assessment to feedback-plus-feedback loops is described. Its extension to the assessment of set-point tracking and cascade control will be provided in Sect. 2.6. All methods presented are illustrated using many examples.

2.1 System Descriptions and Basics

For the description of the methods in this chapter, we assume generic feedback control systems shown in Fig. 2.1, where $r(k)$ is the set point, $u(k)$ the controller output, $e(k)$ the control error, $y(k)$ the process output, and $\varepsilon(k)$ is the unmeasured disturbance. G_c, G_p and G_ε denote the transfer functions of the feedback controller, the process and disturbance dynamics, respectively. The set point is set to zero by convenience, and the disturbances are assumed to be zero mean. If the reference value and/or the mean of the disturbances are not zero, they can be made mean-free by a simple transformation.

Let the system under consideration be described by an ARMAX model (see Fig. 2.1)

$$A(q)y(k) = q^{-\tau}B(q)u(k) + C(q)\varepsilon(k), \tag{2.1}$$

M. Jelali, *Control Performance Management in Industrial Automation*,
Advances in Industrial Control, DOI 10.1007/978-1-4471-4546-2_2,

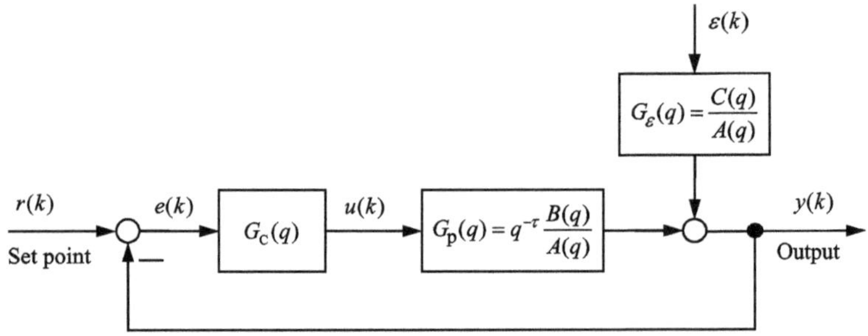

Fig. 2.1 Generic feedback control system structure

where $\varepsilon(k)$ is a zero-mean white noise with the variance σ_ε^2, also referred to as chocks. $A(q)$, $B(q)$ and $C(q)$ are polynomials[1] in q^{-1} of order n, m and p, respectively:

$$\begin{aligned} A(q) &= 1 + a_1 q^{-1} + a_2 q^{-2} + \cdots + a_n q^{-n}, \\ B(q) &= b_0 + b_1 q^{-1} + b_2 q^{-2} + \cdots + b_m q^{-m}, \\ C(q) &= 1 + c_1 q^{-1} + c_2 q^{-2} + \cdots + c_p q^{-p}. \end{aligned} \tag{2.2}$$

τ is an integer number of sampling periods[2] (i.e. the dynamics contain a delay of τ samples), so that the leading term of B is non-zero constant. This means that B is strictly rational or that the input u does not affect the output y immediately, i.e., there is at least one-sample delay ($\tau \geq 1$). Also note that the polynomials A and C are monic, because their leading term is unity.

The noise model of an ARMAX model includes only random steps. For a generalisation of the treatment, an ARIMAX model of the form

$$A(q)y(k) = q^{-\tau} B(q)u(k) + \frac{C(q)}{\Delta}\varepsilon(k) \tag{2.3}$$

may be needed to describe drifting (non-stationary) disturbances. As before, u is the input, y is the output, and ε is the white noise. Δ is the backward difference operator, i.e. $\Delta = 1 - q^{-1}$. ARIMAX models are typically used for the design of model predictive controllers, particularly DMC and GPC.

If the process is assumed to be stable, it can be expressed as the *infinite impulse response* (*IIR*)

$$G_\mathrm{p}(q) = \sum_{i=\tau}^{\infty} h_i q^{-i}, \qquad \lim_{i\to\infty} h_i = 0. \tag{2.4}$$

[1]Following Ljung (1999), q is chosen as an argument of the polynomials rather than q^{-1} (which perhaps would be more natural in view of the right side) in order to be in formal agreement with z-transform and Fourier-transform expressions.

[2]For discrete systems with no time delay, there is a minimum one-sample delay because the output depends on the previous input, i.e. $\tau = 1$.

In practice, the impulse response is truncated at time n_p, called "time-to-steady-state":

$$G_\mathrm{p}(q) \approx \sum_{i=\tau}^{n_\mathrm{p}} h_i q^{-i}, \qquad \lim_{i\to\infty} h_i = 0. \tag{2.5}$$

When considering an ARIMAX model, the disturbance transfer function is marginally stable due to the pole at $q = 1$, so it can be shown that

$$G_\varepsilon(q) = \sum_{i=0}^{\infty} e_i q^{-i}, \tag{2.6}$$

where the coefficients e_i converge to $C(1)/A(1)$. Defining n_ε as the settling time of the disturbance dynamics enables Eq. 2.6 to be expressed as

$$G_\varepsilon(q) = \sum_{i=0}^{n_\varepsilon} e_i q^{-i} + e_{n_\varepsilon} \sum_{i=n_\varepsilon+1}^{\infty} q^{-i}. \tag{2.7}$$

As the "differenced" load transfer function, i.e. ΔG_ε, is stable, it can be written as

$$\Delta G_\varepsilon(q) = \frac{C(q)}{A(q)} = \sum_{i=0}^{\infty} d_i q^{-i} \approx \sum_{i=0}^{n_\varepsilon} d_i q^{-i} \tag{2.8}$$

with $d_0 = e_0 = 1$, $d_i = e_i - e_{i-1}$ for $i = 1, 2, \ldots, n_\varepsilon$.

2.1.1 Note on Input–Output Models

Following the system identification and control performance monitoring literature, the *polynomial operator form* is used throughout this book for the description of input–output models. This makes use of the backwards-shift (or unit delay) operator q^{-1} defined as

$$q^{-1} f(k) = f(k-1).$$

For instance, the difference equation of a linear system

$$y(k) + a_1 y(k-1) + \cdots + a_n y(k-n) = b_0 u(k) + b_1 u(k-1) + \cdots + b_m u(k-m) \tag{2.9}$$

thus becomes

$$\left(1 + a_1 q^{-1} + \cdots + a_n q^{-n}\right) y(k) = \left(b_0 + b_1 q^{-1} + \cdots + b_m q^{-m}\right) u(k). \tag{2.10}$$

This will simply be denoted by

$$A(q) y(k) = B(q) u(k), \tag{2.11}$$

where $A(q)$ and $B(q)$ are the following polynomials, in fact, depending on q^{-1} in the form

$$A(q) = 1 + a_1 q^{-1} + \cdots + a_n q^{-n}, \tag{2.12}$$

$$B(q) = b_0 + b_1 q^{-1} + \cdots + b_m q^{-m}. \tag{2.13}$$

The ratio of both polynomials

$$G(q) = \frac{B(q)}{A(q)} \tag{2.14}$$

is considered as the discrete *transfer operator* of the discrete transfer function (strictly speaking, $G(z)$ should be used in the latter case) of the system. For *time-invariant linear systems*, the forward-shift operator q and the complex variable z defining the z-transform are equivalent. In this case, one can use either one (q is just replaced with z), and the appropriate signification will result from the context; see Ratjen and Jelali (2006).

However, the shift operator q and thus the transfer operator $G(q)$ can be applied for any discrete-time system, thus as well to linear systems with time-varying coefficients (e.g. in the context of adaptive control) or nonlinear systems, where the z-transform and thus the concept of transfer function *do not* apply.

Note that the variable z is analytical: we speak of numerical values z_i of the poles of a transfer function $G(z)$. The operator q does not possess any numerical values; it gives the transfer function $G(q)$, whose mathematical expression is *strictly* identical to $G(z)$.

2.2 Minimum-Variance Control (MVC)

The minimum-variance control (MVC), also referred to as optimal H_2 control and first derived by Åström (1979), is the best possible feedback control for linear systems in the sense that it achieves the smallest possible closed-loop output variance. More specifically, the MVC task is formulated as minimisation of the variance of the error between the set point and the actual output at $k + \tau$, given all the information up to time k:

$$J = E\{[r - y(k+\tau)]^2\} \tag{2.15}$$

or

$$J = E\{y^2(k+\tau)\}, \tag{2.16}$$

when the set point is assumed zero (without loss of generality), i.e., the case of regulation or disturbance rejection is considered. The discrete time delay τ is defined as the number of whole periods of delay in the process, i.e. (Harris 1989)

$$\tau = 1 + f = 1 + \mathrm{int}(T_\mathrm{d}/T_\mathrm{s}), \tag{2.17}$$

where T_d is the (continuous) process delay arising from true process dead time or analysis delay, and T_s denotes the sampling time. f is the number of integer periods of delay.

The design of minimum-variance controller requires a perfect system model and a perfect disturbance model and will result in a complete cancellation of the error (other than measurement noise) one sample time after the system time delay τ. The test for detecting MVC follows immediately (see Sect. 2.3): if the sample autocorrelations of the system output are zero beyond τ, then MVC is being achieved.[3] Further, if there is no process noise, i.e. $\varepsilon(k) = 0$, then MVC is equivalent to a deadbeat controller.

To enable minimisation of Eq. 2.15 with respect to the control input u, first we need to relate the controlled output y to u. When both sides of Eq. 2.1 are multiplied by E_τ and the left side is substituted using the *Diophantine equation*, also known as the *polynomial division identity*,

$$E_\tau(q)A(q) = -q^{-\tau}F_\tau(q) + C(q), \tag{2.18}$$

where

$$E_\tau(q) = e_0 + e_1q^{-1} + e_2q^{-2} + \cdots + e_{\tau-1}q^{-(\tau-1)}, \tag{2.19}$$

$$F_\tau(q) = f_0 + f_1q^{-1} + f_2q^{-2} + \cdots + f_{n-1}q^{-(n-1)}, \tag{2.20}$$

we get the prediction of the output τ steps ahead as

$$y(k+\tau) = \frac{F_\tau(q)}{C(q)}y(k) + \frac{E_\tau(q)B(q)}{C(q)}u(k) + E_\tau(q)\varepsilon(k+\tau). \tag{2.21}$$

The right-hand side of this equation contains the three terms: present and past output signals, present and past control signals and future error signals, respectively. As future terms are not available at time k, only the realisable terms of the optimal output prediction are then given by

$$\hat{y}(k+\tau) = \frac{F_\tau(q)}{C(q)}y(k) + \frac{E_\tau(q)B(q)}{C(q)}u(k). \tag{2.22}$$

Now, the control action is selected to optimise the variance of the output (τ steps ahead), i.e.

$$\begin{aligned}\min_{u(k)} J(k) &= \min_{u(k)} E\{y^2(k+\tau)\} \\ &= \min_{u(k)} E\left\{\frac{F_\tau(q)}{C(q)}y(k) + \frac{E_\tau(q)B(q)}{C(q)}u(k) + E_\tau(q)\varepsilon(k+\tau)\right\}.\end{aligned} \tag{2.23}$$

[3] One should here remember the linear correlation test used for the validation of identified linear models; see Sect. 2.3.

(The set point is first assumed to be zero.)[4] This equation contains past inputs, past outputs and future disturbances. As the disturbance is assumed to be white noise, its future values cannot be correlated with past signals. Therefore, the minimum will be achieved when the sum of the first two components is set to zero:

$$\frac{F_\tau(q)}{C(q)}y(k) + \frac{E_\tau(q)B(q)}{C(q)}u(k) = 0, \tag{2.24}$$

which gives the MVC law

$$u(k) = -\frac{F_\tau(q)}{E_\tau(q)B(q)}y(k). \tag{2.25}$$

The same procedure applied to ARIMAX models (Eq. 2.3) leads to

$$\Delta u(k) = -\frac{F_\tau(q)}{E_\tau(q)B(q)}y(k). \tag{2.26}$$

These control laws imply that, no matter what the system dynamics is, all system poles (included in $A(q)$ and thus $F(q)$) and zeros, included in $B(q)$, are cancelled by MVC. Consequently, the basic MVC design is restricted for stable and minimum-phase systems. In practice, cancelling of system dynamics means to exert aggressive control effort, which may not be tolerated from the operational point of view. Another limitation is the sensitivity against system changes, i.e. the lack of robustness to modelling errors.

For non-minimum-phase systems, i.e. with unstable $B(q)$, MVC can be designed with some (minor) modifications. The unstable zeros are not inverted, similar to the treatment in the IMC design (Morari and Zafiriou 1989). The control law for non-minimum-phase (ARMAX) processes is given by

$$\Delta u(k) = -\frac{S(q)}{R(q)}y(k), \tag{2.27}$$

where S and R are the solution of the Diophantine equation

$$A(q)R(q) = -q^{-\tau}B(q)S(q) + C(q)B_-(q)B_+^{-1}(q). \tag{2.28}$$

The polynomial B is decomposed into a minimum phase part B_- and non-minimum phase part B_+.

From the MVC laws given above it is clear that the main vehicle for calculating minimum-variance controllers is the solution of the Diophantine Eqs. 2.18 and 2.28. For simple cases, it is possible to get solutions; see Example 2.1. However, constructing solutions for Diophantine equations usually requires the use of a software

[4]The basic MVC is designed to solve regulation problems, where the objective is to compensate for stochastic disturbances and not to follow a reference trajectory. However, MVC can be extended to include variations in the reference, as described below.

package. A standard one for this purpose is available from Kwakernaak and Sebek (2000). Another solver is provided by Moudgalya (2007) in form of a MATLAB function xdync.

Using the MVC, the minimum value of the output variance, shortly denoted minimum variance, is achieved:

$$J_{\min}(k) = \min_{u(k)} E\{y^2(k+\tau)\} = E\{E_\tau(q)\varepsilon(k+\tau)\}$$
$$= \left(\sum_{i=0}^{\tau-1} e_i^2\right)\sigma_\varepsilon^2 \equiv \sigma_{\mathrm{MV}}^2, \quad (2.29)$$

where σ_ε^2 is the (disturbance) noise variance. Note that σ_{MV}^2 is the same as the variance of the prediction error $y - \hat{y}$. The achieved output of the closed-loop system under MVC is

$$y(k) = E_\tau(q)\varepsilon(k). \quad (2.30)$$

Note that whereas the controller itself may require the specification of the system model and disturbance model, both are not needed for MVC-based performance assessment, as described below (Sect. 2.4). It is important to stress that the adoption of MVC as a benchmark does not imply that it should be the goal towards which the existing control should be driven, or that it is always practical, desirable, or even possible to implement. Nevertheless, the performance bound set by the MVC is exceeded by all other (linear) controllers; hence, it serves as an appropriate benchmark against which the performance of other controllers may be compared.

The reader is encouraged to consult the textbook by Moudgalya (2007: Chap. 11), including many examples and MATLAB functions (mv for minimum phase systems, mv_nm for non-minimum phase systems). Minimum variance control (placed in a conventional feedback structure) can be viewed in an IMC structure or an SPC structure; see Sect. 3.2. The equivalence between MVC and IMC was revealed by Bergh and MacGregor (1987) to analyse the robustness of MVC. Refer also to Qin (1998), who derived the MVC using the IMC structure.

Example 2.1 Consider the first-order system described by the transfer function

$$y(k) = \frac{q^{-\tau}}{1 + a_1 q^{-1}} u(k) + \frac{1}{1 + a_1 q^{-1}} \varepsilon(k) \quad (2.31)$$

with $a_1 = -0.9$ and the time delay $\tau = 3$. This is an ARMAX model with

$$A(q^{-1}) = 1 + a_1 q^{-1}, \quad n = 1,$$
$$B(q^{-1}) = 1, \quad m = 0,$$
$$C(q^{-1}) = 1, \quad p = 0.$$

The Diophantine Eq. 2.18 takes the form

$$E_3(q^{-1})A(q^{-1}) + q^{-\tau} F_3(q^{-1}) = C(q^{-1}),$$

$$(1+e_1q^{-1}+e_2q^{-2})(1+a_1q^{-1})+f_0q^{-3}=1.$$

Comparing the same powers of q^{-1} gives

$$\begin{aligned}
q^0&: \quad 1=1,\\
q^{-1}&: \quad e_1+a_1=0 \Rightarrow e_1=-a_1,\\
q^{-2}&: \quad e_2+a_1e_1=0 \Rightarrow e_2=a_1^2,\\
q^{-3}&: \quad a_1e_2+f_0=0 \Rightarrow f_0=-a_1^3.
\end{aligned}$$

The closed loop is then given by (Eq. 2.30)

$$y(k)=E_\tau(q^{-1})\varepsilon(k)=(1-a_1q^{-1}+a_1^2q^{-2}+\cdots)\varepsilon(k).$$

In fact, the first three terms will be the same irrespective of the (linear) controller used. The MVC law has the form (Eq. 2.25):

$$u(k)=-\frac{-a_1^3}{1-a_1q^{-1}+a_1^2q^{-2}}y(k).$$

This gives for $a_1=-0.9$:

$$u(k)=-\frac{0.729}{1+0.9q^{-1}+0.81q^{-2}}y(k). \tag{2.32}$$

This control law can also be determined using the function mv from Moudgalya's MATLAB software (Moudgalya 2007: Sect. 11.4).

2.3 Auto-Correlation Test for Minimum Variance

Auto-correlation is a method that is used to determine how data in a time series are related. Auto-correlation-based analysis provides to discover the nature of disturbances acting on the process and how they affect the system by comparing current process measurements patterns with those exhibited in the past during "normal" operation.

A fundamental test for assessing the performance of control loops is to check the auto-correlation of the output samples: the autocorrelation should die out beyond the time delay τ. Using a representative sample of measured output data, the sample auto-correlation can be computed (i.e. estimated). Statistically significant values of the estimated auto-correlations existing beyond the delay provide evidence that the current controller is not minimum variance. Furthermore, if there exist many large auto-correlation values that persist beyond τ, the control performance deviates substantially from the MV performance bound. If only few slightly significant values exist beyond τ, the performance is close to that of MVC. Since the auto-correlations are statistical estimates based on a finite sample of data, they will never be truly zero.

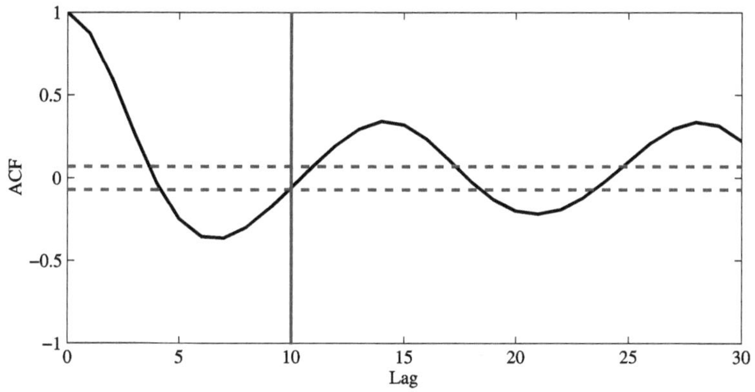

Fig. 2.2 Example of auto-correlation test

Therefore, to assess whether the true auto-correlations $\rho_{yy}(j)$ might be zero or not, their estimated values must be compared to their statistical confidence intervals, e.g. 95 % or 2σ. Box and Jenkins (1970) showed that, if $\rho_{yy}(j)$ is zero for $j \geq \tau$, then the variance is

$$\sigma^2 = \text{var}\{\rho_{yy}(j)\} \approx \frac{1}{N}\left[1 + 2\sum_{i=1}^{\tau-1} \rho_{yy}^2(i)\right], \quad j \geq \tau. \tag{2.33}$$

Therefore, the 95 % confidence interval for $\rho_{yy}(j)$ is $[-2\sigma, \; 2\sigma]$. If most auto-correlation coefficients $\rho_{yy}(j)$ are inside this interval for $j \geq \tau$, the control is roughly achieving minimum variance; otherwise, it is not.

Example 2.2 Figure 2.2 depicts the auto-correlation estimates for a gauge control loop and their 95 % confidence levels. It is observed that the auto-correlation functions are far outside the confidence limits after the time delay of 10. Therefore, we conclude that the control is not achieving minimum variance. Furthermore, the auto-correlation function is oscillatory, indicating that oscillation exists in the original data.

The motivation behind the use of auto-correlation function (ACF) is that it can be easily estimated from plant response data. Moreover, the dynamic response characteristics for data trends can be inferred without having to resort to the more complicated tasks associated with the identification and interpretation of time-series models. For example, a slowly decaying auto-correlation function implies an under-tuned loop, and an oscillatory ACF typically implies an over-tuned loop. For multivariable systems, off-diagonal plots can be used to trace the source of disturbance or the interaction between each process variables. Figure 2.3 shows an example taken from Huang et al. (1999), clearly indicating that the first loop has relatively poor performance while the second loop has very fast decay dynamics and thus good

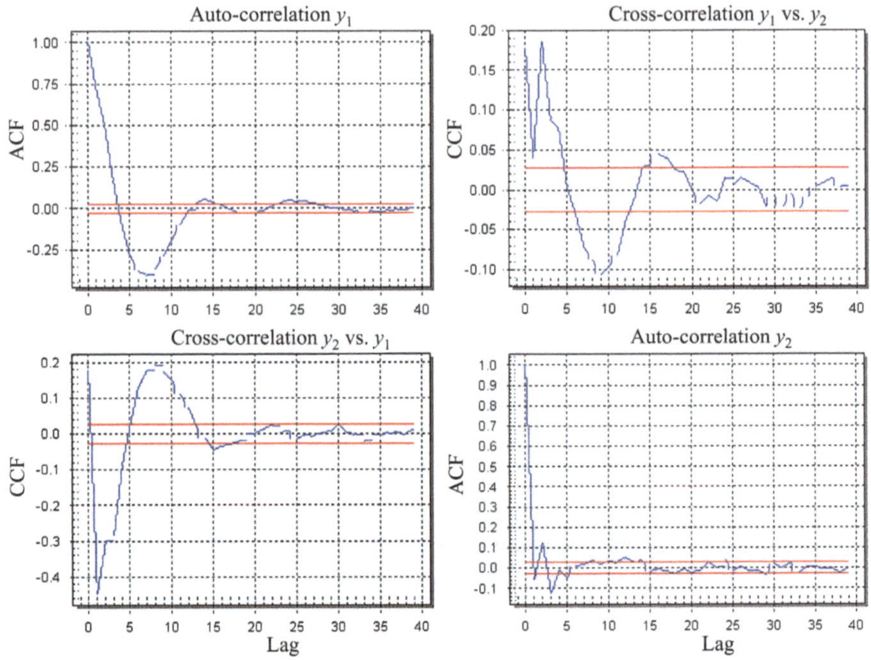

Fig. 2.3 Correlation functions of a multivariate process (Huang et al. 1999)

performance. The off-diagonal subplots indicate interaction between the two loops. Note that the ACF plot of the multivariate system is not necessarily symmetric.

2.4 Minimum-Variance Index/Harris Index

In the following, we present algorithms that will use routine (closed-loop) operating data to assess the performance of control loops against MVC as benchmark. MVC-based assessment first described by Harris (1989) compares the actual system-output variance σ_y to the output variance σ_{MV}^2 as obtained using minimum-variance controller applied to an estimated time-series model from measured output data.

The Harris index is defined as

$$\eta_{\mathrm{MV}} = \frac{\sigma_{\mathrm{MV}}^2}{\sigma_y^2}. \tag{2.34}$$

This index will of course be always within the interval [0, 1], where values close to unity indicate good control with respect to the theoretically achievable output variance. “0” means the worst performance, including unstable control. No matter what the current controller is, we need only the following information about the system:

- Appropriately collected closed-loop data for the controlled variable.
- Known or estimated system time delay (τ).

Moreover, there are two advantages for using this index over a simple error variance metric:

1. Taking the ratio of the two variances results in a metric that is (supposedly) independent of the underlying disturbances—a key feature in an industrial situation, where the disturbances can vary widely.
2. The metric is scale independent, bounded between 0 and 1. This is an important consideration for a plant user, who might be faced with evaluating hundreds or even thousands of control loops.

2.4.1 Estimation from Time-Series Analysis

From the measured (closed-loop) output data, a time-series model, typically of AR/ARMA type, is estimated:

$$y(k) = \frac{\hat{C}(q)}{\hat{A}(q)}\varepsilon(k). \tag{2.35}$$

A series expansion, i.e. impulse-response, of this model gives

$$\begin{aligned} y(k) &= \left(\sum_{i=0}^{\infty} e_i q^{-i}\right)\varepsilon(k) \\ &= \underbrace{\left(e_0 + e_1 q^{-1} + e_2 q^{-2} + \cdots + e_{\tau-1} q^{-(\tau-1)}\right)\varepsilon(k)}_{\text{feedback-invariant}} \\ &\quad + \underbrace{\left(e_\tau q^{-\tau} + e_{\tau+1} q^{-(\tau+1)} + \cdots\right)\varepsilon(k)}_{\text{feedback-varying}}. \end{aligned} \tag{2.36}$$

The first τ impulse response coefficients can be estimated through τ-term polynomial long division, or equivalently via resolution of the Diophantine identity:

$$\hat{C}(q) = \hat{E}_\tau(q)\hat{A}(q) + q^{-\tau}\hat{F}_\tau(q), \tag{2.37}$$

where $\hat{E}_\tau$ is an estimate of E_τ in Eq. 2.18. The *feedback-invariant terms* are not functions of the process model or the controller; they depend only on the characteristics of the disturbance acting on the process.

Since the first τ terms are invariant irrespective of the controller (Fig. 2.4), the minimum-variance estimate corresponding to the feedback-invariant part is given by

$$\sigma_{\mathrm{MV}}^2 = \sum_{i=0}^{\tau-1} e_i^2 \sigma_\varepsilon^2. \tag{2.38}$$

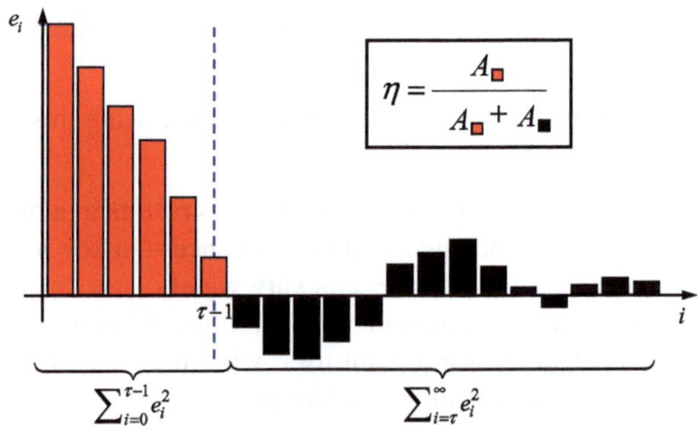

Fig. 2.4 An impulse response showing the contributions to the Harris index

The first coefficient of the impulse response, e_0, is often normalised to be equal to unity.

The estimate of the actual output variance can be directly estimated from the collected output samples using the standard relation in Eq. 1.1. However, it is suggested to use the (already) estimated time-series model also for evaluating the current variance. From the series expansion of the time-series model (Eq. 2.36), we obtain

$$\sigma_y^2 = \sum_{i=0}^{\infty} e_i^2 \sigma_\varepsilon^2. \tag{2.39}$$

Since the noise variance will be cancelled in Eq. 2.34, it is neither needed nor has an effect on the performance index. This compares the sum of the τ first impulse-response coefficients squared to the total sum; see Fig. 2.4.

The performance index η_{MV} corresponds to the ratio of the variance, which could theoretically be achieved under minimum variance control, to the actual variance. η_{MV} is a number between 0 (far from minimum-variance performance) and 1 (minimum-variance performance) that reflects the inflation of the output variance over the theoretical minimum variance bound. As indicated in Desborough and Harris (1992), it is more useful to replace σ_y^2 by the mean-squares error of y to account for offset

$$\eta_{\mathrm{MV}} = \frac{\hat{\sigma}_{\mathrm{MV}}^2}{\mathrm{MSE}} = \frac{\hat{\sigma}_{\mathrm{MV}}^2}{\hat{\sigma}_y^2 + \bar{y}^2}. \tag{2.40}$$

See also Sect. 2.4.2. If η_{MV} is considerably less than 1, re-tuning the controller will yield benefits. If η_{MV} is close to 1, the performance cannot be improved by re-tuning the existing controller; only process or plant changes, such as changes in the location of sensors and actuators, inspection of valves, other control loop components, or even alterations to the control structure can lead to better performance.

Although $\varepsilon(k)$ is unknown, it can be replaced by the estimated innovations sequence. This can be obtained by pre-whitening the system output variable $y(k)$ via time series analysis based on an AR or ARMA model (alternatively a Kalman filter-based innovation model in state-space form); see Box and MacGregor (1974),

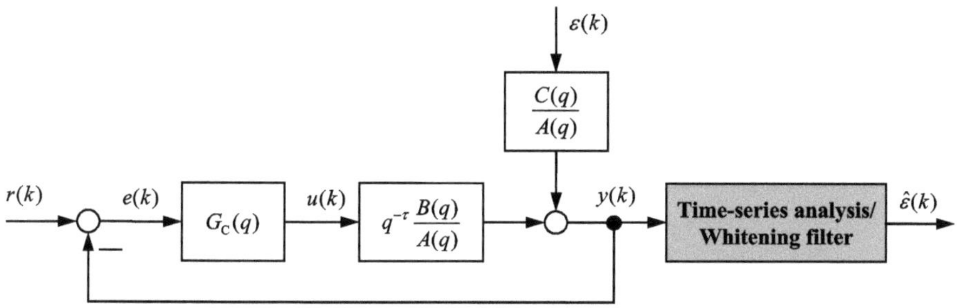

Fig. 2.5 Schematic representation of the white noise or innovation sequence estimation

Söderström and Stoica (1989) and Goodwin and Sin (1984). An estimate for the random chocks is then found, e.g. by inverting the estimated ARMA model

$$\boldsymbol{\varepsilon} = \hat{C}^{-1}(q)\hat{A}(q)\boldsymbol{y}, \tag{2.41}$$

where $\boldsymbol{y}$ is the vector of the output data.

The aim of pre-whitening (or simply whitening) is at tracking back the source of variations in a regulatory closed-loop system to white noise excitation ("the driving force"), as shown in Fig. 2.5. This means reversing the relationship between $y(k)$ and $\varepsilon(k)$. The process of obtaining a "whitening" filter is analogous to time-series modelling, where the final test of the adequacy of the model, i.e. validation, consists of checking if the residuals are "white". These residuals are the estimated white noise sequence. In contrast to time-series modelling, where the estimation of the model is of core interest, the residual or innovation sequence is the main item of interest in the "whitening" process and thus in control performance assessment.

To summarise, the complete algorithm to evaluate the MVC-based (Harris) index and to assess feedback controls contains the steps described in Procedure 2.1.

Procedure 2.1 Performance assessment based the Harris index.

1. Select the time-series-model type and orders.
2. Determine/estimate the system time delay τ.
3. Identify the closed-loop model from collected output samples [ar/arma(x)] (Eq. 2.35).
4. Calculate the series expansion (impulse response) for the estimated model (Eq. 2.36) [dimpulse].
5. Estimate the minimum variance from Eq. 2.38.
6. Estimate the actual output variance from Eqs. 1.1 or 2.39.
7. Compute the (Harris) performance index (Eq. 2.34).

2.4.2 Estimation Algorithms

In this section, some different algorithms are described for the estimation of the Harris index from normal operating data, irrespective of the controller installed on the process. These algorithms do not necessitate the solution of Diophantine equation.

2.4.2.1 Direct Least-Squares Estimation

A simple way to estimate the Harris index η_{MV} from closed-loop routine data is to use linear regression methods, without the necessity of solving any Diophantine equation or performing polynomial long divisions. From Eq. 2.21 the process output under any installed feedback controller $G_c(q)$ can be expressed as

$$y(k) = E_\tau(q)\varepsilon(k+\tau) + q^{-\tau}\frac{F_\tau(q) - B(q)E_\tau(q)G_c(q)}{C(q)}y(k). \tag{2.42}$$

Under the assumption of closed-loop stability, the second term in the previous equation can be approximated by a finite-length (n) AR model:

$$y(k) = \sum_{i=1}^{n} \Theta_i y(k-\tau-i+1) + E_\tau(q)\varepsilon(k) \tag{2.43}$$

with the unknown model parameters Θ_i.

Running k over a range of values and stacking up similar terms yields

$$\boldsymbol{y} = \boldsymbol{X\Theta} + E_\tau(q)\boldsymbol{\varepsilon}(k) \tag{2.44}$$

with

$$\boldsymbol{y} = \begin{bmatrix} y(N) \\ y(N-1) \\ \vdots \\ y(n+\tau) \end{bmatrix}, \qquad \boldsymbol{\Theta} = \begin{bmatrix} \Theta_1 \\ \Theta_2 \\ \vdots \\ \Theta_n \end{bmatrix},$$

$$\boldsymbol{X} = \begin{bmatrix} y(N-\tau) & y(N-\tau-1) & \cdots & y(N-\tau-n+1) \\ y(N-\tau-1) & y(N-\tau-2) & \cdots & y(N-\tau-n) \\ \vdots & \vdots & \cdots & \vdots \\ y(n) & y(n-1) & \cdots & y(1) \end{bmatrix}.$$

The parameter vector $\boldsymbol{\Theta}$ can be estimated with LS method, i.e. by fitting the recorded closed-loop data $\{y_1, y_2, \ldots, y_N\}$ to the model Eq. 2.43. The LS solution follows as

$$\hat{\boldsymbol{\Theta}} = \left(\boldsymbol{X}^{\mathrm{T}}\boldsymbol{X}\right)^{-1}\boldsymbol{X}^{\mathrm{T}}\boldsymbol{y}. \tag{2.45}$$

An estimate of the minimum variance can be determined as the residual mean square error

$$\hat{\sigma}^2_{\mathrm{MV}} = \frac{1}{N-\tau-2n+1}(\boldsymbol{y}-\boldsymbol{X\Theta})^{\mathrm{T}}(\boldsymbol{y}-\boldsymbol{X\Theta}), \tag{2.46}$$

while the actual variance results as

$$\hat{\sigma}^2_y = \frac{1}{N-\tau-n+1}\boldsymbol{y}^{\mathrm{T}}\boldsymbol{y}. \tag{2.47}$$

The Harris index can then be formed as

$$\hat{\eta}_{\mathrm{MV}}(\tau) = \frac{\hat{\sigma}^2_{\mathrm{MV}}}{\hat{\sigma}^2_y} = \frac{N-\tau-n+1}{N-\tau-2n+1}\frac{(\boldsymbol{y}-\boldsymbol{X\Theta})^{\mathrm{T}}(\boldsymbol{y}-\boldsymbol{X\Theta})}{\boldsymbol{y}^{\mathrm{T}}\boldsymbol{y}} \tag{2.48}$$

or

$$\hat{\eta}_{\mathrm{MV}}(\tau) = \frac{\hat{\sigma}^2_{\mathrm{MV}}}{\mathrm{MSE}} = \frac{N-\tau-n+1}{N-\tau-2n+1}\frac{(\boldsymbol{y}-\boldsymbol{X}\Theta)^{\mathrm{T}}(\boldsymbol{y}-\boldsymbol{X\Theta})}{\boldsymbol{y}^{\mathrm{T}}\boldsymbol{y}+(N-\tau-n+1)\bar{\boldsymbol{y}}^2}, \tag{2.49}$$

when the mean square error is used rather than the variance to penalise non-zero steady-state errors; see Eq. 2.40. It is important to note that the signal $y(k)$ has always to be made free from the set point value prior to the index calculation.

Exact distributional properties of the estimated performance indices are complicated and not amenable to a closed-form solution. Desborough and Harris (1992) approximated first and second moments for the estimated performance indices and resorted to a normal theory to develop approximate confidence intervals. Asymptotically, the performance indices are ratios of correlated quadratic forms, and as such the distributions of the performance indices are non-symmetric. Refinements to the confidence intervals developed in Desborough and Harris (1992) can be obtained with little extra computational effort, by resorting to the extensive statistical literature on the distributional properties of quadratic forms (Harris 2004).

2.4.2.2 Online/Recursive Least-Squares Estimation

One advantage of the LS approach is that recursive algorithms to find $\hat{\eta}_{\mathrm{MV}}(k)$ are readably available. An online estimation of the index becomes possible. This is useful to detect change points in control monitoring. Also, if the process is nonlinear and the dynamics are slow enough that the process can be considered locally linear, recursive estimation of the performance index provides a local estimate of the controller performance. Alternatively, $\hat{\eta}_{\mathrm{MV}}(k)$ can be used online as a tuning tool to immediately show whether the tuning changes have improved or degraded control performance (Desborough and Harris 1992). This assumes that the disturbance model does not change significantly.

Typically, recursive LS (RLS) algorithms minimise a cost function of the form

$$V = (\boldsymbol{y}-\boldsymbol{X\Theta})^{\mathrm{T}}\boldsymbol{\Lambda}\,(\boldsymbol{y}-\boldsymbol{X\Theta}), \tag{2.50}$$

where $\boldsymbol{\Lambda}$ is a diagonal matrix with elements $(\lambda, \lambda^2, \ldots, \lambda^N)$. λ is the so-called forgetting factor used to place more emphasis on recent data. An estimate of the MV at time k is given by

$$\sigma^2_{\mathrm{MV}}(k) = \lambda\sigma^2_{\mathrm{MV}}(k-1) + \varepsilon^2(k). \tag{2.51}$$

An estimate of the performance index is computed as

$$\hat{\eta}_{\mathrm{MV}}(k) = \frac{\sigma^2_{\mathrm{MV}}(k)}{\sigma^2_y(k)}, \tag{2.52}$$

where $\sigma^2_y(k)$ is the exponentially weighted moving mean square error

$$\sigma^2_y(k) = \lambda\sigma^2_y(k-1) + y^2(k). \tag{2.53}$$

Instead of an RLS method, a stochastic gradient algorithm, which does not need matrix computations, can be used as well. This has been proposed by Ingimundarson (2002, 2003) for performance assessment of λ-tuned PI controllers.

As stated by Tyler and Morari (1995), the recursive estimation described works well as long as the closed loop is accurately represented by an AR(MA) model. This does not apply for closed-loop models with moving average parameters. An alternative approach to the recursive index estimation is therefore to use a hierarchical method based on data windowing: the data are first broken into segments with similar dynamic properties. Efficient algorithms, such as those proposed by Basseville (1988), can be applied to rapidly detect changes in the closed-loop dynamics. Once a change has been detected, the Harris index can be computed for the largest data segment available with similar dynamics.

In practice, it is often sufficient to use moving windows to study the change in performance of the process over time. Drops or drifts in the performance index can be easily observed in such performance pictures like those shown in Fig. 2.6 ($N = 500$). However, care has to be taken to not use too small data windows; see the guidelines in Sects. 7.1.2 and 7.3.2. In this example, a performance deterioration caused by a big (non-stationary) disturbance appearing at time $k = 2025$ can be observed.

2.4.2.3 Filtering and Correlation Analysis (FCOR) Method

Huang et al. (1997a, 1997b, 1997c, 2000) have developed a method to derive the MVC-based performance index by filtering (i.e. pre-whitening) and subsequent correlation analysis (thus called FCOR) between of the delay-free output and estimated random shocks obtained by a pre-whitening filter. Calculation of the system correlation eliminates the need to determine the impulse response coefficients from the estimated closed-loop transfer function. The FCOR algorithm is presented in this section following Huang and Shah (1999).

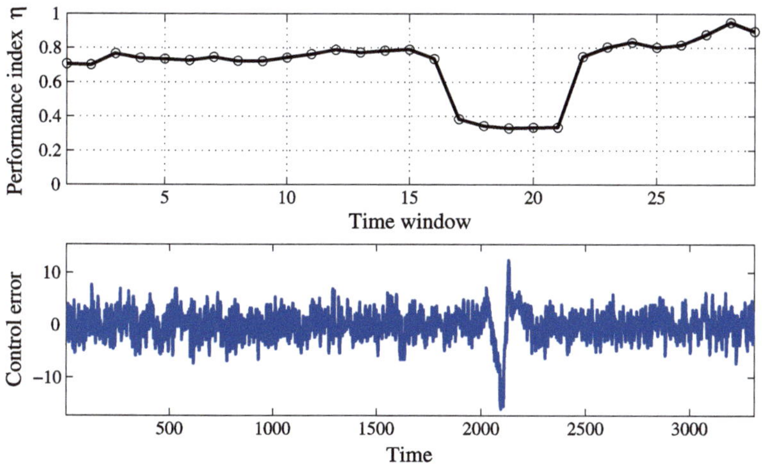

Fig. 2.6 An example of the Harris index trend computed for moving data windows (gauge control loop)

Consider the (stable) closed-loop system described by the infinite-order moving average process in Eq. 2.36. Multiplying this equation by the error terms $\varepsilon(k)$, $\varepsilon(k-1)$, ..., $\varepsilon(k-\tau+1)$ respectively and then taking the expectation of both sides of the equation yields

$$\begin{aligned} r_{y\varepsilon}(0) &= \mathrm{E}\{y(k)\varepsilon(k)\} = e_0\sigma_\varepsilon^2, \\ r_{y\varepsilon}(1) &= \mathrm{E}\{y(k)\varepsilon(k-1)\} = e_1\sigma_\varepsilon^2, \\ r_{y\varepsilon}(2) &= \mathrm{E}\{y(k)\varepsilon(k-2)\} = e_2\sigma_\varepsilon^2, \\ &\vdots \\ r_{y\varepsilon}(\tau-1) &= \mathrm{E}\{y(k)\varepsilon(k-\tau+1)\} = e_{\tau-1}\sigma_\varepsilon^2. \end{aligned} \tag{2.54}$$

Therefore, the minimum variance is

$$\sigma_{\mathrm{MV}}^2 = \sum_{i=0}^{\tau-1} e_i^2\sigma_\varepsilon^2 = \sum_{i=0}^{\tau-1}\left(\frac{r_{y\varepsilon}(i)}{\sigma_\varepsilon^2}\right)^2\sigma_\varepsilon^2 = \sum_{i=0}^{\tau-1} r_{y\varepsilon}^2(i)/\sigma_\varepsilon^2. \tag{2.55}$$

Substituting Eq. 2.55 into Eq. 2.34 leads to the performance index

$$\eta_{\mathrm{MV,cor}} = \sum_{i=0}^{\tau-1} r_{y\varepsilon}^2(i)/(\sigma_y^2\sigma_\varepsilon^2) = \sum_{i=0}^{\tau-1} \rho_{y\varepsilon}^2(i) = \mathbf{Z}^{\mathrm{T}}\mathbf{Z}, \tag{2.56}$$

where $\mathbf{Z}$ is the *cross-correlation coefficient* vector between $y(k)$ and $\varepsilon(k)$ for lags 0 to $\tau-1$ and is denoted

$$\mathbf{Z} := \left[\rho_{y\varepsilon}(0),\ \rho_{y\varepsilon}(1),\ \rho_{y\varepsilon}(2),\ \ldots,\ \rho_{y\varepsilon}(\tau-1)\right]^{\mathrm{T}}. \tag{2.57}$$

The corresponding sampled version of the performance index is therefore given by

$$\hat{\eta}_{\mathrm{MV,cor}} = \sum_{i=0}^{\tau-1} \hat{\rho}_{y\varepsilon}^2(i) = \hat{\mathbf{Z}}^{\mathrm{T}} \hat{\mathbf{Z}}, \tag{2.58}$$

where

$$\hat{\rho}_{y\varepsilon}(l) = \frac{\sum_{k=1}^{M} y(k)\varepsilon(k-l)}{\sum_{k=1}^{M} y^2(k) \sum_{k=1}^{M} \varepsilon^2(k)}.$$

$\varepsilon(k)$ can be determined from pre-whitening of $y(k)$ via time series, as explained in Sect. 2.4.1. The complete FCOR algorithm is described in Procedure 2.2.

Procedure 2.2 Filtering and correlation-based (FCOR) algorithm.

1. Select the time-series-model type and orders.
2. Determine/estimate the system time delay τ.
3. Identify an appropriate closed-loop model from collected output samples $y(k)$.
4. Filter the system output data $y(k)$ from the model to obtain an estimate for the whitened sequence (Eq. 2.41).
5. Calculate the cross-correlation coefficients between $y(k)$ and $\varepsilon(k)$ for lags 0 to $\tau - 1$ from Eq. 2.54.
6. Use Eq. 2.58 to compute the performance index.

Note that FCOR is a general methodology rather than a specific algorithm. It consists of two steps: filtering/whitening and correlation analysis. The advantage of FCOR is its flexibility in the form of the filter and its suitability for MIMO systems due to its simplicity in computation, namely computation of the Markov matrices can be avoided. The filtering step in FCOR can use any algorithm, such as AR, ARMA, state space, or even a nonlinear time series modelling algorithm.

2.4.2.4 Examples

The following examples illustrate the performance assessment results in terms of the Harris index obtained using Procedure 2.2. Some controller tuning rules will be evaluated using the MVC-based assessment introduced above.

Example 2.3 Consider the first-order system from Example 2.1. The MVC is used here just to simulate the process under this ideal controller and to show that the Harris index will take the value of 1 in this case. This is confirmed by the Harris index value given in Table 2.1 (fourth row). In this simulation, $\varepsilon(k)$ was a normally distributed noise with the variance $\sigma_\varepsilon^2 = 0.01$. The Harris index values have been determined from using $N = 1500$ simulated data points ($T_{\mathrm{s}} = 0.5$) and modelling the closed loop by an AR model of order $n = 30$.

Table 2.1 Harris index value for the different controllers

Controller	K_c	$\hat{\eta}$
P1	0.05	0.62
P2	0.25	0.76
P3	0.50	0.37
MVC	–	0.99

The impulse responses for a P-only controller with different gains and for the MVC are illustrated in Fig. 2.7. It can be seen that P1 is a sluggish controller, P2 a well-tuned controller and P3 an aggressive controller. The figure also shows how the impulse response for the MVC dies beyond $\tau = 3$. The first three (controller-invariant) coefficients are marked with circles in the figure. From this example, it can be learned that the Harris index for a well-tuned controller ($\eta = 0.76$) does not always achieve that of MVC ($\eta = 0.99 \approx 1$).

In the following, we consider two simulated processes used by Seborg et al. (2004: Chap. 12) to compare different controller tuning rules in terms of (deterministic) set-point tracking and disturbance rejection. Here we evaluate the stochastic control performance using the Harris index. For all processes, we assume that the process model (without time delay) and disturbance model are identical and the disturbance is normally distributed noise with the variance $\sigma_\varepsilon^2 = 0.01$.

Example 2.4 A blending system with a measurement time delay modelled by

$$G_p(s) = \frac{1.54}{5.93s + 1} e^{-1.07s} \tag{2.59}$$

and controlled by a PI controller is considered. Figure 2.8 illustrates the results gained from modelling the closed loop by an AR model of order $n = 20$ using

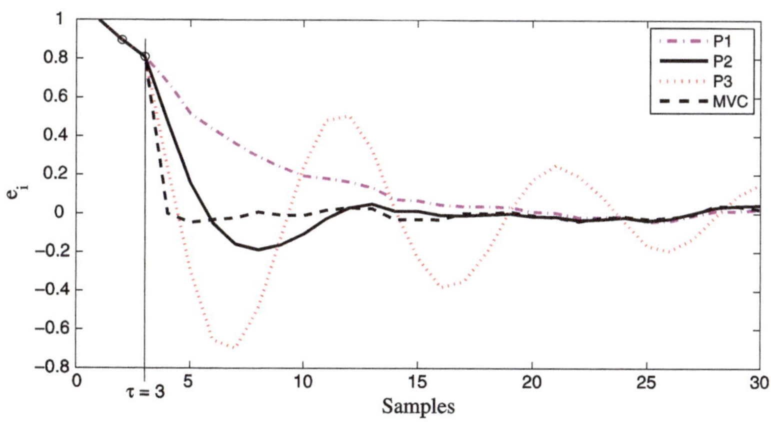

Fig. 2.7 Impulse responses for different controllers

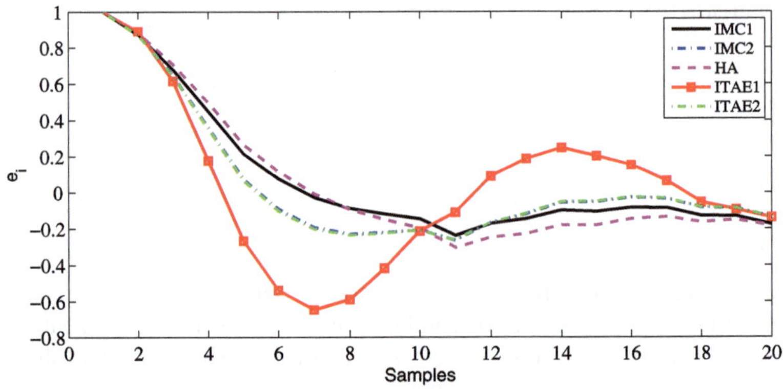

Fig. 2.8 Harris index value for the blending process and different controllers

Table 2.2 Impulse responses for the blending process and different controllers

Controller/tuning rule	Acronym	K_c	T_I	η
IMC ($\lambda = T/3$)	IMC1	1.27	5.93	0.76
IMC ($\lambda = T_d$)	IMC2	1.80	5.93	0.81
Hägglund and Åström	HA	1.10	2.95	0.65
ITAE (disturbance)	ITAE1	2.97	2.75	0.59
ITAE (set point)	ITAE2	1.83	5.93	0.81

Table 2.3 Harris index value for the lag-dominant process and different controllers

Controller/tuning rule	Acronym	K_c	T_I	$\hat{\eta}$
IMC ($\lambda = 1.0$)	IMC1	0.5	100	0.63
IMC ($\lambda = 2.0$) based on integrator approximation	IMC2	0.556	5	0.52
IMC ($\lambda = 1.0$) based on Skogestad's modification	IMC3	0.5	8	0.56
Direct synthesis (disturbance)	DS-d	0.551	4.91	0.52

1500 output samples. IMC and ITAE (set point) yield the best performance, and ITAE (disturbance) the least performance, as a consequence of the most aggressive settings; see Table 2.2. Note that IMC2 and ITAE (set point) have almost identical impulse responses for this example, but this is not true in general.

Example 2.5 This example is a lag-dominant model with $T_d/T = 0.01$:

$$G_p(s) = \frac{100}{100s+1}e^{-s}. \tag{2.60}$$

Table 2.3 contains the results gained from modelling the closed loop by an AR model of order $n = 20$ using 1500 output samples. IMC1 leads to the best per-

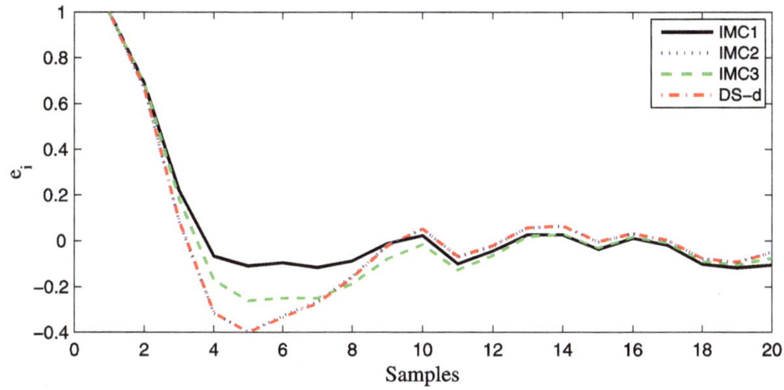

Fig. 2.9 Impulse responses for the lag-dominant process and different controllers

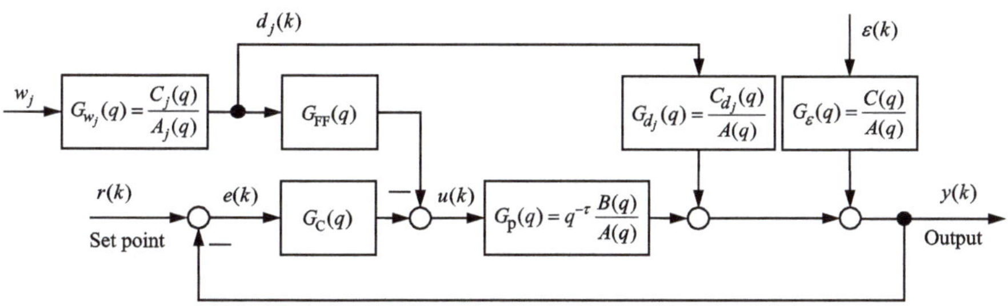

Fig. 2.10 Generic feedback plus feedforward control system structure

formance and both IMC2 and DS-d to the least performance, which have almost identical controller settings and thus almost identical impulse responses, as can be seen in Fig. 2.9.

2.5 Assessment of Feedback/Feedforward Controls

Feedforward control (FFC) should always be introduced to reduce the process variability due to disturbances. An ineffective FFC will contribute to a large variance due to the measured disturbances. Therefore, one additional task in assessing feedback/feedforward control loops is to diagnose whether a poor performance is due to feedback control or feedforward control. This section focuses on the analysis of variance (ANOVA), to quantify major contributions to system-output variance (Desborough and Harris 1993; Huang and Shah 1999). A detailed derivation of the algorithm can also be found by Ratjen and Jelali (2006).

The MV is calculated differently in feedback/feedforward control (Fig. 2.10) than feedback control alone. The major difference is in the estimation of the variance of

Table 2.4 Analysis of variance for feedback plus feedforward control

Disturbance	MV	FB	FF	FB/FF	Total
$\varepsilon(k)$	$\sigma^2_{\mathrm{MV},\varepsilon}$	$\sigma^2_{\mathrm{FB},\varepsilon}$	–	–	$\sigma^2_{y,\varepsilon}$
$w_1(k)$	$\sigma^2_{\mathrm{MV},w_1}$	–	$\sigma^2_{\mathrm{FF},w_1}$	$\sigma^2_{\mathrm{FB/FF},w_1}$	σ^2_{y,w_1}
$\vdots$	$\vdots$	$\vdots$	$\vdots$	$\vdots$	$\vdots$
$w_p(k)$	$\sigma^2_{\mathrm{MV},w_p}$	–	$\sigma^2_{\mathrm{FF},w_p}$	$\sigma^2_{\mathrm{FB/FF},w_p}$	σ^2_{y,w_p}
Total	σ^2_{MV}		$\sigma^2_y - \sigma^2_{\mathrm{MV}}$		σ^2_y

the unmeasured disturbance $\varepsilon(k)$. An ARMAX model of the MISO form

$$y(k) = \frac{C(q^{-1})}{A(q^{-1})}\varepsilon(k) + \sum_{j=1}^{p} \frac{B(q^{-1})}{A(q^{-1})} d_j(k - \tau_j) \tag{2.61}$$

should be used for identifying the closed-loop model to include the effect of p measured disturbances d_j, as opposed to an AR(MA) model only. Then each measured disturbance model is identified as an AR(I)MA time-series model, i.e. $A_j(q)d_j(k) = C_j(q)w_j(k)$, leading to

$$y(k) = \frac{C(q^{-1})}{A(q^{-1})}\varepsilon(k) + \sum_{j=1}^{p} \frac{B_j(q^{-1})C_j(q^{-1})}{A(q^{-1})A_j(q^{-1})} w_j(k - \tau_j). \tag{2.62}$$

Time delays (τ and τ_j) are required and the model orders also need to be determined. The identified closed-loop model is then used to carry out an ANOVA table for the output y based on the time delays in the feedforward and feedback paths. This method can yield valuable information about the sources of variability, provided that all measured disturbances are mutually independent.

The cross-correlation between the measured disturbances (as potential feedforward variables) and the output can be used to determine which of them could be used for FFC. The analysis of variance (Desborough and Harris, 1993) highlights the contribution of the disturbances to the overall variance (Table 2.4):

$$\sigma_y^2 = \sigma^2_{\mathrm{MV},\varepsilon} + \sigma^2_{\mathrm{FB},\varepsilon} + \sum_{j=1}^{n_w} \left(\sigma^2_{\mathrm{MV},w_j} + \sigma^2_{\mathrm{FF},w_j} + \sigma^2_{\mathrm{FB/FF},w_j}\right), \tag{2.63}$$

where

- $\sigma^2_{\mathrm{MV},\varepsilon}$: the minimum variance of the FBC, arising from unmeasured disturbance ε
- $\sigma^2_{\mathrm{FB},\varepsilon}$: the variance due to the non-optimality of the FBC
- $\sum_{j=1}^{p} \sigma^2_{\mathrm{MV},w_j}$: the minimum variance of the FFC, coming from r measured disturbances w_i
- $\sum_{j=1}^{p} \sigma^2_{\mathrm{FF},w_j}$: the variance due to the non-optimality of the FFC

- $\sum_{j=1}^{p} \sigma^2_{\mathrm{FB/FF},w_j}$: the variance due to the non-optimality of the combination FBC/FFC.

The bottom row in Table 2.4 consists of, from left to right, the summation of the minimum variances, the sum of all the variance due to non-optimality of the controller components and the total variance. It is important to note that it is not possible to unambiguously attribute variance inflation to either the feedback controller alone or the feedforward controller alone, hence the column labelled "FF/FB" in the table. Note that if the process is invertible, it is always possible to eliminate the variance inflation due to both this component and the feedforward component using a feedforward controller, regardless of the feedback controller (Desborough and Harris 1993). If one row contains a considerable portion of the total variance in the columns FF and FB + FF, this implies that re-tuning is needed. If only the term FB + FF is large, it can be expected that the feedback controller may handle the disturbance satisfactory. The analysis of variance helps quantify how much the performance of the control loop can be improved, which can be translated in terms of increased product quality and/or material/energy consumption; see Sect. 13.2.3.

The procedure for variance estimation for feedforward/feedback control loops can be outlined as follows. It is demonstrated in simulation studies discussed below. An industrial application of this algorithm is presented in Sect. 16.3.1.

Procedure 2.3 Variance estimation for feedforward/feedback control.

1. Determine or estimate the time delays.
2. Fit an ARMAX model to the closed-loop output samples $y(k)$ and measured disturbance samples $d_i(k)$ as inputs.
3. Fit individual AR(IMA) models to each of the feedforward variables $d_i(k)$.
4. Calculate the series expansions (impulse responses) for the estimated models.
5. Compute the variances as in Table 2.4.

Example 2.6 The system consists of a pure time-delay process affected by output noise and a measurable disturbance. This linear system, adopted from Desborough and Harris (1993), has the structure and parameters illustrated in Fig. 2.11. For the simulation study, the driving noises were Gaussian random signals with the variances σ^2_w and σ^2_ε. A simple integral feedback controller was used as the initial controller.

Five cases will be studied. The first case considers the effect of weak disturbances, i.e. with low disturbance variance. In the second case, we increase the disturbance variance, compared to the first case. In the third case, the disturbance dynamics will be altered so that its average residence time will be significantly shorter. In the first three cases, the system was operated under feedback-only control, so the assessment method will give hints whether the loop should be extended with feedforward control. In the fourth case, a feedforward component will be added to the controller. Finally, the feedback controller will be re-tuned and evaluated again.

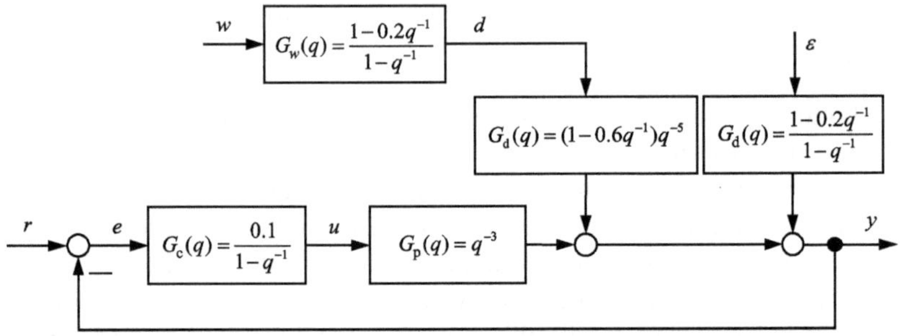

Fig. 2.11 Structure and transfer functions of the considered control loop

Table 2.5 Analysis of variance table [% of total variance] for Case 1

Disturbance	MV	FB	FF	FB/FF	Total
$\varepsilon(k)$	53.4	42.8	–	–	96.3
$w(k)$	0	–	3.0	0.7	3.7
Total	53.4		46.5		100

Case 1. *Weak Disturbances*. The variances of the driving noises were $\sigma_w^2 = 0.1$ and $\sigma_\varepsilon^2 = 1$. A simulation of the system was carried out, and equidistant data were collected at a sampling time $T_s = 0.1$ s. Steady-state operating data with 1000 samples were selected for calculating the ANOVA table; see Table 2.5. From this it can be deduced that the feedback controller ($\sigma_{y,\varepsilon}^2 = 96.3$ %) is far from the minimum achievable variance ($\sigma_{\mathrm{MV},\varepsilon}^2 = 53.4$ %) for the unmeasured noise. There is a major portion of the variance (42.8 %) which can be handled by a feedback tuning. However, the contribution (3.7 %) to the variance from the measured disturbance is small. 3 % can be reduced if an optimal feedforward controller is implemented. There is also a negligible portion of the variance (0.7 %) which can be handled by a combination of feedforward/feedback tuning. The conclusion is that the assessment method suggests that the feedback controller would benefit from improved tuning. However, an implementation of a feedforward controller is not recommended.

Case 2. *Strong Disturbances*. Here the transfer functions of the system are left unchanged, but the noise variance of the disturbance has been increased, i.e. $\sigma_w^2 = \sigma_\varepsilon^2 = 1$. Performing an ANOVA-analysis shows that the measured disturbance is now responsible for 27.7 % of the total variance, from which 22.6 % can be handled by feedforward control, see Table 2.6. This implies that the control loop will benefit from implementing a feedforward controller in this case.

Case 3. *Speed-up of Disturbance Dynamics*. In this case, the disturbance dynamics have changed such that there is no time delay, i.e.

$$G_{\mathrm{d}}(s) = \frac{0.8}{1+4s}. \tag{2.64}$$

Table 2.6 Analysis of variance table [% of total variance] for Case 2

Disturbance	MV	FB	FF	FB/FF	Total
$\varepsilon(k)$	40.5	31.9	–	–	72.3
$w(k)$	0	–	22.6	5.0	27.7
Total	40.5		59.5		100

Table 2.7 Analysis of variance table [% of total variance] for Case 3

Disturbance	MV	FB	FF	FB/FF	Total
$\varepsilon(k)$	61.3	38.1	–	–	99.4
$w(k)$	0	–	0.4	0.2	0.6
Total	61.3		38.7		100

Table 2.8 Analysis of variance table [% of total variance] for Case 4 with static feedforward control

Disturbance	MV	FB	FF	FB/FF	Total
$\varepsilon(k)$	46.1	38.1	–	–	84.2
$w(k)$	0	–	15.4	0.4	15.8
Total	46.1		53.9		100

Table 2.9 Analysis of variance table [% of total variance] for Case 4 with dynamic feedforward control

Disturbance	MV	FB	FF	FB/FF	Total
$\varepsilon(k)$	49.8	41.7	–	–	91.6
$w(k)$	0	–	6.0	2.4	8.4
Total	49.8		50.1		100

The variances of the driving noises are the same as in Case 2. From performing similar simulations and looking at the ANOVA table (Table 2.7), the same conclusions can be drawn as in Case 1.

Case 4. *Effect of Feedforward Control.* The implementation of a simple proportional feedforward controller $G_{\mathrm{FF}} = 0.5$ leads to a 7 % decrease of the feedforward portion of variance, as shown in the ANOVA Table 2.8, compared to the performance in Case 2 (Table 2.6). This is due to the static feedforward which attempts to compensate for changes in the feedforward variable before these changes appear in the output. When now a two-sample delay is included in the feedforward controller, i.e. $G_{\mathrm{FF}} = 0.5q^{-2}$, the feedforward portion of variance decreases up to 6 %, as shown in the ANOVA Table 2.9. From this table it can also be deduced that re-tuning or redesigning the feedback controller will yield the largest return, since it is still far from the minimum-variance performance.

Table 2.10 Analysis of variance table [% of total variance] for Case 5 with dynamic feedforward control and re-tuned feedback control

Disturbance	MV	FB	FF	FB/FF	Total
$\varepsilon(k)$	71.4	14.1	–	–	85.6
$w(k)$	0	–	8.5	6.0	14.4
Total	71.4		28.6		100

Case 5. *Re-tuning of the Feedback Controller*. Just an increase of the proportional controller gain to $K_c = 0.31$ yields substantial performance improvement for the feedback controller, as can be seen in Table 2.10. A further decrease of variance may be only achieved by redesigning the controller.

2.6 Assessment of Set-Point Tracking and Cascade Control

Most of the techniques presented above can be applied to single control loops operating in a regulatory mode, i.e. with constant set point. However, when set-point variations occur frequently, neglecting them will lead to under-estimation of the regulatory performance improvement. Equation 2.30 has to be extended to include the transfer function relating the control error to the set-point changes. The superposition principle gives the resulting closed-loop relation

$$y(k) = \frac{F_\tau(q)}{E_\tau(q)A(q)} q^{-\tau} r(k) + E_\tau(q)\varepsilon(k). \tag{2.65}$$

The estimation procedure of Sect. 2.4 thus has to be modified by estimating an ARMAX model with $r(k)$ as the input signal, similar to the case of feedback plus feedforward control; see Sect. 2.5.

2.6.1 Performance Assessment of Cascade Control Systems

In process control applications, the rejection of load disturbances is often of main concern. To improve the control performance for this task, the implementation of a cascade control system is a good option. Indeed, cascade control is widely used in the process industries and is particularly useful when the disturbances are associated with the manipulated variable or when the final control element exhibits nonlinear behaviour (Shinskey 1996). Therefore, the main criterion to assess cascade control loops is its capability to reject load disturbances. Typical examples of cascade control from the metal processing industry are strip thickness control and flatness control systems; see Chap. 15.

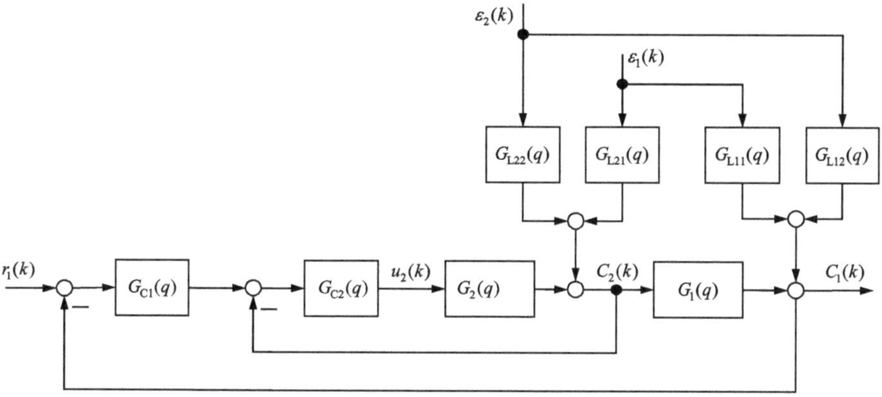

Fig. 2.12 Block diagram of a cascade control system

The minimum achievable variance with cascade control is generally lower than that from single-loop feedback control and can provide useful information on potential performance improvement. The above techniques can directly be used to analyse the performance of the primary loop under the assumption of constant set point, but require modifications for the analysis of the secondary loop. The relationships for the minimum-variance assessment of cascade control systems (Fig. 2.12) are derived following Ko and Edgar (2000).

Subscript 1 in this figure refers to the primary control loop, while subscript 2 refers to the secondary control loop. $C_1(k)$ and $C_2(k)$ are the process outputs of primary loop and the secondary loop, respectively. $C_1(k)$ is the deviation variable from its set point, and $C_2(k)$ the deviation of the secondary output from its steady-state value, which is required to keep the primary output at its set point. $G_1(q) \equiv G_1^*(q)q^{-\tau_1}$ is the process transfer function in the primary loop with time delay equal to τ_1, and $G_1^*(q)$ is the primary process model without any time delay. It is assumed that $G_1(q)$ has a stable inverse, i.e. all zeros lie inside the unit circle. The disturbance filters $G_{L11}(q)$ and $G_{L12}(q)$ are assumed to be rational functions of q^{-1}, and they are driven by zero-mean white noise sequences $\varepsilon_1(k)$ and $\varepsilon_2(k)$, respectively. Similarly, for the secondary loop, we have $G_2(q) \equiv G_2^*(q)q^{-\tau_2}$ as the process transfer function in the secondary loop with time delay equal to τ_2 and $G_2^*(q)$ is the secondary process model without any time delay. $G_2(q)$ is also assumed to be minimum-phase. The combined effect of all unmeasured disturbances to the secondary output is represented as a superposition of disturbance filters $G_{L21}(q)$ and $G_{L22}(q)$ driven by zero-mean white noise sequences $\varepsilon_1(k)$ and $\varepsilon_2(k)$, respectively.

Using block-diagram algebra, from Table 2.11 it can simply be seen that:

$$
\begin{aligned}
C_1(k) &= G_1(q)C_2(k) + G_{L11}(q)\varepsilon_1(k) + G_{L12}(q)\varepsilon_2(k), \\
C_2(k) &= G_2(q)u_2(k) + G_{L21}(q)\varepsilon_1(k) + G_{L22}(q)\varepsilon_2(k),
\end{aligned}
\tag{2.66}
$$

where $u_2(k)$ is the manipulated variable in the secondary control loop. The MVC algorithm for the equation system 2.66 is given by

- *Primary Controller*

$$G_{\mathrm{c1,MV}} = \frac{G_1^*(Q_{22}R_{21} - Q_{21}R_{22}) + (R_{11} + T_1)G_{\mathrm{L22}} - (R_{12} + T_2)G_{\mathrm{L21}}}{(Q_{11} + S_1 q^{-\tau_1})(R_{12} + T_2 + G_1^* R_{22}) - (Q_{12} + S_2 q^{-\tau_1})(R_{11} + T_1 + G_1^* R_{21})}. \tag{2.67}$$

- *Secondary Controller*

$$G_{\mathrm{c2,MV}} = \frac{(Q_{11} + S_1 q^{-\tau_1})(R_{12} + T_2 + G_1^* R_{22}) - (Q_{12} + S_2 q^{-\tau_1})(R_{11} + T_1 + G_1^* R_{21})}{G_1^*[G_{\mathrm{L11}}S_2 - G_{\mathrm{L12}}S_1 + (R_{11}Q_{12} - R_{12}Q_{11})q^{-\tau_2}]}, \tag{2.68}$$

where $Q_{11}(q)$ and $Q_{12}(q)$ are polynomials in q^{-1} of order $\tau_1 + \tau_2 - 1$, $Q_{21}(q)$, $Q_{22}(q)$, S_1(q) and S_2(q) are polynomials in q^{-1} of order $\tau_2 - 1$, and $R_{ij}(q)$ $(i, j = 1, 2)$ are proper transfer functions that satisfy the following Diophantine identities:

$$\begin{aligned}
G_{\mathrm{L11}} &= Q_{11} + R_{11} q^{-\tau_1-\tau_2},\\
G_{\mathrm{L12}} &= Q_{12} + R_{12} q^{-\tau_1-\tau_2},\\
G_{\mathrm{L21}} &= Q_{21} + R_{21} q^{-\tau_2},\\
G_{\mathrm{L22}} &= Q_{22} + R_{22} q^{-\tau_2},\\
G_1^* Q_{21} &= S_1 + T_1 q^{-\tau_2},\\
G_1^* Q_{22} &= S_2 + T_2 q^{-\tau_2}.
\end{aligned} \tag{2.69}$$

The primary output $C_1(k)$ under this optimal control algorithm is an MA process of order $\tau_1 + \tau_2 - 1$

$$C_1(k) = \left[Q_{11}(q) + S_1(q)q^{-\tau_1}\right]\varepsilon_1(k) + \left[Q_{12}(q) + S_2(q)q^{-\tau_1}\right]\varepsilon_2(k), \tag{2.70}$$

and the MV of $C_1(k)$ is

$$\sigma^2_{C_1,\mathrm{MV}} = \mathrm{trace}\left\{\left(\sum_{i=0}^{\tau_1+\tau_1-1} \boldsymbol{N}_i^{\mathrm{T}} \boldsymbol{N}_i\right)\boldsymbol{\Sigma}_\varepsilon\right\}, \tag{2.71}$$

where $\boldsymbol{N}_i$ $(i = 0, 1, \ldots, \tau_1 + \tau_2 - 1)$ are defined as the coefficient matrices of the matrix polynomial $[(Q_{11} + S_1 q^{-\tau_1})(Q_{12} + S_2 q^{-\tau_1})]$, and $\boldsymbol{\Sigma}_\varepsilon$ is the variance-covariance matrix of the white noise vector $[\varepsilon_1(k)\varepsilon_2(k)]^{\mathrm{T}}$. The derivation of the above relationships can be found by Ko and Edgar (2000). Since the polynomials Q_{11}, Q_{12}, S_1, S_2 in Eq. 2.70 are all feedback-invariant, the expression for the primary output under MV cascade control can be estimated from the first $\tau_1 + \tau_2 - 1$ MA coefficients of the closed-loop transfer functions relating $\varepsilon_1(k)$ to $C_1(k)$ and $\varepsilon_2(k)$ to $C_1(k)$. No joint identification of the process dynamics and the disturbance model is needed.

The closed-loop transfer functions in this case can be obtained from the first row of the transfer function matrix estimated via *multivariate time-series analysis* of

$[C_1(k), C_2(k)]^{\mathrm{T}}$. For this analysis, an AR model [arx setting an empty input] can be used efficiently with its computational speed. Alternatively, a state space model can be estimated via the prediction error method [pem] or a subspace identification method [n4sid]; see Sect. 7.2.6. The sample variance-covariance matrix of the residual vectors thus provides an estimate of the variance and the covariance elements of the innovation sequences. The closed-loop impulse-response coefficients can then be determined via simple correlation analysis between the output variables and the estimated innovations sequences, or by solving a suitable Diophantine identity concerning the estimated parameter matrix polynomial,

$$\begin{aligned}\sigma^2_{C_1,\mathrm{MV}} &= \mathrm{var}\big\{\big(h_{10} + h_{11}q^{-1} + \cdots + h_{1,\tau_1+\tau_1-1}q^{-(\tau_1+\tau_1-1)}\big)\varepsilon_1 \\ &\quad + \big(h_{20} + h_{21}q^{-1} + \cdots + h_{2,\tau_1+\tau_1-1}q^{-(\tau_1+\tau_1-1)}\big)\varepsilon_2\big\} \\ &= \mathrm{trace}\left\{\left(\sum_{i=0}^{\tau_1+\tau_1-1} \boldsymbol{N}_i^{\mathrm{T}}\boldsymbol{N}_i\right)\boldsymbol{\Sigma}_\varepsilon\right\}. \end{aligned} \tag{2.72}$$

The MV performance index for the cascade control system is defined as

$$\eta = \frac{\sigma^2_{C_1,\mathrm{MV}}}{\sigma^2_{C_1}}. \tag{2.73}$$

Naturally, the question arises if the MV can be calculated by just applying univariate analysis on $C_1(k)$. Indeed, this would lead to similar results, but only in the case where the net disturbance effect driven by $\varepsilon_2(k)$ is negligible. Otherwise, the estimated performance index from univariate analysis is higher than the estimated performance index value obtained through multivariate analysis. Thus, univariate analysis of cascade control loops would yield an over-estimate of the control performance.

Example 2.7 We consider the example of a process described by Eqs. 2.66 with (Ko and Edgar 2000)

$$\begin{aligned} G_1(q) &= \frac{q^{-2}}{1-0.9q^{-1}}, & G_{\mathrm{L}11}(q) &= \frac{1}{1-0.8q^{-1}}, & G_{\mathrm{L}12}(q) &= \frac{q^{-1}}{1-0.1q^{-1}}, \\ G_2(q) &= \frac{q^{-1}}{1-0.5q^{-1}}, & G_{\mathrm{L}21}(q) &= \frac{q^{-1}}{1-0.2q^{-1}}, & G_{\mathrm{L}22}(q) &= \frac{1}{1-0.3q^{-1}} \end{aligned} \tag{2.74}$$

to illustrate how the assessment procedure of cascade control systems works in detail. For this system, the primary process time delay is two samples ($\tau_1 = 2$), and the secondary process time delay is one sample ($\tau_2 = 1$). The process is subjected to disturbances in the form of white noise sequences $\{\varepsilon_1(k), \varepsilon_2(k)\}$ with unity variance-covariance matrix $\boldsymbol{\Sigma}_\varepsilon = \left[\begin{smallmatrix}1.0 & 0.1\\ 0.1 & 1.0\end{smallmatrix}\right]$. A closed-loop simulation was performed using a

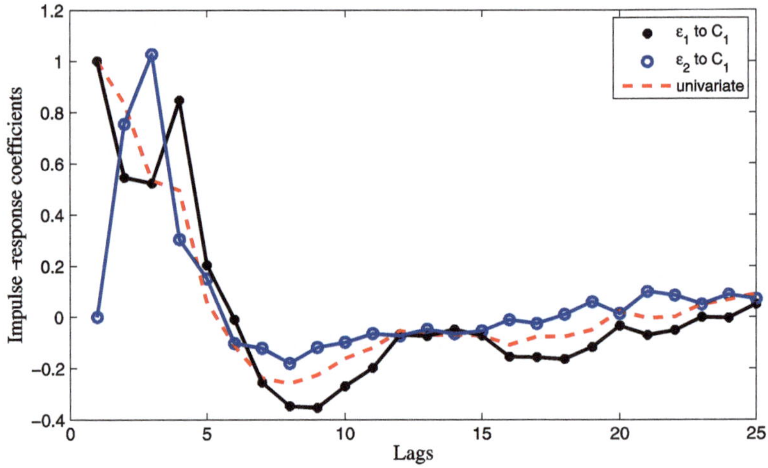

Fig. 2.13 Closed-loop impulse responses from simulated data

PI controller for the primary loop and P-only controller for the secondary loop. The transfer functions for the controllers used are

$$G_{c1}(q) = \frac{0.48 - 0.4q^{-1}}{1 - q^{-1}}; \qquad G_{c2}(q) = 0.7.$$

A data set of 2000 samples for the primary and secondary outputs was collected and a multivariate AR model of 25th order was fitted to the gathered data. The estimated variance-covariance matrix of the white noise sequence is $\hat{\boldsymbol{\Sigma}}_\varepsilon = \left[\begin{smallmatrix} 1.36 & 0.43 \\ 0.43 & 1.30 \end{smallmatrix}\right]$. From this model the estimated closed-loop impulse responses have been obtained, as shown in Fig. 2.13.

The estimated minimum variance by multivariate analysis is (Eq. 2.72)

$$\begin{aligned}\sigma^2_{C_1,\mathrm{MV}} = \mathrm{trace}\left\{\left(\begin{bmatrix}1\\0\end{bmatrix}\begin{bmatrix}1 & 0\end{bmatrix} + \begin{bmatrix}0.48\\0.81\end{bmatrix}\begin{bmatrix}0.48 & 0.81\end{bmatrix},\right.\right. \\ \left.\left. + \begin{bmatrix}0.49\\1.03\end{bmatrix}\begin{bmatrix}0.49 & 1.03\end{bmatrix}\right)\begin{bmatrix}1.36 & 0.43\\0.43 & 1.30\end{bmatrix}\right\} \approx 5.01. \qquad (2.75)\end{aligned}$$

This gives the estimated performance index (Eq. 2.73)

$$\eta_{C_1,\mathrm{multi}} = \frac{\sigma^2_{C_1,\mathrm{MV}}}{\sigma^2_{C_1}} = \frac{5.01}{7.28} \approx 0.69. \qquad (2.76)$$

For a comparison, univariate analysis using an AR model of 25th order fitted to the primary response data $C_1(k)$ has also been carried out. The obtained univariate closed-loop impulse response is also illustrated in Fig. 2.13. The estimated minimum variance by the univariate analysis is (Eq. 2.38)

$$\sigma^2_{\hat{C}_1,\mathrm{MV}} = \left(1.0^2 + 0.836^2 + 0.534^2\right)(2.822) = 5.596.$$

This yields the estimated performance index as

$$\eta_{C_1,\mathrm{uni}} = \frac{\sigma^2_{\hat{C}_1,\mathrm{MV}}}{\sigma^2_{\hat{C}_1}} = \frac{5.596}{7.28} \approx 0.79. \tag{2.77}$$

The estimated performance index by univariate analysis is (13 %) higher than the estimated performance index value obtained through a multivariate analysis. Thus, univariate analysis of cascade control loops would yield an over-estimate of the controller's performance. Note that analysis of the inner loop yields an estimated performance index of

$$\eta_{C_2,\mathrm{uni}} = \frac{2.76}{2.61} \approx 0.92,$$

indicating very good performance.

Just to theoretically confirm the results achieved above, we now calculate the minimum variance from the full knowledge of the process and disturbance models by solving Diophantine Eqs. 2.69 using

$$\frac{1}{1-aq^{-1}} = \sum_{i=0}^{\infty} a^i q^{-i}. \tag{2.78}$$

This gives specifically for the considered case:[5]

$$\frac{1}{1-0.8q^{-1}} = 1 + 0.8q^{-1} + 0.64q^{-2} + 0.512q^{-3} = Q_{11} + R_{11}q^{-3},$$

$$\frac{q^{-1}}{1-0.1q^{-1}} = q^{-1}\left(1 + 0.1q^{-1} + 0.01q^{-2}\right) = Q_{12} + R_{12}q^{-3},$$

$$\frac{q^{-1}}{1-0.2q^{-1}} = q^{-1}(1) = Q_{21} + R_{21}q^{-1},$$

$$\frac{1}{1-0.3q^{-1}} = 1 + 0.3q^{-1} = Q_{22} + R_{22}q^{-1},$$

$$\frac{1}{1-0.9q^{-1}}Q_{21} = S_1 + T_1q^{-1},$$

$$\frac{1}{1-0.9q^{-1}}Q_{22} = S_2 + T_2q^{-1},$$

[5] Remaining error terms are ignored here for simplicity.

where Q_{11} and Q_{12} are polynomials in q^{-1} of order 2, Q_{21}, Q_{22}, S_1 and S_2 are constants, and Q_{ij} and T_i $(i, j = 1, 2)$ are proper transfer functions. The solution of these identities yields

$$\begin{aligned}
Q_{11} &= 1 + 0.8q^{-1} + 0.64q^{-2}, & R_{11} &= 0.512,\\
Q_{12} &= q^{-1} + 0.1q^{-2}, & R_{12} &= 0.01,\\
Q_{21} &= 0, & R_{21} &= 1,\\
Q_{22} &= 1, & R_{22} &= 0.3,\\
S_1 &= 0, & T_1 &= 0,\\
S_2 &= 1, & T_2 &= 0.9.
\end{aligned}$$

Thus, from Eq. 2.70, the primary output $C_1(k)$ under minimum-variance cascade control is given by

$$C_1(k) = \left[1 + 0.8q^{-1} + 0.64q^{-2}\right]\varepsilon_1(k) + \left[q^{-1} + 1.1q^{-2}\right]\varepsilon_2(k). \tag{2.79}$$

The minimum variance follows as (Eq. 2.72)

$$\begin{aligned}
\sigma^2_{C_1,\mathrm{MV}} = \operatorname{trace}\Bigg\{\Bigg(&\begin{bmatrix}1\\0\end{bmatrix}\begin{bmatrix}1 & 0\end{bmatrix} + \begin{bmatrix}0.8\\1\end{bmatrix}\begin{bmatrix}0.8 & 1\end{bmatrix}\\
&+ \begin{bmatrix}0.64\\1.1\end{bmatrix}\begin{bmatrix}0.64 & 1.1\end{bmatrix}\Bigg)\begin{bmatrix}1 & 0.1\\0.1 & 1\end{bmatrix}\Bigg\} \approx 4.56,
\end{aligned} \tag{2.80}$$

which is close to the value in Eq. 2.75, estimated only from the simulated closed-loop data and the knowledge of process time delays.

2.6.2 Assessment of Different Tuning Strategies

The conventional strategy to tune a cascade control loop is to first tune the secondary controller with the primary controller in the manual mode. Then the primary controller is transferred to automatic, and it is tuned. If the secondary controller is re-tuned for some reason, usually the primary controller must also be re-tuned. It has been devised by Seborg et al. (2004) to tune the slave loop tighter than the master loop to get improved stability characteristics and thus allow larger values of K_{c1} to be used in the primary control loop. Note that the presence of an integrator in the secondary loop is not strictly necessary since the null steady-state error can be assured by the primary loop controller. If integral action is employed in both the master and the slave controllers, the integrator windup should be carefully handled. A typical approach is to stop the integration of the primary controller when the output of the secondary controller attains its limits (Visioli 2006).

Principally, any (appropriate) tuning method can be applied to both controllers. However, there are some tuning methods explicitly tailored for cascade control systems, such as relay feedback simultaneous tuning, IMC-based simultaneous tuning and SPC-based simultaneous tuning. These will not be described here, but the reader is referred to Visioli (2006: Chap. 9) and the references included therein. As the usual sequential tuning is time-consuming, simultaneous tuning methods should be preferred.

Example 2.8 Consider the cascade system with the transfer functions

$$G_1(s) = \frac{1}{(s+1)^3}, \qquad G_2(s) = \frac{1}{s+1}. \tag{2.81}$$

This system was used by Åström and Hägglund (2006) to show the improved performance for deterministic load disturbance rejection of cascade control, compared with conventional PI control. The parameters of the latter controller were $K_\mathrm{c} = 0.37$ and $T_\mathrm{I} = 2.2$. For cascade control, a P controller with $K_\mathrm{c} = 5$ was placed in the secondary loop, and a PI controller with $K_\mathrm{c} = 0.55$ and $T_\mathrm{I} = 1.9$ in the primary loop. A high-gain controller is possible in the secondary loop, as its response to the control signal is quite fast.

We now evaluate the performance of both control systems in terms of stochastic disturbance rejection using the MV index. The hypothetical disturbances dynamics and variances are assumed to be the same as in Example 2.7. Under these circumstances, the computed performance index values were $\eta_{1,\mathrm{uni}} = 0.62$ for the conventional controller and $\eta_{1,\mathrm{uni}} = 0.85$, $\eta_{1,\mathrm{multi}} = 0.77$ for the cascade control. These results reveal the increase of stochastic performance achieved by the cascade control.

Example 2.9 An evaluation and comparison of the cascade-control tuning methods mentioned above in the context of CPM is provided using the following example:

$$G_1(s) = \frac{e^{-4s}}{(5s+1)^2}, \qquad G_2(s) = \frac{e^{-0.2s}}{s+1}. \tag{2.82}$$

The different PID controller settings derived by Visioli (2006) are given in Table 2.11. This also contains achieved values of the performance indices, the MV index $\eta_{2,\mathrm{uni}}$ for the inner loop and the MV index $\eta_{1,\mathrm{multi}}$ for the outer loop based on the multivariable performance analysis. The results confirm again the need for multivariable analysis; otherwise the performance is overestimated.

All tuning methods yields similar and satisfactory, but not excellent, control performance in the primary loop, despite the excellent (stochastic) performance in the inner loop. Therefore, there is still improvement potential for the primary loop from stochastic performance view point. If, for instance, the primary controller is significantly detuned, i.e., λ is increased, the variance is increased, however, only at the expense of increased rise time.

Table 2.11 Used controller parameters and assessment results

	Primary controller			Secondary controller			**Minimum-variance assessment**		
	K_{c1}	T_{I1}	T_{D1}	K_{c2}	T_{I2}	T_{D2}	$\eta_{2,\mathrm{uni}}$	$\eta_{1,\mathrm{uni}}$	$\eta_{1,\mathrm{multi}}$
Initial tuning	1.0	12.0	0	0.5	4.0	0	**0.82**	**0.72**	**0.64**
Relay feedback tuning	1.18	19.0	4.75	0.56	2.14	0	**0.95**	**0.77**	**0.70**
IMC-based tuning 1	0.94	10.0	1.89	3.31	1.06	0.07	**0.81**	**0.71**	**0.63**
IMC-based tuning 2	0.22	8.30	0.56	2.48	1.04	0.03	**0.84**	**0.89**	**0.83**
SPC-based tuning	1.5	8.22	0.91	3.17	1.05	0	**0.91**	**0.78**	**0.58**

From this example we also learn that tuning cascade control should always be driven towards maximising the Harris index (calculated using multivariable analysis) of the primary loop, when the variance is the main point. Thereby, it is not necessary to maximise the Harris index for the secondary loop. This conclusion seems to be not in agreement with the conventional approach of tuning cascade controllers. A similar conclusion was also pointed out by Teo et al. (2005) from their experience on another example.

2.7 Summary and Conclusions

Performance assessment based on minimum variance control, as the standard method for evaluating controllers, has been presented in detail. Besides batch calculation, the performance index can also be computed recursively, enabling the use of control charts for online monitoring of changes in controller performance. The following advantages of MV benchmarking contributed to its popularity and usage in the majority of CPM applications:

- Metrics based on MVC are the main criteria used in stochastic performance assessment, providing a direct relationship between the variance of key variables and product quality or energy/material consumption, which are correlated with financial benefits.
- MV benchmarking is easy to apply and implement and remains valuable as an absolute bound on performance against which real controllers can be compared. Performance monitoring should always include at least a look at the Harris index, as a first pass-assessment layer to bring obvious problems to immediate attention.
- Considering the MVC lower bound in setting performance targets will ensure that overly optimistic and conservative performance targets are avoided. MVC information can also be beneficial in incentive studies.

However, one should be aware about some serious drawbacks:

- A well-functioning loop in the process industry has frequently variance well above the minimum variance. Also industrial controllers (usually of the PID-type) do not have always a chance to match the MVC performance.

- Even though, MV control action can lead to highly undesirable, aggressive control and poor robustness.

Principally, MVC-based assessment is useful irrespective of the type of controller installed at the plant. However, tailored versions of MV assessment, such as those for feedback-plus-feedforward control and cascade control, can also be applied when the controller structure is known. Both control strategies are of widespread use in the process industry. The analysis of variance for feedback-plus-feedforward control helps quantify how much the performance of the control loop can be improved by re-tuning the feedforward component or introducing such a component if not yet implemented. For cascade control, it was shown that multivariate performance analysis should be generally applied, since univariate analysis may yield over-estimated loop performance, thus giving misleading conclusions. Also, tuning cascade control should always be driven towards maximising the Harris index (calculated using multivariable analysis) of the primary loop, when the variance is the main point.

Chapter 3
User-Specified Benchmarking

Minimum-variance benchmarking only considers the most fundamental performance limitation of a control loop owing to the existence of time delays. In practice, however, there are many other limitations on the achievable control performance, such as constraints on controller order, structure and action. A controller showing poor performance relative to MVC *is not necessarily* a *poor* controller. Further analysis using a more realistic performance benchmark is usually required.

Many researchers have introduced modified/extended versions of the Harris index to include design specifications of the user (such as the rise time and settling time) and take into account time delays in the system, leading to more realistic performance indices, referred to as user-specified benchmarks (in terms of user-specified closed-loop dynamics). Unlike the Harris index, extended (user-specified) performance indices do not provide information about how close the current performance is to optimal performance. An extended performance index rather reflects how well the control loop is doing compared to design specifications in terms of a specified prediction horizon, settling time, overshoot, or other parameters.

Section 3.1 provides a general setting for user-specified performance assessment. In Sect. 3.2, the framework of IMC-achievable performance assessment is presented. Section 3.3 deals with the very popular extended-horizon approach. A special performance index based on desired pole location is described in Sect. 3.4. The simplest method of performance assessment is to use criterion values extracted from historical data windows where the controller is believed to perform well. This technique known as historical benchmarking is given in Sect. 3.5. Section 3.6 deals with assessment methods based on reference models.

3.1 General Setting

Consider again the impulse-response representation of the closed-loop dynamics in Eq. 2.36 and rewrite it as

$$y(k) = \left[E_\tau(q) + q^{-\tau} G_\mathrm{R}(q)\right]\varepsilon(k). \tag{3.1}$$

M. Jelali, *Control Performance Management in Industrial Automation*,
Advances in Industrial Control, DOI 10.1007/978-1-4471-4546-2_3,

Since the first term of this equation is, as before, control invariant, the user has only the option to specify G_R as a stable and proper transfer function. This desired behaviour of the closed loop has to be achieved by a correspondingly designed or tuned controller. Many ways exist to achieve the desired behaviour, such as the specification of the closed-loop rise time, settling time, decay ratio, overshoot, frequency characteristics, robust performance measures, etc. Therefore, the side effect of user-specified performance assessment is that no clear guidelines exist on which measure should be specified to get optimal performance and how a certain choice actually affects the performance and robustness of the control system at hand (Huang and Shah 1997). It remains something subjective and arbitrary. Nevertheless, there are some interesting user-specified benchmarking methods that are worth consideration in practice, to be discussed below.

User-specified performance indices are generally defined as

$$\eta_{\mathrm{user}} = \frac{J_{\mathrm{user}}}{J_{\mathrm{act}}}, \tag{3.2}$$

where J_{user} is the corresponding value of the user-specified performance measures. Note that performance index can now take values *higher* than unity, thereby indicating that the current controller is doing better than required, i.e. $J_{\mathrm{act}} < J_{\mathrm{user}}$. In the stochastic performance framework, the variance of the user-specified benchmark control can be calculated from Eq. 3.1, to give the "variance-related" performance index

$$\eta_{\mathrm{user}} = \frac{\sigma^2_{\mathrm{user}}}{\sigma^2_y}. \tag{3.3}$$

At this point, note that the term "user-specified performance" does not imply that users can specify the desired control performance arbitrary, without considering physical limitations. For instance, it is useless to specify closed-loop response within the time-delay period, as it is control invariant. Also, users cannot desire to cancel non-minimum phase zeros.

3.2 IMC-Achievable Performance Assessment

Internal Model Control (IMC), first proposed by Frank (1974) for process control, is a class of model-based control design proven to have good performance and robustness properties against parameter uncertainties. It is well known that the IMC scheme is equivalent to the celebrated Yula–Kucera parameterisation or Q-parameterisation of all stabilising controllers (Youla et al. 1976a, 1976b) for a stable system. In a famous series of papers (García and Morari 1982, 1985a, 1985b; Economou et al. 1986; Economou and Morari 1986; Rivera et al. 1986), Morari and his co-workers greatly expanded on the IMC-design methods and placed the methodology in a sound theoretical framework. García and Morari (1982, 1985a, 1985b) provided a unifying review on IMC and further extended it to multivariable systems. Developments of IMC in the case of nonlinear systems have been

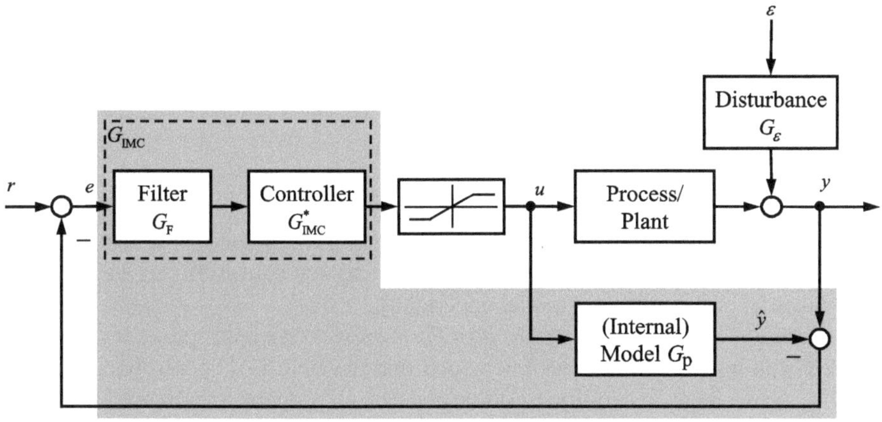

Fig. 3.1 Structure of internal model control

proposed, mainly for continuous-time models, by Economou et al. (1986), Calvet and Arkun (1988) and Henson and Seborg (1991), but also for particular classes of discrete-time models by Alvarez (1995). The interested reader may also wish to consult Morari and Zafiriou (1989) for thorough information on the topic.

The IMC structure is shown in Fig. 3.1. It generally consists of three parts:

- A (forward) prediction model (G_p) of the process.
- A controller (G^*_IMC) based on the invertible part of the process model.
- A (low-pass) filter (G_F).

In the IMC scheme, the role of the system models is emphasised: it is used directly as element within the feedback loop. The filter is used to introduce robustness in the IMC structure in the face of modelling errors (by appropriate reducing of the loop gain) and to smooth out noisy and/or rapidly changing signals, and thus reduces the transient response of the IMC. With nonlinear systems and models, the filter must be designed empirically, usually as a first-order or second-order system.

3.2.1 IMC Design

The IMC-design procedure, although sub-optimal in a general (norm) sense, provides a reasonable balance between performance and robustness. It consists of two basic steps (Morari and Zafiriou 1989):

1. *Nominal Performance Design*. The system model is factorised into one ("bad" all-pass) part (G_p^+), which contains all non-minimum-phase elements (i.e. all

RHP zeros and time delays)[1] and one "good" minimum-phase and invertible part (G_p^-):

$$G_p = G_p^+ G_p^-. \tag{3.4}$$

The IMC controller is then usually set to

$$G_{IMC}^* = \left(G_p^-\right)^{-1}, \tag{3.5}$$

which is physically realisable, i.e. stable and causal, in contrast to the inversion of G_p^+, which is non-realisable, i.e. unstable and non-causal. In this step, no care is taken of constraints and model uncertainty.

2. *Ensuring Robust Stability and Robust Performance.* An appropriate low-pass filter G_F is introduced and designed such that the final IMC controller ($G_{IMC} = G_{IMC}^* G_F$) is now, in addition to stable and causal, proper (in the sense that it does not require signal derivatives), i.e., G_{IMC} must be finite, as its argument (s/q) goes to infinity.[2] Note that the controller properness (usually semi-properness) is not a must, but an option commonly used. The inclusion of the filter provides the roll-off necessary for robustness and milder action of the manipulated variables and implies that the controller is detuned to sacrifice performance for robustness, i.e., "optimal control" is no longer obtained. Usually, the filter parameters are fixed up to the filter gain λ used for tuning (determining the speed of response and the robustness of the closed-loop system).

For linear (time-discrete) systems, a possible factorisation of the process model is (Moudgalya 2007)

$$G_p(z) = z^{-\tau} \frac{B^s(z) B^-(z) B^{nm+}(z)}{A(z)}, \tag{3.6}$$

where (Fig. 3.2)

- $B^s(z)$ is the factor of $B(z)$ with roots inside the unit circle and with positive real parts.
- $B^-(z)$ denotes the factor of $B(z)$ with roots that have negative real parts, irrespective their position from the unit circle.
- $B^{nm+}(z)$ refers to that part of $B(z)$ containing non-minimum zeros of $B(z)$ with positive real parts.

With this factorisation, the IMC becomes

$$G_{IMC}^*(z) = \frac{A(z)}{B^s(z) B_{steady}^-(z) B_{reversed}^{nm+}(z)}, \tag{3.7}$$

where $B_{steady}^-(z)$ is the steady-state equivalent of $B^-(z)$, i.e.

$$B_{steady}^-(z) = B^-(z)|_{z=1},$$

[1] The inversion of RHP zeros and time delays leads to unstability and non-causality, respectively.

[2] In the linear case, this means that the transfer-function denominator order is equal or greater than nominator order.

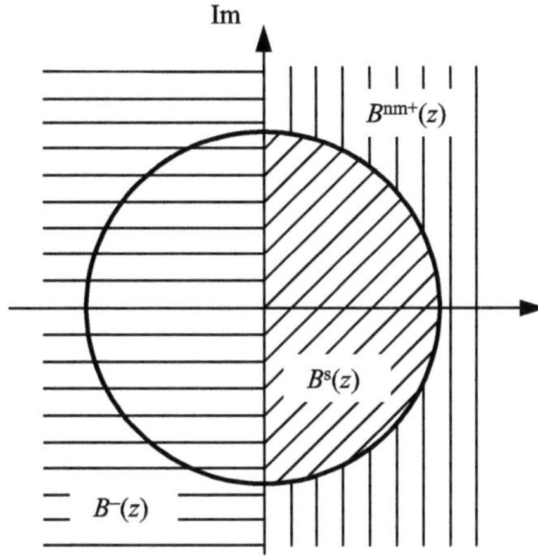

Fig. 3.2 Dividing z-plane region into good, bad and non-minimum phase parts

and $B^{nm+}_{reversed}(z)$ is $B^{nm+}(z)$ with reversed coefficients, i.e.

$$B^{nm+}_{reversed}(z) = B^{nm+}(z)|_{reversed\ coefficients}.$$

For instance, $B^{nm+}(z) = 1 - 1.3z^{-1} \Rightarrow B^{nm+}_{reversed}(z) = z^{-1} - 1.3$.

Three attractive characteristics of IMC are stated as follows:

- *Inherent Stability*. If the controller and the process are input–output stable and a perfect model of the process is available, the closed-loop system is input–output stable. Should the system be not input–output stable, but stabilisable by feedback, IMC can still be applied.
- *Perfect Control*. If the controller is an exact model inverse and the closed-loop system is stable, then an ideal (i.e., error-free) control performance is achieved (i.e. $y = r$). In practice, however, a perfect model/control can never be obtained, and the infinite gain required by perfect control would lead to sensitivity problems under model uncertainties. Hence, a suitably designed filter is introduced to reduce high gains and provide robustness in the IMC scheme.
- *Zero Offset*. If the controller is an exact model inverse and the closed-loop system is stable with an ideal/optimal controller, then offset-free control is attained for asymptotically constant inputs.

Note that the IMC structure can be rearranged to get an equivalent standard feedback controller, where

$$G_c = \frac{G_{IMC}}{1 - G_p G_{IMC}} = \frac{G_F G^*_{IMC}}{1 - G_p G_F G^*_{IMC}}. \tag{3.8}$$

Note also that the Smith-predictor control (SPC; Fig. 3.3), a very popular and effective time-delay compensator, can be rearranged to get an equivalent IMC structure, where

$$G_c = \frac{G_{SPC}}{1 + G_{SPC} G^*_p}. \tag{3.9}$$

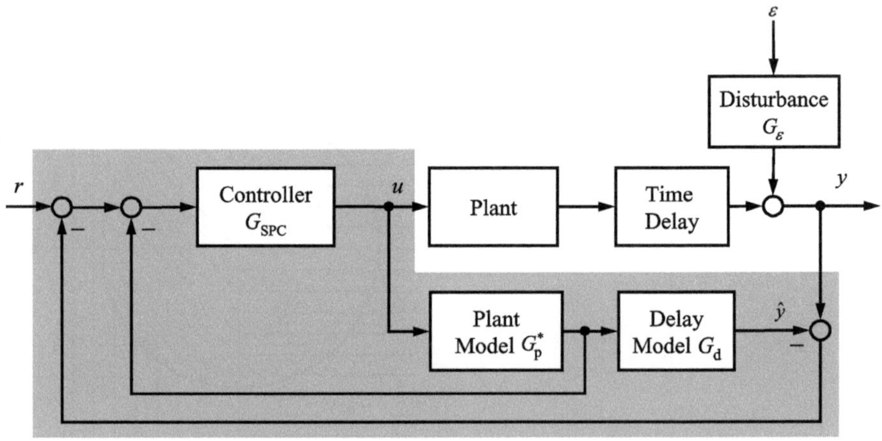

Fig. 3.3 Block diagram of Smith-predictor control

From this representation many aspects like stability properties of the SPC can be inferred, as reported by Lee et al. (1996), Litrico and Georges (1999) and Sunan et al. (2002). Note that G_{SPC} is usually a PID controller.

3.2.2 IMC Benchmark

IMC is a popular technique for controller design in the process industry, including special variants, such as Dahlin's controller and lambda (λ)-tuning. The strength of the IMC approach is that it provides a good trade-off between performance and robustness to process changes and model-plant mismatch, while reducing the controller-tuning parameters to a single coefficient, the filter constant λ. λ is also equivalent to the closed-loop time constant. These properties provide enough motivation to consider IMC-achievable performance as a performance assessment benchmark.

Consider the process described by the model

$$y(k) = G_p u(k) + G_\varepsilon \varepsilon(k) = \frac{B(q)}{A(q)} q^{-\tau} u(k) + \frac{C(q)}{A(q)} \varepsilon(k). \tag{3.10}$$

From IMC block diagram (Fig. 3.1) we can derive the general relationship

$$y(k) = \frac{1}{1+G_c G_p} G_\varepsilon \varepsilon(k) = \frac{1 - G_{IMC} G_p}{1 - G_{IMC} G_p + G_{IMC} \hat{G}_p} G_\varepsilon \varepsilon(k). \tag{3.11}$$

If perfect process modelling, i.e., $\hat{G}_p = G_p$, is assumed, the process output is given by

$$y(k) = G_\varepsilon \varepsilon(k) - G_p G_{IMC} G_\varepsilon \varepsilon(k) = G_\varepsilon \varepsilon(k) - q^{-\tau} G_p^* G_{IMC} G_\varepsilon \varepsilon(k), \tag{3.12}$$

where G_p^* is the delay-free part of the process model G_p.

Huang and Shah (1997) proposed to specify G_{R} in Eq. 3.1 as

$$G_{\mathrm{R}} = \frac{1 - G_{\mathrm{F}}}{R_\tau}, \tag{3.13}$$

where G_{F} is the (stable and proper) IMC filter, and R_τ the relational proper transfer function defined by the Diophantine equation

$$G_\varepsilon(q) = E_\tau(q) + q^{-\tau} R_\tau(q). \tag{3.14}$$

A low-pass filter of the form

$$G_{\mathrm{F}}(q) = \frac{(1-\alpha)}{1-\alpha q^{-1}}, \quad \alpha = \mathrm{e}^{-(T_{\mathrm{s}}/\lambda)}, \tag{3.15}$$

with a specified closed-loop time constant λ, is usually included to ensure a trade-off between performance and modelling errors. Substituting Eq. 3.14 into 3.12 yields

$$y(k) = E_\tau \varepsilon(k) + q^{-\tau}[R_\tau - \tilde{G}_{\mathrm{p}} \tilde{G}_{\mathrm{IMC}} G_\varepsilon]\varepsilon(k). \tag{3.16}$$

Comparing with Eq. 3.1 leads to

$$G_R = R_\tau - \tilde{G}_{\mathrm{p}} G_{\mathrm{IMC}} G_\varepsilon \quad \Rightarrow \quad G_{\mathrm{IMC}} = \frac{R_\tau - G_R}{G_{\mathrm{p}}^* G_\varepsilon}. \tag{3.17}$$

From Eqs. 3.13 and 3.17 we get the IMC controller

$$G_{\mathrm{IMC}} = \frac{G_{\mathrm{F}} R_\tau}{G_{\mathrm{p}}^* G_\varepsilon}. \tag{3.18}$$

Since this controller is stable and proper, the closed-loop response specified in Eq. 3.1 is achievable. Substituting back into Eq. 3.16 gives

$$y(k) = E_\tau \varepsilon(k) + q^{-\tau} R_\tau (1 - G_{\mathrm{F}})\varepsilon(k). \tag{3.19}$$

From this it easy to see that if $G_{\mathrm{F}} = 1$, we get the MVC response.

To further simplify the assessment task, let us assume the disturbance model being of the random-walk type, i.e.

$$G_\varepsilon(q) = \frac{1}{1-q^{-1}}. \tag{3.20}$$

This can be expanded using long division as

$$G_\varepsilon(q) = \underbrace{1 + q^{-1} + q^{-2} + \cdots}_{E_\tau} + q^{-\tau} \underbrace{\frac{1}{1-q^{-1}}}_{R_\tau}. \tag{3.21}$$

Substituting R_τ into Eq. 3.19 gives

$$y(k) = E_\tau \varepsilon(k) + q^{-\tau} \frac{\alpha}{1-\alpha q^{-1}} \varepsilon(k). \tag{3.22}$$

The *best IMC-achievable* variance of y for a desired closed-loop time constant is therefore

$$\sigma_{\mathrm{IMC}}^2 = \sigma_{\mathrm{MV}}^2 + \frac{\alpha^2}{1-\alpha^2}\sigma_\varepsilon^2 = \tau\sigma_\varepsilon^2 + \frac{\alpha^2}{1-\alpha^2}\sigma_\varepsilon^2, \tag{3.23}$$

which is, of course, larger than the minimum variance σ_{MV}^2. The variance σ_{IMC}^2 can be used to assess the performance degradation due to the requirement of robustness in the controller design. For a given desired closed-loop time constant λ, the best achievable performance should be σ_{IMC}^2. Equation 3.23 also reveals an important conflict between minimum variance and robustness: a larger α results in more robust control but dramatically increases the IMC-achievable variance. The IMC-performance index can now be introduced as

$$\eta_{\mathrm{IMC}} = \frac{\hat{\sigma}_{\mathrm{IMC}}^2}{\hat{\sigma}_y^2} = \frac{\sum_{i=0}^{\tau-1} e_i^2\sigma_\varepsilon^2 + \frac{\alpha^2}{1-\alpha^2}\sigma_\varepsilon^2}{\sum_{i=0}^{\infty} e_i^2\sigma_\varepsilon^2} = \frac{\sum_{i=0}^{\tau-1} e_i^2 + \frac{\alpha^2}{1-\alpha^2}}{\sum_{i=0}^{\infty} e_i^2}. \tag{3.24}$$

Note that only routine operating data and the knowledge/estimation of the time delay are needed to estimate the index η_{IMC}.

Example 3.1 We illustrate the IMC-achievable performance assessment with the model of a paper machine given by Åström and Wittenmark (1997):

$$y(k) = \frac{0.63q^{-3}}{1-0.37q^{-1}}u(k) + \frac{1}{1-q^{-1}}\varepsilon(k). \tag{3.25}$$

The noise ε is assumed to have unity variance. The IMC design for this system is obtained as

$$G_{\mathrm{IMC}} = \frac{1-0.37q^{-1}}{0.63}\frac{(1-\alpha)}{1-\alpha q^{-1}}.$$

Let the desired closed-loop time constant be $\lambda = 5T_{\mathrm{s}}$ with $T_{\mathrm{s}} = 1$ s, corresponding to $\alpha \approx 0.82$. The IMC-achievable variance can be calculated from Eq. 3.23 as $\sigma_{\mathrm{IMC}}^2 = 5.0525\sigma_\varepsilon^2$. Consider the process under integral control

$$G_{\mathrm{c}} = \frac{0.1}{1-q^{-1}}.$$

The closed-loop system was simulated and 5000 data points were recorded.

A truncated AR model is estimated from the gathered data and used for calculating the impulse response coefficients and the Harris index. The output variance with the integral controller is $\sigma_{\mathrm{I}}^2 = 6.3868$. Figure 3.4 shows the impulse response with the integral controller and IMC. Of course, the closed-loop dynamics with IMC requires complete knowledge of the process and disturbance models. This is, however, not needed for computing the IMC-achievable performance index. The IMC-achievable performance index and Harris index were computed as $\eta_{\mathrm{IMC}} = 0.79$ and $\eta_{\mathrm{MV}} = 0.47$, respectively. The conclusion is that the controller achieves good performance compared to the IMC performance, although it is far from MVC performance (indicated by the Harris index).

3.3 Extended Horizon Approach

To consider user specifications and/or to avoid requiring the loop time delay for calculating the performance index, Desborough and Harris (1992) and Thornhill

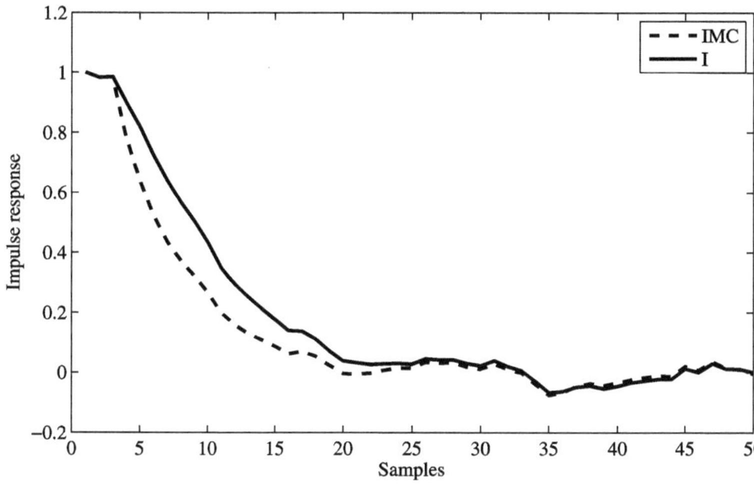

Fig. 3.4 Closed-loop impulse responses from paper machine simulation

et al. (1999) proposed the use of the extended horizon performance index (EHPI). The general expression to calculate the performance index is the following:

$$\eta_b = \frac{\sum_{i=0}^{b-1} e_i^2}{\sum_{i=0}^{\infty} e_i^2}, \tag{3.26}$$

where b is the prediction horizon. If b equals the time delay of the system, η_b is identified to be the Harris index η (Eq. 2.34). When η_b is calculated for b larger than the time delay, it is referred to as an *extended horizon performance index*. In this case, an interpretation is that η_b gives the portion of the variance that can be accounted for with a b-step-ahead predictor.

For a general b, the index tells us how large the portion of the variance of the control signal comes from noise older than b samples. Figure 2.4 can again be used for a visual interpretation of the contributions to η_b, just replacing τ with b. A rise in η_b means that impulse-response coefficients older than b have decreased compared to the first b ones. As a consequence, the benchmark is no longer minimum-variance control, and the quality of control in the interval between the time delay and the prediction horizon is not assessed.

Thornhill et al. (1999) suggested a practical approach, which can be applied when prior knowledge of the time delay is not available: the performance indices are calculated over a range of time-delay values and are known as extended horizon performance indices. A plot of η vs. time-delay values (Fig. 3.5) helps one in selecting an appropriate value of the prediction horizon, to be used instead of the time delay for performance assessment. A good choice falls on the region where the CPI does not vary rapidly, i.e. flat area (Thornhill et al. 1999). Our experience with this method suggests to take the prediction horizon somewhere near the first inflexion point (Jelali 2006a).

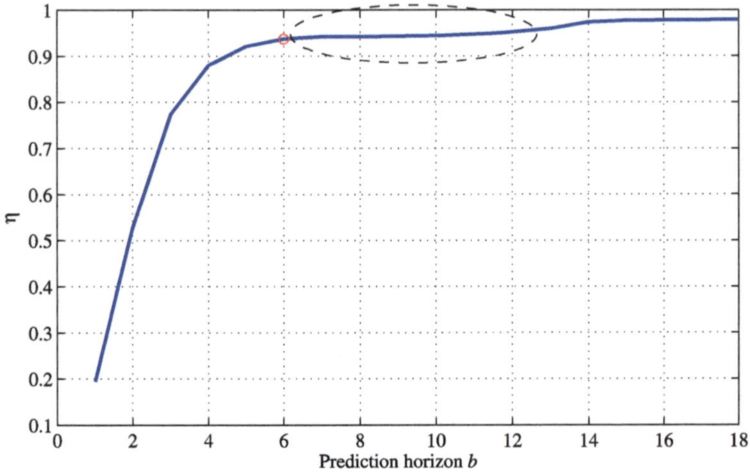

Fig. 3.5 Prediction horizon plot for finding a good choice of the prediction horizon. In this case, the loop has a time delay $\tau = 6$ (Loop: thickness control, coil no. 90)

This approach avoids the time consuming determination of time delays (Sect. 7.3.1) and regards the prediction horizon as an engineering criterion, representing a demand made on the control loop: the predictable (and thus controllable) components of the control error should be dealt with within the specified time horizon b.

Ingimundarson and Hägglund (2005) also suggested to monitor the EHPI. However, they recommend to select the prediction horizon and an alert limit not according to the type of loop, as done by Thornhill et al. (1999), but from the tuning of the loop, i.e. to fulfil design specifications. At each performance-evaluation instant, an index estimate is compared to the alert limit, say 0.8. If the estimate falls below the alert limit, the loop is not rejecting disturbances as it should. This information is highly desirable in the cases where plant staff is interested in knowing whether the control performance is acceptable, but not how far it is from an optimum.

Ingimundarson (2003) suggested to choose the prediction horizon as $b = \tau + h$ for self-regulating systems and $b = \tau + 2h$ for integrating systems. h is a specified (desired) number of samples for the closed-loop response (corresponding to λ, the time constant of the continuous-time model), particularly used when the (PI) controller is set by λ-tuning. This is a widely used method in the pulp & paper and chemical industries. Note that even in the case where the time delay is known or properly estimated, the prediction horizon approach is useful to get more realistic assessments.

Moreover, prediction horizon plots give an opportunity for a cross-check on the choice of sampling interval. In cases where the choice of sampling interval has been clear from a consideration of the estimated impulse response, it has always been found that the best prediction horizons are typically between three and 10 sampling intervals. Therefore, it is recommended that if the selected prediction horizon exceeds 10 sample intervals, then an increase in the sampling interval should be considered (Thornhill et al. 1999).

3.4 Performance Index Based on Desired Pole Locations

Inspired by the observations of Kozub and Garcia (1993), Horch and Isaksson (1999) proposed a modified performance index based on desired pole locations. Instead of comparing the actual variance to MV, which corresponds to placing all closed-loop poles at the origin, they compare the actual variance to the variance that would be obtained when placing all closed-loop poles *but one* at the origin. The pole not placed at the origin would determine the closed-loop speed and bandwidth, according to its location. The choice of the closed-loop pole (μ) can be based on either control-design guidelines (robustness margins) or additionally available system knowledge, such as the lowest system time constant.

The Horch–Isaksson index is defined by

$$\eta_{\mathrm{HI}} = \frac{\sigma^2_{\mathrm{MV,mod}}}{\sigma^2_y}, \tag{3.27}$$

where the modified minimum variance $\sigma^2_{\mathrm{MV,mod}}$ is now calculated as

$$\begin{aligned}\sigma^2_{\mathrm{MV,mod}} &= \left(\sum_{i=0}^{\tau-1} h_i^2 + h_{\tau-1}^2 \mu^2 \sum_{i=0}^{\infty} \mu^{2i}\right)\sigma_e^2 = \left(\sum_{i=0}^{\tau-1} h_i^2 + h_{\tau-1}^2 \frac{\mu^2}{1-\mu^2}\right)\sigma_e^2 \\ &= \sigma^2_{\mathrm{MV}} + h_{\tau-1}^2 \frac{\mu^2}{1-\mu^2}\sigma_e^2 = \sigma^2_{\mathrm{MV}} + \sigma^2_{\mu}.\end{aligned} \tag{3.28}$$

This variance has two parts: the minimum variance and the contribution from the first-order decay (corresponding to the specified pole μ). Thus, the assessment algorithm for this modified index is the same as that for the Harris index (Sect. 2.4), but replacing the minimum variance by its modified version (Eq. 3.28) in the performance-index calculation (Eq. 3.27); see Procedure 3.1. Comparing Eq. 3.28 with Eq. 3.23 shows that the Horch–Isaksson index is equivalent to the IMC-achievable index.

Procedure 3.1 Performance assessment based on the Horch–Isaksson index.

1. Preparation. Select the time-series-model type and orders.
2. Determine/estimate the system time delay τ.
3. Identify the closed-loop model from collected output samples.
4. Calculate the series expansion (impulse response) for the estimated model (Eq. 2.36).
5. Estimate the minimum variance from Eq. 2.38.
6. Calculate the modified minimum variance from Eq. 3.28.
7. Estimate the actual output-variance from Eqs. 1.1 or 2.39.
8. Compute the (Horch–Isaksson) performance index (Eq. 3.27).

3.5 Historical or Reference Benchmarks

In practice, it is often decided to specify "benchmark" criteria corresponding to assessment values extracted from historical data during a time period when the control system was "doing its job well" from the viewpoint of control or maintenance engineers. Such criteria have been introduced under different terms, such as *baselines* (Gerry 2002), *historical data benchmarks* (HIS), *reference data set benchmarks* (Patwardhan et al. 1998; Huang et al. 1999; Gao et al. 2003), or *reference distributions* (Li et al. 2004).

This approach requires a priori knowledge that the performance was good during certain time period according to some expert assessment. For the selected input and output data, the historical benchmark index is defined as the ratio

$$\eta_{\text{his}} = \frac{J_{\text{his}}}{J_{\text{act}}}, \tag{3.29}$$

where J_{his} is the value of the selected performance criterion (typically the variance), extracted from historical data, and J_{act} the actual value of the criterion, to be extracted from measured process data under the installed controller.

Historical benchmarking techniques do not require a process model or knowledge of process delay and therefore are suitable for monitoring time-varying and nonlinear processes as well. They only need a window of the control-error data collected during a representative period defined as having good control by the user, and these data are used to build the reference value of a performance index. Once the reference value is built, only the control-error data are needed to run the control performance monitor.

However, one should be careful when applying HIS and EHPI benchmarking, as they may be too subjective and rely too much on the current performance situation. Often, the controller is "felt" to work satisfactorily although it is badly performing compared with other benchmarks.

3.6 Reference-Model/Relative Performance Index

The methodology proposed by Li et al. (2003) is based on a reference model that specifies the required closed-loop behaviour and generates the performance index as the ratio of reference and actual value of a performance metric (Fig. 3.6). This is called the *relative performance index* and defined as

$$\eta_{\text{RP}} = \frac{M(e_{\text{ref}})}{M(e)}, \tag{3.30}$$

where M is a performance metric to be selected, such as the mean squared error (MSE), the mean absolute error (MAE), or recursive versions of them, like the exponentially weighted moving average of squared error (EWMASE):

$$M_k = \lambda M_{k-1} + (1-\lambda)e_k^2. \tag{3.31}$$

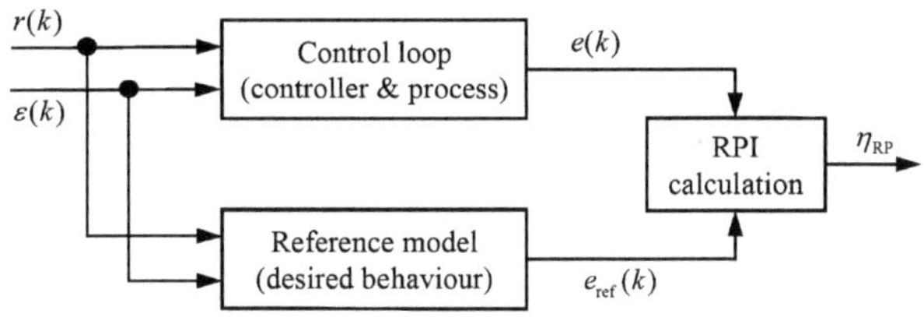

Fig. 3.6 Principle of the relative performance monitor

The RPI provides a measure of the relative performance of the control loop and the reference model, and $1 - \eta_{\mathrm{RP}}$ represents the improvement potential in control-loop performance if re-tuning the loop to the reference-model level of performance. The RPI value can be interpreted as follows (Li et al. 2003):

1. An RPI value close to 1.0 means that the control-loop performance is close to that of the reference model.
2. An RPI value $\ll$1.0 indicates that the loop performance is much worse than that of the reference model, and something should be done on the control loop, such as re-estimating the plant model, re-tuning the controller parameters, checking for valve stiction, etc. In this case the selected performance metric, such as MSE or EWMASE, may be reduced by $100(1 - \eta_{\mathrm{RP}})$ % under similar input conditions if the control loop is restored to reach the same performance level represented by reference model.
3. An RPI value $>$1.0 implies that the control loop has a better performance than the reference model. If the RPI values are much greater than 1.0, the current performance is better than that of the reference model, and an update of the reference model might be useful.

It is not intended to present the RPI computation algorithm in detail here; for this, the reader should refer to Li et al. (2003). Below, the main features are briefly described.

3.6.1 Selection of the Reference Model

In principle, there are many ways to specify the reference model. However, it is not practical at all to use more complicated reference models than a first-order or second-order model with a unity gain. Thus, one or two free parameters have to be selected depending on the application at hand. Always, the user has to specify these parameters, which necessitates corresponding skills or a priori knowledge about how well the controller should actually work.

3.6.2 Disturbance Estimation

A key factor in the calculation of the RPI is the estimation of the unmeasured disturbances. As for the reference model, it is useful and sufficient in most cases to use a simple model for the disturbance estimation. Li et al. (2003) propose to choose a first-order ARX model to describe the plant and then estimate the model parameters so that the deviation between the model output and the measured process output is minimised. The resulting deviation provides an estimate of the acting disturbance. The disturbance estimation process can be formulated using either a batch or recursive variant of least-square parameter estimation. The latter version has the advantage that it can handle time-variant and nonlinear systems commonly found in the process industry.

3.6.3 Selection of Critical Index Value

To automatically flag poor performance, a critical or threshold value $\eta_{\mathrm{RP,crit}}$ of the RPI must be specified. A good choice of the threshold, e.g. 0.75, depends not only on the noise level of the data but also on the user's tolerance level on the deviation of a control-loop performance from that of the reference model. Statistical tests like the F-test could also be used to establish values for controller-monitor flagging. However, this method for determining the threshold would add complexity.

3.6.4 Index Properties

Although the RPI may be a useful index for control performance assessment, a lot of options and parameters must be specified by the (experienced) user. This makes the index much more involved than calculating the MV index. A strength of using the RPI, which should be highlighted here, is that its values have been observed to be in agreement with time-domain criteria, usually used to assess set-point tracking performance. This is not the case for minimum variance benchmarks, which are more suited for the assessment of stochastic disturbance-rejection performance. Therefore, this RPI characteristic complements those of the MV index. Also, the RPI is able to differentiate oscillations caused internally and those induced externally. This also complements the MV index, so that a good strategy may be to simultaneously calculate both indices. When the MV index signals poor control performance and the RPI does not, external oscillatory disturbances should be concluded to occur.

3.7 Summary and Conclusions

The general setting and different methods of user-specified performance assessment have been presented. Although these approaches may be useful in many situations, their main dilemma is that the specifications are arbitrary in some way, and it is not always clear how such specifications affect the closed-loop dynamics, e.g. in terms of performance optimisation and robustness (Huang and Shah 1999). For each specification type (settling time, decay ratio, overshoot, desired variance, or even reference model, etc.), there is usually an infinite number of possibilities that can be considered, but no general guidelines exist on which option is the best to get performance closest to optimal control. Often, the decision will remain up to an experienced user or control engineer.

Of particular interest remains the extended horizon approach which is useful to apply when no information is available about the time delay. Historical benchmarking can also be attractive due to its simplicity, but must be considered with care, because of the subjective definition of the benchmark values.

Note also that Rhinehart (1995) and Venkataramanan et al. (1997) proposed and applied a statistical test (called r-statistic), which detects deviations from set-point, regardless of the output-noise amplitude.

Chapter 4
Advanced Control Performance Assessment

This chapter deals with extensions of the MV benchmark, which need substantially more information about the plant than just the time delay. An extension of the MV benchmark is the approach of generalised MV (GMV) benchmarking, minimising a weighted sum of the control error and control effort; see Sect. 4.1. A more general but rigorous extension is the linear-quadratic Gaussian (LQG) benchmark presented in Sect. 4.2. Both benchmarks are useful when more information on controller performance, such as how much can the output variance be reduced without significantly affecting the controller output variance, is needed (Shah et al. 2001), or for cases where actuator wear is a concern.

Model predictive control (MPC) technology has been widely implemented in many process industries, such as the chemical, petrochemical, metallurgical and pulp & paper industries, over the past three decades. Therefore, it is also important to use tailored techniques for the assessment of MPC performance. Section 4.3 provides an overview of these methods. It is particularly shown how to use routine operating data to distinguish between poor performance due to plant–model mismatch and that due to improper tuning of the MPC controller. Moreover, performance measures which estimate potential benefit from re-identification of the process model or re-tuning of the controller are introduced. This is essential in MPC monitoring, as a process model is a substantial component of the MPC controller.

4.1 Generalised Minimum-Variance Control (GMVC) Benchmarking

A straightforward extension of the MV benchmark by considering control action penalisation leads to the more flexible approach of generalised MV (GMV) benchmarking suggested by Grimble (2002). This control performance assessment technique is described in this section.

M. Jelali, *Control Performance Management in Industrial Automation*,
Advances in Industrial Control, DOI 10.1007/978-1-4471-4546-2_4,

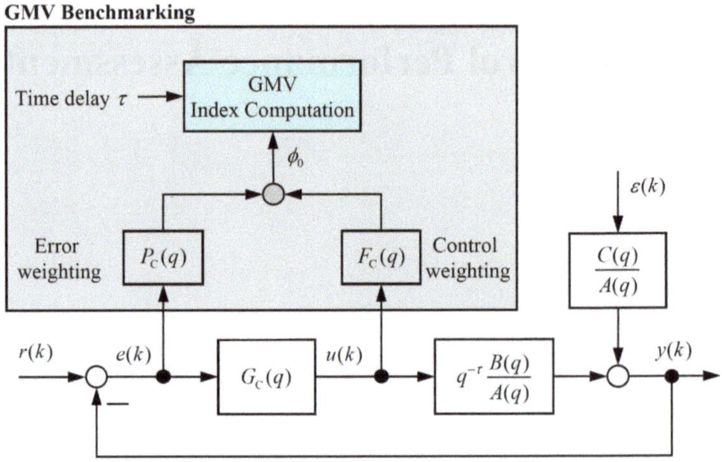

Fig. 4.1 Principle of GMV benchmarking

4.1.1 GMV Control

The derivation of the GMV control law is simpler than that of the LQG control law, to be discussed in Sect. 4.2. By analogy with MV control law, the GMV control algorithm can be defined as one that minimises the following cost function (Grimble 2002a):

$$J_{\mathrm{GMV}} = \mathrm{E}\{\phi_0^2\}, \tag{4.1}$$

where ϕ_0 is the "generalised" output signal (Fig. 4.1),

$$\phi_0(k) := P_c e(k) + F_c u(k), \tag{4.2}$$

and P_c and F_c denote appropriate weighting functions:

$$P_c = \frac{P_{cn}}{P_{cd}}, \qquad P_{cd}(0) = 1, \qquad P_{cn}(0) \neq 1; \tag{4.3}$$

$$F_c = \frac{F_{cn}}{F_{cd}}, \qquad F_{cd}(0) = 1, \qquad F_{cn} = F_{c\tau} q^{-\tau}. \tag{4.4}$$

The error weighting P_c usually includes an integral term. The control weighting F_c is defined to include the time delay, as the control signal affects the output with a τ-step delay. Unlike the related LQG control law, the price to be paid for the "simplicity" is that the dynamic weightings cannot be chosen arbitrary. Rather, the restricting assumption imposed on the dynamic weightings must be fulfilled to ensure the stability of the closed-loop system:

$$D_c := P_{cn} F_{cd} B - F_{ck} P_{cd} A \text{ must be stable.} \tag{4.5}$$

4.1.2 Selection of Weightings

The GMV benchmark algorithm needs a set of dynamic error and control weights to compute the performance index. These weights act as design parameters that specify the type of optimal controller required (regulatory performance, tracking/disturbance rejection, level of robustness): the user is required to know and specify the optimal performance requirements for the control loop under assessment. Guidelines for the selection of the weightings can be found by Grimble and Uduehi (2001) and Grimble and Majecki (2004).

In general, the frequency dependence of the weightings can be used to weight different frequency ranges in the error and control signals. According to Grimble and Majecki (2004), the standard procedure is that the error weighting P_c normally includes an integral term:

$$P_c(q) = \frac{P_{cn}}{1 - q^{-1}}. \tag{4.6}$$

P_{cn} may be constant or have the form $(1 - \alpha q^{-1})$, where $0 < \alpha < 1$ is a tuning parameter; the larger α, the sooner integral action is "turned off". This term leads to integral action in the controller. The general effect of introducing integral error weighting is, however, to introduce high gain into the controller at low frequencies.

The control weighting F_c is chosen as a constant, or as a lead term, i.e.

$$F_c = \rho \quad \text{or} \quad F_c(q) = \rho\left(1 - \gamma q^{-1}\right). \tag{4.7}$$

This weighting provides one mechanism of ensuring the controller rolls-off in high frequencies and does not amplify the measurement noise. ρ and γ are tuning parameters. An additional scalar may be used to balance the steady-state variances of the error and control signals. Controller roll-off at high frequencies is naturally included in LQG or H_2 designs by using a measurement noise model. When such a model is absent, GMV and LQG designs can give a too high gain at high frequencies. An example of the weightings

$$P_c(q) = \frac{1 - 0.85q^{-1}}{1 - q^{-1}}, \qquad F_{c\tau}(q) = -0.015\left(1 - 0.1q^{-1}\right)$$

is shown in Fig. 4.2.

If the considered system is controlled by a PID controller or some other well-defined classical control structure, Uduehi et al. (2007b) suggest to select the ratio of the error weighting to the control weighting equal to the aforementioned controller as starting choice of GMV cost function weighting. Then some slight adjustment, usually reducing the value of the control weighting, should follow.

Despite these guidelines, it should be clear that the full utilisation of the dynamic weighting is only possible when full knowledge of the process model linearised around the working point is available. This makes the application of GMV benchmarking more demanding than other methods. However, the model does not need

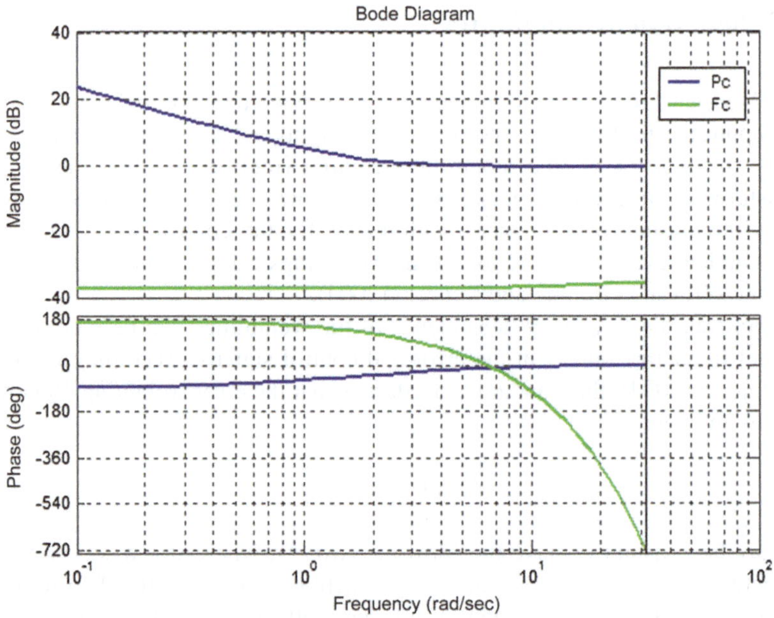

Fig. 4.2 Example of selection of GMV weighting functions (Grimble and Majecki 2005)

to be very accurate, i.e., a simple first-order or second-order approximation often suffices. See Uduehi et al. (2007b) for details.

While selecting the dynamic weightings, one has to be aware of the restriction stated on the weightings (Eq. 4.5). Note that the benchmarking algorithm will still return a controller performance index even if the condition is not satisfied. This, however, will involve the assessment against an inadmissible controller, effectively under-estimating the controller performance index.

4.1.3 GMVC with Static Weightings

Selecting static weightings, i.e.

$$J = \mathrm{E}\{y^2(k+\tau) + \rho u^2(k)\} \quad \text{or} \quad J = \mathrm{E}\{y^2(k+\tau) + \rho(\Delta u)^2(k)\}, \tag{4.8}$$

leads to a simple version of GMVC. An approximate solution to this problem is given by MacGregor (1977):

$$u(k) = -\frac{\alpha_0 F_\tau(q)}{\alpha_0 G_\tau(q) + \rho C(q)} y(k) \quad \text{or} \quad \Delta u(k) = -\frac{\alpha_0 F_\tau(q)}{\alpha_0 G_\tau(q) + \rho C(q)} y(k), \tag{4.9}$$

where

$$G_\tau(q) = E_\tau(q)B(q), \qquad \alpha_0 = \text{constant term of } \frac{G_\tau(q)}{C(q)}.$$

The MATLAB function gmv from Moudgalya (2007: Chap. 11) can be used for the calculation of the GMV control law in Eq. 4.9. Note, however, that it is not required to apply the GMV control law on the process to determine the GMV performance index (Sect. 4.1.4).

4.1.4 Assessment Index and Procedure

The GMV performance index is defined as the ratio

$$\eta_{\rm GMV} = \frac{J_{\rm GMV}}{J_{\rm act}}, \tag{4.10}$$

where $J_{\rm GMV}$ is the value of the cost function under GMV, and $J_{\rm act}$ its the actual value under the installed controller. As for the MV index, there is no need to implement the GMV for calculation of the index.

The GMV control law has the same type of structural characteristics as the MV control law. Thus, exactly the *same* procedure for computing the Harris index (Sect. 2.4) can be applied for the estimation of GMV performance index from data, but replacing the output signal y (or control error e) by the *fictitious signal* ϕ_0. This fact is a unique feature of GMV benchmarking, compared to other advanced designs and could motivate the application of this benchmarking method in practice. The drawback is, again, the weighting selection process, which is not simple at all, unless static weightings are used.

Example 4.1 Consider again the first-order system in Eq. 2.31, now with $a_1 = -0.8$ and $\tau = 2$. The example comes from Uduehi et al. (2007a). A PID controller of the form

$$G_{\rm c}(q) = \frac{1 - 0.4q^{-1}}{(1 - q^{-1})(1 + 0.5q^{-1})}$$

is initially adopted. The process output obtained by setting a unity noise variance is shown in Fig. 4.3 (only 150 samples are plotted). For performance assessment, 3000 samples have been recorded from simulation.

The dynamic weightings for GMV benchmarking have been chosen as

$$P_{\rm c}(q) = \frac{1 - 0.2q^{-1}}{1 - q^{-1}}, \qquad F_{\rm c}(q) = -q^{-2}.$$

The performance indices for the installed controller result to be $\eta_{\rm MV} = 0.29$ and $\eta_{\rm GMV} = 0.41$. Both indices indicate unsatisfactory (stochastic) performance. The

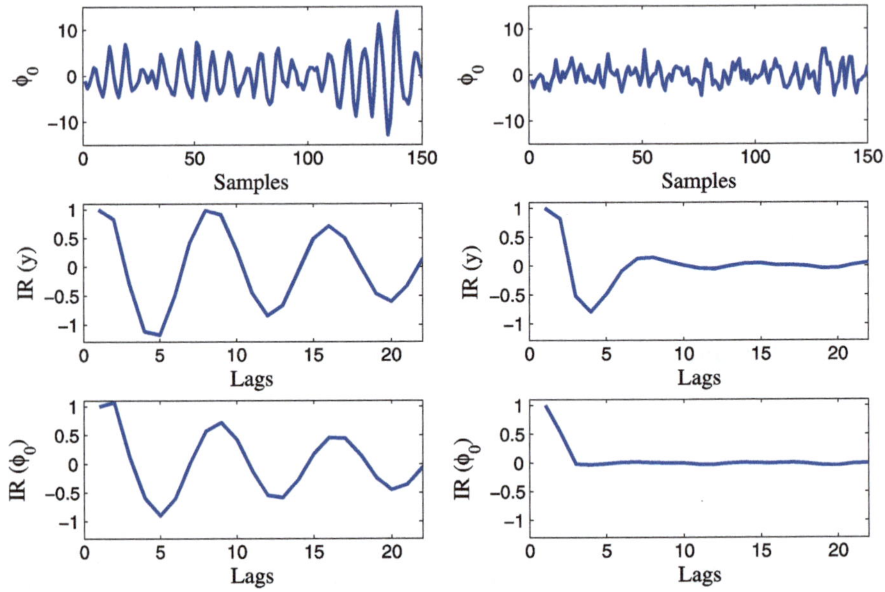

Fig. 4.3 "Generalised" output signal ϕ_0 and its impulse responses with the PID controller (*left*) and the GMV controller (*right*)

impulse responses show that the controller is too aggressive. When the analytically derived GMV controller

$$G_c(q) = \frac{2.44 - 1.44q^{-1}}{(1-q^{-1})(2+1.8q^{-1})}$$

is used, one gets $\eta_{MV} = 0.57$, which signals fair but not yet optimal controller performance. The same conclusion can be stated from looking at the impulse response (IR(y)). However, the GMV index $\eta_{GMV} = 0.98$ indicates maximum performance, which is confirmed by a look at the impulse response (IR(ϕ_0)). This shows how the GMV criterion provides a means of balancing error and control variances and makes the benchmark more realistic.

4.2 Linear-Quadratic Gaussian (LQG) Benchmarking

The linear-quadratic Gaussian (LQG) benchmark was proposed by Huang and Shah (1999) as an alternative to or the next step after applying the MV benchmark, when the latter indicates poor performance. As for MVC, this benchmark does not require that an LQG controller be implemented for the given process. Rather the benchmark provides the performance bound for any linear controller in terms of the weighted input *and* output variance. This is useful when we are interested in knowing how

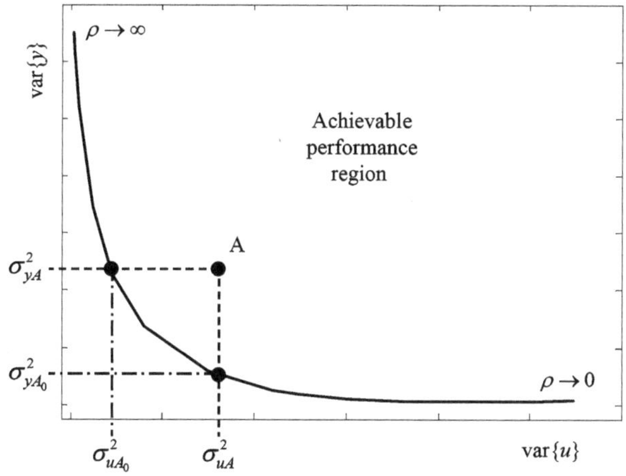

Fig. 4.4 LQG trade-off curve with a typical performance point A of a controller showing improvement potential, i.e. reduction of input and output variance

far away the control performance is from the "best" achievable performance with the same control effort. In mathematical form, this means that the solution of the following problem may be of interest:

$$\text{Given that } \mathrm{E}\{u^2\} \leq \alpha, \text{ what is the lowest achievable } \mathrm{E}\{y^2\}? \tag{4.11}$$

The solution, i.e. the achievable performance, is given by the *trade-off curve*, also known as the *performance limit curve* in Fig. 4.4. This curve can be generated from solving the H_2/LQG problem (Kwakernaak and Sivan 1972; Harris 1985; Boyd and Barratt 1991; Grimble 2006a, 2006b), where the H_2/LQG objective function is defined by:[1]

$$\begin{aligned} J(\rho) &= \mathrm{E}\{y^2(k)\} + \rho \mathrm{E}\{u^2(k)\}, \\ J_{\mathrm{LQG}} &= \mathrm{var}\{y(k)\} + \rho \mathrm{var}\{u(k)\}. \end{aligned} \tag{4.12}$$

By varying the *move suppression* weight ρ, various optimal solutions of $\mathrm{E}\{y^2\}$ and $\mathrm{E}\{u^2\}$ can be calculated. Thus a curve with the optimal output variance as ordinate and the incremental manipulated variable variance as the abscissa can be plotted from these calculations. Boyd and Barratt (1991) have shown that any linear controller can only operate in the region above this curve. Consequently, the trade-off curve defines the limit of performance of all linear controllers, as applied to a linear time-invariant process, including the minimum variance control law. Although the LQG design can handle tracking problems, the approach is generally applied mainly for disturbance rejection.

Depending on the current performance level under the installed controller, one can decide whether the control system can be improved. Suppose that the actual measured performance of a loop with the implemented controller (usually not LQG)

[1]This assumes a system model of the ARMAX type. If a model of the ARIMAX type, i.e. with integrating disturbance term, is considered, $u(k)$ has to be replaced by $\Delta u(k)$.

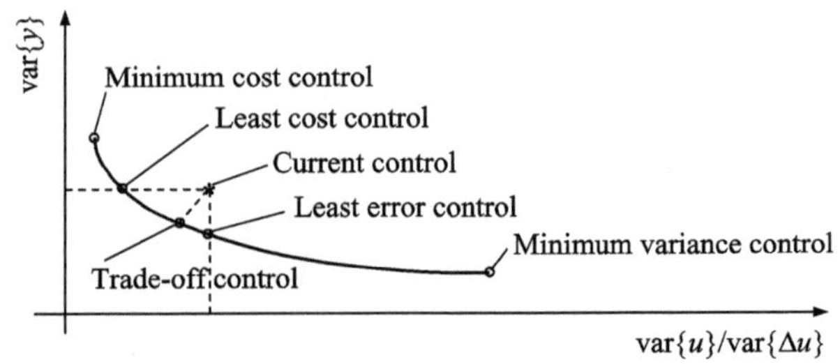

Fig. 4.5 LQG trade-off curve with several optimal controllers

is given by the point $\mathrm{A} = (\sigma^2_{u\mathrm{A}}, \sigma^2_{y\mathrm{A}})$. From the plot it is clear that one of the following is possible, by switching over, if necessary, to an LQG controller: (i) one can achieve the same output variance ($\sigma^2_{y\mathrm{A}}$) for smaller control effort ($\sigma^2_{u\mathrm{A}_0}$); (ii) for the same control effort ($\sigma^2_{u\mathrm{A}}$), one can achieve smaller output variance ($\sigma^2_{y\mathrm{A}_0}$). Nevertheless, working at the limits of this trade-off curve may not give a robust solution (Moudgalya and Shah 2004).

In total, five optimal controllers may be identified from the trade-off curve shown in Fig. 4.5. They are explained as follows (Huang and Shah 1999):

- *Minimum-Cost Control.* This is an optimal controller identified at the left end of the trade-off curve. The minimum-cost controller is optimal in the sense that it offers an offset-free control performance with the minimum possible control effort. It is worthwhile pointing out that this controller is different from the open-loop mode since an integral action is guaranteed to exist in this controller.
- *Least-Cost Control.* This optimal controller offers the same output error as the current or existing controller but with the least control effort. So if the output variance is acceptable but actuator variance has to be reduced, then this represents the lowest achievable manipulative action variance for the given output variance.
- *Least-Error Control.* This optimal controller offers least output error for the same control effort as the existing controller. If the input variance is acceptable but the output variance has to be reduced, then this represents the lowest achievable output variance for the given input variance.
- *Trade-off Control.* This optimal controller can be identified by drawing the shortest line to the trade-off curve from the existing controller; the intersection is the trade-off control. Clearly, this trade-off controller has performance between the least cost control and the least error control. It offers a trade-off between reductions of the output error and the control effort.
- *Minimum-Error (Variance) Control.* This is an optimal controller identified at the right end of the trade-off curve. The minimum-error controller is optimal in the sense that it offers the minimum possible error. Note that this controller may be different from the traditional minimum-variance controller due to the existence of integral action.

LQG benchmark is more realistic than MVC, but it still represents an unattainable standard, even for MPC. Also, much more information on the process is required, i.e. measurement of the manipulating variable(s) and/or process/disturbance models. The uncertainty in the estimated model then has to be "mapped" onto the LQG curve, in which case it would become a fuzzy trade-off curve. Alternately the uncertainty region can be mapped into a region around the current performance of the controller relative to the LQG curve; see Patwardhan and Shah (2002). Moreover, the use of an LQG benchmark for performance assessment is much more complicated and leads to a higher computational burden (a state estimator and the solution of algebraic Riccati equations, as shown below) than the traditional methods based on the MVC. Serious critical issues related to LQG-based performance assessment can be found in Kozub (2002). All these demanding requirements and drawbacks make the calculation and use of—at least—the standard LQG benchmark not always practical.

4.2.1 Classical LQG Framework

An alternative approach to the design of pole-placement controllers is to specify an optimisation index to be minimised. This underlies the concept of linear-quadratic regulator (LQR), obtained by minimising a quadratic index. When the states required are estimated by a Kalman filter, the estimator–regulator combination is known as the LQG controller. The LQG-control problem is to find the first input $\boldsymbol{u}_1$ of the sequence $\boldsymbol{u}_f = (\boldsymbol{u}_1, \ldots, \boldsymbol{u}_N)$ that minimises the quadratic cost function J over the horizon N,

$$J = \sum_{k=1}^{N} \hat{\boldsymbol{y}}_k^{\mathrm{T}} \boldsymbol{Q} \hat{\boldsymbol{y}}_k + \boldsymbol{u}_k^{\mathrm{T}} \boldsymbol{R} \boldsymbol{u}_k, \tag{4.13}$$

where $\hat{\boldsymbol{y}}_k$ is the k-step-ahead optimal predicted output given past inputs and outputs and future inputs up to time k. When setting $\boldsymbol{Q} = \boldsymbol{I}$ and $\boldsymbol{R} = \rho \boldsymbol{I}$, Eq. 4.13 becomes Eq. 4.12. The classical approach of solving the LQG problem consists of three stages: (1) system identification of a state-space model, (2) Kalman-filter design and (3) LQ-controller design. For details, readers are referred to standard texts on this topic, such as Kwakernaak and Sivan (1972) and Anderson and Moore (1991).

MATLAB's Control System Toolbox provides the functions lqry or dlqr to design an LQG controller and kalman or dlqe to design a Kalman filter. If the state transition matrix is singular, e.g. owing to time delays, then the function dlqe2 in the MATLAB MPC Toolbox can be used. Moreover, the function lqgreg is available to form the feedback regulator.

4.2.2 Polynomial Domain Approach

An equivalent polynomial approach can also be formulated to design the LQG controller using transfer functions; see Grimble and Johnson (1988), Åström and Wittenmark (1997). An excellent recent reference is Grimble (2006a). The polynomial approach has the advantage that it does not require concepts, such as states, controllability and observability. Moreover, in the polynomial approach, complicated noise models and different kinds of weightings of the signals in the performance measure can easily be handled (Moudgalya and Shah 2004). We found that the polynomial approach is much easier for generating LQG performance limit curves.

Polynomial solutions are based on spectral factorisation, for which good numerical methods exist, e.g. Kucera (1979) and Harris and Davis (1992). The required LQG controllers can be computed, for instance, using the MATLAB function doflq (Kammer et al. 1996) or the function lqg (Moudgalya 2007: Chap. 13).

4.2.3 LQG Framework as Special Case of MPC

As recommended by Huang and Shah (1999), it is more useful to solve the LQG problem using the MPC approach, specifically via an infinite general-predictive control (GPC) solution, i.e. with an infinite prediction horizon and an infinite control horizon. Nevertheless, in practice, a finite value of the prediction horizon is usually sufficient to achieve the approximate infinite horizon LQG solution via the GPC approach. This means also that the trade-off curve for MPC always lies above that for LQG; see Sect. 4.3.

4.2.4 Subspace-Based LQG Design

In state-space-based controllers, the state is used as an intermediate variable between the past and the future to find the control inputs. Indeed, in the case of an LQG controller, the state estimate that minimises the prediction criterion (Kalman filter) is calculated from the past inputs and outputs. In a second step, the future inputs that minimise the control criterion (LQ controller) are then found from the state. This idea is also clearly present in subspace system identification, where the Kalman filter state sequence serves as interface between the past and future through the projection of the future outputs $\boldsymbol{Y}_f$ into the past inputs $\boldsymbol{U}_f$ and outputs $\boldsymbol{Y}_p$, and future inputs $\boldsymbol{U}_f$.

The big advantage in subspace identification is that the Kalman filter states can be found without the knowledge of the system parameters $\boldsymbol{A}$, $\boldsymbol{B}$, $\boldsymbol{C}$, $\boldsymbol{D}$, $\boldsymbol{K}$ and $\boldsymbol{S}$, and thus without having to solve the Kalman filter equations. This property is exploited to straightforwardly calculate the control law. Details of this method are given by Favoreel et al. (1999). Here, we only show the steps of both LQG-design approaches in Fig. 4.6 for comparison. The main conclusion is that the three steps involved in the classical approach are *short-circuited* and replaced by a QR and an SVD of matrices *directly* constructed out of input and output data gathered. Therefore, the subspace-based approach for LQG design should always be preferred.

So far the LQG design can be efficiently performed based on subspace identification, provided that the first n_{IR} impulse-response coefficients of the controller are

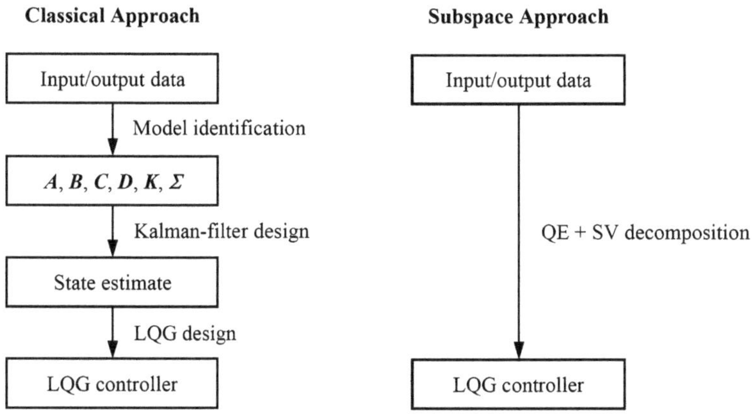

Fig. 4.6 Steps of both LQG-design methods

known. This can, however, be a serious limitation in the industrial practice, where the parameterisation of the controller is often not easy to discover. In such a situation, one might try to identify the controller parameters from data, as proposed by Bezergianni and Georgakis (2003). However, the best way should be to apply methods, such as that proposed by Kadali and Huang (2002a, 2002b) which does not need the knowledge of the controller at all. This approach needs however set-point excitation for proper closed-loop identification of the subspace matrices.

4.2.5 Generation of the LQG Performance Limit Curve

For the construction of the LQG performance limit curve, an LQG controller minimising the objective function in Eq. 4.12 is computed as[2]

$$u(k) = -\frac{S(q)}{R(q)} y(k), \tag{4.14}$$

where for simplicity the controller set point is assumed constant at its initial steady-state value, i.e. $r = 0$. Inserting this control law into the system description in Eq. 2.1 gives the closed-loop relationships:

$$y(k) = \frac{C(q)R(q)}{A(q)R(q) + q^{-\tau}B(q)S(q)}\varepsilon(k), \tag{4.15}$$

$$u(k) = \frac{C(q)S(q)}{A(q)R(q) + q^{-\tau}B(q)S(q)}\varepsilon(k). \tag{4.16}$$

[2]The more general form of this control law is $R(q)u(k) = -S(q)y(k) + T(q)r(k)$, referred to as the RST regulator.

Applying Parseval's theorem to Eqs. 4.15 and 4.16 yields the variances of the process output and input as

$$\sigma_y^2 = \mathrm{var}\{y(k)\} = \frac{\sigma_\varepsilon^2}{2\pi j}\oint_{|z|=1}\left|\frac{C(q)R(q)}{A(q)R(q)+q^{-\tau}B(q)S(q)}\right|^2\frac{\mathrm{d}z}{z}, \tag{4.17}$$

$$\sigma_u^2 = \mathrm{var}\{\Delta u(k)\} = \frac{\sigma_\varepsilon^2}{2\pi j}\oint_{|z|=1}\left|\frac{C(q)S(q)}{A(q)R(q)+q^{-\tau}B(q)S(q)}\right|^2\frac{\mathrm{d}z}{z}. \tag{4.18}$$

These equations can be numerically solved using a contour integration algorithm, such as that given by Åström (1979). In MATLAB, the function covar can be used for the evaluation of such integrals. Without loss of generality, the white noise sequence is supposed to have unity variance, i.e., $\mathrm{var}\{\varepsilon(k)\} = 1$. If this is not the case, the system description can always be scaled to satisfy this assumption. Instead of evaluating the Parseval integrals, one can simulate the closed loop and compute the variances from the calculated signals. Obviously, these will not exactly match the real variances owing to the finite simulation time. It is essential to note that the variance calculation requires the knowledge of the disturbance dynamics, which is usually not needed for controller design. The procedure for generating the LQG performance limit curve is summarised below.

Procedure 4.1 Construction of the LQG trade-off curve.

1. Determine a system model in the form of Eq. 2.1.
2. By varying ρ, compute a series of LQG control laws (Eq. 4.14).
3. Solve the Parseval integral Eqs. 4.17 and 4.18 using contour integration, or simulate the closed loop, to give the variances of y and u for each ρ value.
4. Plot $\mathrm{var}\{y(k)\}$ vs. $\mathrm{var}\{u(k)\}$ to provide the trade-off curve.

For the MIMO case, the objective function of the LQG control is expressed as

$$J_{\mathrm{LQG}} = \mathrm{E}\{\boldsymbol{y}^{\mathrm{T}}(k)\boldsymbol{W}\boldsymbol{y}(k)\} + \rho\mathrm{E}\{\boldsymbol{u}^{\mathrm{T}}(k)\boldsymbol{R}\boldsymbol{u}(k)\}, \tag{4.19}$$

where $\boldsymbol{W}$ is the output weighting, which should be selected so that it reflects the relative importance of the individual outputs. Similarly, the control weighting $\boldsymbol{R}$ has to be chosen according to the relative cost of the individual inputs. As in the SISO case, by varying ρ, different LQG control laws can be computed. Then solving the Parseval integral equations or simulating the closed loop provides the H_2 norms $\mathrm{E}\{\boldsymbol{y}^T(k)\boldsymbol{W}\boldsymbol{y}(k)\}$ and $\mathrm{E}\{\boldsymbol{u}^T(k)\boldsymbol{R}\boldsymbol{u}(k)\}$ which can be used to plot the trade-off curve.

4.2.6 LQG Assessment Using Routine Operating Data

As mentioned above, the construction of the LQG trade-off curve necessitates the complete knowledge of the plant and disturbance models, i.e. G_{p} and G_ε. A disturbance model can be determined from fitting a time series model, e.g. an AR model,

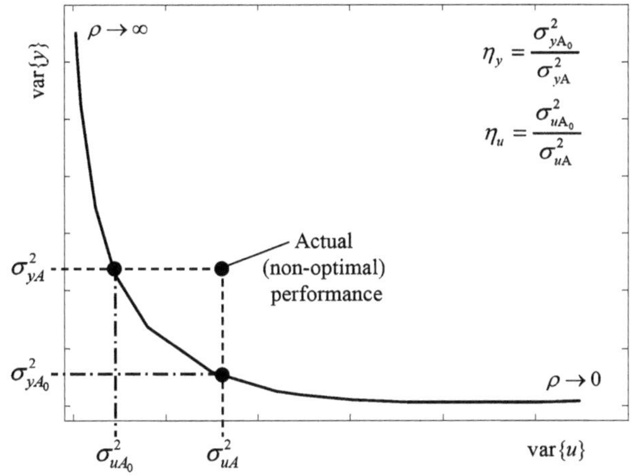

Fig. 4.7 LQG trade-off curve with the definition of LQG performance indices

to routine operating data, as for the computation of the Harris index. However, the estimation of the plant model, when not available, is much more involved. The addition of a dither signal to the set point or controller output is generally required to introduce a source of external excitation into the feedback loop; see Sect. 6.1.3. Note that the injection of dither signals is highly undesirable and seldom allowed in industrial practice.

To assess the performance of an installed controller relative to the LQG benchmark, the following performance indices are defined (see Fig. 4.7):

$$\eta_y = \frac{\sigma^2_{y,\mathrm{LQG}}}{\sigma^2_y} = \frac{\sigma^2_{y\mathrm{A}0}}{\sigma^2_{y\mathrm{A}}}, \tag{4.20}$$

$$\eta_u = \frac{\sigma^2_{u,\mathrm{LQG}}}{\sigma^2_u} = \frac{\sigma^2_{u\mathrm{A}0}}{\sigma^2_{u\mathrm{A}}}. \tag{4.21}$$

η_y and η_u vary between 0 and 1. If η_y is equal to 1 for the given input variance, then the controller is giving optimal performance with respect to the process variance. If not, then the controller is non-optimal, and there is scope for improvement in terms of process response without affecting the input variance. Similarly, if η_u is equal to 1 for the given output variance, then the controller is giving optimal performance with respect to the input variance. If not, then the controller is non-optimal, and there is scope to reduce input variance without affecting the output variance. Using these two measures, one can see how far the control performance is from the LQG benchmark. For specific values η_y and η_u, there is a potential of decreasing the output variance and input variance by $100(1-\eta_y)$ % and $100(1-\eta_u)$ %, respectively. Note that the performance measures do have inherent variability. Harris (2004) has given the sampling distribution statistical properties of some related quadratic-type performance indices.

The whole procedure for assessing the performance of control systems against the LQG benchmark is given as follows.

Procedure 4.2 Construction of the LQG trade-off curve.

1. Estimate a disturbance model G_ε from normal operating data.
2. When G_p is not available, try to identify at least a rough estimate for G_p from normal operating data using the method by Ko and Edgar (1998) or that by Julien et al. (2004); see Sect. 7.2.7. If this is not feasible, i.e., in the case the time delay is too short relative to the disturbance dynamics, estimate a process model G_p from closed-loop operating data under the injection of a dither signal to the set point or controller output. If the insertion of such a signal is not possible/allowed, it is indispensable to perform open-loop tests for the (re)estimation of G_p.
3. Generate the LQG trade-off curve using Procedure 4.1.
4. Evaluate the trade-off curve to determine the variances $\sigma^2_{y\mathrm{A}_0}$ and $\sigma^2_{u\mathrm{A}_0}$.
5. Calculate the performance indices (Eqs. 4.20 and 4.21) to assess the performance and deduce the possible improvement potential.

4.2.6.1 Confidence Limits for Variances

It is useful to give confidence limits for the variances $\sigma^2_{\Delta y}$ and σ^2_u calculated using measured data from control loop under the installed controller. If the data $y(k)$ and $\Delta u(k)$ were uncorrelated Gaussian white noise sequences, an approximate joint 95 % confidence region would be defined by the intervals (Montgomery and Runger 1992; Julien et al. 2004):

$$\frac{(N-1)\sigma^2_{\Delta u}}{\chi^2_{\alpha/4,N-1}} \le \hat{\sigma}^2_{\Delta u} \le \frac{(N-1)\sigma^2_{\Delta u}}{\chi^2_{1-\alpha/4,N-1}}, \tag{4.22}$$

$$\frac{(N-1)\sigma^2_{y}}{\chi^2_{\alpha/4,N-1}} \le \hat{\sigma}^2_{y} \le \frac{(N-1)\sigma^2_{y}}{\chi^2_{1-\alpha/4,N-1}}, \tag{4.23}$$

where $\chi^2_{\alpha/4,N-1}$ and $\chi^2_{1-\alpha/4,N-1}$ are the upper and lower $\alpha/4$ percentage points of the chi-square distribution with $N-1$ degrees of freedom, respectively, and $\alpha = 0.05$. The limits calculated from the aforementioned equations give a rectangle around the variance values calculated from data. However, these confidence limits are ideal values, i.e., they do not take into account the correlation between $y(k)$ and $\Delta u(k)$ usually induced by feedback control.

Example 4.2 Reconsider Example 2.3. For this system, the LQG trade-off curve has been generated using the MATLAB function lqg from Moudgalya (2007: Chap. 13) for ρ in the range 0.01–100; the closed loop has been simulated using the function cl. From the curve in Fig. 4.8 it can be seen that the performance of controllers P1 lies on the trade-off curve, thus P1 is optimal relative to the LQG benchmark, but the controller will be sluggish, as the weighting parameter is too high. P3 (not shown in the figure) is far from the trade-off curve, i.e. far from optimum.

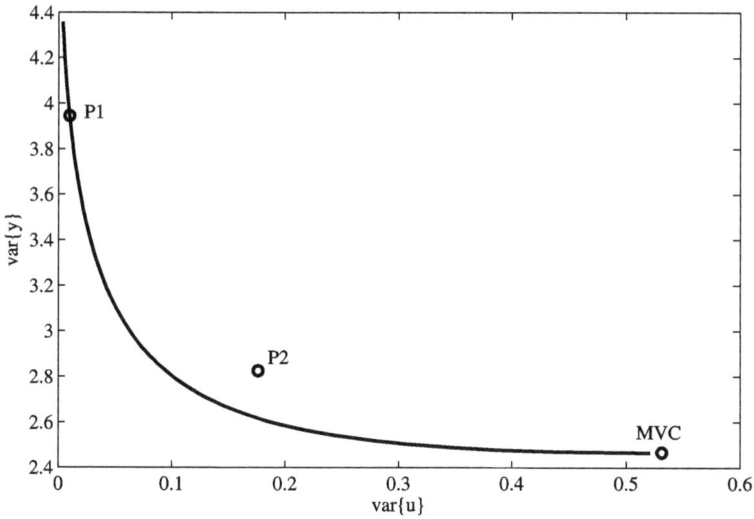

Fig. 4.8 LQG performance curve and performance points of the considered controllers (Example 4.2)

The performance indices for P2 have been calculated as

$$\eta_{y,\mathrm{P2}} = \frac{2.617}{2.825} \approx 0.93, \qquad \eta_{u,\mathrm{P2}} = \frac{0.0946}{0.1766} \approx 0.54.$$

Although the controller performance is close to optimal with respect to the process output variance (93 % of the optimal), the performance index with respect to the input variance is only 0.54. This indicates a maximum possible scope of 46 % to reduce the input variance without increasing the output variance; see Fig. 4.8. It can also be concluded that the measures of achievable performance with LQG as benchmark are more realistic than those obtained relative to the MVC, when control action cannot be allowed to exceed certain level.

The performance point for MVC is obtained, as expected, for $\rho \to 0$. We have calculated the LQG control law for $\rho = 0.001$ to get

$$u(k) = -\frac{0.7283}{1 + 0.8991q^{-1} + 0.8092q^{-2}} y(k)$$

just to confirm that it is very close to the exact transfer function of the MVC in Eq. 2.32.

Example 4.3 This example taken from Hugo (2006) has an integrating disturbance:

$$y(k) = -\frac{0.08q^{-3}}{1 - 0.92q^{-1}} u(k) + \frac{1}{1 - q^{-1}} \varepsilon(k), \qquad \mathrm{var}\{\varepsilon\} = 0.02. \tag{4.24}$$

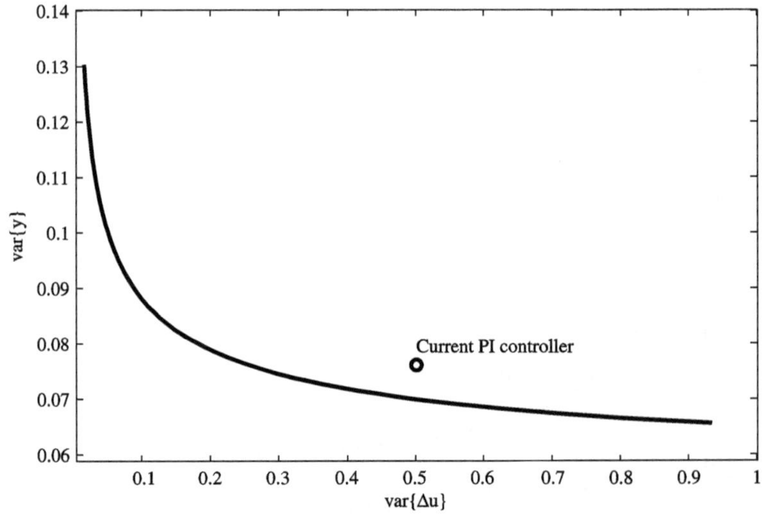

Fig. 4.9 LQG performance limit curve and performance point of the current controller (Example 4.3)

A PI controller with $K_c = 3.33$ and $T_I = 6.94$ is used to close the loop. These tuning parameters were determined using the ITAE tuning rules for load disturbances. For this system, the minimum variance index were found to be 0.74, indicating good performance, but also that a time-delay compensator would improve control performance. This was expected as the ITAE tuning rules should give a response that is close to minimum variance for PI controllers. The LQG trade-off curve has been constructed for ρ in the range 0.006–2.0 and is shown in Fig. 4.9.

The performance indices according to Eqs. 4.20 and 4.21 have been calculated:

$$\eta_y = \frac{0.0698}{0.0761} \approx 0.92, \qquad \eta_u = \frac{0.248}{0.502} \approx 0.49.$$

Again, the LQG benchmark is more realistic, indicating near optimal performance with respect to the output variance (92 % of the optimal). However, the performance index with respect to the input variability is only 0.49. Hence there is a maximum possible scope of 51 % to reduce the input variance without increasing the output variance.

4.3 Model Predictive Control (MPC) Assessment

A major innovation in the field of process control over the last millennium (since the 1970s) has been the development of MPC. Model predictive control or receding horizon control (RHC) is a form of control, in which the current control action

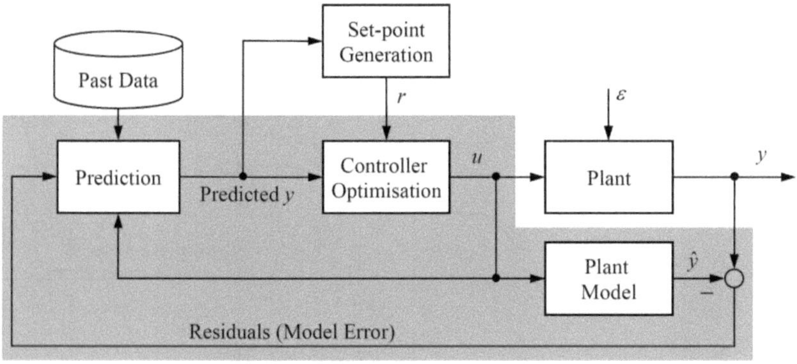

Fig. 4.10 Schematic block diagram for model predictive control

is obtained by solving on-line, at each sampling instant, a finite horizon open-loop optimal control problem, using the current state of the plant as the initial state. The optimisation carried out based on a prediction model yields an optimal control sequence, and the first control in this sequence is applied to the plant; see Fig. 4.10. This is its main difference from conventional control which uses a pre-computed control law.

MPC has wide popularity in many process industries, owing to its general way of coping with process-control problems. It can be recognised that the most important feature of model predictive control is its ability to handle constraints explicitly in the design. Nowadays, MPC algorithms are the secondly most used algorithms in the process industry—besides PID control. With over 4500 industrial installations, model predictive control (MPC) is currently the most widely implemented advanced process control technology for process plants (Qin and Badgwell 2003).

4.3.1 Basic Principle and Properties

The methodology of all the controllers belonging to the MPC family is characterised by the following strategy, represented in Fig. 4.11, referred to as "*receding/moving horizon control*":

1. *Prediction*. The future outputs for the *prediction horizon* N_2 are predicted at each instant k using the plant model. These predicted outputs $\hat{y}(k+i|k)$ for $i = 1, 2, \ldots, N_2$ depend on past inputs and past outputs, and on the future control signals $u(k+i|k)$, $i = 0, \ldots, N_u - 1$, which are those to be sent to the system and to be calculated. $N_u \leq N_2$ denotes the *control horizon*, within which the control signal is changed.
2. *Optimisation*. The set of future control signals is calculated by minimising a determined criterion in order to keep the system as close as possible to the reference trajectory $r(k+L)$ (which can be the set point itself or an approximate of it). An

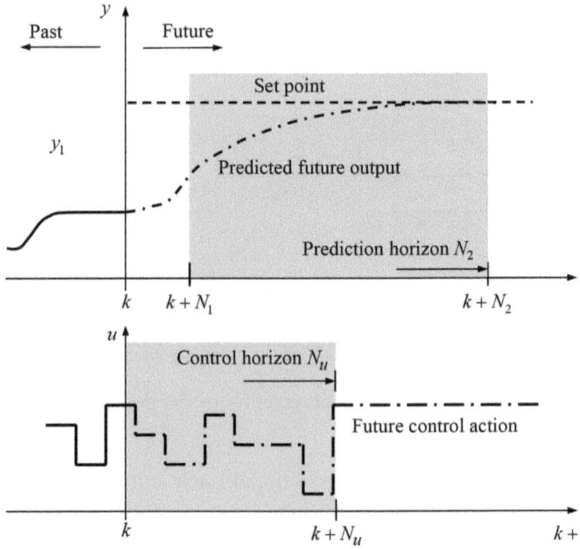

Fig. 4.11 Principle of MPC Strategy

explicit solution can be obtained if the criterion is quadratic, the model is linear and there are no constraints, otherwise an iterative optimisation method has to be used. Some assumptions about the structure of the future control law are also made in some cases, such as that it will be constant from a given instant.

3. *Applied Control Signal.* Not the entire control-input sequence, but only the first element $u(k|k)$ is applied to the system, whereas the next control signal values calculated are rejected. This is because at the next sampling instant $y(k+1)$ is already known and step 1 is repeated with the new value and all the sequences are brought up to date. Also, the set point may change over the next intervals. Thus, the $u(k+1|k+1)$ is calculated (which in principle will be different to the $u(k+1|k)$ because of the new information available) using the *receding horizon concept.*
4. *Model Update.* Current measurement information is used to estimate the (unmeasured) disturbances by making assumptions about their nature and thus to adapt the prediction model. The simplest approach (used in most MPC methods) is to assume that the disturbance is constant as the difference between the actual and estimated output, and to keep this unchanged during the prediction horizon.

Observe the IMC structure in Fig. 4.10, which is inherent for most MPC schemes.

MPC is the only control methodology born in industry and it has made a significant impact on industrial control engineering. It has so far been applied mainly in chemical and petro-chemical industry, but the benefits of MPC are currently being increasingly discovered in other sectors of the process industry.

The penetration of MPC into industrial practice has been motivated by the following facts:

- Its underlying idea is intuitive and easy to understand.

- It can be used to control a great variety of processes, from those with relatively simple dynamics to other more complex ones, including "difficult" systems, such as those containing long time delays or non-minimum-phase ones. It is thus much more powerful than PID control.
- Its basic formulation extends to multivariable plants with almost no major modifications.
- It intrinsically compensates for (dominant) time delays.
- Feedforward control for disturbance compensation can be integrated in MPC schemes in a straightforward way.
- The extension of MPC to the treatment of constraints is conceptually simple, and these can be systematically included already during the design phase and not ad hoc during commissioning.
- It is very useful when future references, e.g. for path following, set-point tracking or batch-process control, are known.
- Operational and economical criteria may be incorporated in the objective function.

Predictive control comes in a large variety of shapes using a lot of different names, such *model predictive heuristic control* (*MPHC*) (Richalet et al. 1978), *dynamic matrix control* (*DMC*) (Cutler and Ramaker 1980), *quadratic DMC* (*QDMC*) (Garcia and Morshedi 1986; Prett and Garcia 1988) and *generalised predictive control* (*GPC*) (Clarke et al. 1987a, 1987b), just to mention a few. However, the underlying ideas are the same in all methods, it is the details that distinguish them from one another. The various MPC techniques are thoroughly presented in many standard text books, e.g. by Soeterboeck (1992), Camacho and Bordons (1999) and Maciejowski (2001). Contributed books on the subject have published by Allgöwer and Zheng (2000) and Kouvaritakis and Cannon (2001). Notable reviews of MPC theory include those of García et al. (1989), Ricker (1991), Morari and Lee (1991), Rawlings and Muske (1993), Rawlings et al. (1994), Mayne (1997) and Lee and Cooley (1997). Excellent reviews of industrial MPC technology are provided by Qin and Badgwell (1997, 2003).

MPC systems should be monitored on a regular basis to ensure that their performance does not degrade owing to changes in the process, instrumentation, or process conditions, including disturbances. If the performance becomes significantly worse, re-tuning the controller or re-identifying the process model may be required. An important motivation for introducing MPC is that it facilitates process operation closer to targets and limiting constraints. Thus, an evaluation of MPC performance should include measures to check whether these objectives have been achieved (Seborg et al. 2004).

A standard objective function often used in MPC is (assuming zero set point)

$$J = \sum_{i=N_1}^{N_2} \hat{y}^2(k+i) + \rho \sum_{i=1}^{N_u} \Delta u^2(k+i-1), \tag{4.25}$$

where N_1 and N_2 are the minimum and maximum prediction horizons, and N_u the control horizon. $\hat{y}(k+i)$ is an optimal i-steps-ahead forecast of the controlled variable to be formulated via the process model.

From accumulated experience of applying MPC algorithms in the last decades, a number of engineering rules have been identified to obtain appropriate values of MPC tuning parameters for good control performance. Some of these guidelines are recalled:

- *Prediction Horizons.* N_1 and N_2 mark the limits of the time interval in which it is desired for the process variable to follow the reference. Typically, N_1 is chosen to be zero. However, for systems with time delay, there is no reason to select N_1 less than the number of time delay samples, since the output will only begin to evolve passed this time. N_2 is usually chosen approximately as the settling time in samples. Smaller values of N_2 make the control action more aggressive. Nothing is gained when costing future errors that cannot be influenced by future control actions.
- *Control Horizon.* N_u should be less than N_2. Typically, N_u is taken to be about one-half to one-third of $N_2 - N_1$ for plants having large time constants, as is usual for chemical processes. N_u should be greater than or equal to the number of unstable poles in the process to guarantee the stability of infinite horizon MPC. Large N_u tends to make control more aggressive. An advantage of a large N_u is that it allows one detect constraint violations before they are reached, averages the control objective over time and handles unknown variable time delays.
- *Control Weighting.* ρ is to be selected large or small, depending on whether less or more aggressive control is desired. Increasing ρ makes control more damped; in the opposite direction, decreasing ρ makes the control action more aggressive and the control response faster. For multivariable systems, the parameter ρ (which is then a matrix) becomes especially useful to weight the different control efforts. Note that the parameters ρ and N_u are strongly related to each other.

Theses rules provide only a basis for tuning MPC controllers. In practice, the parameters are often selected per trial and error, a procedure that can be time-consuming, since the parameters are dependent on each other. Note also that not all parameter combinations guarantee a stable controller. Tuning methods for MPC have been proposed by some researchers, e.g. Clarke and Scattoloni (1991), MacIntosh et al. (1991) and Rawlings and Muske (1993).

Many approaches exist for the benchmarking of MPC systems; see Schäfer and Çinar (2002) for an overview. Some methods are briefly discussed in the following. Recall that in MPC, the process model and the optimisation are substantial components of the (online) controller design; see Qin and Badgwell (1997). In contrast to the assessment of PID controllers, where no model is directly involved, a fundamental question related to the performance assessment of model-based control, particularly MPC, is whether detected poor control performance is due to bad controller tuning or inaccurate modelling.

4.3.2 Constrained Minimum-Variance Control

An approach based on constrained minimum-variance controller has been proposed by Ko and Edgar (2001a) for the assessment of MPC. First, a constrained MVC is designed using the receding horizon concept, the same method of constrained optimisation applied in MPC design. Subsequently, the achievable MV performance bounds in constrained MPC systems via disturbance model identification and (constrained) closed-loop simulation with the constrained MVC are estimated. Knowledge of the process model is thus required to develop the constrained MVC. It is also assumed that the process model has a stable inverse. When the constraints become inactive, the proposed approach naturally recovers unconstrained MV performance bounds.

4.3.3 Design-Case MPC Benchmarking

The so-called design-case benchmarking, recommended by Patwardhan et al. (1998), Shah et al. (2001) and Gao et al. (2003), evaluates the controller performance using a criterion commensurate with the actual design objective(s) and then compares it with the achieved performance:

$$\begin{aligned}\eta_{\mathrm{MPC}} &= \frac{J_{\mathrm{opt}}}{J_{\mathrm{act}}}\\ &= \frac{\sum_{j=1}^{N_2} \hat{e}^{\mathrm{T}}(k+j-N_2)Q\hat{e}(k+j-N_2) + \sum_{j=1}^{N_u} \Delta u^{*\mathrm{T}}(k+j-N_2)R\Delta u^{*}(k+j-1)}{\sum_{j=1}^{N_2} e^{\mathrm{T}}(k+j-N_2)Qe(k+j-N_2) + \sum_{j=1}^{N_u} \Delta u^{\mathrm{T}}(k+j-N_2)R\Delta u(k+j-1)},\end{aligned} \tag{4.26}$$

where $\hat{e}$ is the estimated control error, Δu^{*} the optimal control moves, and e and Δu the measured values of outputs and inputs, respectively, at corresponding sampling instants. Q and R are weightings of relative importance of controlled and manipulated variables. The performance index will be equal to unity when the achieved performance exactly meets the design requirements. The actual output may differ significantly from the predicted output due to inadequacy of the model structure, nonlinearities, modelling uncertainty, etc. The inputs will differ from design values in part due to the receding horizon nature of the MPC law.

Another benchmarking method (very similar to the design-case method) referred to as the expectation-case approach has been proposed by Zhang and Henson (1999): the actual performance is compared (online) to the expected performance (judged to be satisfactory) obtained when controller actions are implemented on the process model *instead* of the plant:

$$\eta_{\mathrm{MPC}} = \frac{J_{\mathrm{exp}}}{J_{\mathrm{act}}} = \frac{\sum_{j=1}^{N_2} \hat{e}^{\mathrm{T}}(k+j-N_2)Q\hat{e}(k+j-N_2)}{\sum_{j=1}^{N_2} e^{\mathrm{T}}(k+j-N_2)Qe(k+j-N_2)}. \tag{4.27}$$

The performance is then assumed to be generated by an ARMA process

$$\Delta\eta_{\mathrm{MPC}}(k) = \frac{\hat{A}(q^{-1})}{\hat{C}(q^{-1})}\eta_{\mathrm{MPC}}(k), \tag{4.28}$$

whose parameters are estimated via standard identification techniques and used to detect statistically significant changes in controller performance. The estimated noise variance is used to compute 95 % confidence limits on $\Delta\eta_{\mathrm{MPC}}(k)$.

The advantage of using any of the above criteria for the purpose of performance assessment is that it is a measure of the deviation of the controller performance from the (user-specified) design or expected performance. Thus, a low performance index truly indicates changes in the process or the presence of disturbances, resulting in sub-optimal control. The estimation of the indices in Eqs. 4.26 and 4.27 does not involve any time series analysis or identification: (i) the design objective (J_{opt} or J_{exp}) is calculated by the controller at every instant, i.e. from the optimisation step in MPC, which implies that the online predicted values of $\hat{e}$ and Δu^* have to be available for the design-case approach; (ii) only the measured input and output data are needed to find the achieved performance (J_{act}). Moreover, the design objective approach has no restrictive assumptions and thus can deal with the multivariate and constrained nature of MPC.

Nevertheless, the application of the aforementioned methods is limited by the fact that installed MPC routines may not return the design value of the cost index for user inspection. In addition, it must be recognised that a low performance index could be caused by errors in the plant and/or disturbance models. Thus a further task is to distinguish between both possible causes of poor performance.

4.3.4 Infinite-Horizon Model Predictive Control

The derivation of MPC control laws is usually based on a system model of the ARIMAX type (Eq. 2.3), typically (as in DMC) presuming the unmeasured disturbance be a random walk, i.e.

$$A(q)y(k) = q^{-\tau}B(q)u(k) + \frac{1}{\Delta}\varepsilon(k). \tag{4.29}$$

The objective of MPC is to minimise the cost function in Eq. 4.25. For $N_1 = 1$, $N_u = N_2$ and $N_2 \to \infty$, Eq. 4.25 converges to the LQG objective function, i.e.

$$\frac{1}{N_2}J_{\mathrm{MPC}} \to J_{\mathrm{LQG}} = \mathrm{var}\{y(k)\} + \rho\,\mathrm{var}\{\Delta u(k)\}. \tag{4.30}$$

The solution is known as the *infinite-horizon MPC controller*. This controller is usually adopted for MPC assessment although finite-horizon MPC will be found in practice, owing to the following facts:

- If the prediction horizon N_2 is at least equal to the plant time-to-steady-state, as always recommended, a finite-horizon MPC will be virtually identical to the infinite-horizon MPC and provides a reasonable approximation of several commercial MPC systems (Huang and Shah 1999; Julien et al. 2004).
- The infinite-horizon LQG algorithm endows the algorithm with powerful stabilising properties. For the case of a perfect model, it was shown to be stabilising for any reasonable linear plant as long as ρ is positive (Mayne et al. 2000; Qin and Badgwell 2003).
- The use of MPC benchmarks commensurate with the actually designed controller requires that details of the installed MPC controller (used disturbance model, parameters N_1, N_2, N_u, etc.) to be available from MPC software supplier. This is, however, often more a hope than reality, due to know-how protection reasons.

4.3.4.1 Generation of MPC Performance Limit Curve

As for LQG, the MPC performance limit curve can be simply constructed by plotting $\mathrm{var}\{y\}$ vs. $\mathrm{var}\{\Delta u\}$ for values of ρ in the range $[0, \infty)$. Obviously, this generally requires the knowledge of the complete system model. When assessing the performance of MPC, this assumption is nothing special having in mind that a system model is explicitly integrated in the design of a model predictive controller. Note that many commercial MPC systems use a random walk to model the disturbances. Moreover, the (infinite-horizon) MPC controller has to be expressed in the RST form of Eq. 4.14, so that the MPC performance curve can be computed by solving Eqs. 4.17 and 4.18. For this purpose, any solution to the LQG optimisation problem (Sect. 4.2) can be adopted.

However, during subsequent months and years of control-system operation, the controller performance is not expected to fall on the original MPC performance curve due to changes in process/plant characteristics. Therefore, a periodical performance assessment of MPC by re-calculating the performance curve is highly recommended, ideally when it can be generated from routine operating data. If the current performance point does not fall on or near the MPC curve, this is a strong indication of mismatch in the plant and/or disturbance models. If a significant portion of this deviation can be attributed to error in the plant model, then additional response testing should be carried out before commissioning is resumed (Julien et al. 2004).

The MATLAB function gpc_Nc from Moudgalya (2007: Chap. 12) can be used to express a GPC controller with specified parameters N_1, N_p, N_c and ρ in an RST form.

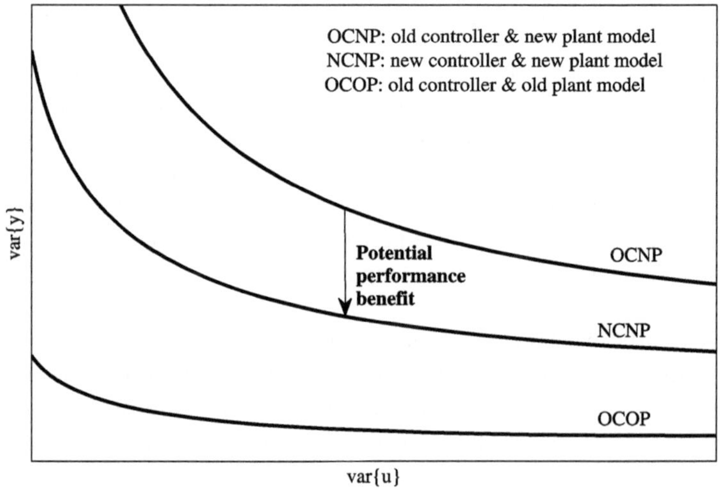

Fig. 4.12 MPC performance curves

4.3.4.2 Assessing the Potential Performance Improvement of MPC

It has been shown by Julien et al. (2004)[3] that it is possible *under certain circumstances* to identify a model of the plant dynamics only from normal operating data, i.e. without the injection of a dither signal, as is usual in closed-loop identification, but with the knowledge of the time delay; see Sect. 7.2.7 for details. Once an updated process model $\hat{G}_p$ has been generated using this method, a new performance curve can be obtained. Using the new model and the old, i.e. installed controller, the so-called *old controller & new process model* (*OCNP*) performance curve can be constructed, again by solving Eqs. 4.17 and 4.18. Inspection of the OCNP curve may indicate that it is possible to re-tune the existing controller to a desirable performance region despite the presence of plant–model mismatch, eliminating the need for a new identification experiment. Performance indices like those defined by Eqs. 4.20 and 4.21 can be calculated to quantify the possible improvement potential.

On the other hand, one can design a new MPC controller based on the new model and then generate the so-called *new controller & new process model* (*NCNP*) performance curve, also referred to as *MPC-achievable* performance curve. The distance between the NCNP curve and the OCNC curve indicates the performance deficiency due to the plant–model mismatch; see Fig. 4.12.

The OCNP curve shows how the performance would be affected were the controller to be re-tuned by varying the move suppression coefficient, ρ. Hence, the OCNP and NCNP curves enable the user to differentiate between variance inflation

[3]In this section, MATLAB routines kindly provided by Julien et al. (2004) have been used for the computation of infinite-horizon MPCs and the corresponding performance curves.

caused by tuning vs. plant–model mismatch. The potential benefit of re-identifying the plant can then be weighed against the time and expense associated with a new response test.

4.3.4.3 MPC Assessment Procedure

A complete procedure for assessing the performance, diagnosing possible reasons for poor performance and quantifying improvement potential of MPC-controlled loops is given as follows.

Procedure 4.3 MPC performance assessment and diagnostic (Julien et al. 2004).

1. Construct the OCOP performance curve for the MPC regulator in the form of Eq. 4.14, based on the initial plant and disturbance models identified prior to commissioning.
2. Using routine operating data, compute the input ($\sigma^2_{\Delta u \mathrm{A}_0}$) and output ($\sigma^2_{y\mathrm{A}_0}$) sample variances for the existing controller and compare its performance to the OCOP curve. If the curve does not intersect the joint confidence region for ($\sigma^2_{\Delta u \mathrm{A}_0}, \sigma^2_{y\mathrm{A}_0}$) (Eqs. 4.22 and 4.23), one can conclude that there is process–model mismatch.
3. Use routine operating data to re-estimate the process and disturbance model when feasible, as described in Sect. 7.2.7. This is usually successful when the time delay is at least 15 % of the disturbance settling time and exceeds the duration of any initial wrong-way transients in the disturbance impulse response by at least four intervals.
4. Construct the OCNP performance curve based on the new system model and the MPC controller designed for the original model. This curve should pass through the confidence region for ($\sigma^2_{\Delta u \mathrm{A}_0}, \sigma^2_{y\mathrm{A}_0}$) if the closed-loop identification has been successful. If not, then stop: an identification experiment must be carried out to estimate G_ε and G_p.
5. Compute the NCNP performance curve based on the updated system model and a new MPC controller designed for the updated plant model: this curve summarises the closed-loop performance one could expect the MPC strategy to exhibit if it were rebuilt based on data collected during a new response test.

This procedure is limited to cases where the process time delay is "large" relative to the settling time of the disturbance, as only in this case the process model can be estimated from routine operating data (Sect. 7.2.7). Moreover, it should be recognised that large data sets may be required to ensure that the process models converge, and so this method will evaluate an "average" controller performance over the range of disturbances encountered. Note also that the extension of this method to multivariate MPC systems with constraint needs further research.

Example 4.4 We illustrate the MPC performance assessment technique with the paper machine model of Example 3.1. The noise ε is assumed to have the variance

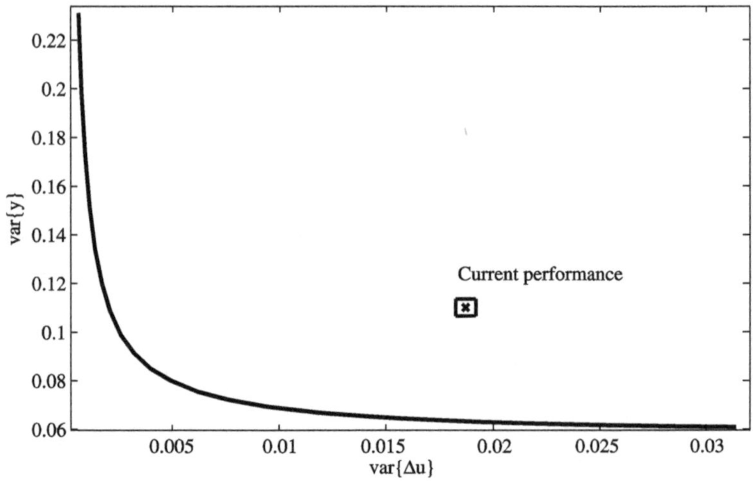

Fig. 4.13 Old controller & old process curve and installed controller performance (Example 4.4)

$\sigma_\varepsilon^2 = 0.02$. The assessment calculations are based on data sets containing 10000 observations of u and y, generated by simulated closed-loop model ($T_s = 1$). The adopted initial controller is an infinite-horizon MPC regulator determined for the "old process" (Eq. 3.25) with a penalty on move of $\rho = 0.5$ as ("old controller")

$$G_c(q) = \frac{1 - 0.37q^{-1}}{1.1764 - 0.7037q^{-1} + 0.1573q^{-2} - 0.63q^{-3}}. \tag{4.31}$$

We consider the scenario where the disturbance dynamics has changed so that the "new process" is given by

$$y(k) = \frac{0.63q^{-3}}{1 - 0.37q^{-1}}u(k) + \frac{1}{(1 - q^{-1})(1 - 0.3q^{-1})}\varepsilon(k). \tag{4.32}$$

Steps 1 and 2 of Procedure 4.3 are applied to generate the performance curves and evaluate the controller performance. Looking at the OCOP curve in Fig. 4.13 indicates significant degradation in performance. However, it is still not known whether the performance degradation is due to changes in the plant and/or disturbance dynamics.

As in Step 3 of the procedure, the disturbance dynamics is estimated from routine operating data and turns out that the disturbance impulse response does not settle out before the time delay, but extrapolation can be applied to yield Fig. 4.14.

As the extrapolated coefficients fit well with the true ones, an estimate of the plant dynamics G_p is identified from the pre-filtered input–output data (Sect. 7.2.7; Eq. 7.37). Figure 4.15 compares the step response of the estimated process model with that of the true one and indicates good agreement.

Using the estimated process model, Steps 4 and 5 of Procedure 4.3 can be carried out and lead to the OCNP and NCNP performance curves shown in Fig. 4.16.

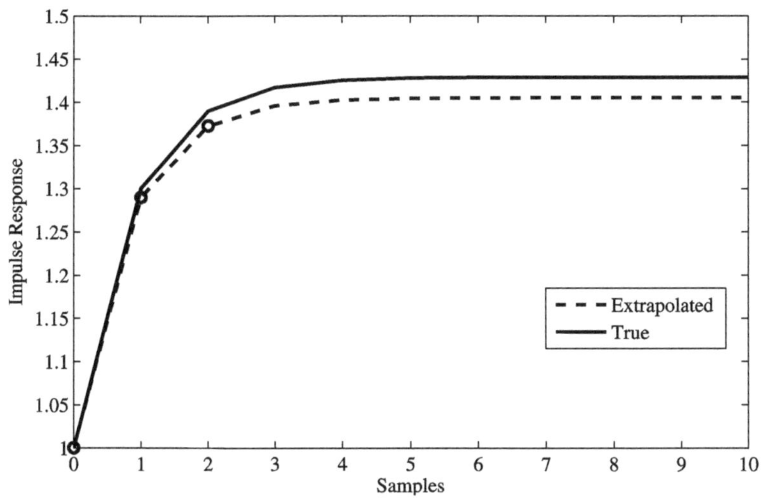

Fig. 4.14 True and extrapolated impulse response of disturbance model (Example 4.4; changed disturbance dynamics)

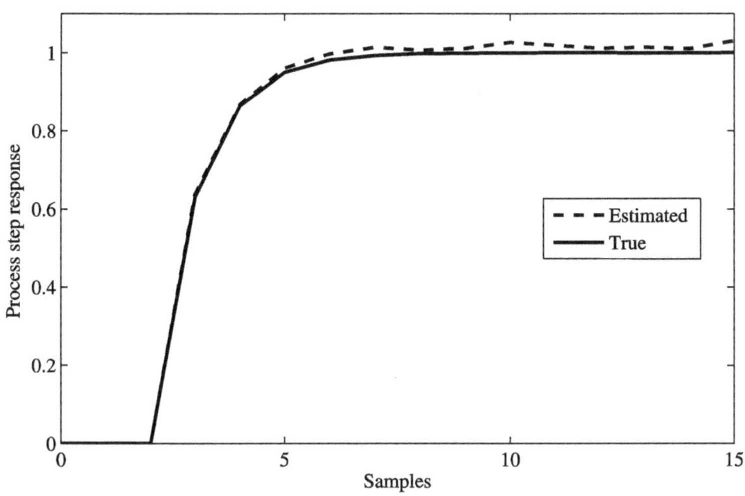

Fig. 4.15 True and estimated step response of process model (Example 4.4; changed disturbance dynamics)

Both curves fall together and pass through the confidence region (rectangle) for the actual variances. This implies that no real benefit would be gained by carrying out a new response test and re-design of the MPC controller. This is indeed the right conclusion, as we only changed the disturbance dynamics.

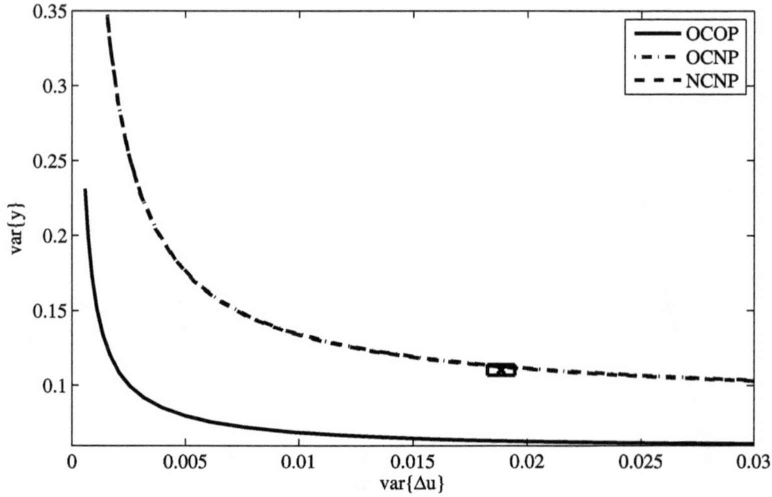

Fig. 4.16 Updated performance curves (Example 4.4; changed disturbance dynamics)

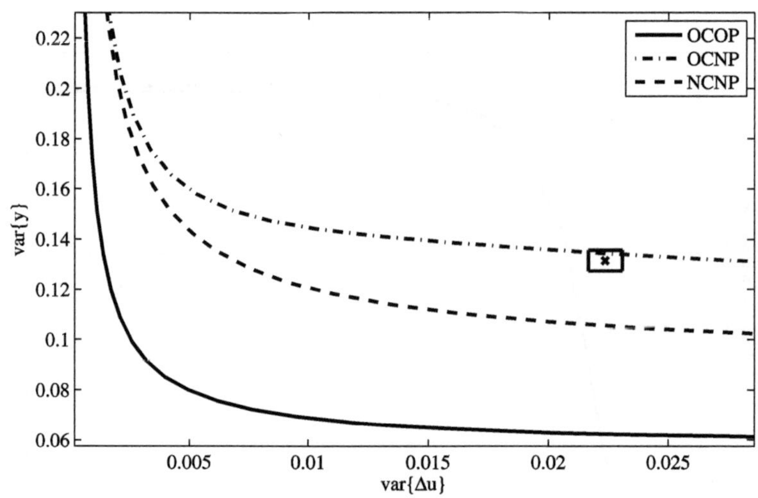

Fig. 4.17 Updated performance curves (Example 4.5; changed plant and disturbance dynamics)

Example 4.5 Consider again the paper machine model in Eq. 3.25. In this case, both the plant and disturbance dynamics are changed so that the "new process" is expressed by

$$y(k) = \frac{0.63q^{-3}}{(1-0.37q^{-1})(1-0.35q^{-1})}u(k) + \frac{1}{(1-q^{-1})(1-0.3q^{-1})}\varepsilon(k).$$

The updated performance curves are shown in Fig. 4.17.

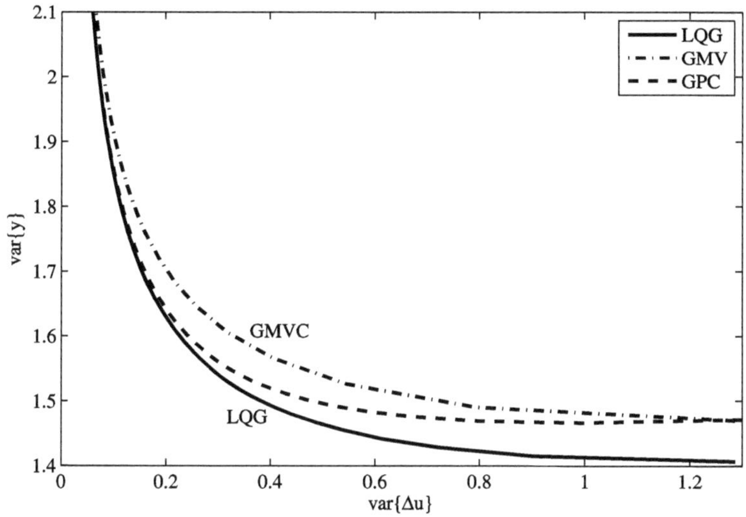

Fig. 4.18 Performance curves for LQG, GMVC and GPC of MacGregor's viscosity problem

The clear distance between the OCNP and NCNP curves indicate the potential from performance improvement by re-testing the plant and re-designing the MPC controller. For instance, if the current level of the input variance is maintained, there is a maximum possible scope of

$$\frac{0.1314 - 0.1056}{0.1314} \times 100 \approx 20\ \%$$

to reduce the output variance.

Example 4.6 The viscosity control problem presented by MacGregor (1977) is considered:

$$y(k) = \frac{0.51 + 1.21q^{-1}}{1 - 0.44q^{-1}}u(k-1) + \frac{1}{1-q^{-1}}\varepsilon(k). \tag{4.33}$$

The noise ε is assumed to have unity variance. For this system, we construct the performance curves for LQG, GMVC and GPC ($N_2 = N_u = 2$), shown in Fig. 4.18. It can seen how the GMVC and GPC curves lie significantly above the LQG curve and fall together for smaller ρ. The MVC may be adopted if the large input variance is acceptable. A possible comparison strategy of the other controllers is to determine the output variances for comparable input variances, as listed in Table 4.1. This shows that LQG achieves a smaller output variance than GMVC or GPC. In view of this criterion, we would prefer the LQG controller, but this is not a general suggestion.

Note that as the weighting parameter ρ is reduced to zero, the performance of an LQG controller becomes that of the MVC, while satisfying closed-loop stability.

Table 4.1 Comparison of MVC, GMVC, LQG and GPC controllers for MacGregor's viscosity problem

	MVC	GMVC; $\rho = 1.0$	LQG; $\rho = 1.4$	GPC; $\rho = 2.8$
σ_y^2	1.4070	1.7194	1.6452	1.6591
$\sigma_{\Delta u}^2$	1.2994	0.1867	1.1878	0.1864

In contrast, closed-loop systems with GMVC and GPC become unstable for non-minimum phase processes when the control effort goes to zero (Moudgalya 2007). This is the reason why in Fig. 4.18 the GMVC and GPC performance curves move upwards as $\rho \to 0$. In view of this, the LQG benchmark is also the recommended assessment method.

4.3.5 Assessing the Effect of Constraints

The most important feature of MPC is that it provides a very flexible approach to incorporate constraints into the computation of optimal control inputs. It is clear that the performance of a constrained loop is different from that of the unconstrained loop (when the constraints are active). Most commercial MPC packages use quadratic program (QP) solutions to calculate the optimal input sequence, and the resulting control law is generally nonlinear. Performance assessment of MPC loops with constraints is thus more complicated.

To evaluate the effect of the constraint imposed on the process, the installed controller performance may be compared to that of the unconstrained MPC controller. The performance gap between both controllers is only due to the constraints. Obviously, the actual variance in process output converges to that in the unconstrained case, when the constraints become inactive. Therefore, if the constrained MPC loop works near the unconstrained controller benchmark, it indicates that the constraints have little effect on the system performance. If the constrained MPC loop works far from the unconstrained controller benchmark, one can conclude that the constraints greatly affect the loop performance. Recall that the design-case MPC assessment approach, introduced in Sect. 4.3.2, is well suited for evaluating MPC with constraints.

4.3.6 Economic Performance Assessment of MPC

MPC performance assessment with respect to constraints and economical benefits is an active CPM field. Economic performance assessment provides a systematic methodology founded on the use of APC strategies, such as MPC, to determine the economic benefit potential and to evaluate the performance of MPC applications.

This assessment is carried out using available information about normal MPC operation, including the base case process operation data, the steady-state gain matrix of the process model, the constraint high and low limits for the input and output variables, the economic objective function and the mean operating points for the input and output variables. Economic performance assessment should give answers to the two questions:

1. Which controlled variable (CV)/manipulated variable (MV) is the most sensitive in terms of economic potential?
2. How much does this CV/MV need to be tuned or changed to achieve a given desired benefit potential?

In most processes, the economic benefit is derived from operating quality variables at or near their optimal operating points, which often are located near the upper or lower constraint boundaries. However, in most cases, MPC applications cannot achieve this economic optimum due to the following factors:

1. Process variability forces the key quality variables to back off from their optimal values.
2. Conservative or too tight setting of constraints reduces the degrees of freedom to operate the process.

Thus, MPC performance could be improved by simply adjusting the current operating point, subject to the availability of a degree of freedom, by the following corresponding measures:

1. Reducing process variability, e.g., if the variability of one or some of the controlled variables can be reduced to certain percentage by tuning the lower level regulatory control loops.
2. Relaxing a constraint limit on certain process variables.

In some situations, both measures can be simultaneously applied. The reduction in variability or relaxation of a limit provides the opportunity to shift the current operating point closer to its optimal operating point and consequently improve yield (and thus financial profit); see Fig. 1.3.

Beside economic performance assessment, the economic performance assessment of MPC usually includes sensitivity analysis and tuning guidelines. Sensitivity analysis aims to check the impact of variability reduction or constraint relaxation on the benefit potential. The tuning guidelines provide suggestions on how and where to tune the control-system operation to achieve the desired target benefit potential.

Economic performance assessment was originally developed by Xu et al. (2006a, 2006b, 2007, 2008) by extending the linear matrix inequality (LMI) method to a deterministic process and implementing a minimum variance basis for process variability tuning. This method has been extended by Agarwal et al. (2007a, 2007b) and Akande et al. (2009) to include both deterministic and stochastic processes. The LMI-based economic performance assessment approach has been extensively applied to different industrial processes, for example the hydrogen unit by Lee et al. (2008). Moreover, Zhao et al. (2009) proposed a stochastic solution that uses the

LQG benchmark for variability tuning. The basic principles and equations of these methods will be reviewed in the following.

4.3.6.1 Linear Matrix Inequality Performance Analysis

LMI-based performance assessment of MPC aims to minimise the following economic objective function (Xu et al. 2007; Lee et al. 2008):

$$\min \frac{1}{N}\sum_{k=1}^{N} J_{\mathrm{eco}}(k), \tag{4.34}$$

where N is the total number of data points collected, and $_{\mathrm{eco}}(k)$ is defined as

$$\begin{aligned} J_{\mathrm{eco}}(k) = {} & \sum_{i=1}^{p}\left[b_{y,i}(k)\bar{y}_i + a_{y,i}^2(k)\big(\bar{y}_i - y_{s,i}(k)\big)^2\right] \\ & + \sum_{j=1}^{m}\left[b_{u,j}(k)\bar{u}_j + a_{u,j}^2(k)\big(\bar{u}_j - u_{s,j}(k)\big)^2\right]. \end{aligned} \tag{4.35}$$

$a_{y,i}$ is the quadratic coefficient for the ith CV, $b_{y,i}$ is the linear coefficient for the ith CV, $a_{u,j}$ is the quadratic coefficient for the jth MV, $b_{u,j}$ is the linear coefficient for the jth MV, $y_{s,i}$ is the set point (target) value for the ith CV, $u_{s,j}$ is the set point (target) value for the jth MV, $(\bar{y}, \bar{u})$ is the optimal mean operating point to be determined, p is the number of CV, and m is the number of MV. The optimisation is performed subject to the following constraints:

$$\sum_{j=1}^{m} K_{ij}\Delta\bar{u}_j \quad \forall i = 1, 2, \ldots, p, \tag{4.36}$$

$$\begin{aligned} & L_{y,i}(k) - \lambda_{y,i}\alpha_{y,i}(k) + 2\sigma_{y,i}(1+\nu_i) \le \bar{y}_i \le H_{y,i}(k) + \mu_{y,i}\alpha_{y,i}(k) - 2\sigma_{y,i}(1+\nu_i) \\ & \quad \forall i = 1, 2, \ldots, p, \end{aligned} \tag{4.37}$$

$$\begin{aligned} & L_{u,j}(k) - \lambda_{u,j}\alpha_{u,j}(k) + 2\sigma_{u,j} \le \bar{u}_j \le H_{u,j}(k) + \mu_{u,j}\alpha_{u,j}(k) - 2\sigma_{u,j} \\ & \quad \forall j = 1, 2, \ldots, m, \end{aligned} \tag{4.38}$$

where K_{ij} is the (i, j) entry of the gain matrix $\boldsymbol{K}$,

$$\begin{aligned} \bar{y}_i &= \bar{y}_{0,i} + \Delta\bar{y}_i, \quad \bar{y}_{0,i} = \frac{1}{N}\sum_{k=1}^{N} y_{0,i} \quad \forall i = 1, 2, \ldots, p, \\ \bar{u}_j &= \bar{u}_{0,j} + \Delta\bar{u}_j, \quad \bar{u}_{0,j} = \frac{1}{N}\sum_{k=1}^{N} u_{0,j} \quad \forall j = 1, 2, \ldots, m. \end{aligned} \tag{4.39}$$

$\bar{y}_0$ and $\bar{u}_0$ are the current mean operating points of the controlled and manipulated variables, y and u are the routine operating process data, $\Delta\bar{y}_0$ and $\Delta\bar{u}_0$ are the proposed shift in the mean operating conditions, $L(k)$ is the lower limit (LL) for the given variable, $H(k)$ is the upper limit (UL) for the given variable, α is half the constraint region for the given variable, σ is the standard deviation for the variable, ν is the percent change in variability, μ is the percent change in UL, and λ is the percent change in LL. The last three variables are specified by the user.

Based on the above equations, the following special cases can be distinguished (Xu et al. 2007; Lee et al. 2008):

1. *Base Case Operation.* The base case potential, $J_{\text{eco},0}$, is calculated by replacing $\bar{y}_i$ with $\bar{y}_{0,i}$ and $\bar{u}_j$ with $\bar{u}_{0,j}$. $J_{\text{eco},0}$ serves as a reference for the calculation of benefit potentials.
2. *Ideal Case Operation.* In this scenario, the disturbance influence is not considered, and a strict steady-state operation is assumed. There is no backoff due to the disturbances and the constraint limits are kept unchanged. This case is characterised by $\nu_i = -1$, $\lambda_{y,i} = \mu_{y,i} = \lambda_{u,j} = \mu_{u,j} = 0$ $\forall i = 1, 2, \ldots, p$ and $\forall j = 1, 2, \ldots, m$, to give the ideal benefit $J_{\text{eco,ideal}}$, and hence the *ideal benefit potential* is found from $\Delta J_{\text{eco,ideal}} = J_{\text{eco,ideal}} - J_{\text{eco},0}$.
3. *Current Case Operation.* In this scenario, the present level of disturbance is taken into account. The purpose is to move the operating point as close as possible to the ideal optimum. This case is characterised by $\nu_i = 0$, $\lambda_{y,i} = \mu_{y,i} = \lambda_{u,j} = \mu_{u,j} = 0$ $\forall i = 1, 2, \ldots, p$ and $\forall j = 1, 2, \ldots, m$, to give the current benefit $J_{\text{eco,current}}$, and hence the *current benefit potential* is found from $\Delta J_{\text{eco,current}} = J_{\text{eco,current}} - J_{\text{eco},0}$. Note that if the constraints in Eq. 4.37 (respectively 4.38) become unfeasible, then they are replaced by $\bar{y}_i = \bar{y}_{0,i}$ or $\Delta\bar{y}_i = 0$ (respectively $\bar{u}_j = \bar{u}_{0,j}$ or $\Delta\bar{y}_i = 0$).
4. *Reducing Variability Scenario.* In the current case operation, no action is taken to reduce the variability of the controlled variables, and the existing benefit potential is obtained by shifting mean values only. If the variability of one or some of the controlled variables can be reduced to certain percentage ($\nu_i = \nu_{\text{V},i}$) by tuning the lower level regulatory control loops, then the backoff can also be reduced, which will allow further mean values' shifting in the direction of the ideal operating point. If a significant increase of the benefit potential is observed by variability reduction on one or more of the controlled variables, it is worth tuning the corresponding regulatory control loops. If the optimal operating point by variability reduction is denoted as $\bar{y}_{\text{V},i}, \bar{u}_{\text{V},j}$ with the corresponding objective function as $J_{\text{eco,V}}$, then the *optimal benefit potential by reducing variability*, $\Delta J_{\text{eco,V}}$, can be obtained by $\Delta J_{\text{eco,V}} = J_{\text{eco,V}} - J_{\text{eco},0}$. Note that $\lambda_{y,i} = \mu_{y,i} = \lambda_{u,j} = \mu_{u,j} = 0$ $\forall i = 1, 2, \ldots, p$ and $\forall j = 1, 2, \ldots, m$.
5. *Relaxing Constraint Scenario.* The constraint limits of both controlled and manipulated variables are usually determined in the design phase when an MPC application is commissioned, but some of them may be allowed to adjust afterwards. This leaves an alternative way to improve the MPC economic performance. For the quality variables, the constraint relaxation may allow their mean values to be moved closer to the ideal optimal operating point even if there is

no reduction of variability, i.e. $v_i = 0$. For the constrained variables, however, the increase on their operating ranges will allow more variability to be transferred from the quality variables to them, which may create room for the quality variables to reduce their variability. If the optimal result due to the constraint relaxation is represented by $\bar{y}_{\mathrm{C},i}, \bar{u}_{\mathrm{C},j}$ with the corresponding objective function as $J_{\mathrm{eco,C}}$, then the *optimal benefit potential by relaxing constraints*, $\Delta J_{\mathrm{eco,C}}$, is given by $\Delta J_{\mathrm{eco,C}} = J_{\mathrm{eco,C}} - J_{\mathrm{eco,0}}$.

Note that constraint relaxation can be done more easily than variability reduction because it does not require controller parameters tuning, while variability reduction involves the tuning of controller settings such that the variability of one or more specific process-output variables can be reduced.

4.3.6.2 Sensitivity Analysis

The aim of sensitivity analysis is to quantify the importance of variability reduction or constraint relaxation of different variables in terms of their contributions to the benefit potential. This should find out which process variables should be given the priority for economic performance improvement. Two separate algorithms exist in the literature, one for constraint tuning and the other for variability tuning.

The first algorithm is used to determine sensitivity analysis for *constraint tuning*. In this case, the following set of equations is solved (Lee et al. 2008):

$$\min_{\bar{y_i},\bar{u_j},\lambda_{y,i},\mu_{y,i},\mu_{u,j}} \left[\sum_{i=1}^{p} (\lambda_{y,i} + \mu_{y,i}) + \sum_{j=1}^{m} (\lambda_{u,j} + \mu_{u,j}) \right] \tag{4.40}$$

subject to

$$\begin{cases} \bar{\lambda}_{y,i} \ge \lambda_{y,i} \ge 0, & flag_{\lambda_{y,i}} = 1, \\ \lambda_{y,i} = 0, & flag_{\lambda_{y,i}} = 0, \end{cases} \tag{4.41}$$

$$\begin{cases} \bar{\mu}_{y,i} \ge \mu_{y,i} \ge 0, & flag_{\mu_{y,i}} = 1, \\ \mu_{y,i} = 0, & flag_{\mu_{y,i}} = 0, \end{cases} \tag{4.42}$$

$$\begin{cases} \bar{\lambda}_{u,j} \ge \lambda_{u,j} \ge 0, & flag_{\lambda_{u,j}} = 1, \\ \lambda_{u,j} = 0, & flag_{\lambda_{u,j}} = 0, \end{cases} \tag{4.43}$$

$$\begin{cases} \bar{\mu}_{u,j} \ge \mu_{u,j} \ge 0, & flag_{\mu_{u,j}} = 1, \\ \mu_{u,j} = 0, & flag_{\mu_{u,j}} = 0, \end{cases} \tag{4.44}$$

$$\frac{J_{\mathrm{eco,0}} - J_{\mathrm{eco}}}{\Delta J_{\mathrm{eco,ideal}}} = R_C, \tag{4.45}$$

$$L_{y,i}(k) - \lambda_{y,i}\alpha_{y,i}(k) + 2\sigma_{y,i} \le \bar{y}_i \le H_{y,i}(k) + \mu_{y,i}\alpha_{y,i}(k) - 2\sigma_{y,i}$$
$$\forall i = 1, 2, \ldots, p, \tag{4.46}$$
$$L_{u,j}(k) - \lambda_{u,j}\alpha_{u,j}(k) + 2\sigma_{u,j} \le \bar{u}_j \le H_{u,j}(k) + \mu_{u,j}\alpha_{u,j}(k) - 2\sigma_{u,j}$$
$$\forall j = 1, 2, \ldots, m, \tag{4.47}$$

where $\bar{\lambda}$ represents the given constant upper bound for the variable λ, $\bar{\mu}$ represents the given constant upper bound for the variable μ, *flag* is a Boolean variable that represents whether or not the bounds of the given variable can be changed, and R_C is the given target benefit potential ratio with $0 \le R_C \le 1$.

The second algorithm is used to perform sensitivity analysis for *variability tuning*. In this case, the following set of equations is solved (Lee et al. 2008):

$$\min_{\bar{y}_i, \bar{u}_i, \nu_i} \left(\sum_{i=1}^{p} |\nu_i| \right) \tag{4.48}$$

subject to

$$\begin{cases} \underline{\nu}_i \le \nu_i \le 0, & \mathit{flag}_{\nu_i} = 1, \\ \nu_i = 0, & \mathit{flag}_{\nu_i} = 0, \end{cases} \tag{4.49}$$

$$\frac{J_{\text{eco},0} - J_{\text{eco}}}{\Delta J_{\text{eco,ideal}}} = R_C, \tag{4.50}$$

$$L_{y,i}(k) - \lambda_{y,i}\alpha_{y,i}(k) + 2\sigma_{y,i}(1 + \nu_i)$$
$$\le \bar{y}_i \le H_{y,i}(k) + \mu_{y,i}\alpha_{y,i}(k) - 2\sigma_{y,i}(1 + \nu_i) \quad \forall i = 1, 2, \ldots, p, \tag{4.51}$$
$$L_{u,j}(k) - \lambda_{u,j}\alpha_{u,j}(k) + 2\sigma_{u,j} \le \bar{u}_j \le H_{u,j}(k) + \mu_{u,j}\alpha_{u,j}(k) - 2\sigma_{u,j}$$
$$\forall j = 1, 2, \ldots, m, \tag{4.52}$$

where $\underline{\nu}$ represents the given constant lower bound for the variable ν.

4.3.6.3 LQG Benchmark Approach

Instead of taking the mean operating point to describe the quality variables (as in Eq. 4.34), it is more useful to consider their probability distribution. Such a stochastic-based approach will provide better estimates of the appropriate variables to consider. In this case, the optimisation is performed by modifying Eq. 4.34 to include the probability of each of the potential future scenarios, i.e.

$$\min \frac{1}{N} \sum_{k=1}^{N} \mathrm{E}\{J_{\text{eco}}(k)\}, \tag{4.53}$$

where E is the expectation operator with the other information being the same as for the deterministic case. To solve this stochastic problem, Agarwal et al. (2007a) proposed to divide the operating region for each quality variable into a finite number of smaller operating regions. Then for each subdivided region, the continuous-valued stochastic economic objective function is obtained by discretisation of each region. This discrete value is then solved for the quality variable assuming that in the subdivided region the variables are normally distributed. However, this approach suffers from two problems:

1. The discretisation the region into smaller subregions yields a loss in precision that can be detrimental in obtaining accurate results.
2. The number of subregions that are created can quickly grow limiting its use in high-dimensional problems.

A closed-form solution was given by Akande et al. (2009) for the stochastic minimum-variance benchmark case. It has been shown that if the cost function is linear, then the stochastic optimisation will give the same result as the deterministic optimisation. On the other hand, for a quadratic cost function, the results will be different.

Another approach to the stochastic problem is to consider the LQG benchmark for identifying the achievable minimum variance subject to input variance constraints. This approach is a more realistic benchmark than minimum-variance control when evaluating constraint control, such as MPC. However, this approach requires a complete plant model and can only address input variance constraints, i.e. only soft and not hard constraints can be handled (Huang and Kadali 2008).

The first step in using the LQG approach is to formulate the MPC problem as an LQG problem. One approach to this problem was proposed by Zhao et al. (2009), where the goal is to

$$\begin{aligned} \min_{\bar{y}_i,\bar{u}_j,\sigma_Y^2,\sigma_U^2} \mathrm{E}\Bigg\{ & \sum_{i=1}^{p}\big[b_{y,i}(k)\bar{y}_i + a_{y,i}^2(k)\big(\bar{y}_i - y_{s,i}(k)\big)^2\big] \\ & + \sum_{j=1}^{m}\big[b_{u,j}(k)\bar{u}_j + a_{u,j}^2(k)\big(\bar{u}_j - u_{s,j}(k)\big)^2\big]\Bigg\} \end{aligned} \tag{4.54}$$

subject to the constraints given in Eqs. 4.36–4.38, where σ_Y^2 and σ_U^2 are calculated as follows:

$$\begin{aligned} \sigma_Y^2 &= \sum_{i=1}^{p} q_i \sigma_{y,i}^2, \\ \sigma_U^2 &= \sum_{j=1}^{m} r_j \sigma_{u,j}^2. \end{aligned} \tag{4.55}$$

q_i is the diagonal entry of the weighting matrix $\boldsymbol{Q}$ for the outputs, r_j is the diagonal entry of the weighting matrix $\boldsymbol{R}$ for the inputs, $\sigma_{y,i}^2$ is the variance for the ith output, and $\sigma_{u,j}^2$ is the variance for the jth input. The trade-off curve can then be determined for different values of ρ by solving Eqs. 4.12 and 4.54. From this curve it is possible to obtain the desired values. In general, it is difficult to obtain a tractable solution if the quadratic terms are included. Therefore, Zhao et al. (2009) proposed solving the case where only the linear terms are considered, i.e.

$$\min_{\bar{y}_i, \bar{u}_j, \sigma_Y^2, \sigma_U^2} \mathrm{E}\left\{ \sum_{i=1}^{p} c_y^{(i)} \bar{y}_i - \sum_{j=1}^{m} c_u^{(j)} \bar{u}_j \right\} \tag{4.56}$$

subject to the constraints

$$\begin{aligned}
\Delta \bar{y}_i &= \sum_{j=1}^{m} (K_{ij} \Delta \bar{u}_j), \\
\bar{y}_i &= \bar{y}_{i,0} + \Delta \bar{y}_i, \\
\bar{u}_j &= \bar{u}_{j,0} + \Delta \bar{u}_j, \\
y_{i,\min} + z_{\alpha/2}\sigma_{y,i} &\le \bar{y}_i \le y_{i,\max} - z_{\alpha/2}\sigma_{y,i}, \\
u_{j,\min} + 3\sigma_{u,j} &\le \bar{u}_j \le u_{j,\max} - 3\sigma_{u,j}, \\
\sigma_{y,i} &\ge 0, \\
\sigma_{u,j} &\ge 0,
\end{aligned} \tag{4.57}$$

where σ_Y^2 and σ_U^2 are calculated as in Eq. 4.55, $z_{\alpha/2}$ is the z-score at the α-level, and Δ represents a shift in the process. Using this approach, the LQG-curve is then calculated.

4.4 Summary and Conclusions

This chapter has provided advanced methodologies for controller performance assessment. The main feature of these important methods is to minimise a weighted sum of the set-point error and the control effort, and thus avoid excessive control action that can result from minimum-variance control. For the assessment purpose, a performance curve is constructed by plotting the variance of the process variable against that of the (differenced) input over a range of values of the move suppression parameter. Such a trade-off curve is particularly valuable when assessing the performance model-predictive controllers. A formal procedure was described (Sect. 4.3.4.3) which utilises routine operating data to update the plant and disturbance models for MPC. Although not universally applicable, the method provides a useful way to determine when it becomes worthwhile to invest in re-identification of the plant dynamics and re-commissioning of MPC. Moreover, LQG benchmarking (with its performance curve) remains the standard against which other

Table 4.2 Summary of requirements of different control benchmarking methods

Method	HIS	EHPI	MV	GMV	LQG	MPC
Time delay			√	√	√	√
Control error		√	√	√	√	√
Control input	(√)			√	√	√
Weightings					√	√
Process/disturbance model					√	(√)
Controller structure						√

Table 4.3 Summary of benefits of different control benchmarking methods (Required computational burden: Low: $*$ – high: $*****$)

Method	HIS	EHPI	MV	GMV	LQG	MPC
Control benchmark	√	√	√	√	√	√
Limitation of actuator energy				√	√	√
Reflection of controller structure	(√)	(√)				√
Required computational burden	*	**	**	***	*****	*****

controllers should be compared, when the penalisation of control effort is important.

Despite these nice features, there are enough reasons for not using advanced benchmarking techniques, for instance:

1. Advanced assessment methods normally require a full system model to calculate the performance indices. Although it is possible under certain circumstances to estimate a process model, a disturbance model, or both from closed-loop data, this method can be complicated and rely on an adequate signal-to-noise ratio in the data set.
2. If a move suppression weight (LQG, MPC) or even more complex dynamic weightings (GMV) are needed, the user must specify them and decide whether the current controller variance is sufficiently close to a somewhat detuned controller to be acceptable. This is not a trivial task having in mind that the output variance is a very nonlinear function of the move suppression parameter. Moreover, a priori knowledge and/or simulations are usually necessary to select proper design parameters of advanced benchmarking techniques.
3. Such involved methods are only applicable and useful for the assessment of model-based controllers, which are used in maximally 10 % of industrial control loops.

Inspired by Clegg (2002), the benchmarking methods HIS, EHPI, MV, GMV, LQG and MPC are compared in terms of parameters/data requirements and benefits; see Table 4.2 and Table 4.3. It can be concluded that, usually, calculating more sophisticated and realistic benchmarks requires more prior knowledge and data and is computationally expensive. On the other hand, using historical benchmarks (which do not require model identification) is the easiest approach but must be taken with care, as it is too subjective and may be misleading.

Chapter 5
Deterministic Controller Assessment

Systems in the process industries normally have at least one stochastic source acting on the system process and are operated around a constant operating point for long periods of time or within the batches. Also, for most of these systems, reducing the output variances leads to improved product quality, reduced energy/material consumption and thus higher efficiency and productivity. Therefore, for such systems, considering MV and related benchmarks is quite natural and useful.

However, systems in some other industrial fields, such as power and servo industry and robotics, show references and disturbances tending to be more deterministic rather than stochastic. In these cases, the processes are operated with frequent changes in the references and hence output levels. For instance, power plants have to follow a daily load program, which is fixed as a function of typical load demand and energy market requirements (Uduehi et al. 2007c). So, for such systems, deterministic assessment techniques are needed.

This chapter presents three methods for control performance assessment based on deterministic criteria: settling time and IAE indices gained from set-point response data (Sect. 5.2), the idle index for detection of sluggish control (Sect. 5.3) and the area index for evaluating deterministic load-disturbance rejection performance (Sect. 5.4). These techniques are compared and discussed using simulation examples in Sect. 5.4.3.

5.1 Performance Metrics

It has been a tradition in (PID) control engineering to judge the control performance based on step changes in set points or load disturbances. When analysing a set-point response, the criteria used to describe how well the process responds to the change can include the rise time, settling time, decay ratio, overshoot and steady-state error (Fig. 5.1, Table 5.1).

These criteria can be used both as specifications for tuning/commissioning of control loops as well as for documenting changes in performance due to the adjustment of the controller or process parameters. Typically, designers select one of the

M. Jelali, *Control Performance Management in Industrial Automation*,
Advances in Industrial Control, DOI 10.1007/978-1-4471-4546-2_5,

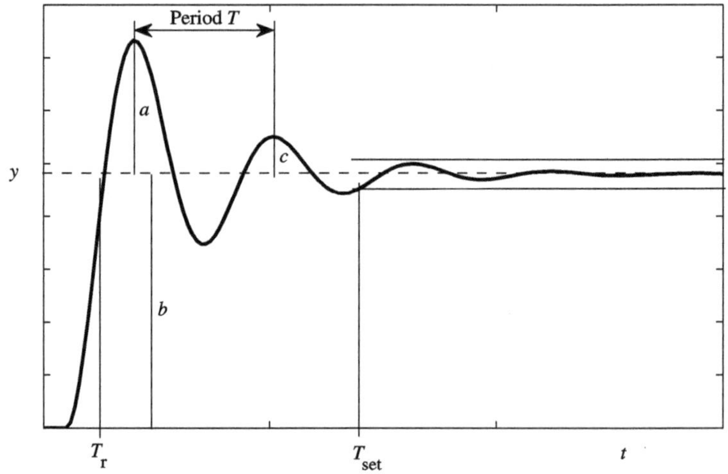

Fig. 5.1 Set-point response features

Table 5.1 Typical set-point response criteria

Criterion	Definition and Interpretation
Rise time T_r	Inverse of the largest slope of the step response or the time it takes for the step response to change from 10 to 90 % of its steady-state value. A large rise time may be the result of a sluggish controller.
Settling time T_{set}	Time it takes before the step response remains within $p\%$ (commonly $p = 1, 2$, or 5 %) of its steady-state value. Time spent outside the desired level generally relates to undesirable product. Therefore, a short settling time is sought.
Decay ratio $d = c/a$	Ratio between two consecutive maxima of the error. A large decay rate is associated with an aggressive controller, and visible oscillations are present in the set-point response. The smaller the decay rate, the faster the oscillations will be dampened. Traditionally, a quarter amplitude damping (i.e., $d = 1/4$) has been used. This value is, however, often too high.
Overshoot $\alpha = 100a/b$	Ratio between the difference between first peak and the steady-state value of set-point response. An aggressive controller can increase the amount of overshoot associated with a set-point change. Commonly, an overshoot of 8 to 10 % is specified. In many situations, it is desirable, however, to have an over-damped response with no overshoot.
Steady-state error	Steady-state control error. This is always zero for a controller with integral action.

above metrics and define optimal control as the tunings that achieve the minimum value of the criteria.

The set-point response criteria explained above are based on a single point of the response curve. Other closed-loop performance metrics include the integral of error criteria, which focus on deviation from set point, and thus characterise the

Table 5.2 Typical integral error criteria

Criterion	Formula	Comment
Integral of the squared error (ISE)	$\int_0^\infty e^2(t)\,\mathrm{d}t$	Very aggressive criterion because squaring the error term provides a greater punishment for large error.
Integral of the absolute value of the error (IAE)	$\int_0^\infty \lvert e(t)\rvert\,\mathrm{d}t$	Tends to produce controller settings that are between those for the ITAE and ISE criteria.
Integral of the time-weighted absolute error (ITAE)	$\int_0^\infty t\lvert e(t)\rvert\,\mathrm{d}t$	Most conservative of the error criteria; the multiplication by time gives greater weighting to error that persists over a longer passage of time.
Integral of multiplied absolute error (ITNAE)	$\int_0^\infty t^n\lvert e(t)\rvert\,\mathrm{d}t$	Most conservative of the error criteria; the multiplication by time gives greater weighting to error that persists over a longer passage of time.
Quadratic error (QE)	$\int_0^\infty [e^2(t)+\rho u^2(t)]\,\mathrm{d}t$	Standard criterion used for optimal control design. ρ is a weighting factor.

entire closed-loop response curve. It is common to use some features of the control error, typically extrema (e.g. maximum error, time where maximum error occurs, asymptotes, areas); see Table 5.2. Shinskey (1996) argues that the IAE value is a good economic performance measure because the size and length of the error in either direction is proportional to lost revenue.

5.2 Controller Assessment Based on Set-Point Response Data

Performance indices based on set-point data have been developed by Swanda and Seborg (1997, 1999) to characterise the performance of PID-type feedback control loops. Index values are determined to indicate the transition point from satisfactory control to unsatisfactory control for various control objectives. The method will be outlined in the following.

5.2.1 Normalised Criteria

As mentioned in Sect. 5.1, two traditional performance criteria are the settling time T_{set} and the integral of absolute errors IAE. However, the absolute values of T_{set} and IAE give little indication of control loop performance without relation to the process dynamics. The rationale of Swanda and Seborg's approach is to compare the achieved performance with that of a PI controller tuned with the IMC rule based on a FOPTD process model:

$$G(s) = \frac{K_{\rm p}{\rm e}^{-T_{\rm a}s}}{Ts+1}, \tag{5.1}$$

where $K_{\rm p}$ is the static process gain, $T_{\rm a}$ the apparent time delay, and T is the (apparent) time constant or lag. The term "apparent" is used to emphasise that the parameters are approximate. $\tau_{\rm a}$ is a simple mean to characterise the net time delay, right-half-plane zeros and system order. In this context, two important performance indications, namely, the normalised versions of the settling time $T_{\rm set}$ and the IAE are considered:

$$T^*_{\rm set} = \frac{T_{\rm set}}{T_{\rm a}}, \tag{5.2}$$

$$IAE_{\rm d} = \frac{IAE}{|\Delta r|T_{\rm a}}, \tag{5.3}$$

where Δr is the size of the set-point step change. Both criteria are related to each other by (Swanda and Seborg 1999):

$$IAE_{\rm d} \approx \frac{T^*_{\rm set}}{2.30} + 0.565 \quad \text{for } T^*_{\rm set} \geq 3.30. \tag{5.4}$$

Under this condition, the corresponding gain margin $A_{\rm m}$ and phase margin $\varphi_{\rm m}$ can also be expressed as functions of $T^*_{\rm set}$ by

$$A_{\rm m} = \frac{\pi}{2} IAE_{\rm d} = \frac{\pi}{2}\left(\frac{T^*_{\rm set}}{2.30} + 0.565\right), \tag{5.5}$$

$$\varphi_{\rm m} = \frac{\pi}{2} - \frac{1}{IAE_{\rm d}} = \frac{\pi}{2} - \frac{1}{\frac{T^*_{\rm set}}{2.30} + 0.565}. \tag{5.6}$$

These relationships have been derived by fitting the parameters to the analytical solutions for different models controlled by an IMC-PI controller and for a settling time defined at $y = 0.9\Delta r$. Swanda and Seborg (1999) claimed that they are accurate enough and applicable to other process models. Given $T^*_{\rm set}$, Eqs. 5.5 and 5.6 can be used to throw light on the performance and robustness of the control system. These equations clearly indicate that if $T^*_{\rm set}$ is large, then the gain and phase margins are large, leading to better robustness at the cost of poor performance.

5.2.2 Assessment Methodology

Optimal $T^*_{\rm set}$ and $\text{IAE}_{\rm d}$ values have been determined by Swanda and Seborg (1999) for different representative models and serve as benchmarks for control system performance. Controller settings which minimise $T^*_{\rm set}$ and $\text{IAE}_{\rm d}$ criteria were determined using the MATLAB Optimization Toolbox.

Table 5.3 Swanda and Seborg's performance classes for PI control

Class	Dimensionless settling time T^*_{set}	Overshoot α [%]
High performance	$T^*_{set} \leq 4.6$	–
Fair/Acceptable performance	$4.6 < T^*_{set} \leq 13.3$	–
Excessively sluggish	$T^*_{set} > 13.3$	≤ 10
Aggressive/Oscillatory	$T^*_{set} > 13.3$	> 10

To quantify how far a PI controller is from the best achievable performance and to identify poorly performing control loops, different performance classes are defined, as given in Table 5.3. For a particular class, both conditions in Table 5.3 should be met. However, this definition can be relaxed to a single bound if one performance index is favoured over another (Swanda and Seborg 1999).

The assessment strategy can be interpreted as follows:

- An overshoot value of 10 % is used to distinguish between the excessively sluggish and poorly tuned controllers. A characteristic of a detuned controller is that it has little or no overshoot. No overshoot is a definitive indication of sluggish control. Therefore, if $\alpha > 10$ % and the upper bounds of T^*_{set} and IAE_d are exceeded, then the controller is considered to be poorly tuned. In fact, an $\alpha \leq 10$ % bound can be applied regardless of the values of the normalised performance indices.
- If the best achievable performance of PI control is desired, then the controller should be re-tuned if the calculated index values are outside those for the high-performance class. Furthermore, determining if a controller has the best achievable performance is useful, because if this ideal limit does not meet manufacturing specifications, then re-tuning the PI controller will not solve the problem. In this situation, a more advanced controller, such as a model predictive controller, would need to be considered.

Furthermore, the approximate relationships in Eqs. 5.5 and 5.6 can be used to determine robustness benchmarks for the current level of performance. For example, if $T^*_{set} = 5$, the corresponding benchmark values for A_m and φ_m, are 4.3 and 69°. If it is determined that the A_m and φ_m values are significantly less than the benchmarks, then the controller has a poor performance-robustness trade-off and re-tuning of the controller is advisable.

The assessment procedure described here can be summarised in the following procedure.

Procedure 5.1 Performance assessment based on dimensionless settling time.

1. Carry out a set-point step experiment with the closed loop.
2. Identify the values of apparent time delay T_a, settling time T_{set} and overshoot α from collected output data.
3. Calculate the normalised settling time T^*_{set} value.
4. Use Table 5.3 to assess the control performance.
5. Calculate the corresponding values of gain margin A_m and phase margin φ_m (Eqs. 5.5 and 5.6, respectively) and assess the performance-robustness trade-off.

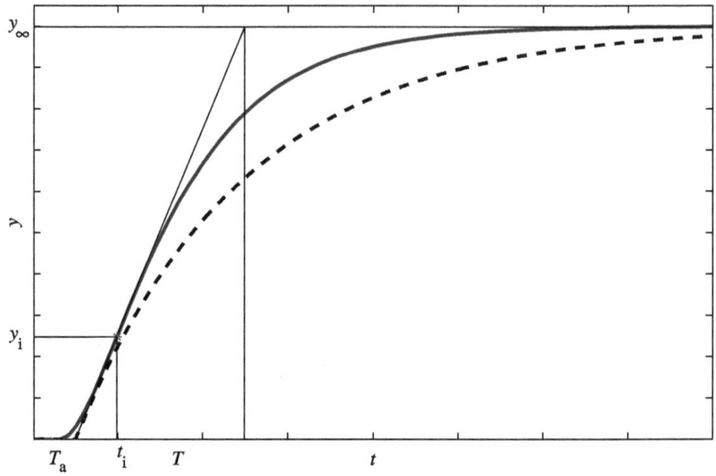

Fig. 5.2 Approximation of a step response using the tangent method

5.2.3 Determination of Time Delay from Step Response

A key point in the performance assessment based on dimensionless settling time is the determination of the apparent time delay τ_a. For this purpose, the step response is approximated by a FOPTD (Eq. 5.1) or a second-order-plus-time-delay (SOPTD) model

$$G(s) = \frac{K_p e^{-T_a s}}{\frac{s^2}{\omega_0^2} + \frac{2D}{\omega_0}s + 1}, \tag{5.7}$$

depending on the damping behaviour. Pragmatically, one can first try a SOPTD approximation. If the damping coefficient value is higher than unity, a FOPTD model may be sufficient.

5.2.3.1 FOPTD Approximation

When the system response is over-damped, it is sufficient to approximate the step response by a FOPTD. There are many methods to generate such approximations. One of the first methods, the tangent method, was described by Ziegler and Nichols (1942) for systems with essentially monotone step responses; see Fig. 5.2.

The tangent method for obtaining the time constant suffers from using only a single point to estimate the time constant. Use of several points from the response may provide a better estimate. Strejc (1959) proposed to use two data points $A(t_1, y_1)$ and $B(t_2, y_2)$, e.g. $A(t_{20\,\%}, y_{20\,\%})$ and $B(t_{85\,\%}, y_{85\,\%})$ of the measured reaction curve to

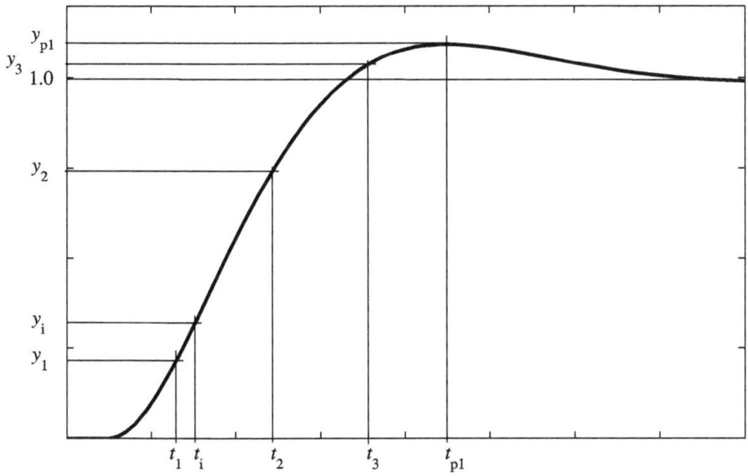

Fig. 5.3 Typical step-response data points used for approximation of an under-damped second-order-plus-time-delay process

give the model parameters:

$$\hat{T} = \frac{t_2 - t_1}{\ln(\frac{y_\infty - y_1}{y_\infty - y_2})}, \tag{5.8}$$

$$\hat{T}_a = t_1 + \hat{T}\ln\left(1 - \frac{y_1}{y_\infty}\right) \quad \text{or} \quad \hat{T}_a = t_2 + \hat{T}\ln\left(1 - \frac{y_2}{y_\infty}\right). \tag{5.9}$$

5.2.3.2 SOPTD Approximation

This is the right option when the step response is under-damped. There are many methods available in the literature to perform this approximation, e.g. Yuwana and Seborg (1982), Rangaiah and Krishnaswamy (1994, 1996). This technique belongs to those involving allowing $y(t)$ and its model approximation to intersect at two to five points, including the point of inflection (t_i, y_i) and the first peak (t_{p1}, y_{p1}), as shown in Fig. 5.3. In the context of controller assessment based on set-point response data, we found that the three-point approximation method by Rangaiah and Krishnaswamy (1996) is the most reliable approach.

The maximum point is used to calculate the maximum overshoot M_{p1} and the damping coefficient D as (Huang and Chou 1994)

$$M_{p1} = \frac{y_{p1} - y_\infty}{y_\infty}, \qquad \hat{D} = \sqrt{\frac{\ln^2 M_{p1}}{\pi^2 + \ln^2 M_{p1}}}, \tag{5.10}$$

respectively.

Now that an estimate for D is available, the task reduces to a two-parameter problem of estimating τ_a and T so as to achieve a good model fit. For this purpose, the step response is normalised to a gain of unity ($y^*(t)$) and with respect to T, i.e. $t^* = (t - T_\mathrm{a})/T$. The three-point method (Rangaiah and Krishnaswamy 1996) requires values of $(t^*_{\mathrm{p1}}, y^*_{\mathrm{p1}})$, $(t^*_\mathrm{i}, y^*_\mathrm{i})$ and (t^*_2, y^*_2) and performing the following procedure.

Procedure 5.2 Estimating second-order plus time delay parameters using the three-point method of Rangaiah and Krishnaswamy (1996).

1. Locate the first peak $(t_{\mathrm{p1}}, y_{\mathrm{p1}})$ from the measured system response and estimate the damping ratio from Eq. 5.10.
2. Compute y^*_i from an analytical expression derived by Huang and Clements (1982)
$$y^*_\mathrm{i} = 1 - \frac{1}{\sqrt{1-\hat{D}^2}} \exp\left(-\frac{\hat{D}}{\sqrt{1-\hat{D}^2}} \arctan\frac{\sqrt{1-\hat{D}^2}}{\hat{D}}\right) \sin\left(2\arctan\frac{\sqrt{1-\hat{D}^2}}{\hat{D}}\right), \tag{5.11}$$
read off the corresponding t_i from the measured system response, and calculate
$$t^*_\mathrm{i} = \frac{1}{\sqrt{1-\hat{D}^2}} \arctan\frac{\sqrt{1-\hat{D}^2}}{\hat{D}}. \tag{5.12}$$
3. Estimate y^*_2 from the empirical relation
$$y^*_2 = 1.8277 - 1.7652\hat{D} + 0.6188\hat{D}^2, \tag{5.13}$$
read off the corresponding t_2 from the measured system response, and calculate
$$t^*_2 = 3.4752 - 1.3702\hat{D} + 0.1930\hat{D}^2. \tag{5.14}$$
4. Evaluate $\hat{T}$ from
$$\hat{T} = \frac{t_2 - t_\mathrm{i}}{t^*_2 - t^*_\mathrm{i}}. \tag{5.15}$$
An estimate for the time delay is then obtained from
$$\hat{T}_\mathrm{a} = t_\mathrm{i} - t^*_\mathrm{i}\hat{T}. \tag{5.16}$$

Note that this approximation method is particularly suited for under-damped processes, i.e., $0.4 < D < 0.8$. Outside this range, the following methods could be used: (i) for $D < 0.4$, i.e., when oscillations are significant, the method by Yuwana and Seborg (1982) may be performed; (ii) for $D > 0.8$, i.e., when the response is sluggish, the method by Rangaiah and Krishnaswamy (1994) is recommended.

5.2.3.3 Optimisation-Based Approximation

The main problem of the application of most FOPTD- and SOPTD-approximation methods is their high sensitivity to noise in the measured output signal. Also, the techniques will fail when the step response is not complete, i.e., the steady state is not reached, or even when the system input is not an ideal step, e.g. a steep ramp. A method that works well despite the presence of noise is to fit a FOPTD or SOPTD to the measured response data based on an optimisation algorithm, e.g. Nelder–Mead minimisation using the routine fminsearch of the MATLAB Optimization Toolbox. This means to identify the parameters of a FOPTD or SOPTD model, which minimise the objective function

$$V(\boldsymbol{\Theta}) = \sum_{k=1}^{N} \left[y(k) - \hat{y}(k|\boldsymbol{\Theta}) \right]^2 \tag{5.17}$$

with $\boldsymbol{\Theta} = [K_\mathrm{p}, T, T_\mathrm{a}]^\mathrm{T}$ for a FOPTD model and $\boldsymbol{\Theta} = [K_\mathrm{p}, T, D, T_\mathrm{a}]^\mathrm{T}$ for a SOPTD model. It is then recommended to use the estimated response data for the calculation of the settling time and overshoot, rather than considering the measured (noisy) response data.

5.2.4 Application Examples

5.2.4.1 Simulation Example

We consider a process represented by the model ($T_\mathrm{s} = 1$ s)

$$y(s) = \frac{5\mathrm{e}^{-9s}}{(10s+1)(s+1)} u(s) + \frac{1}{(10s+1)(s+1)} \varepsilon(s), \tag{5.18}$$

controlled by a PI controller with the settings $K_\mathrm{c} = 0.144$ and $T_\mathrm{I} = 6$ s. Some approximation methods have been applied to give the SOPTD model parameters in Table 5.4. Overall, the optimisation-based method seems to yield the best approximation (Fig. 2.4). Having in mind the aforementioned advantages, it should always be the preferred method. Note that the obtained models approximate the closed-loop behaviour and should not be confused with the parameters of process model in Eq. 5.18.

Figure 5.4 also contains the set-point-response-based performance assessment results. Whatever the approximation method is used, the results clearly indicate oscillatory/aggressive controller tuning (look at Table 5.3). The corresponding gain and phase margin are $A_\mathrm{m} \approx 16$ and $\varphi_\mathrm{m} \approx 85°$, respectively, indicating high robustness.

Table 5.4 Set-point-response approximation and assessment results for the simulation example

Approximation method	Estimated SOPTD parameters				Assessment results		
	K_p	ω_0	D	T_a	T^*_{set}	IAE_{d}	α
Yuwana and Seborg (1982)	1.0	0.13	0.11	11.0	25.7	11.5	67.5
Rangaiah and Krishnaswamy (1996)	1.0	0.18	0.09	13.0	20.5	9.3	74.7
Optimisation-based	1.0	0.11	0.12	6.6	23.1	10.4	69.5

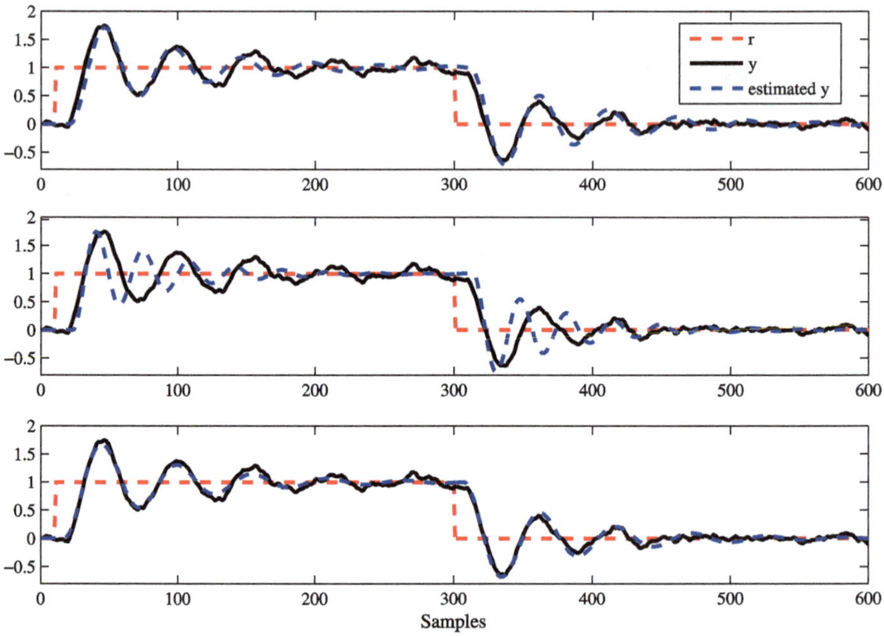

Fig. 5.4 Simulated and estimated step responses used for performance assessment of the control loop in the simulation example (*top*: Yuwana and Seborg (1982); *middle*: Rangaiah and Krishnaswamy (1996); *bottom*: optimisation-based)

5.2.4.2 Industrial Example

The set-point assessment technique is applied to data shown in Fig. 5.5, which were gathered from a flow control loop in a pulp mill. The set point was changed stepwise while the gain K_c of the PI controller was changed from 0.2 to 0.04 at time 840 s and to 0.35 at time 1690 s. The integral time was $T_I = 9$ s in all cases. It is known from the loop that the time delay varies rather much (Horch and Stattin 2002). FOPTD models have been estimated by the optimisation function `fminsearch`, and the estimated step responses (Fig. 5.6) were used to calculate the settling times and overshoots. In contrast, the approximation methods in Sect. 5.2.3.1 fail, as the step responses are too noisy. The performance assessment results (Procedure 5.1) are summarised in Table 5.5 for the different data segments with different controller

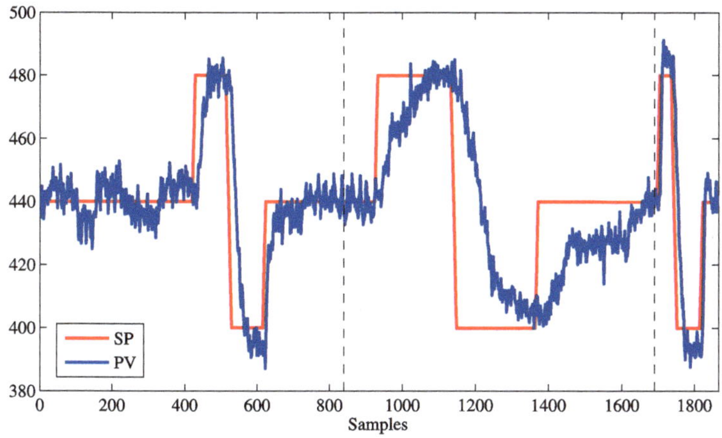

Fig. 5.5 Data from a flow control loop. The instants of controller re-tuning are denoted by *vertical dashed lines*

Table 5.5 Set-point-response-based assessment results for the flow control loop

Data segment no.	T_a	T^*_{set}	IAE_{d}	α	Assessment
1	14.8	3.0	2.1	0	High performance
2	26.7	6.0	3.3	0	Fair performance
3	7.8	2.8	2.0	0	High performance

settings. The assessment results found are in good agreement with the knowledge about the control loop.

Note that similar results have been achieved by Horch and Stattin (2002), who identified Kautz models to compute the settling times and overshoots of the closed-loop step responses and Laguerre models to estimate the time delays; see also Sects. 7.2.5 and 7.3.1.

5.3 Idle Index for Detecting Sluggish Control

Very conservative tuning, usually with fixed controller settings, is often established in the process industry due to lack of time to optimise controllers during commissioning of control systems. When operating conditions change, this results in sluggish control and thus unnecessarily large and long deviations from the set point remain. Therefore, it is often beneficial to detect sluggish control loops using tailored methods for this purpose. This section reviews the idle-index technique proposed by Hägglund (1999) and discusses some key aspects when the method is applied to real-world data.

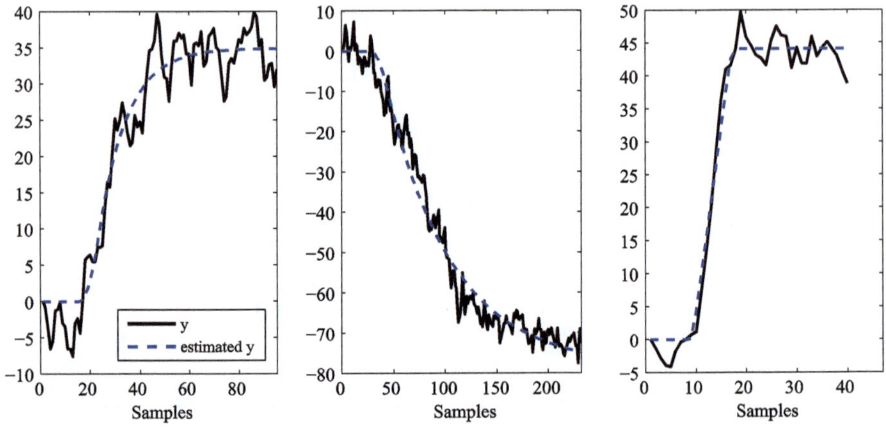

Fig. 5.6 Measured and estimated step responses used for performance assessment of the flow control loop. Compared to Table 5.5, the data are mean-centred for each window prior to identification

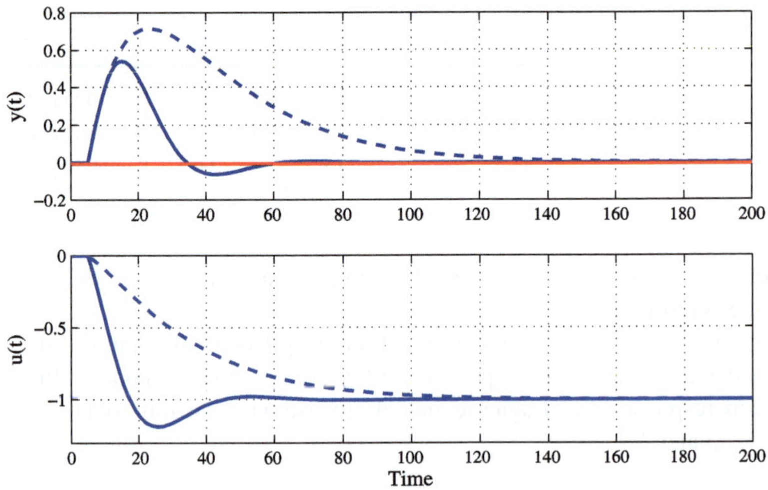

Fig. 5.7 Good (*solid*) and sluggish (*dash*) control of load disturbances

5.3.1 Characterisation of Sluggish Control

Figure 5.7 shows two responses to load disturbances in the form of step changes at the process input. One response is good, with a quick recovery with small undershoot. The second response, however, is very sluggish. Both responses have an initial phase where the two signals go in opposite directions, i.e. $\Delta u \Delta y < 0$, where Δu and Δy are the increments of control signal and the process output, u and y, respectively. Characteristic for the sluggish response is that, after this initial phase, a very long time period occurs, where the correlation between the two signal increments is positive.

5.3.2 Idle Index

Hägglund (1999, 2005) presented a method to detect sluggish control loops by using the so-called *idle index*. It assesses the time for a loop needed to recover from a stepwise load disturbance. The idle index describes the relation between times of positive and negative correlation between the control signal and the process output increments, Δu and Δy, respectively.

The idle index (I_i) is defined as

$$I_\mathrm{i} = \frac{t_\mathrm{pos} - t_\mathrm{neg}}{t_\mathrm{pos} + t_\mathrm{neg}} \tag{5.19}$$

for loops with a positive gain, and

$$I_\mathrm{i} = \frac{-t_\mathrm{pos} + t_\mathrm{neg}}{t_\mathrm{pos} + t_\mathrm{neg}} \tag{5.20}$$

for loops with a negative gain.

To form the index, the time periods when the correlations between the signal increments are positive and negative, respectively, are calculated first. The following quantities are updated at every sampling instant:

$$\begin{aligned} t_\mathrm{pos} &= \begin{cases} t_\mathrm{pos} + T_\mathrm{s} & \text{if } \Delta u \Delta y > 0, \\ t_\mathrm{pos} & \text{if } \Delta u \Delta y \le 0, \end{cases} \\ t_\mathrm{neg} &= \begin{cases} t_\mathrm{neg} + T_\mathrm{s} & \text{if } \Delta u \Delta y < 0, \\ t_\mathrm{neg} & \text{if } \Delta u \Delta y \ge 0, \end{cases} \end{aligned} \tag{5.21}$$

where T_s is the sampling time.

I_i is bounded in the interval $[-1, 1]$. A positive value of I_i close to 1 means that the control is sluggish. Values close to 0 indicate that the controller tuning is reasonably good. Negative values of I_i close to -1 may imply well-tuned control, but can also be obtained for oscillatory control as well. Therefore, it is necessary to combine the idle index calculation with an oscillation-detection procedure to make the right decision. Also, the idle index tool requires that the set point is available to exclude periods when the excitation is caused by set-point changes.

5.3.3 Practical Conditions and Parameter Selection

When the methodology described in Sect. 5.3.2 has to be applied in practical cases, there are many technical problems to be solved. It was Kuehl and Horch (2005) who revealed these issues and suggested appropriate data-processing techniques.

5.3.3.1 P-Only Control

It is evident that the idle index is not designed to reasonably assess the control performance of loops with P-only controllers (luckily not very frequent in the process industries). This becomes clear when considering the controller equation for P-control $\Delta u = -K_c \Delta y$. All increments of y and u will be of different sign, no matter how fast or sluggish the controller performance may be. The idle index for data from loops with pure P-control will hence always have a value close to -1.

5.3.3.2 Need for Signal Filtering

The procedure is very sensitive to noise, since increments of the signals are studied. Without proper filtering, the index *completely fails*. To find a suitable filter time-constant, it is necessary to have some information about the process dynamics. When the analysis is done offline, non-causal (zero-phase) filtering [filtfilt] should be applied.

5.3.3.3 Low-Pass Filtering

Low-pass filtering is the simplest and most common choice. To keep sudden changes in the signals, it is necessary to permanently supervise the deviation between the input x and output x_f of the filter. As soon as the deviation exceeds a certain limit ε, this should be interpreted as a relevant change in the variable and the filter is re-initialised with the current signal value:

$$x_f(k) = x(k) \quad \text{if } \left|x(k) - x_f(k)\right| > \varepsilon_0. \tag{5.22}$$

Kuehl and Horch (2005) proposed the heuristic rule $0.4\sigma_\varepsilon < \varepsilon_0 < 0.6\Delta d$, where σ_ε is the standard deviation of the noise, and Δd the typical size of load disturbances. A noise-level estimate can be determined as the standard deviation of the prediction error resulting from fitting an AR(MA)X model to the data.

5.3.3.4 Regression Filtering

Regression filtering is also a simple method for suppressing noise. First, the data set is split up into segments containing single load disturbances by using a simple disturbance detection algorithm that begins collecting data once the disturbance is larger than a given threshold ε, as mentioned above, and stops when reaching the consecutive disturbance. There, the next segment begins. An ordinary regression is then performed within each segment by fitting a polynomial [polyfit & poylval] of default order $n = 10$ to the data in the least-squares sense. The polynomial order surely is subject to further optimisation. Yet, orders higher than 10 tend to overfit the data, and very small orders produce rather slurry results (Kuehl and Horch 2005). This procedure acts very similar to the low-pass filter with re-initialisation but suppresses noise more rigorously.

5.3.3.5 Wavelet Denoising

Wavelet denoising is a much more involved method but does not show any big advantage compared with the other techniques in the considered context. For details of this approach, consult Kuehl and Horch (2005) and the references cited therein.

5.3.3.6 Exclusion of Steady-state Data

The performance indications drawn from the idle index assume that the load disturbances are step changes or at least abrupt changes—a reasonable assumption in many situations, since load changes are often caused by sudden changes in production. However, if the load disturbances are varying slowly, the idle index may become positive and close to one even in situations where the control is not sluggish (Hägglund 2005). To avoid this, it might be advantageous to calculate the idle index only during periods when there are abrupt load changes, which may be accomplished using load-detection procedures; see Hägglund (1995) and Hägglund and Åström (2000).

Calculations of the idle index near the steady state should thus be avoided, when the signal-to-noise ratio is small. A simple way to ensure this is to perform the calculations only when

$$|e| > e_0, \tag{5.23}$$

where e_0 is a threshold based on a noise-level estimate or fixed to a few percent of the signal range. The exclusion of steady-state data can also be done by the algorithm proposed by Cao and Rhinehart (1995).

5.3.3.7 Signal Quantisation

Quantisation can be described as

$$x_{\mathrm{quant}} = q\,\mathrm{round}\left(\frac{x_{\mathrm{f}}}{q}\right), \tag{5.24}$$

where x_{quant} is the quantified signal, x_{f} the filtered signal, and q the quantisation level. The choice of q is suggested to follow the heuristic rule $q = [0.05\text{–}0.1]y_{\max}$, where $y_{\max}$ is the maximum change of the output variable due the disturbance step (Kuehl and Horch 2005).

In our experience, quantisation should not be applied when the signals are properly filtered, e.g. using a zero-phase filter with appropriately selected cut-off frequency. In such a case, quantisation would lead to an artificial increase of the idle index and thus may be misleading; see also the example in Sect. 5.3.4. This finding is a consequence of the fact that the choice of the quantisation level is a matter of how much noise is left in the signals.

5.3.3.8 Combined Methods

A number of methods can be combined for handling noise and avoiding misleading results. A suitable data pre-processing procedure proposed by Kuehl and Horch (2005) looks as follows:

1. Filtering with
   (a) re-initialised low-pass filter or
   (b) linear regression with re-initialisation or
   (c) wavelet denoising.
2. Exclusion of steady-state data.
3. Quantisation.

5.3.4 Illustrative Example

A FOPTD model described as $G(s) = 1/(10s + 1)\mathrm{e}^{-5s}$ is perturbed with a single stepwise load disturbance of amplitude 1 at the process entry. White noise with a variance of 0.002 has been added to the process output. The process is controlled with a sluggishly tuned PI controller with $K_{\mathrm{c}} = 0.8$ and $T_{\mathrm{I}} = 30.0$. The process output y and controller output u (Fig. 5.8) are subject to an idle index evaluation. In a first step, zero-phase filtering and linear regression, as described in Sect. 5.3.3, are applied, respectively. The results[1] can be seen in Fig. 5.9 and Fig. 5.10. Finally, the processed data is quantised as described in Sect. 5.3.3.7 with the same quantisation interval chosen as $q = 0.001$ for all data sets.

[1]MATLAB software implementing this method was provided by A. Horch and was used for this example.

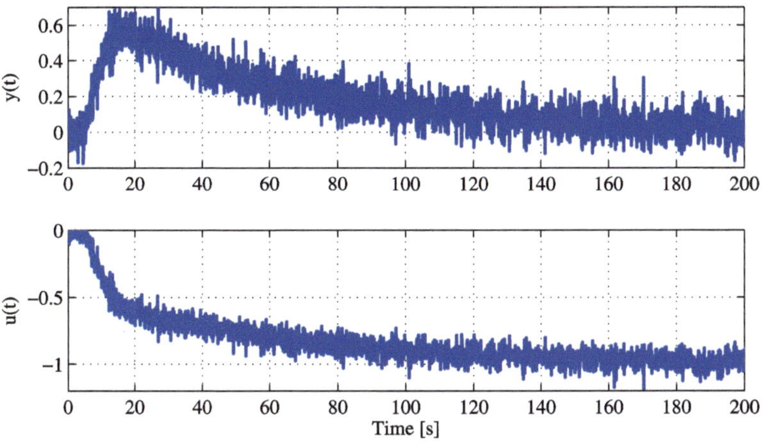

Fig. 5.8 Simulated data corrupted with noise

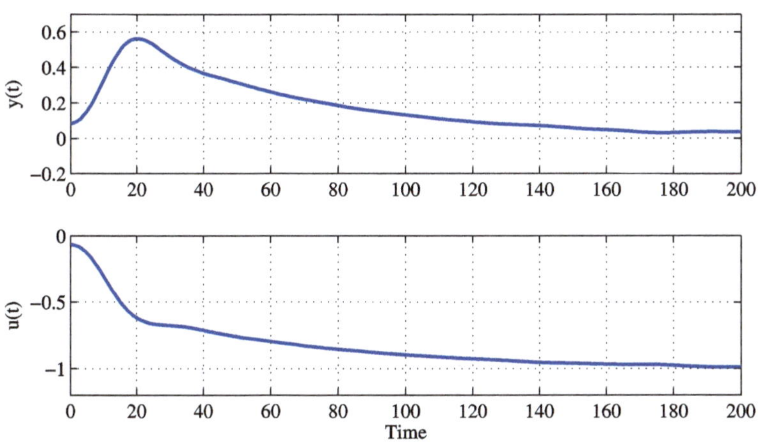

Fig. 5.9 Simulated data after zero-phase filtering

For some steps in the signal processing chain, the corresponding idle index value has been calculated. This example confirms that the idle index calculation completely fails when using noisy data. A comparison to the value in the noise-free case then reveals improvements due to signal processing. All idle index values are summarised in Table 5.6. Apparently, quantisation does not really help in this example, bearing in mind the danger of using a higher q than required. The best performing method is zero-phase filtering. However, this method can only be applied offline, as the filter is non-realisable.

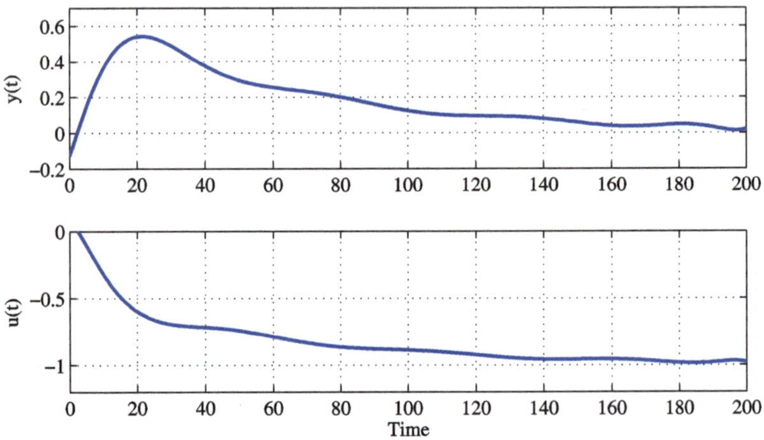

Fig. 5.10 Simulated data after regression filtering

Table 5.6 Idle index values after some stages of data processing

Condition/processing method	Idle index value
No noise	**0.82**
Noisy	−0.99
Zero-phase filtering ($\omega_c = 0.008$)	**0.72**
Regression filtering ($n = 10$)	**0.56**
Low-pass filtering	−0.96
Regression filtering & quantisation with $q = 0.001$	0.28
Low-pass filtering & quantisation with $q = 0.001$	−0.11

5.4 Assessment of Load Disturbance Rejection Performance

The aim of the methodology proposed by Visioli (2005) is to verify, by evaluating an abrupt load disturbance response, if the tuning of the adopted PI controller guarantees good load-disturbance rejection performance. The IAE criterion is thus used to ensure a low error magnitude and a stable response, i.e. low settling time, at the same time (Shinskey 1996).

5.4.1 Methodology

Visioli's technique is based on the analysis of the control signal when an abrupt load disturbance occurs on the process, and it aims to estimate a generalised damping index of the closed-loop system. The performance index proposed is called the *area index* (*AI*) and is based on the control signal $u(t)$ that compensates for a step load

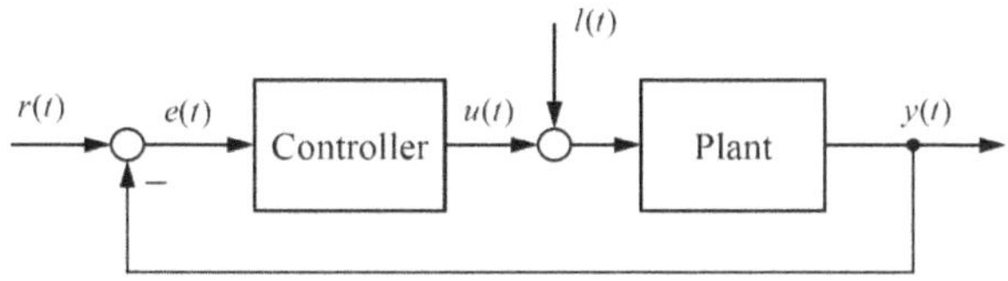

Fig. 5.11 Closed loop with acting load disturbance $l(t)$

disturbance occurring on the process; see Fig. 5.11. The value of the area index then decides whether it can be deduced if the control loop is too oscillatory.

The area index is calculated as the ratio between the maximal value of the determined areas (Fig. 5.12) and the sum of them, excluding the area A_0, i.e. the area between the time instant in which the step load disturbance occurs and the first time instant at which $u(t)$ attains u_0. Formally, the area index is defined as

$$I_\mathrm{a} := \begin{cases} 1 & \text{if } N < 3, \\ \frac{\max(A_1,\ldots,A_{N-2})}{\sum_{i=1}^{N-1} A_i} & \text{elsewhere,} \end{cases}$$
$$A_i = \int_{t_i}^{t_{i+1}} \left|u(t) - u_0\right| \mathrm{d}t, \quad i = 0, 1, \ldots, N-1, \tag{5.25}$$

where u_0 denotes the new steady-state value achieved by the control signal after the transient load disturbance response, t_0 the time instant in which the step load disturbance occurs, $t_1, \ldots, t_{N-1}$ the subsequent time instants, and t_N the time instant in which the transient response ends and the manipulated variable attains its steady-state value u_0. From a practical point of view, the value of t_N can be selected as the minimum time after that $u(t)$ remains within a $p\%$ (e.g. 1 %) range of u_0.

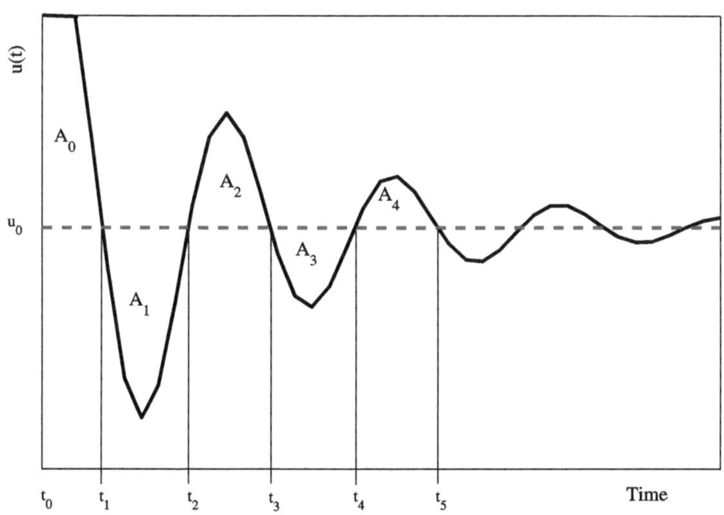

Fig. 5.12 Significant parameters for determining the area index

Table 5.7 Visioli's performance-assessment rules for PI controllers

	$I_i < -0.6$ (low)	$I_i \in [-0.6, 0]$ (medium)	$I_i > 0$ (high)
$I_a > 0.7$ (high)	K_c too low	K_c too low, T_I too low	K_c too low, T_I too high
$I_a \in [0.35, 0.7]$ (medium)	**K_c ok, T_I ok**	K_c too low, T_I too low	–
$I_a < 0.35$ (low)	K_c too high and/or T_I too low	T_I too high	T_I too high

The area index can be combined with other indices to assess the performance of PI controllers. Based on the results obtained, the rules presented in Table 5.7 have been devised by Visioli (2005) to assess the tuning of the PI settings. The value of the area index is considered to be low if it is less than 0.35, medium if it is $0.35 < I_a < 0.7$ and high if it is greater than 0.7. The value of the idle index is considered to be low if it is less than −0.6, medium if it is $-0.6 < I_i < 0$ and high if it is greater than zero.

From Eq. 5.25 it can be deduced that the value of the area index is always in the interval (0, 1]. Also, Visioli (2005) showed that the index can be related to the damping factor D of the closed-loop transfer function from the load disturbance signal $l(t)$, acting at the process input, to the controller output. It has been concluded that the more the value of AI approaches zero, the more the control loop is oscillatory, whilst the more the value of AI approaches unity, the more the control loop is sluggish. Therefore, a well-tuned controller gives a medium value of AI.

When both the values of the area index and of the idle index are low, it is not possible to take decision based on both indices. In this situation, it is convenient to evaluate the following output index:

$$I_o := \begin{cases} 0 & \text{if } N < 1, \\ \frac{\sum_{i=1}^{N-1} A_{i,\text{negativ}}}{\sum_{i=1}^{N-1} A_i} & \text{elsewhere} \end{cases} \tag{5.26}$$

with

$$A_i = \int_{t_i}^{t_{i+1}} \left| y(t) - y_0 \right| \mathrm{d}t \quad i = 0, 1, \ldots, N-1. \tag{5.27}$$

In case $I_o < 0.35$ it can be concluded that both the proportional gain and integral time constant values are too high. Otherwise, the oscillatory response is caused by a too high value of K_c and/or a too low value of T_I. In the latter case, experience suggested to decrease the value of the proportional gain anyway (Visioli 2005).

5.4.2 Practical Conditions

As all control-error-area-based methods, the area index determination is extremely sensitive to noisy signals. As the area index is (usually) determined off-line, a standard filtering procedure can be applied before calculating the different areas. Alternatively, the concept of noise band, successfully applied in industry (Shinskey 1996), can be sufficient. This means to discard those areas A_i whose value is less than a pre-defined threshold from the analysis, because they are actually due to the noise. This threshold can be determined by considering the control signal for a sufficiently long time interval when the process is at an equilibrium point and by determining the maximum area between two consecutive crossings with respect to its steady state value. The latter can be calculated as the mean value of the control signal itself in the considered time interval.

Another aspect that has to be taken into account is that the area index is significant only when an abrupt load change occurs, i.e., when the load change is fast enough with respect to the dynamics of the complementary sensitivity function. Thus, the method has to be applied only in these situations, e.g., when a sudden change in the production occurs or, obviously, when a step signal is deliberately added to the manipulated variable for this purpose. Otherwise, a higher value of the index might result (Visioli 2005). To verify that this condition applies or to detect those phases with significant load-disturbance changes, many methods exist, which can be applied, e.g. Hägglund (1995) and Hägglund and Åström (2000).

Among the different indices proposed in the literature, the merit of the area index is to provide an indication on in which direction the controller parameters have to be re-tuned. This will be used as the basis for a new iterative PI-controller tuning procedure presented in Sect. 14.4.1.

5.4.3 Illustrative Example

The following fourth-order process is considered:

$$G_{\mathrm{p}}(s) = \frac{1}{(s+1)^4}. \tag{5.28}$$

The optimal tuning has been given by Visioli (2005) as $K_{\mathrm{c}} = 1.65$ and $T_{\mathrm{I}} = 4.15$, leading to $I_{\mathrm{i}} = -0.80$ and $I_{\mathrm{a}} = 0.35$. Looking at Table 5.7 confirms that this tuning is optimal (K_{c} ok, T_{I} ok). In the first case, only the integrator time is increased to $T_{\mathrm{I}} = 9.0$. The resulting values of the indices are $I_{\mathrm{i}} = -0.16$ and $I_a = 0.03$, indicating that T_{I} is too high, which is the right conclusion. The case tried next is $K_{\mathrm{c}} = 2.7$ and $T_{\mathrm{I}} = 9.0$, giving $I_{\mathrm{i}} = -0.91$ and $I_{\mathrm{a}} = 0.24$. This signals that either K_{c} is too high and/or T_{I} is too low.

The same conclusion is obtained when we consider the tuning $K_{\mathrm{c}} = 2.2$ and $T_{\mathrm{I}} = 3.1$ (case 3), giving $I_{\mathrm{i}} = -0.75$ and $I_{\mathrm{a}} = 0.15$. It appears that the idle index and area index alone are not able to distinguish between the two last cases. However, the calculated output index values 0.05 (case 2) and 0.43 (case 3) indicate that both K_{c} and T_{I} are too high in case 2, and K_{c} is too high and/or T_{I} is too low in case 3. In this latter case, the value of K_{c} should be decreased anyway. The results of the whole case study are summarised in Table 5.8.

Table 5.8 Assessment results for the illustrative example

Case	K_c	T_I	I_i	I_a	I_o	IAE	Assessment
0	1.65	4.15	−0.80	0.35	–	2.79	K_c ok, T_I ok (optimal tuning).
1	1.65	9.00	−0.16	0.03	–	5.44	T_I too high.
2	2.70	9.00	−0.91	0.24	0.05	3.60	both K_c and T_I too high.
3	2.20	3.10	−0.75	0.15	0.43	5.24	K_c too high or T_I too low; decrease K_c anyway.

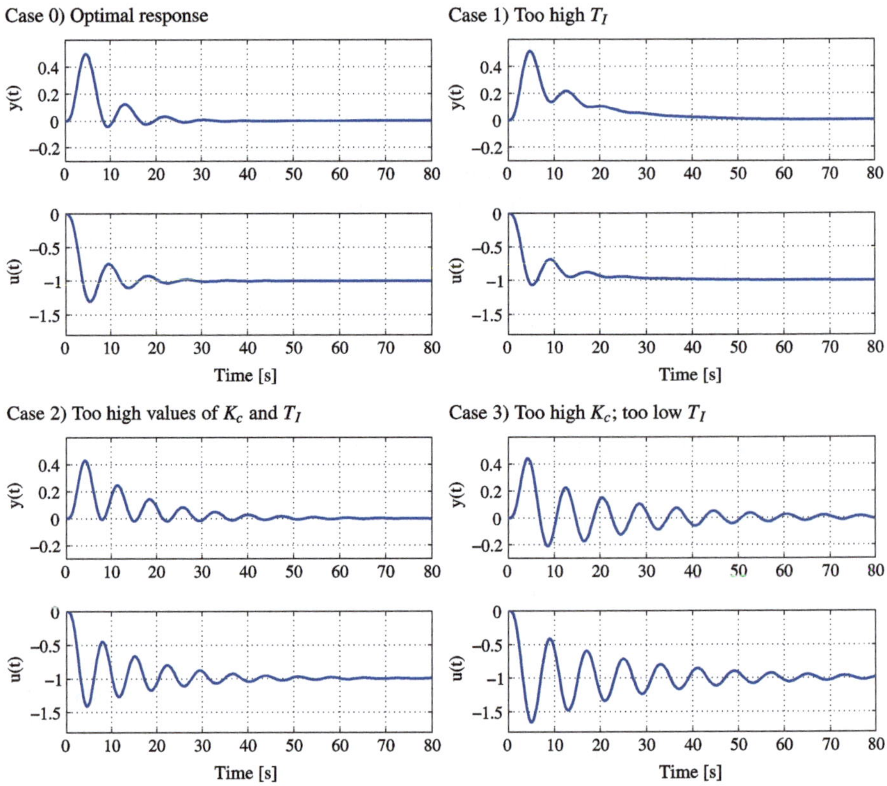

Fig. 5.13 Load disturbances responses for the different cases of the illustrative example

The unit step load disturbance responses, together with the corresponding manipulated variable signals, are plotted in Fig. 5.13. Note the difference between the step responses (y) in case 2 and case 3, which is characterised by the output index, which is based on computing the ratio of the sum of negative output areas to the overall sum of areas.

Table 5.9 Performance assessment results for example P1

	Minimum-variance assessment	Assessment based on set-point response data			Assessment of load-disturbance rejection performance	
Index	η	T^*_{set}	IAE_{d}	α	I_{a}	I_{i}
IMC1	0.76	3.6	2.4	0.0	1.0	−0.08
	Good performance	High performance			K_{c} too low, T_{I} too low	
IMC2	0.81	3.4	2.3	4.3	0.96	−0.63
	High performance	High performance			K_{c} too low	
Hägglund	0.65	27.5	12.2	18.6	0.92	−0.46
and Åström	Good performance	Oscillatory/aggressive			K_{c} too low, T_{I} too low	
ITAE	**0.59**	**9.1**	**4.6**	**61.3**	**0.68**	**−0.77**
(disturbance)	**Fair performance**	**Fair/acceptable performance**			**K_{c} ok, T_{I} ok**	
ITAE	0.81	3.4	2.3	4.8	0.95	−0.63
(Set point)	High performance	High performance			K_{c} too low	

5.5 Comparative Simulation Studies

The three approaches minimum-variance assessment (Sect. 2.4), deterministic assessment of set-point tracking (Sect. 5.2) and load-disturbance rejection (Sect. 5.4) are now applied and compared on simulation examples. For this purpose, three different processes are considered: the loop from Example 2.4 (P_1) and two loops discussed by Visioli (2006): the process in Eq. 5.28 (P_3) and that having the transfer function (P_2)

$$G_{\mathrm{p}}(s) = \frac{1}{10s+1}e^{-5s}. \tag{5.29}$$

The results are presented in Tables 5.9, 5.10 and 5.11.

The calculated performance indices confirm the known fact that IMC tuning is more suitable for set-point tracking than disturbance rejection. It is observed that PI controllers tuned for top load-rejection performance exhibit large values of overshoot and decay ratio, which results in long settling times. The optimal values for Visioli's indices are achieved when applying the ITAE (disturbance) tuning rule. It is expected that this rule gives a response that is very close to minimum variance for PI controllers.

A main lesson to be learned from the results of this example is that the control objective, including the expected type of disturbances, of the loop must guide the selection of the right assessment method. In other words, when assessing a controller with the three approaches, one can directly see for what purpose the control loop has really been tuned. Also, different tunings for the same objective can be compared to identify the best one during controller commissioning.

Table 5.10 Performance assessment results for example P2

	Minimum-variance assessment	Assessment based on set-point response data			Assessment of load-disturbance rejection performance	
Index	η	T^*_{set}	IAE_{d}	α	I_{a}	I_{i}
$K_{\mathrm{c}}=1.86$, $T_{\mathrm{I}}=20.0$	0.86	7.2	3.8	12.8	0.14	−0.12
	High performance	Fair/acceptable performance			T_{I} too high	
$K_{\mathrm{c}}=1.86$, $T_{\mathrm{I}}=10.36$	**0.82**	**7.0**	**3.7**	**38.3**	**0.61**	**−0.75**
	High performance	**Fair/acceptable performance**			**K_{c} ok, T_{I} ok**	
$K_{\mathrm{c}}=3.0$, $T_{\mathrm{I}}=20.0$	0.56	27.5	12.2	18.6	0.20	−0.90
	Good performance	Oscillatory/aggressive			Either K_c too high or T_I too low	

Table 5.11 Performance assessment results for example P3

	Minimum-variance assessment	Assessment based on set-point response data			Assessment of load-disturbance rejection performance	
Index	η	T^*_{set}	IAE_{d}	α	I_{a}	I_{i}
$K_{\mathrm{c}}=1.65$, $T_{\mathrm{I}}=4.15$	**0.89**	**9.3**	**4.7**	**27.6**	**0.35**	**−0.78**
	High performance	**Fair/acceptable performance**			**K_{c} ok, T_{I} ok**	
$K_{\mathrm{c}}=1.2$, $T_{\mathrm{I}}=2.0$	0.79	8.9	4.6	51.3	0.40	−0.51
	Good performance	Fair/acceptable performance			K_c too low, T_I too low	

5.6 Summary and Conclusions

Three deterministic methods for performance assessment have been presented, discussed and compared in this chapter. The first technique assesses the performance of PI controllers from closed-loop response data for a set-point step change. For this purpose, two dimensionless performance indices, the normalised settling time and the normalised integral of the absolute value of the error, are used. The methodology identifies poorly performing control loops, such as those that are oscillatory or excessively sluggish. This technique also provides insight concerning the performance-robustness trade-off inherent in the IMC tuning method and analytical relationships between the dimensionless performance indices, the gain margin and the phase margin. To work properly, it is necessary for the method to have accurate estimates of the apparent time delay, the settling time and the overshoot from step response. Methods for this purpose have been presented with the conclusion that it is recommended to identify the parameters from fitting a FOPTD or SOPTD model to the step response to avoid problems with noisy signals.

The idle index is an apparently simple indicator for sluggish control. It evaluates controller action due to significant, step-wise load disturbances with a focus on the transient behaviour of the control loop. However, in practical situations, where the signals are noisy and show different behaviour (steady-state, transients), the idle index completely fails. Therefore, careful pre-processing of the data, such as steady-state detection, filtering and signal quantisation, can be necessary. A set of techniques have been described to perform these tasks. Despite these pre-treatment measures, the existence of *distinct load step disturbances* is still decisive for the capability of detecting sluggish loops using the idle index. Moreover, a problem associated with the idle index is that a negative value close to -1 may be obtained both from a well-tuned loop or from an oscillatory loop. Thus, an oscillation detection technique (Chap. 8) has be combined with the idle index method to get the right indication.

The idle index method can be improved by considering additional indices, namely, the area index and the output index. This combination provides an efficient way to assess the tuning of PI controllers with respect to load disturbance rejection performance. It has been shown that the three indices together give valuable indication on how PI controller parameters, i.e. proportional gain and integral time, have to be modified to achieve better performance. Note that the same practical issues to be considered for the computation of the idle index are also relevant for the calculation of the area index. The method is particularly sensitive to noise, thus pre-filtering is essential.

From the comparative study presented we concluded that the control objective, including the expected type of disturbances, of the loop must guide the selection of the right assessment method. In other words, when assessing a controller with the different methods, one can directly see for what purpose the control loop has really been tuned. Also, different tunings for the same objective can be compared to pick up the best one during controller commissioning.

Chapter 6
Minimum-Variance Assessment of Multivariable Control Systems

The simplest approach for assessing multivariable control systems is to split the problem into p MISO (or SISO) systems and then apply single-loop assessment methods. This strategy, however, would only indicate the potential for performance improvement by adjusting the individual loops. Since the loops can be coupled, a multivariable control strategy can further reduce process variations; thus, only multivariable assessment can provide the right measure of performance improvement potential in the general case. In this chapter, methods for multivariable minimum variance benchmarking will be presented.

The chapter is organised as follows. Section 6.1 introduces the interactor matrix and gives ways to determine or estimate it. In Sect. 6.2, it is shown how to use the interactor matrix to derive the multivariable variant of MVC. Section 6.3 presents the FCOR algorithm (Huang et al. 1997a, 1997b), as the most known algorithm for assessing MIMO control systems based on routine operating data and the knowledge of the interactor matrix. As the interactor matrix is hard to determine, and thus control assessment based on it is difficult, an assessment procedure that does not require the interactor matrix is proposed in Sect. 6.4. Numerous examples will be given to illustrate how the methods work.

6.1 Interactor Matrix: Time-Delay Analogy

The introduction of the interactor matrix is important not only because it solves the multivariable minimum-variance control problem (Sect. 6.1.3), but also it provides a basic tool to seek for the performance assessment index for multivariable processes (Sect. 6.3).

6.1.1 Definition and Special Forms

The delay structure of a multivariable plant has a direct effect on the minimum achievable variance and becomes in some sense the so-called *interactor matrix*

M. Jelali, *Control Performance Management in Industrial Automation*,
Advances in Industrial Control, DOI 10.1007/978-1-4471-4546-2_6,

(Wolovich and Falb 1976; Goodwin and Sin 1984). The interactor matrix is defined for any $r \times m$ proper, rational polynomial transfer-function matrix $\boldsymbol{G}_\mathrm{p}$ as the *unique*, non-singular $r \times r$ *lower left triangular* polynomial matrix $\boldsymbol{D}$ that satisfies the conditions

$$\begin{aligned}
&\text{(i)}\ \det \boldsymbol{D}(q) = q^n, \\
&\text{(ii)}\ \lim_{q^{-1}\to 0} \boldsymbol{D}(q)\boldsymbol{G}_\mathrm{p}(q) = \boldsymbol{K}, \quad \boldsymbol{K} \text{ finite and full rank,}
\end{aligned} \tag{6.1}$$

where n is the number of infinite zeros of $\boldsymbol{G}_\mathrm{p}$. $\boldsymbol{D}$ can be written in the Markov parameter representation as

$$\boldsymbol{D}(q) = \boldsymbol{D}_0 q^\tau + \boldsymbol{D}_1 q^{\tau-1} + \cdots + \boldsymbol{D}_v q^{\tau-v}, \tag{6.2}$$

where τ denotes the order of the interactor matrix and is unique for a given transfer-function matrix, v is the relative degree of the interactor matrix, i.e., the difference between the maximum and minimum power of q in $\boldsymbol{D}$, and $\boldsymbol{D}_i$ $(i = 0, 1, \ldots, v)$ are the coefficients matrices.

The interpretation of the interactor is that it is a part of the transfer-function matrix that is feedback-invariant and therefore constitutes a fundamental performance limitation in the system. This is equivalent to the role that the time delay plays in SISO systems; however, the multivariable case is normally more complex due to interactions between loops.

The general interactor matrix has a full matrix and is not unique. The uniqueness is the result of restricting the interactor matrix to be lower left triangular. Some important special forms of the (general) interactor matrix can be distinguished:

- *Simple Interactor Matrix*. If all the delays are equal, then the simple interactor matrix is obtained:

$$\boldsymbol{D}(q) = q^\tau \boldsymbol{I}_r, \tag{6.3}$$

  and this is the direct equivalent of the scalar time delay.
- *Diagonal Interactor matrix*. In the next simple case, the interactor may have a diagonal structure:

$$\boldsymbol{D}(q) = \operatorname{diag}\big(q^{\tau_1}, q^{\tau_2}, \ldots, q^{\tau_r}\big), \tag{6.4}$$

  where τ_i is the minimum delay between all the inputs and output i $(i = 1, 2, \ldots, r)$.
- *Unitary Interactor Matrix*. A particularly useful form is the unitary interactor matrix (Peng and Kinnaert 1992) which satisfies

$$\boldsymbol{D}^\mathrm{T}\big(q^{-1}\big)\boldsymbol{D}(q) = \boldsymbol{I}. \tag{6.5}$$

  The weighted unitary interactor matrix has the form

$$\boldsymbol{D}_w^\mathrm{T}\big(q^{-1}\big)\boldsymbol{D}_w(q) = \boldsymbol{W}, \tag{6.6}$$

where $\boldsymbol{W} > 0$ is a symmetric weighting matrix. The important property of a unitary matrix is that it does not change the spectral properties of a filtered signal, i.e. $\|\boldsymbol{D}\boldsymbol{x}\|_2 = \|\boldsymbol{x}\|_2$. In particular, the variance of the filtered signal remains the same as that of the original signal. This property will be exploited later in this chapter (Sect. 6.2) to derive the MIMO MV control law.

The knowledge of the interactor matrix is a prerequisite for standard controller performance assessment algorithms. The interactor matrix can be calculated from the plant transfer-function matrix, e.g. using the algorithm given by Rogozinski et al. (1987), or estimated from plant data, as suggested by Huang and Shah (1999).

Generally, the algorithms for determining the interactor matrix require a priori knowledge of the entire transfer matrix, which can be gained from open-loop identification. To weaken these requirements, Shah et al. (1987) have suggested (i) estimating the first few Markov parameters from closed-loop data via a (relatively high-frequency) dither signal or set-points changes and (ii) factorising the interactor directly from the estimated Markov parameters.

6.1.2 Recursive Determination of Unitary Interactor Matrices

When the process transfer function $\boldsymbol{G}_\mathrm{p}$ is known, which is a strong requirement in the CPM context, a unitary interactor matrix can be determined using different algorithms available in the literature. One common feature of these algorithms is to use the Markov parameter representation of $\boldsymbol{G}_\mathrm{p}$,

$$\boldsymbol{G}_\mathrm{p} = \sum_{i=0}^{\infty} \boldsymbol{G}_i q^{-(i+1)}, \tag{6.7}$$

and that of the interactor matrix (Eq. 6.2). From Eq. 6.1 we can write

$$\begin{aligned}
&\boldsymbol{D}_0\boldsymbol{G}_0 = 0,\\
&\boldsymbol{D}_1\boldsymbol{G}_0 + \boldsymbol{D}_0\boldsymbol{G}_1 = 0,\\
&\vdots\\
&\boldsymbol{D}_{\tau-1}\boldsymbol{G}_0 + \cdots + \boldsymbol{D}_1\boldsymbol{G}_{\tau-2} + \boldsymbol{D}_0\boldsymbol{G}_{\tau-1} = \boldsymbol{K},
\end{aligned}$$

or in matrix form

$$[\boldsymbol{D}_\tau, \ldots, \boldsymbol{D}_1] \begin{bmatrix} \boldsymbol{G}_0 & 0 & 0 & \ldots & 0 \\ \boldsymbol{G}_1 & \boldsymbol{G}_0 & 0 & \cdots & 0 \\ \vdots & \vdots & \ddots & \ddots & \vdots \\ \boldsymbol{G}_{\tau-2} & \boldsymbol{G}_{\tau-3} & \cdots & \ddots & \vdots \\ \boldsymbol{G}_{\tau-1} & \boldsymbol{G}_{\tau-2} & \cdots & \cdots & \boldsymbol{G}_0 \end{bmatrix} = [\boldsymbol{K},\ 0,\ \ldots,\ 0],$$

or simply

$$D'G' = K', \tag{6.8}$$

where G' is a block-Toeplitz matrix. Solving these algebraic equations gives the general solution of the interactor matrix. However, there is no unique solution for Eq. 6.8, and a direct inversion will not be always possible. This fact implies that an "optimal" (application-dependent) solution is sought; thus different algorithms have been proposed, e.g. by Rogozinski et al. (1987), Paplinski and Rogozinski (1990), Peng and Kinnaert (1992) and Bittanti et al. (1994).

In what follows, the method suggested by Huang and Shah (1999) is presented. It applies for systems having a full rank $n \times m$ and proper rational polynomial transfer function matrix, i.e. $\mathrm{rank}(G_p) = \min(n, m)$ and $\lim_{q^{-1}\to 0} G_p(q) < \infty$.

Procedure 6.1 Computation of the unitary interactor matrix.

1. Construct the matrices G' and K'.
2. Compute a singular value decomposition (SVD) of G' as

$$G' = U\Sigma V^T = \begin{bmatrix} U_1 & U_2 \end{bmatrix} \begin{bmatrix} \Sigma_r & 0 \\ 0 & 0 \end{bmatrix} \begin{bmatrix} V_1^T \\ V_2^T \end{bmatrix}, \tag{6.9}$$

   where $[U_1,\ U_2]$ and $[V_1,\ V_2]^T$ are orthogonal matrices, the columns of U_2 span the null space of G' in the sense that $U_2^T G' = \mathbf{0}$, Σ_r is a full rank diagonal matrix, and the rows of V_1^T span the row space of G'.
3. Check if

$$\mathrm{rank}(G') \geq \mathrm{rank}(K') = \mathrm{rank}(K) = \min(n, m) \tag{6.10}$$

   and if each row of K' is within the row space spanned by V_1^T or orthogonal to the row space spanned by V_2^T, i.e.

$$K'V_2 = 0. \tag{6.11}$$

   These conditions, when fulfilled, also determine the order of the interactor matrix $\tau = \min(n, m)$. Note that the condition in Eq. 6.1 can be simplified by writing

$$K'V_2 = [K,\ 0,\ \ldots,\ 0] \begin{bmatrix} V_{21} \\ V_{22} \\ \vdots \\ V_{2\tau} \end{bmatrix} = KV_{21}, \tag{6.12}$$

   where V_{21} is the upper partition of V_2 with its row dimension the same as the column dimension of G_p. Thus, the aforementioned condition is equivalent to

$$KV_{21} = 0, \tag{6.13}$$

or even

$$V_{21} = 0 \tag{6.14}$$

if $\boldsymbol{K}$ (or $\boldsymbol{G}_\mathrm{p}$) is a square matrix or is an $n \times m$ non-square matrix with $n > m$.

4. If the conditions are not satisfied, expand the block-Toeplitz matrix by adding more Markov parameters until they are satisfied.
5. Construct a block matrix of the first τ Markov parameters as

$$\boldsymbol{\Lambda} = \begin{bmatrix} \boldsymbol{G}_0 \\ \boldsymbol{G}_1 \\ \vdots \\ \boldsymbol{G}_\tau \end{bmatrix}. \tag{6.15}$$

6. Apply the recursive algorithm of Rogozinski et al. (1987) and Peng and Kinnaert (1992), to give the interactor matrix $\boldsymbol{D}$.

Care has to be taken when calculating the numerical rank for the case where rows or columns of a matrix are very small, i.e. very close to zero. This will result in singular values very close to zero. Thus, a regularisation of the matrix $\boldsymbol{G}'$ is recommended. In our experience, the aforementioned procedure is difficult to automate so that it is always suggested to carefully inspect the results of each step.

Example 6.1 Consider a 2×2 multivariable process with a diagonal transfer function matrix and the same time delay of two samples ($\tau = 2$) for both input/output channels:

$$\boldsymbol{G}_\mathrm{p}(q) = \begin{bmatrix} \frac{q^{-2}}{1-0.1q^{-1}} & 0 \\ 0 & \frac{2q^{-2}}{1-0.4q^{-1}} \end{bmatrix}. \tag{6.16}$$

Since

$$\lim_{q^{-1}\to 0} q^2 \boldsymbol{G}_\mathrm{p} = \begin{bmatrix} 1 & 0 \\ 0 & 2 \end{bmatrix} =: \boldsymbol{K}$$

is a constant full-rank matrix, the process has a *simple* interactor matrix of the form $\boldsymbol{D} = q^2 \boldsymbol{I}$.

Example 6.2 The system considered here has a diagonal transfer function matrix, but the input/output channels have now different time delays ($\tau_1 = 2$, $\tau_2 = 3$):

$$\boldsymbol{G}_\mathrm{p}(q) = \begin{bmatrix} \frac{q^{-2}}{1-0.1q^{-1}} & 0 \\ 0 & \frac{2q^{-3}}{1-0.4q^{-1}} \end{bmatrix}. \tag{6.17}$$

Since

$$\lim_{q^{-1}\to 0} \begin{bmatrix} q^2 & 0 \\ 0 & q^3 \end{bmatrix} \boldsymbol{G}_\mathrm{p} = \begin{bmatrix} 1 & 0 \\ 0 & 2 \end{bmatrix} =: \boldsymbol{K}$$

is a constant full-rank matrix, the process has a *diagonal* interactor matrix of the form $\boldsymbol{D}(q) = \mathrm{diag}(q^2,\ q^3)$.

Example 6.3 The process transfer function is changed to

$$\boldsymbol{G}_{\mathrm{p}}(q) = \begin{bmatrix} \frac{q^{-1}}{1-0.1q^{-1}} & \frac{q^{-1}}{1-0.1q^{-1}} \\ \frac{q^{-1}}{1-0.3q^{-1}} & \frac{2q^{-1}}{1-0.4q^{-1}} \end{bmatrix}. \tag{6.18}$$

It can easily be seen that the system still has a simple interactor matrix $\boldsymbol{D} = q\boldsymbol{I}$, as we can write

$$\lim_{q^{-1}\to 0} \begin{bmatrix} q & 0 \\ 0 & q \end{bmatrix} \boldsymbol{G}_{\mathrm{p}} = \begin{bmatrix} 1 & 0 \\ 0 & 2 \end{bmatrix} =: \boldsymbol{K},$$

which is a constant full-rank matrix.

Example 6.4 Consider a 2×2 multivariable process with the transfer function matrix (Huang and Shah 1998)

$$\boldsymbol{G}_{\mathrm{p}}(q) = \begin{bmatrix} \frac{q^{-1}}{1-0.4q^{-1}} & \frac{K_{12}q^{-2}}{1-0.1q^{-1}} \\ \frac{0.3q^{-1}}{1-0.1q^{-1}} & \frac{q^{-2}}{1-0.8q^{-1}} \end{bmatrix}, \tag{6.19}$$

where K_{12} controls the extent of interaction among the controlled variables. In this example, $K_{12} = 1$ is taken. There is no chance to find a simple or diagonal interactor matrix for this system, i.e. it has a general interactor matrix. Using Eq. 2.78 leads to the Markov parameter representation

$$\begin{aligned} \boldsymbol{G}_{\mathrm{p}}(q) &= \begin{bmatrix} q^{-1}(1+0.4q^{-1}+\cdots) & q^{-2}(1+0.1q^{-1}+\cdots) \\ 0.3q^{-1}(1+0.1q^{-1}+\cdots) & q^{-2}(1+0.8q^{-1}+\cdots) \end{bmatrix} \\ &= \begin{bmatrix} 0 & 0 \\ 0 & 0 \end{bmatrix} q^{0} + \begin{bmatrix} 1 & 0 \\ 0.3 & 0 \end{bmatrix} q^{-1} + \begin{bmatrix} 0.4 & 1 \\ 0.3 & 1 \end{bmatrix} q^{-2} + \cdots. \end{aligned}$$

Skipping the first step $\boldsymbol{G}' = \boldsymbol{G}_0$, which is obviously rank deficient, the block-Toeplitz matrix is formed as

$$\boldsymbol{G}' = \begin{bmatrix} 1 & 0 & 0 & 0 \\ 0.3 & 0 & 0 & 0 \\ 0.4 & 1 & 1 & 0 \\ 0.3 & 1 & 0.3 & 0 \end{bmatrix}.$$

Applying the SVD (Eq. 6.9) gives

$$\boldsymbol{U} = \begin{bmatrix} -0.2205 & -0.9320 & -0.0146 & -0.2873 \\ -0.0662 & -0.2796 & -0.0044 & 0.9578 \\ -0.7919 & 0.1964 & -0.5782 & 0 \\ -0.5656 & 0.1211 & 0.8157 & 0 \end{bmatrix},$$

$$\boldsymbol{\Sigma} = \begin{bmatrix} 1.8154 & 0 & 0 & 0 \\ 0 & 0.9832 & 0 & 0 \\ 0 & 0 & 0.4094 & 0 \\ 0 & 0 & 0 & 0 \end{bmatrix} \quad \Rightarrow \quad \boldsymbol{\Sigma}_r = \begin{bmatrix} 1.8154 & 0 & 0 \\ 0 & 0.9832 & 0 \\ 0 & 0 & 0.4094 \end{bmatrix},$$

$$\boldsymbol{V} = \begin{bmatrix} -0.4004 & -0.9163 & -0.0060 & 0 \\ -0.7478 & 0.3229 & 0.5801 & 0 \\ -0.5297 & 0.2367 & -0.8145 & 0 \\ 0 & 0 & 0 & 1.0 \end{bmatrix} \quad \Rightarrow \quad \boldsymbol{V}_{21} = \begin{bmatrix} 0 \\ 0 \end{bmatrix}.$$

Therefore, both necessary conditions in Eqs. 6.10 and 6.14 are satisfied, giving the order of the interactor matrix $\tau = \min(n, m) = 2$. The block matrix of the first two Markov parameters can be constructed as

$$\boldsymbol{\Lambda} = \begin{bmatrix} \boldsymbol{G}_0 \\ \boldsymbol{G}_1 \\ \boldsymbol{G}_2 \end{bmatrix} = \begin{bmatrix} 0 & 0 \\ 0 & 0 \\ 1 & 0 \\ 0.3 & 0 \\ 0.4 & 1 \\ 0.3 & 1 \end{bmatrix}.$$

Applying the factorisation algorithm of Rogozinski et al. (1987) yields the interactor matrix

$$\boldsymbol{D}(q) = \begin{bmatrix} -0.9578q & -0.2873q \\ 0.2873q^2 & -0.9578q^2 \end{bmatrix}.$$

It can be easily verified that the condition in Eq. 6.5 is satisfied.

6.1.3 Estimation from Closed-Loop Identification

It is a well known fact that the delay structure or the interaction matrix is "feedback-control invariant" (Wolovich and Falb 1976). This implies that although the Markov parameters of the open and closed-loop transfer function matrix are different, their linear combination yields the same interactor matrix; see also Huang and Shah (1999) for the mathematical proof. Exploiting this result, the interactor matrix of an open-loop transfer function can be estimated from the closed-loop data, provided that one is allowed to insert a dither signal $\boldsymbol{w}(k)$ to the set points $\boldsymbol{r}(k)$ or to the controller outputs $\boldsymbol{u}(k)$; see Fig. 6.1. A dither signal should be high-frequency random, ideally a white noise. The magnitude of the dither signal should be selected such that it has a very weak effect on the process output relative to the existing process disturbances.

In practice, however, the injection of such dither signals will be seldom allowed. In such situations, a series of simple changes of the set point may be conducted instead, and parametric models, e.g. ARX, ARMAX or PEM models, may be fitted to the gathered data. When using the armax function of the MATLAB identification toolbox, the MIMO system has to be split into p MISO subsystems, and the

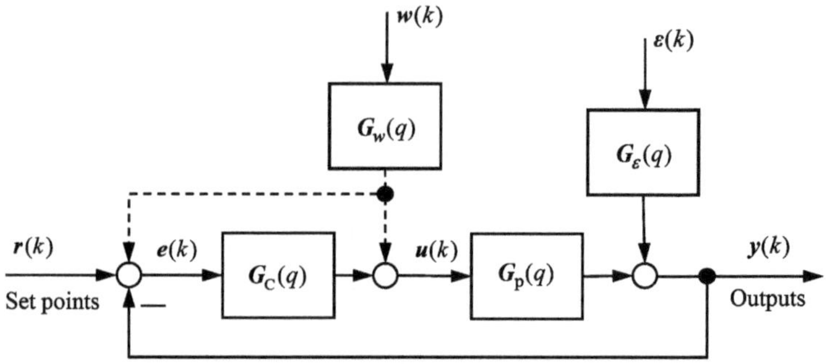

Fig. 6.1 Diagram of control loop with dither-signal injection

identification task is carried out for each output separately. The MATLAB function `dimpulse` can then be applied to each submodel. The resulting coefficients are combined to form the Markov parameter representation required for the determination of the interaction matrix using the technique already described in Sect. 6.1.2. Note that the identification task primarily seeks to estimate the closed-loop transfer function, i.e. from $\boldsymbol{w}$ to $\boldsymbol{y}$.

Example 6.5 We consider the multivariable control system (Fig. 6.1) with the following transfer functions and covariance matrices (Huang and Shah 1999):

$$\boldsymbol{G}_{\mathrm{P}}(q) = \begin{bmatrix} \frac{q^{-1}}{1-0.4q^{-1}} & \frac{K_{12}q^{-2}}{1-0.8q^{-1}} \\ \frac{0.3q^{-1}}{1-0.1q^{-1}} & \frac{q^{-2}}{1-0.8q^{-1}} \end{bmatrix}, \qquad \boldsymbol{G}_{\varepsilon}(q) = \begin{bmatrix} \frac{1}{1-0.5q^{-1}} & -\frac{0.6}{1-0.5q^{-1}} \\ \frac{0.5}{1-0.5q^{-1}} & \frac{1}{1-0.5q^{-1}} \end{bmatrix}, \tag{6.20}$$

$$\boldsymbol{G}_{\mathrm{c}}(q) = \begin{bmatrix} \frac{0.5-0.2q^{-1}}{1-0.5q^{-1}} & 0 \\ 0 & \frac{0.25-0.2q^{-1}}{(1-0.5q^{-1})(1+0.5q^{-1})} \end{bmatrix}.$$

The set point and K_{12} are assumed to be zero. The disturbances are white random noises with unit variance. The dither signal is a two-dimensional white noise sequence with the variances of 0.05 and 0.07, passed through the transfer function $\boldsymbol{G}_w$, which is selected as the discrete version of a high-pass filter $s/(s+1)$ for a sampling time $T_{\mathrm{s}} = 0.1$ s. For the sake of comparison, we will present not only the results achieved from identified models, but also those calculated using the knowledge of the exact transfer matrices.

A data set of 3000 samples has been used for the identification. The first three Markov-parameter matrices are calculated as

$$\hat{\boldsymbol{G}}' = \begin{bmatrix} \hat{\boldsymbol{G}}_0 \\ \hat{\boldsymbol{G}}_1 \\ \hat{\boldsymbol{G}}_2 \end{bmatrix} = \begin{bmatrix} 0.0239 & 0.0516 \\ -0.0127 & -0.1407 \\ 1.0357 & 0.0517 \\ 0.2997 & -0.0920 \\ -0.1060 & -0.1443 \\ -0.0830 & 0.8524 \end{bmatrix}, \qquad \boldsymbol{G}' = \begin{bmatrix} 0 & 0 \\ 0 & 0 \\ 1 & 0 \\ 0.3 & 0 \\ 0.4 & 0 \\ 0.03 & 1 \end{bmatrix}.$$

The deviations between the estimated and the theoretical values are mainly due to the non-sufficient excitation through the dither signal. Increasing the frequency of the high-pass filter reduces these deviations. For the determination of the interactor matrix, it is important to omit singular values very close to zero, i.e. less than a threshold a. A rule of thumb is to use $a = 2/\sqrt{N}$, where N is the data length. This did not help here, thus $a = 0.16$ has been used to give

$$\hat{\boldsymbol{G}}' = \begin{bmatrix} 0 & 0 \\ 0 & 0 \\ 1.0363 & 0.0234 \\ 0.2975 & 0.0067 \\ 0.0102 & -0.1355 \\ -0.0646 & 0.8538 \end{bmatrix}.$$

Despite the deviations between the identified and theoretical block-Toeplitz matrix entries, using the algorithm in Sect. 6.1.2 yields an estimate of the interaction matrix

$$\hat{\boldsymbol{D}}(q) = \begin{bmatrix} -0.9612q & -0.2759q \\ 0.2759q^2 & -0.9612q^2 \end{bmatrix},$$

which is very close to the theoretical interaction matrix

$$\boldsymbol{D}(q) = \begin{bmatrix} -0.9578q & -0.2873q \\ 0.2873q^2 & -0.9578q^2 \end{bmatrix}.$$

Example 6.6 Consider the system with the transfer functions given by Huang and Shah (1999)

$$\boldsymbol{G}_{\mathrm{p}}(q) = \begin{bmatrix} \frac{q^{-2}}{1-0.4q^{-1}} & \frac{2q^{-2}}{1-0.5q^{-1}} \\ \frac{q^{-1}}{1-0.1q^{-1}} & \frac{q^{-2}}{1-0.2q^{-1}} \end{bmatrix}, \qquad \boldsymbol{G}_{\varepsilon}(q) = \begin{bmatrix} \frac{2}{1-0.9q^{-1}} & \frac{1}{1-0.3q^{-1}} \\ \frac{1}{1-0.4q^{-1}} & \frac{2}{1-0.5q^{-1}} \end{bmatrix}, \qquad (6.21)$$

$$\boldsymbol{G}_{\mathrm{c}}(q) = \begin{bmatrix} 0.2 & 0 \\ 0 & 0.2 \end{bmatrix}.$$

All other parameters are selected as in Example 6.5. The same procedure applied here leads to the estimate of the interaction matrix

$$\hat{\boldsymbol{D}}(q) = \begin{bmatrix} -0.5700q & -0.8216q \\ -0.8216q^2 & -0.5700q^2 \end{bmatrix},$$

which shows remarkable deviations from the theoretical matrix

$$\hat{\boldsymbol{D}}(q) = \begin{bmatrix} 0 & q \\ q^2 & 0 \end{bmatrix}.$$

If the variance of the dither signals is increased to $\sigma_w^2 = 0.2$, we get

$$\hat{\boldsymbol{D}}(q) = \begin{bmatrix} -0.0602q & -0.9982q \\ 0.9982q^2 & -0.0602q^2 \end{bmatrix},$$

which now agrees well with the theoretical result. This example shows the importance of selecting the right dither signal for the estimation process. Recall that the dither signal should not affect the process output a lot, compared to the disturbances "routinely" occurring during the normal process operation.

6.2 Interactor-Matrix-Based Minimum-Variance Control Law

The minimum-variance control law can easily be extended to the multivariable case if the unitary interactor matrix is known, which is a strong requirement in CPM. Consider first a MIMO process

$$\begin{aligned} \boldsymbol{y}(k) &= \boldsymbol{G}_{\mathrm{p}}(q)\boldsymbol{u}(k) + \boldsymbol{G}_{\varepsilon}(q)\boldsymbol{\varepsilon}(k) \\ &= \boldsymbol{D}^{-1}(q)\tilde{\boldsymbol{G}}_{\mathrm{p}}(q)\boldsymbol{u}(k) + \boldsymbol{G}_{\varepsilon}(q)\boldsymbol{\varepsilon}(k), \quad \tilde{\boldsymbol{G}}_{\mathrm{p}}(q) := \boldsymbol{D}(q)\boldsymbol{G}_{\mathrm{p}}(q) \end{aligned} \tag{6.22}$$

with a general *unitary* interactor matrix $\boldsymbol{D}$. Multiplying both sides of Eq. 6.22 by $q^{-\tau}\boldsymbol{D}$ yields

$$\begin{aligned} q^{-\tau}\boldsymbol{D}(q)\boldsymbol{y}(k) &= q^{-\tau}\tilde{\boldsymbol{G}}_{\mathrm{p}}(q)\boldsymbol{u}(k) + q^{-\tau}\boldsymbol{D}(q)\boldsymbol{G}_{\varepsilon}(q)\boldsymbol{\varepsilon}(k) \\ &= q^{-\tau}\tilde{\boldsymbol{G}}_{\mathrm{p}}(q)\boldsymbol{u}(k) + \tilde{\boldsymbol{G}}_{\varepsilon}(q)\boldsymbol{\varepsilon}(k), \end{aligned} \tag{6.23}$$

where $\tilde{\boldsymbol{G}}_{\varepsilon} := q^{-\tau}\boldsymbol{D}\boldsymbol{G}_{\varepsilon}$ is a proper transfer function matrix. By defining the interactor-filtered output $\tilde{\boldsymbol{y}} = q^{-\tau}\boldsymbol{D}\boldsymbol{y}$, Eq. 6.23 can be transformed to a process with a *simple* interactor matrix, i.e.

$$\tilde{\boldsymbol{y}}(k) = q^{-\tau}\tilde{\boldsymbol{G}}_{\mathrm{p}}(q)\boldsymbol{u}(k) + \tilde{\boldsymbol{G}}_{\varepsilon}(q)\boldsymbol{\varepsilon}(k). \tag{6.24}$$

Substituting the Diophantine identity

$$\tilde{\boldsymbol{G}}_{\varepsilon} = q^{-\tau}\boldsymbol{D}\boldsymbol{G}_{\varepsilon} = \underbrace{\tilde{\boldsymbol{F}}_0 + \tilde{\boldsymbol{F}}_1 q^{-1} + \cdots + \tilde{\boldsymbol{F}}_{\tau-1} q^{-(\tau-1)}}_{\tilde{\boldsymbol{F}}} + q^{-\tau}\tilde{\boldsymbol{R}}, \tag{6.25}$$

where $\tilde{\boldsymbol{R}}$ is the remaining proper and rational transfer function matrix, into Eq. 6.24 leads to

$$\tilde{\boldsymbol{y}}(k) = \tilde{\boldsymbol{G}}_{\mathrm{p}}(q)\boldsymbol{u}(k-\tau) + \tilde{\boldsymbol{R}}(q)\boldsymbol{\varepsilon}(k-\tau) + \tilde{\boldsymbol{F}}\boldsymbol{\varepsilon}(k). \tag{6.26}$$

The last term in this equation cannot be affected by the control action, i.e.

$$\mathrm{var}\{\tilde{\boldsymbol{y}}\} = \mathrm{E}\{\tilde{\boldsymbol{y}}\tilde{\boldsymbol{y}}^{\mathrm{T}}\} \geq \mathrm{var}\{\tilde{\boldsymbol{F}}\varepsilon\}.$$

Therefore,

$$\mathrm{E}\{\tilde{\boldsymbol{y}}^{\mathrm{T}}\tilde{\boldsymbol{y}}\} \geq \mathrm{trace}\big(\mathrm{var}\{\tilde{\boldsymbol{F}}\varepsilon\}\big).$$

The minimum variance control is achieved when the sum of the first two terms on the right-hand side of Eq. 6.26 is set to zero, i.e.

$$\tilde{\boldsymbol{G}}_{\mathrm{p}}(q)\boldsymbol{u}(k-\tau) + \tilde{\boldsymbol{R}}(q)\boldsymbol{\varepsilon}(k-\tau) = 0.$$

This gives

$$\boldsymbol{u}(k) = -\tilde{\boldsymbol{G}}_{\mathrm{p}}^{-1}(q)\tilde{\boldsymbol{R}}(q)\boldsymbol{\varepsilon}(k). \tag{6.27}$$

Substituting Eq. 6.27 into 6.26 yields

$$\tilde{\boldsymbol{y}}(k) = \tilde{\boldsymbol{F}}(q)\boldsymbol{\varepsilon}(k). \tag{6.28}$$

Therefore,

$$\boldsymbol{\varepsilon}(k) = \tilde{\boldsymbol{F}}^{-1}(q)\tilde{\boldsymbol{y}}(k). \tag{6.29}$$

Substituting Eq. 6.29 into Eq. 6.27 gives the MV control law

$$\boldsymbol{u}(k) = -\tilde{\boldsymbol{G}}_{\mathrm{p}}^{-1}(q)\tilde{\boldsymbol{R}}(q)\tilde{\boldsymbol{F}}^{-1}(q)\tilde{\boldsymbol{y}}(k) = -\tilde{\boldsymbol{G}}_{\mathrm{p}}^{-1}(q)\tilde{\boldsymbol{R}}(q)\tilde{\boldsymbol{F}}^{-1}(q)q^{-\tau}\boldsymbol{D}(q)\boldsymbol{y}(k). \tag{6.30}$$

So far, the optimal control law, which minimises the LQ objective function of the interactor-filtered output

$$\tilde{J} = \mathrm{E}\{\tilde{\boldsymbol{y}}^{\mathrm{T}}\tilde{\boldsymbol{y}}\}, \tag{6.31}$$

has been found.

However, as already mentioned in Sect. 6.1.1, if $\boldsymbol{D}$ is a *unitary* interactor matrix (Eq. 6.5), the variance is "filter-invariant". Thus, the control law in Eq. 6.30 also minimises the original LQ objective function of the original output

$$J = \mathrm{E}\{\boldsymbol{y}^{\mathrm{T}}\boldsymbol{y}\}, \tag{6.32}$$

and even $\tilde{J} = J$ holds. It can also be proven that this control law is unique, output-ordering invariant and scaling invariant. However, the value of the minimum variance itself depends on the type of interaction matrix used (Ettaleb 1999). Note that the MV control will not be required for calculating the performance index. Also it is not needed to be implemented at the plant at all.

6.3 Assessment Based on the Interactor Matrix

There exist many approaches/algorithms to assess the performance of multivariable systems from routine-operating data. Examples are the FCOR approach (Huang et al. 1997a, 1997b; Huang and Shah 1999), the spectral-factorisation-based approach (Harris et al. 1996a) and the admissible minimum-variance and minimum ISE control approaches for multivariable processes with unstable zeros (Tsiligiannis and Svoronos 1989). In this section, we describe the computationally simple FCOR performance assessment algorithm, largely based on the descriptions by Huang et al. (1997a, 1997b) and Huang and Shah (1999).

As for the SISO case, but now the interactor-filtered, routine operating data (under control) $\tilde{\boldsymbol{y}}(k)$ are modelled by a multivariate MA process

$$\begin{aligned}\tilde{\boldsymbol{y}} - \mathrm{E}\{\tilde{\boldsymbol{y}}\} &= \underbrace{\tilde{\boldsymbol{F}}_0 + \tilde{\boldsymbol{F}}_1\boldsymbol{\varepsilon}(k-1) + \cdots + \tilde{\boldsymbol{F}}_{\tau-1}\boldsymbol{\varepsilon}(k-(\tau-1))}_{\tilde{\boldsymbol{\varepsilon}}(k)} \\ &\quad + \underbrace{\tilde{\boldsymbol{L}}_0\boldsymbol{\varepsilon}(k-\tau) + \tilde{\boldsymbol{L}}_1\boldsymbol{\varepsilon}(k-(\tau+1)) + \cdots}_{\tilde{\boldsymbol{w}}(k-\tau)}, \end{aligned} \tag{6.33}$$

which leads to an estimate of the white noise $\boldsymbol{\varepsilon}(k)$ (pre-whitening). Then the MV term, $\tilde{\boldsymbol{\varepsilon}}(k) = \tilde{\boldsymbol{F}}\boldsymbol{\varepsilon}(k)$, consists of the first τ terms of this MA model and, thus, can be separated from time-series analysis of the data and used as benchmark measure of multivariate MVC.

From Eq. 6.33, the covariance between $\tilde{\boldsymbol{y}}(k)$ and the white noise sequence at lag i (for $i < \tau$) is given by

$$\boldsymbol{\Sigma}_{\tilde{y}\varepsilon}(i) = \mathrm{E}\{\tilde{\boldsymbol{y}}(k)\boldsymbol{\varepsilon}^{\mathrm{T}}(k-i)\} = \tilde{\boldsymbol{F}}_i\boldsymbol{\Sigma}_{\boldsymbol{\varepsilon}}, \quad \boldsymbol{\Sigma}_{\boldsymbol{\varepsilon}} = \mathrm{E}\{\boldsymbol{\varepsilon}\boldsymbol{\varepsilon}^{\mathrm{T}}\}. \tag{6.34}$$

From

$$\begin{aligned}\tilde{\boldsymbol{y}}(k)|_{\mathrm{mv}} &= q^{-\tau}\boldsymbol{D}(q)y(k)|_{\mathrm{mv}} \\ &= \tilde{\boldsymbol{\varepsilon}}(k) = \tilde{\boldsymbol{F}}_0\boldsymbol{\varepsilon}(k) + \tilde{\boldsymbol{F}}_1\boldsymbol{\varepsilon}(k-1) + \cdots + \tilde{\boldsymbol{F}}_{\tau-1}\boldsymbol{\varepsilon}\big(k-(\tau-1)\big)\end{aligned}$$

one can obtain

$$\boldsymbol{y}(k)|_{\mathrm{mv}} = q^{\tau}\boldsymbol{D}^{-1}(q)\big[\tilde{\boldsymbol{F}}_0\boldsymbol{\varepsilon}(k) + \tilde{\boldsymbol{F}}_1\boldsymbol{\varepsilon}(k-1) + \cdots + \tilde{\boldsymbol{F}}_{\tau-1}\boldsymbol{\varepsilon}\big(k-(\tau-1)\big)\big]. \tag{6.35}$$

For the unitary matrix $\boldsymbol{D}(q)$, the property in Eq. 6.5 gives

$$\boldsymbol{D}^{-1}(q) = \big(\boldsymbol{D}_0 q^{\tau} + \cdots + \boldsymbol{D}_{\tau-1}q\big)^{-1} = \boldsymbol{D}_0^{\mathrm{T}}q^{-\tau} + \cdots + \boldsymbol{D}_{\tau-1}^{\mathrm{T}}q^{-1}.$$

Substituting this into Eq. 6.35 yields

$$\begin{aligned}\boldsymbol{y}(k)|_{\mathrm{mv}} &= q^{\tau}\big(\boldsymbol{D}_0^{\mathrm{T}}q^{-\tau} + \cdots + \boldsymbol{D}_{\tau-1}^{\mathrm{T}}q^{-1}\big) \\ &\quad \times \big(\tilde{\boldsymbol{F}}_0 + \tilde{\boldsymbol{F}}_1 q^{-1} + \cdots + \tilde{\boldsymbol{F}}_{\tau-1}q^{-(\tau-1)}\big)\boldsymbol{\varepsilon}(k).\end{aligned} \tag{6.36}$$

Multiplication and grouping the coefficients for each q^{-i}, having in mind that any term with positive power in q must be zero (to ensure causality), leads to

$$\boldsymbol{y}(k)|_{\mathrm{mv}} = \big(\boldsymbol{E}_0 + \boldsymbol{E}_1 q^{-1} + \cdots + \boldsymbol{E}_{\tau-1}q^{-(\tau-1)}\big)\boldsymbol{\varepsilon}(k). \tag{6.37}$$

Equation 6.36 can also be transformed in the compact matrix form

$$[\boldsymbol{E}_0,\ \boldsymbol{E}_1,\ \ldots,\ \boldsymbol{E}_{\tau-1}] = [\boldsymbol{D}_0^{\mathrm{T}},\ \boldsymbol{D}_1^{\mathrm{T}},\ \ldots,\boldsymbol{D}_{\tau-1}^{\mathrm{T}}]\begin{bmatrix} \tilde{\boldsymbol{F}}_0 & \tilde{\boldsymbol{F}}_1 & \cdots & \tilde{\boldsymbol{F}}_{\tau-1} \\ \tilde{\boldsymbol{F}}_1 & \vdots & \vdots & \\ \vdots & \vdots & & \\ \vdots & \tilde{\boldsymbol{F}}_{\tau-1} & & \\ \tilde{\boldsymbol{F}}_{\tau-1} & & & \end{bmatrix}. \tag{6.38}$$

The matrices $\boldsymbol{E}_i$ correspond to the Markov parameters, and the variance under MVC can be given as

$$\begin{aligned} \boldsymbol{\Sigma}_{\mathrm{mv}} &= \mathrm{var}\{\boldsymbol{y}(k)|_{\mathrm{mv}}\} = \boldsymbol{E}_0\boldsymbol{\Sigma}_{\boldsymbol{\varepsilon}}\boldsymbol{E}_0^{\mathrm{T}} + \cdots + \boldsymbol{E}_{\tau-1}\boldsymbol{\Sigma}_{\boldsymbol{\varepsilon}}\boldsymbol{E}_{\tau-1}^{\mathrm{T}} =: \boldsymbol{X}\boldsymbol{X}^{\mathrm{T}}, \\ &\text{where } \boldsymbol{X} = \left[\boldsymbol{E}_0\boldsymbol{\Sigma}_{\boldsymbol{\varepsilon}}^{1/2},\ \boldsymbol{E}_1\boldsymbol{\Sigma}_{\boldsymbol{\varepsilon}}^{1/2},\ \ldots,\ \boldsymbol{E}_{\tau-1}\boldsymbol{\Sigma}_{\boldsymbol{\varepsilon}}^{1/2}\right]. \end{aligned} \tag{6.39}$$

The similarity to the SISO case (Eq. 2.38) can be clearly observed. From Eq. 6.34 we have

$$\tilde{\boldsymbol{F}}_i = \boldsymbol{\Sigma}_{\tilde{y}\varepsilon}(i)\boldsymbol{\Sigma}_{\boldsymbol{\varepsilon}}^{-1}. \tag{6.40}$$

Substituting this in Eq. 6.39 yields

$$\begin{aligned} \boldsymbol{X} = {} & [\boldsymbol{D}_0^{\mathrm{T}},\ \boldsymbol{D}_1^{\mathrm{T}},\ \ldots,\ \boldsymbol{D}_{\tau-1}^{\mathrm{T}}] \\ & \times \begin{bmatrix} \boldsymbol{\Sigma}_{\tilde{y}\varepsilon}(0)\boldsymbol{\Sigma}_{\boldsymbol{\varepsilon}}^{-1/2} & \boldsymbol{\Sigma}_{\tilde{y}\varepsilon}(1)\boldsymbol{\Sigma}_{\boldsymbol{\varepsilon}}^{-1/2} & \cdots & \boldsymbol{\Sigma}_{\tilde{y}\varepsilon}(\tau-1)\boldsymbol{\Sigma}_{\boldsymbol{\varepsilon}}^{-1/2} \\ \boldsymbol{\Sigma}_{\tilde{y}\varepsilon}(1)\boldsymbol{\Sigma}_{\boldsymbol{\varepsilon}}^{-1/2} & \vdots & \vdots & \\ \vdots & \vdots & & \\ \vdots & \boldsymbol{\Sigma}_{\tilde{y}\varepsilon}(\tau-1)\boldsymbol{\Sigma}_{\boldsymbol{\varepsilon}}^{-1/2} & & \\ \boldsymbol{\Sigma}_{\tilde{y}\varepsilon}(\tau-1)\boldsymbol{\Sigma}_{\boldsymbol{\varepsilon}}^{-1/2} & & & \end{bmatrix}. \end{aligned} \tag{6.41}$$

The multivariable performance index can now be determined by

$$\eta = \frac{\text{minimum variance}}{\text{actual variance}} = \frac{\mathrm{E}\{\boldsymbol{y}^{\mathrm{T}}\boldsymbol{y}\}_{\mathrm{mv}}}{\mathrm{E}\{\boldsymbol{y}^{\mathrm{T}}\boldsymbol{y}\}} = \frac{\mathrm{trace}(\boldsymbol{\Sigma}_{\mathrm{mv}})}{\mathrm{trace}(\mathrm{E}\{\boldsymbol{y}^{\mathrm{T}}\boldsymbol{y}\})} = \frac{\mathrm{trace}(\boldsymbol{X}\boldsymbol{X}^{\mathrm{T}})}{\mathrm{trace}(\boldsymbol{\Sigma}_{\boldsymbol{y}})}. \tag{6.42}$$

Often, one is interested in comparing the variance-covariance matrix of the actual output with the variance-covariance matrix of the ideal output under MVC. For this purpose, the performance indices of individual outputs are obtained from the diagonal elements

$$\begin{aligned} [\eta_{y_1},\ \ldots,\ \eta_{y_p}]^{\mathrm{T}} &= \mathrm{diag}\left(\boldsymbol{\Sigma}_{\mathrm{mv}}\tilde{\boldsymbol{\Sigma}}_{\boldsymbol{y}}^{-1}\right) \\ &= \mathrm{diag}\left(\boldsymbol{X}\boldsymbol{X}^{\mathrm{T}}\tilde{\boldsymbol{\Sigma}}_{\boldsymbol{y}}^{-1}\right), \quad \tilde{\boldsymbol{\Sigma}}_{\boldsymbol{y}} = \mathrm{diag}(\boldsymbol{\Sigma}_{\boldsymbol{y}}). \end{aligned} \tag{6.43}$$

The individual output performance indices represent the performance of each output with respect to the ideal output under MVC. If an offset exists in the process output, then the output variance $\boldsymbol{\Sigma}_y$ should be replaced by the output mean-square error $\mathrm{E}\{(\boldsymbol{y}-\boldsymbol{y}_{\mathrm{ref}})(\boldsymbol{y}-\boldsymbol{y}_{\mathrm{ref}})^{\mathrm{T}}\}$ in the above calculations of the performance indices. This is however not necessary, as usually the measured output signals are mean-centred prior to the index calculation.

In summary, the implementation of the FCOR algorithm consists of the following steps.

Procedure 6.2 MIMO FCOR algorithm (Huang et al., 1997a, 1997b).

1. Filter the (mean-centred) process output data $\boldsymbol{y}(k)$ by an appropriate time-series model to obtain the "whitened" sequence $\boldsymbol{\varepsilon}(k)$ (Eq. 2.41).
2. Form an a priori knowledge or estimate of the interactor matrix $\boldsymbol{D}(q)$ and transform $\boldsymbol{y}(k)$ to the interactor-filtered form: $\tilde{\boldsymbol{y}}(k)=q^{-\tau}\boldsymbol{D}(q)\boldsymbol{y}(k)$.
3. Calculate the covariance $\boldsymbol{\Sigma}_{\tilde{y}\varepsilon}$ between $\tilde{\boldsymbol{y}}(k)$ and $\boldsymbol{e}(k)$ up to lag $\tau-1$ and the auto-covariances $\boldsymbol{\Sigma}_y$ and $\boldsymbol{\Sigma}_\varepsilon$ using the $\boldsymbol{y}(k)$ and $\boldsymbol{\varepsilon}(k)$ sequences, respectively, and the offset $\delta=\mathrm{E}\{\boldsymbol{y}(k)\}-\boldsymbol{y}_{\mathrm{ref}}$ if any.
4. All required information is now available to calculate the performance indices using Eqs. 6.41, 6.42 and 6.43.

Example 6.7 Consider again the system from Example 6.4 with $K_{12}=0.5$ and the disturbance transfer matrix

$$\boldsymbol{G}_\varepsilon(q)=\begin{bmatrix}\frac{1}{1-0.5q^{-1}} & \frac{-q^{-1}}{1-0.6q^{-1}}\\ \frac{q^{-1}}{1-0.7q^{-1}} & \frac{1}{1-0.8q^{-1}}\end{bmatrix}. \tag{6.44}$$

With the unitary interactor matrix determined in Example 6.4, we compute the Diophantine identity (Eq. 6.25) as

$$\tilde{\boldsymbol{G}}_\varepsilon=q^{-\tau}\boldsymbol{D}\boldsymbol{G}_\varepsilon=\underbrace{\begin{bmatrix}-0.9578q^{-1} & -0.2873q^{-1}\\ 0.2873-0.8142q^{-1} & -0.9578-1.0536q^{-1}\end{bmatrix}}_{\tilde{\boldsymbol{F}}}+q^{-2}\tilde{\boldsymbol{R}}.$$

The minimum-variance term can be written as (Eq. 6.33)

$$\tilde{\boldsymbol{\varepsilon}}(k)=\tilde{\boldsymbol{F}}\varepsilon(k)=\left\|\begin{bmatrix}0 & 0\\ 0.2873 & -0.9578\end{bmatrix}+q^{-1}\begin{bmatrix}-0.9578 & -0.2873\\ -0.9291 & -1.0536\end{bmatrix}\right\|\boldsymbol{\varepsilon}(k).$$

Therefore (Eq. 6.35),

$$\boldsymbol{y}(k)|_{\mathrm{mv}}=q^{\tau}\boldsymbol{D}^{-1}\tilde{\boldsymbol{\varepsilon}}(k)=\left(\begin{bmatrix}1.0 & -0.0\\ -0.0 & 1.0\end{bmatrix}+q^{-1}\begin{bmatrix}-0.2670 & -0.3028\\ 0.8899 & 1.0092\end{bmatrix}\right)\boldsymbol{\varepsilon}(k).$$

This explicit expression can always be estimated from routine operating data under any feedback control with a priori knowledge of the unitary interactor matrix. If we assume the disturbances be unit-variance noises, i.e. $\boldsymbol{\Sigma}_{\boldsymbol{\varepsilon}} = \mathrm{E}\{\boldsymbol{\varepsilon}\boldsymbol{\varepsilon}^{\mathrm{T}}\} = \boldsymbol{I}$, then the minimum variance can be calculated as (Eq. 6.39)

$$\boldsymbol{\Sigma}_{\mathrm{mv}} = \mathrm{var}\{\boldsymbol{y}(k)|_{\mathrm{mv}}\} = \begin{bmatrix} 1.1629 & -0.5431 \\ -0.5431 & 2.8104 \end{bmatrix}$$

and with the quadratic performance measure (H_2 norm) as (Eq. 6.42)

$$\mathrm{E}\{\boldsymbol{y}^{\mathrm{T}}\boldsymbol{y}\}_{\mathrm{mv}} = \mathrm{trace}(\boldsymbol{\Sigma}_{\mathrm{mv}}) = 3.9733.$$

A data set of 3000 samples has been used to determine the performance indices for the system output controlled by the multi-loop MV controller based on two single loops without interaction compensation:

$$\boldsymbol{G}_{\mathrm{c}}(q) = \begin{bmatrix} K_{\mathrm{c}}\frac{0.5-0.2q^{-1}}{1-0.5q^{-1}} & 0 \\ 0 & \frac{0.25-0.2q^{-1}}{(1-0.5q^{-1})(1+0.5q^{-1})} \end{bmatrix} \tag{6.45}$$

with $K_{\mathrm{c}} = 1$. The multivariable performance index has been calculated as (Eq. 6.42)

$$\eta = \frac{\mathrm{trace}(\boldsymbol{\Sigma}_{\mathrm{mv}})}{\text{actual variance}} = \frac{3.9733}{6.2805} \approx 0.63,$$

and the individual output performance indices as (Eq. 6.43)

$$\begin{bmatrix} \eta_{y_1} \\ \eta_{y2} \end{bmatrix} = \mathrm{diag}\left\{\begin{bmatrix} 1.1629 & -0.5431 \\ -0.5431 & 2.8104 \end{bmatrix}\begin{bmatrix} 2.2175 & 0 \\ 0 & 4.0630 \end{bmatrix}^{-1}\right\} \approx \begin{bmatrix} 0.52 \\ 0.69 \end{bmatrix}.$$

The performance assessment is then carried out for increasing K_{12}, and the results obtained for the overall index and for the individual output indices are illustrated in Fig. 6.2. In this example, when $K_{12} \to 0$, all performance indices converge to a similar value of about 0.6 owing to the weak interaction. However, if the interaction increases, the indices clearly diverge from each other. The control performance quickly deteriorates with increasing K_{12}, as indicated by the decrease of the overall index η as well as the index η_{y1}. The performance degradation is due to the fact that the interaction part is not compensated by the multi-loop controller. It appears that the performance of y_1 is much more sensitive to the interaction measure K_{12} than that of y_2. One can conclude that there is enough incentive to re-tune the controller or implement advanced multivariable control.

Now, the interaction factor is set to $K_{12} = 6$, and the controller gain K_{c} (Eq. 6.45) is varied with the objective to investigate its influence on the performance indices. The obtained results are shown in Fig. 6.3. It can be seen that a maximum overall (MIMO) index of about 0.28 can be achieved for $K_c \approx 1.15$. This indicates that the controller structure limits the achievable performance, and thus a re-design of the controller is needed in this case. It is also observed that the influence of the controller gain on both outputs is "unbalanced": the individual performance index for y_2 is much higher than that of y_1. This implies that the improvement work has

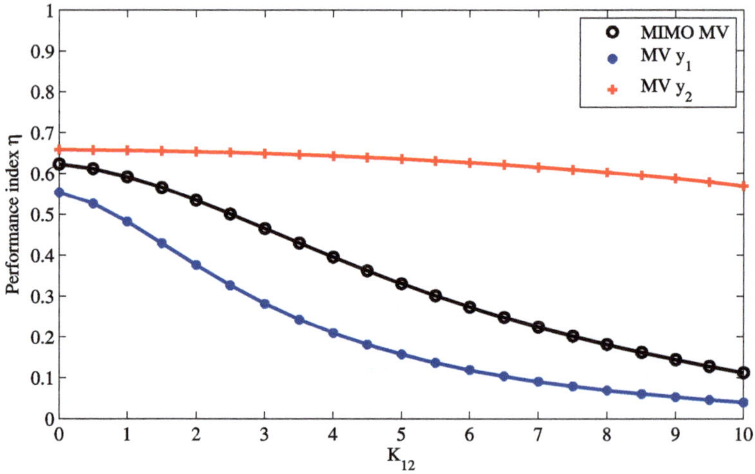

Fig. 6.2 Performance assessment of a MIMO process under multi-loop MVC vs. interaction factor K_{12}

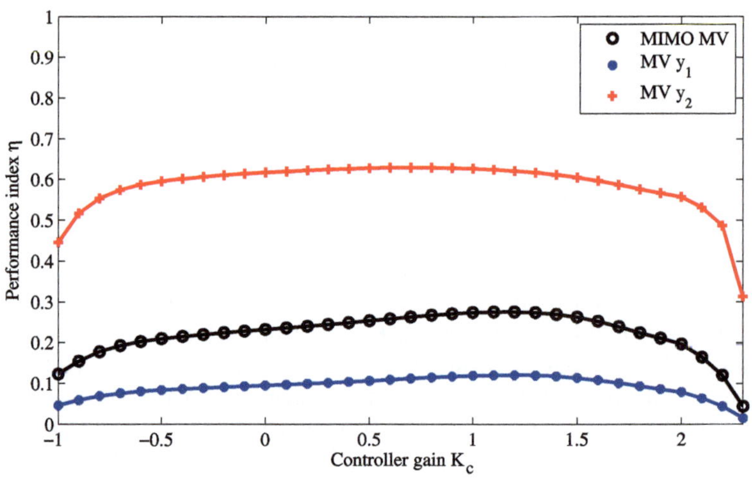

Fig. 6.3 Performance assessment of a MIMO process under multi-loop MVC vs. controller gain K_c

to be focused on loop 2, but the implementation of a multivariable controller will be the best solution.

6.4 Assessment Without Knowledge of the Interactor Matrix

A major problem of the assessment of MIMO control systems described so far is the need for the interactor matrix, which is not unique and generally requires the complete knowledge of the entire transfer matrix of the system at hand. For this reason, it is highly desirable to get around the problem and to obtain bounds on the controller performance index by using easily obtained plant information. Some methods and procedures for performance assessment without knowledge of the interactor matrix will be briefly described in this section. Generally, these approaches will lead to suboptimal but practical performance benchmarks, which can be estimated from routine operating (closed-loop) data with a minimum of a priori knowledge.

It was Ettaleb (1999) who first proposed to define an absolute lower bound on the achievable value for each output variance, thus avoiding the need to identify the interactor matrix. Ko and Edgar (2001b) developed a method which integrates the calculation of interactor matrix and the estimation of performance index without explicitly calculating the interactor matrix. However, this approach still needs to have the Markov matrices of the system transfer function matrix as a priori.

Huang et al. (2006) developed a truly model-free and interactor-free method to calculate the performance index directly from input/output data without the knowledge of interactor matrix or Markov matrices. With this algorithm, there is no need to know interactor matrix, Markov parameters or transfer function matrices. The only information needed is two sets of data, one open-loop experiment data and one closed-loop routine operating data.

Recently, it has been proven by Xia et al. (2006) that, given a diagonal delay matrix $\boldsymbol{D}_{\mathrm{i/o}} = \mathrm{diag}(\tau_{i,\min})$ with the order τ, the least conservative lower bound of $J_{\mathrm{lmv}} \leq J_{\mathrm{mv}}$ for the minimum variance can be found by calculating the MV performance index using $\boldsymbol{D} = \boldsymbol{D}_{\mathrm{i/o}} = \mathrm{diag}(\tau_{i,\min})$, which yields a significant simplification of the performance assessment of MIMO control systems. Similar approaches to simplify the MVC-based performance assessment of multivariable systems can be found by Huang et al. (2006). Particularly, two data-driven subspace algorithms have been proposed to compute the optimal prediction errors and closed-loop potentials.

Based on these works, we propose a practical method and procedure to assess the performance of the multivariable control systems, requiring only closed-loop data from normal process operation and the knowledge of time delays.

6.4.1 Lower Bound of MIMO Performance Index

The MIMO system is decomposed into p MISO systems. An absolute measure of the controller performance can be built by defining the performance index associated with each output $y_i(k)$ as

$$\eta_{y_i} = \frac{\sigma^2_{y_i}|_{\mathrm{mv}}}{\sigma^2_{y_i}}, \tag{6.46}$$

where $\sigma_{y_i}^2$ is the actual variance of the output y_i, and $\sigma_{y_i}^2|_{\mathrm{mv}}$ is the minimum achievable output variance of the MISO system (see Eq. 2.38)

$$\sigma_{y_i}^2|_{\mathrm{mv}} = \sum_{k=0}^{\tau_{i,\min}-1} \boldsymbol{f}_{i,k}\boldsymbol{\Sigma}_{\boldsymbol{\varepsilon}}\boldsymbol{f}_{i,k}^{\mathrm{T}}, \tag{6.47}$$

where $f_{i,k}$ are the impulse-response coefficients (Markov parameters) associated with the output y_i and τ_{ij} is the minimum delay, i.e.

$$\tau_{i,\min} = \min_{1\le j\le m} \tau_{ij} \tag{6.48}$$

is used. Therefore, the methods and algorithms presented in Sect. 2.4 for the SISO case can be applied to estimate η_{y_i} from routine operating data, setting $\tau = \tau_{ij}$. Note that to know whether those bounds defined in Eq. 6.47 are all achievable, it is necessary to determine the interactor matrix $\boldsymbol{D}$ and check whether it is diagonal, i.e. $\boldsymbol{D} = \mathrm{diag}(\tau_{1,\min}, \ldots, \tau_{\mathrm{p},\min})$.

The MIMO (overall) performance index is then determined by

$$\eta = \frac{\sum_{i=1}^{p}\sigma_{y_i}^2|_{\mathrm{mv}}}{\sum_{i=1}^{p}\sigma_{y_i}^2} = \frac{\sum_{i=1}^{p}\sum_{k=0}^{\tau_{i,\min}-1} \boldsymbol{f}_{i,k}\boldsymbol{\Sigma}_{\boldsymbol{\varepsilon}}\boldsymbol{f}_{i,k}^{\mathrm{T}}}{\sum_{i=1}^{p}\sigma_{y_i}^2}, \tag{6.49}$$

where $\boldsymbol{f}_k$ is the kth row of Markov-parameter matrix. This gives for the special case of 2×2 multivariable process:

$$\eta = \frac{\sum_{k=0}^{\tau_{1,\min}-1} f_{1,k}^2\Sigma_{11} + \sum_{k=0}^{\tau_{2,\min}-1} f_{1,k}^2\Sigma_{22} + \sum_{k=0}^{\tau_{2,\min}-1} f_{1,k}f_{2,k}(\Sigma_{12}+\Sigma_{21})}{\sigma_{y_1}^2+\sigma_{y_2}^2} \tag{6.50}$$

with $\boldsymbol{\Sigma}_{\boldsymbol{\varepsilon}} = \begin{bmatrix}\Sigma_{11} & \Sigma_{12}\\ \Sigma_{21} & \Sigma_{22}\end{bmatrix}$.

If the covariance matrix $\boldsymbol{\Sigma}_{\varepsilon}$ is diagonal, Eqs. 6.47 and 6.49 can be further simplified as

$$\sigma_{y_i}^2|_{\mathrm{mv}} = \sum_{k=0}^{\tau_{i,\min}-1} f_k^2\sigma_{\varepsilon_i}^2, \tag{6.51}$$

$$\eta = \frac{\sum_{i=1}^{p}\sigma_{y_i}^2|_{\mathrm{mv}}}{\sum_{i=1}^{p}\sigma_{y_i}^2} = \frac{\sum_{i=1}^{p}\sum_{k=0}^{\tau_{i,\min}-1} f_{i,k}^2\sigma_{\varepsilon_i}^2}{\sum_{i=1}^{p}\sigma_{y_i}^2}, \tag{6.52}$$

respectively. For a 2×2 multivariable process, we have

$$\eta = \frac{\sum_{k=0}^{\tau_{1,\min}-1} f_{1,k}^2\sigma_{\varepsilon_1}^2 + \sum_{k=0}^{\tau_{2,\min}-1} f_{2,k}^2\sigma_{\varepsilon_2}^2}{\sigma_{y_1}^2+\sigma_{y_2}^2}. \tag{6.53}$$

6.4.2 Upper Bound of MIMO Minimum Variance Benchmark

Looking again at Eqs. 6.23 and 6.35, it can be easily seen that setting a simple interactor matrix, i.e. $\boldsymbol{D} = q^{\tau}\boldsymbol{I}$, would significantly simplify the calculation of the MV performance index, provided that the order τ of the interactor matrix (equivalent to the time delay in the univariate case) is known. Otherwise, we may assume that τ is equal to the largest time delay. Then, the variance of the first τ terms may be used to represent the "minimum variance" as in the univariate case, i.e.

$$\boldsymbol{y}(k)|_{\text{umv}} = \left(\boldsymbol{F}_0 + \boldsymbol{F}_1 q^{-1} + \cdots + \boldsymbol{F}_{\tau-1} q^{-(\tau-1)}\right)\boldsymbol{\varepsilon}(k). \tag{6.54}$$

Strictly speaking, this represents the minimum-variance τ-step ahead prediction error (Huang et al. 2005a, 2005b). Thus, an upper bound $J_{\text{umv}} \geq J_{\text{mv}}$ for the minimum variance follows for all unitary matrices of order τ.

Although this bound is not necessarily achievable, it does provide us with an estimate of the maximum potential to improve the control performance. This is similar to the minimum variance control benchmark that is not necessarily achievable but delivers an estimated maximum potential of improvement. Huang et al. (2005b) have shown that the difference between J_{umv} and J_{mv} depends on the relative degree v of the interaction matrix $\boldsymbol{D}$ (Eq. 6.2): the higher v, the bigger the difference between J_{umv} and J_{mv}. When v is known, an upper bound on difference between J_{umv} and J_{mv} can be calculated without the knowledge of the interaction matrix:

$$J_{\text{umv}} - J_{\text{mv}} \leq \text{trace}\left\{\boldsymbol{F}_{\tau-v}\boldsymbol{\Sigma}_{\boldsymbol{\varepsilon}}\boldsymbol{F}_{\tau-v}^{\text{T}} + \cdots + \boldsymbol{\Sigma}_{\boldsymbol{\varepsilon}}\boldsymbol{F}_{\tau-1}^{\text{T}}\right\}. \tag{6.55}$$

6.4.3 Recommended Procedure for the Assessment of MIMO Control Systems

A practical procedure suggested here for control loop performance assessment of multivariable controllers consists of the following steps.

Procedure 6.3 MIMO assessment without interaction matrix.

1. Determine the (minimum) delays between process inputs and outputs.
2. Fit a MIMO AR(MA) model to the process output data, to obtain the estimated noise sequences $\varepsilon_i(k)$ and the covariance matrix $\boldsymbol{\Sigma}_{\varepsilon}$.
3. Determine the estimated lower output variance bounds of different outputs by use of Eq. 6.47.
4. Calculate the actual variances $\sigma_{y_i}^2$ of the output y_i.
5. Determine the (lower bounds of) performance indices associated with different outputs (Eq. 6.46) to know what the achievable limit for each individual output.
6. Determine the lower bound of overall performance index (Eq. 6.49), i.e. assuming a diagonal interaction matrix $\boldsymbol{D} = \text{diag}(\tau_{1,\text{min}}, \ldots, \tau_{\text{p,min}})$.

Table 6.1 Transfer functions of the considered processes

No.	Process model $\boldsymbol{G}_\mathrm{p}$	Disturbance model $\boldsymbol{G}_\varepsilon$	Controller $\boldsymbol{G}_\mathrm{c}$
P1	$\begin{bmatrix} \frac{q^{-1}}{1-0.4q^{-1}} & \frac{K_{12}q^{-2}}{1-0.8q^{-1}} \\ \frac{0.3q^{-1}}{1-0.1q^{-1}} & \frac{q^{-2}}{1-0.8q^{-1}} \end{bmatrix}$ $K_{12}=0$	$\begin{bmatrix} \frac{1}{1-0.5q^{-1}} & -\frac{0.6}{1-0.5q^{-1}} \\ \frac{0.5}{1-0.5q^{-1}} & \frac{1}{1-0.5q^{-1}} \end{bmatrix}$	$\begin{bmatrix} \frac{0.5-0.2q^{-1}}{1-0.5q^{-1}} & 0 \\ 0 & \frac{0.25-0.2q^{-1}}{(1-0.5q^{-1})(1+0.5q^{-1})} \end{bmatrix}$
P2	Same transfer functions as in 1 but with $K_{12}=4$.		
P3	Same transfer functions as in 1 but with $K_{12}=10$.		
P4	$\begin{bmatrix} \frac{q^{-2}}{1-0.4q^{-1}} & \frac{2q^{-2}}{1-0.5q^{-1}} \\ \frac{q^{-1}}{1-0.1q^{-1}} & \frac{q^{-2}}{1-0.2q^{-1}} \end{bmatrix}$	$\begin{bmatrix} \frac{2}{1-0.9q^{-1}} & \frac{1}{1-0.3q^{-1}} \\ \frac{1}{1-0.4q^{-1}} & \frac{2}{1-0.5q^{-1}} \end{bmatrix}$	$\begin{bmatrix} K_\mathrm{c} & 0 \\ 0 & K_\mathrm{c} \end{bmatrix} = \begin{bmatrix} 0.2 & 0 \\ 0 & 0.2 \end{bmatrix}$
P5	$\begin{bmatrix} \frac{q^{-(\tau-1)}}{1-0.4q^{-1}} & \frac{0.5q^{-\tau}}{1-0.1q^{-1}} \\ \frac{0.3q^{-(\tau-1)}}{1-0.4q^{-1}} & \frac{q^{-\tau}}{1-0.8q^{-1}} \end{bmatrix}$ $\tau=2$	$\begin{bmatrix} \frac{1}{1-0.5q^{-1}} & \frac{q^{-1}}{1-0.6q^{-1}} \\ \frac{q^{-1}}{1-0.7q^{-1}} & \frac{1}{1-0.8q^{-1}} \end{bmatrix}$	$\begin{bmatrix} \frac{0.5-0.2q^{-1}}{1-0.5q^{-1}} & 0 \\ 0 & \frac{0.25-0.2q^{-1}}{(1-0.5q^{-1})(1+0.5q^{-1})} \end{bmatrix}$
P6	Same transfer functions as in 5 but with $\tau=3$.		
P7	Same transfer functions as in 5 but with $\tau=7$.		

7. Determine the upper bound of overall performance index (Eq. 6.54), i.e. assuming a diagonal interaction matrix $\boldsymbol{D}=q^{\tau}\boldsymbol{I}$, where τ is an a priori known value of the order of the interaction matrix or (otherwise) the largest time delay.

The big advantage of this approach is that it does not need any knowledge of process models or experimentation with the process. Only normal operating data and the knowledge of minimum time delays are required. Also, all steps of the procedure can be performed in a small amount of computation. Several examples will be given below to demonstrate the efficiency and practicality of the proposed approach, which has been implemented[1] in MATLAB. Note that it has some common features with the assessment procedures suggested by Ettaleb (1999), Huang et al. (2006) and Xia et al. (2006).

Example 6.8 In this study, a series of processes (Table 6.1) are considered and simulated to demonstrate the calculation of the true MV benchmark and both performance bounds presented in Sect. 6.4. In all cases, the noise excitation $\boldsymbol{\varepsilon}(k)$ is a two-dimensional normal-distributed white noise sequence with $\boldsymbol{\Sigma}_\varepsilon=\boldsymbol{I}$. A set of 3000 samples of the closed-loop output was used for each simulation. All calculations are performed using both the theoretical models and estimated models from routine operating data. The results are presented in Table 6.2. It can be deduced that the estimated indices are in good agreement with their theoretical counterparts for all considered processes and simulation scenarios. Note that for P4, a relatively strong

[1] The basic algorithms were implemented by Martina Thormann and Heinrich Ratjen, see Ratjen (2006).

Table 6.2 Theoretical and estimated minimum variance index and its bounds

	Theoretical			Estimated		
	$\eta_{\text{interactor}}$	η_{lower}	η_{upper}	$\eta_{\text{interactor}}$	η_{lower}	η_{upper}
P1	0.93	0.82	1.0	0.98	0.87	0.99
P2	0.45	0.40	0.50	0.49	0.44	0.50
P3	0.10	0.09	0.11	0.08	0.07	0.08
P4	0.60	0.55	0.60	0.62	0.57	0.62
P5	0.61	0.32	0.79	0.69	0.40	0.78
P6	0.74	0.62	0.78	0.83	0.70	0.85
P7	0.54	0.53	0.54	0.56	0.53	0.56

dither signal with the variance of 0.2 was required (see Example 6.6) for the estimation of the interaction matrix. For P5–P7, it was necessary to omit values of the Markov parameters which are less $a = 0.1$ for the determination of the interaction matrix. η_{lower} and η_{upper} can be regarded as practical indices that give sufficient assessment of the MIMO controllers, having in mind that the interactor is not needed at all.

Example 6.9 The system considered in this example is adopted from Xia et al. (2006). It is a slightly modified version of the control system from Example 6.5, i.e. Eq. 6.20, but with

$$\boldsymbol{G}_{\text{p}}(q) = \begin{bmatrix} \frac{q^{-(\tau-1)}}{1-0.4q^{-1}} & \frac{q^{-\tau}}{1-0.1q^{-1}} \\ \frac{0.3q^{-(\tau-2)}}{1-0.1q^{-1}} & \frac{q^{-(\tau-1)}}{1-0.8q^{-1}} \end{bmatrix}. \tag{6.56}$$

When assuming full knowledge of the plant, a corresponding unitary interactor matrix can be determined as

$$\boldsymbol{D}(q) = \begin{bmatrix} 0.9578q^{\tau-1} & -0.2873q^{\tau-2} \\ -0.28736q^{\tau} & 0.9578q^{\tau-1} \end{bmatrix}.$$

The order of the interactor matrix is equal to the largest time delay of the system, however generally this does not have to be the case.

For increasing values of τ, the MIMO performance index (using the interaction matrix) and its bounds (which do not need the interaction matrix) have been computed. The obtained estimates of the performance indices are presented in Table 6.3 and shown graphically in Fig. 6.5). The upper bound values are very close to the real ones (Table 6.3 and Fig. 6.4), confirming that the corresponding benchmark is valuable for providing an estimate for the performance improvement potential. The difference between the curves decreases rapidly with the increasing time delay (or interactor order) τ. Little difference is observed between the upper and lower bound when τ is larger than 5. This is because, for a stable disturbance model, the Markov coefficient matrices that determine this difference become smaller and smaller. All

Table 6.3 True and estimated values of the minimum variance index and its bounds

τ	3	4	5	6	7	8	9	10
η_{upper}	0.81	0.76	0.71	0.62	0.58	0.56	0.55	0.46
$\hat{\eta}_{\text{upper}}$	0.87	0.80	0.73	0.66	0.59	0.51	0.52	0.46
$\eta_{\text{interactor}}$	0.77	0.73	0.70	0.62	0.55	0.53	0.52	0.44
$\hat{\eta}_{\text{interactor}}$	0.73	0.73	0.70	0.64	0.58	0.51	0.51	0.46
η_{lower}	0.55	0.64	0.64	0.58	0.54	0.53	0.52	0.43
$\hat{\eta}_{\text{lower}}$	0.54	0.64	0.65	0.61	0.56	0.50	0.51	0.46

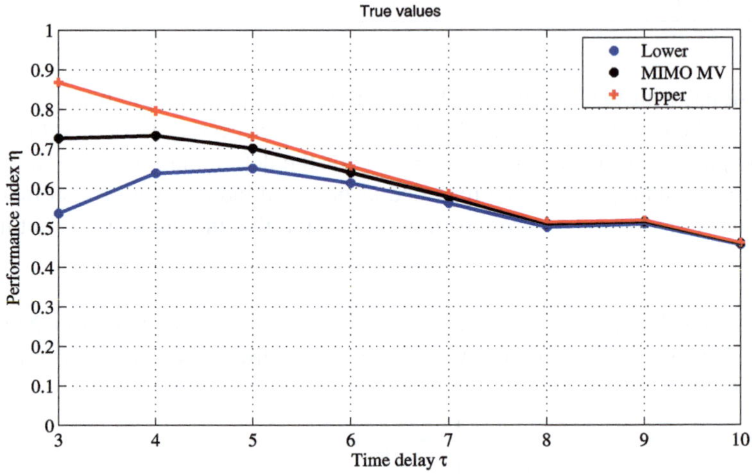

Fig. 6.4 True values of the minimum variance index and its bounds

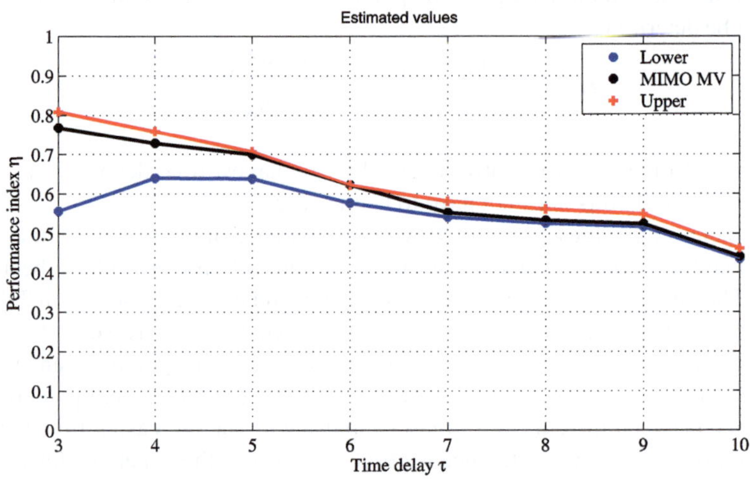

Fig. 6.5 Estimated values of the minimum-variance index and its bounds

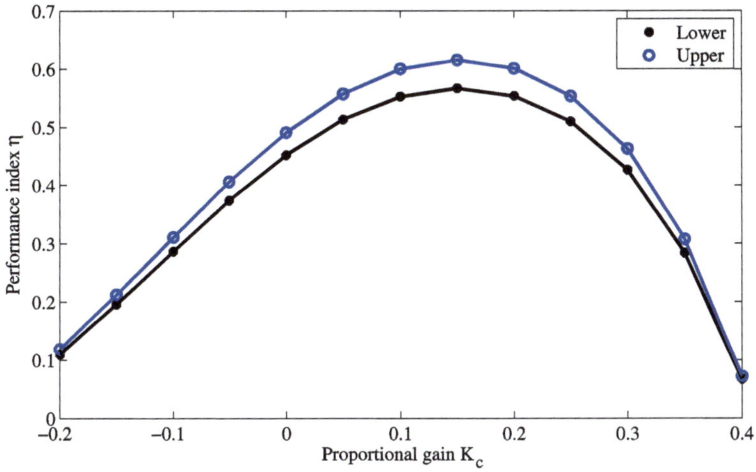

Fig. 6.6 Minimum-variance index bounds vs. the proportional gain K_c for P4

indices correctly indicate performance deterioration of this fixed controller with increasing time delay due to the increased time-delay restriction. Therefore, there is enough incentive to improve the loop performance by re-tuning the controller or implementing advanced multivariable control with a time-delay compensation feature.

What can be learned from this example is that the consideration of the easily computable lower and upper bounds as performance measures suffices to get a clear performance figure, and the estimation of the interaction matrix seems to be practically dispensable. Note that similar performance assessment results have been achieved by Xia et al. (2006) for this example but under consideration of other approaches for the calculation of the lower and upper bounds.

Example 6.10 In this study, we reconsider the system P4 from Table 6.1 but vary the controller proportional gain K_c. The objective is to compare the controller performance for different settings. For each K_c value, both bounds of the MV index were calculated. From the plot in Fig. 6.6 it can be deduced that the optimal controller gain is about 0.15. With increasing gain above this gain, the process becomes more and more oscillatory up to instability for gains higher than 0.4. For this example, the MV index and its upper bound are identical, since the system has a simple interaction matrix $q^2\boldsymbol{I}_2$, and thus $v = 0$. Also for this example, the lower and upper indices give sufficient assessment of the system.

6.5 Summary and Conclusions

MVC benchmarking in the multivariable case has been shown to be formally similar but much more involved than in the univariate case, as it normally requires the knowledge or estimation of the interactor matrix. Although many methods for its factorisation or estimation exist, this task remains very difficult to handle and needs a complete process model or at least its first Markov parameters. Therefore, estimating a lower bound and an upper bound of the minimum achievable variance was recommended instead of the minimum variance itself. Both indices are easily computable and can be found from routine operating data and require only the knowledge of the time delays between each pair of system inputs and outputs. This represents a much weaker assumption than requiring complete process information or the highly undesirable external excitation of the process to estimate the interaction matrix.

Chapter 7
Selection of Key Factors and Parameters in Assessment Algorithms

The performance assessment algorithms presented in the previous chapters contain many options and parameters that must be specified by the user. These factors substantially affect the accuracy and acceptability of the results of assessment exercises. A fundamental basis for performance assessment is to record and carefully inspect suitable closed-loop data. Pre-processing operations, which are suggested and those which should be strictly avoided, will be given in Sect. 7.1. The first decision in control performance assessment is the choice of a (time-series) model structure for describing the net dynamic response associated with the control error. There are different possible structures and different possible identification techniques. The most widely used of them are briefly described in Sect. 7.2. Particularly for MV and GMV benchmarking, it is decisive to properly select or estimate the parameters time delay and model orders. This topic is discussed in Sect. 7.3. Some of the basic models and identification techniques are compared in Sect. 7.4, concerning assessment accuracy and computational load, to provide suggestions of the best suited approaches to be applied in practice.

7.1 Data Pre-processing

An attractive property of the MV benchmarking is that the performance index can be estimated from routine operating data without additional experiments, provided that the system time delay is known or can be estimated with sufficient accuracy. No matter what the current controller is, measuring the controlled variable suffices for performance evaluation based on a closed-loop model. For many other methods, the measurement of the manipulated variable is necessary, as open-loop and closed-loop models are needed to be identified when not available. Operating data must also include the set point, when it is varying, to give the control error.

The first basic step in CPM is data acquisition and preparation for analysis. Comprehensive data needed for the development and implementation of performance evaluation systems are usually available on different sources of the considered

M. Jelali, *Control Performance Management in Industrial Automation*,
Advances in Industrial Control, DOI 10.1007/978-1-4471-4546-2_7,

plant(s). Thus, much time has to be spent to properly collect the data in different ways, i.e. using several communication networks in the plants to interface the process/automation computers/controllers. Sometimes access to signals needed has to be implemented, or even new sensors must be added, but this is seldom the case.

A number of pre-processing techniques should then be applied prior to the assessment task to ensure that the data samples are free from undesired noise or trends, outliers and other corruptions. When ever possible, the data should be first inspected visually to detect corruptions or errors, such as outliers, clipped saturation or quantisation effects. If the data are not evenly or regularly sampled, they need interpolation to make it evenly sampled before any numerical analysis can proceed.

7.1.1 Selection of Sampling Interval

Basically, the data sampling time for CPM should be the same as the controller sampling time. In practice, however, it is recommended to properly down-sample the data to save computation time. Also, Thornhill et al. (1999) stated that the choices of the sampling interval and the number of terms in the model are *not* independent of one another because they both influence the total time span captured by the autoregressive terms. The strategy proposed by Thornhill et al. (1999) is to choose the length of the AR model to be $n = 30$ for all types of control loops and to adjust the sampling interval individually for each loop. The suggestion is to select the sampling interval such that a typical *closed-loop impulse response is fully captured within 30 samples.*

It is worth noting that the performance index itself is not sensitive to the sampling interval, provided that the prediction horizon, b, is adjusted accordingly. That is, if b is 10 sample intervals for 1 s data, it would be 5 sample intervals for 2 s data. The reason for optimising of the sampling interval relates to the diagnosis of the likely cause of poor loop performance. In a spectral analysis, for instance, the spectral features are properly resolved only when the sampling interval is correctly adjusted. The estimated closed-loop impulse response can serve as a diagnostic tool, and its ability to resolve key features is influenced by the sampling interval.

It is necessary to avoid both over-sampling and under-sampling. If the data are sampled too frequently, the transient part of the closed-loop impulse response does not settle within the 30 samples. If the data are under-sampled, the closed-loop impulse response is seen to settle within just a few samples and is not adequately captured because interesting features may be missed between samples (Thornhill et al. 1999).

Example 7.1 The data considered here come from a thickness control loop in a cold rolling mill. Figure 7.1 shows how the sampling time affects the disturbance impulse response. It can be concluded that an appropriate sampling time should be selected not higher than 100 ms to avoid over-sampling, but also not lower than 50 ms to prevent under-sampling.

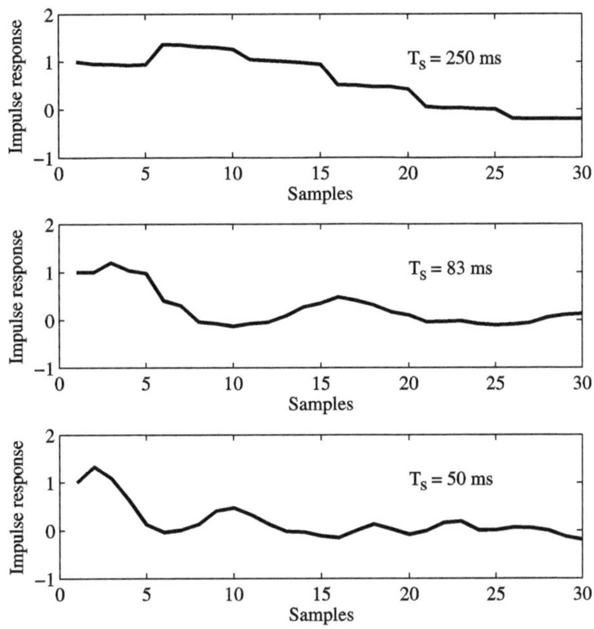

Fig. 7.1 Effect of sampling time on the impulse response (Loop: thickness control)

7.1.2 Selection of Data Length

The data ensemble length has clear effect on the statistical confidence in the performance-index value, which improves as the data ensemble length increases. When data of the control error are considered, it is not necessary for the loop to stay at the same set point throughout the period of data recording, but it is desirable that the loop characteristics remain unchanged. Data episodes during instrument recalibration episode or known plant disturbances such as feed switches or partial trips have, therefore, to be avoided (Thornhill et al. 1999).

The effect of data ensemble length has been assessed using the confidence limits given by Desborough and Harris (1992):

$$\sigma_{\hat{\eta}}^2 \approx \frac{4}{N}(1-\hat{\eta})^2 \left\{ \sum_{k=1}^{\tau-1} \left[\rho_y(k) - \rho_\varepsilon(k)\right]^2 + \sum_{k=\tau}^{\infty} \left[\rho_y(k)\right]^2 \right\}, \tag{7.1}$$

which simplifies to

$$\sigma_{\hat{\eta}}^2 \approx \frac{4}{N}(1-\hat{\eta})^2 \sum_{k=\tau}^{\infty} \left[\rho_y(k)\right]^2 \tag{7.2}$$

for systems without time delay, i.e. $\tau = 1$. $\rho_y(k)$ and $\rho_\varepsilon(k)$ stand for the auto-correlations of y and ε, respectively. The statistical property in Eq. 7.1 also enable us to examine the explicit dependence that time delay and auto-correlation structure of the process have on the uncertainty associated with the estimated index.

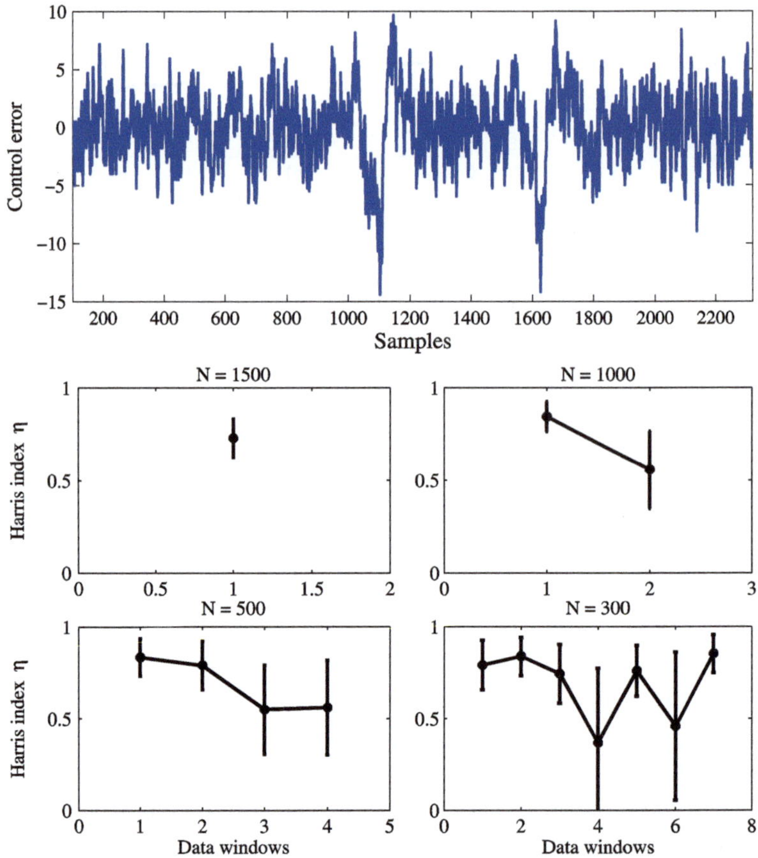

Fig. 7.2 *Top*: assessed control error trend; *bottom*: Effect of data set length on the standard deviations of the Harris index (Loop: thickness control, coil no. 33)

As can be seen from the above equation, short data segments will increase the standard deviations of the statistical estimates. On the other hand, long data segments lead to lower standard deviations. However, too long data sets can give misleading results when many different response characteristics are juxtaposed into one long data set (Kozub 1996). It is agreed by many researchers that a good balance between statistical confidence and the steadiness of the loop characteristics is achieved with a data ensemble between 1000 and 2000, say, 1500 samples.

Example 7.2 We consider data measured from a strip-thickness control loop to show the effect of the data length N. The upper panel in Fig. 7.2 shows the time trend of the control error for 2334 samples. When all data points are used, the Harris index is 0.724 with a standard deviation of 0.101. The lower four subplots in Fig. 7.2 show the index values and the standard deviations when shorter data ensembles are used. For instance, in the lower right-hand plot the data ensembles are 300 points each

($\sigma = 0.102$–0.406). They have considerable variability, and the error bars, which represent the standard deviations, are quite large. By contrast, the standard deviations for data ensemble of 1500 points are somewhat smaller ($\sigma = 0.102$). Particularly in the subplot for $N = 300$, it can be clearly seen that shorter sequences are more responsive to changes in the loop's characteristics. Look at the disturbance episodes between the samples 1020 and 1150 and between the samples 1550 and 1675. Thus, there is a trade-off between confidence in the index value and its sensitivity to features in the data, as also demonstrated by Thornhill et al. (1999) on data from different refinery loops.

The conclusion is that $N = 1500$ is the recommended choice. Certainly, one may use shorter data ensembles of 1000 or even 500 samples, but only at the price of a broader confidence interval for the performance index. The use of data ensembles of less than 500 samples is not recommended at all because the scatter during normal running is relatively large and the standard deviations are high.

7.1.3 Removing of Outliers, Detrending and Pre-filtering

Various unexpected events, such as abnormal pulses, temporary sensor failures, transmitter failures, non-stationary trends or process disturbances, can corrupt the raw data samples. Such bad data should be removed from the data set.

7.1.3.1 Outlier Elimination

Outliers are observations that do not follow the statistical distribution of the bulk of the data and consequently may lead to erroneous results with respect to statistical analysis. They tend to appear as spikes in the sequence of prediction error and will hence give large contributions to the loss function. However, removing certain points or time intervals from the data set is possible only for static data, otherwise it will give rise to transients. Therefore, it may be preferable for dynamic systems to carry out the experiment again to get a better data set. Methods for cleaning data from outliers can be found by Rousseeuw and Leroy (1987), Davies and Gather (1993), Perarson (2002), Liu et al. (2004) and references included therein. As will be shown in Sect. 10.3.4, outlier removing is essential for nonlinearity detection based on surrogate data. However, the presence of outliers does not affect much the value of the Harris index.

7.1.3.2 Detrending

It is important to remove slowly drifting signals to make it more stationary. The MATLAB function detrend can be used for this purpose: it fits a straight line through the data and removes this line from the points. When working with FFT in MATLAB, it is always recommended to detrend the data first. Particularly, nonlinearity assessment based on higher-order statistics (Sect. 10.2) requires the elimination of non-stationary trends to avoid misleading results.

7.1.3.3 Pre-filtering or Down-Sampling

This is usually necessary to avoid aliasing and to remove noise, periodic disturbances, offsets and drifts from the measured signals. Analogue anti-aliasing (low-pass) filters should be used prior to the sampling. The bandwidth of filters should be smaller than the sampling frequency. The rule of thumb governing the design of the filter is that the upper frequency should be about *twice* the desired system bandwidth and the lower frequency should be about *one-tenth* the desired bandwidth. For data down-sampling, MATLAB offers the function resample.

7.1.3.4 Scaling

Particularly in MIMO systems, it is common to have inputs and outputs of different amplitude ranges. Such a diversity in amplitudes can make the model estimation ill-conditioned, which deteriorates the precision of the dynamic response. It is highly recommended to normalise all signals to the same scale and variance. Indeed, scaling makes the estimation algorithm numerically robust, leads to faster convergence and simply tends to give better model quality. Typically, the measured input $u(k)$ and output $y(k)$ sequences are scaled to zero mean and unity variance by

$$u^*(k) = \frac{u(k) - \mathrm{E}\{u(k)\}}{\sigma_u} = \frac{u(k) - \bar{u}}{\sigma_u}, \tag{7.3}$$

$$y^*(k) = \frac{y(k) - \mathrm{E}\{y(k)\}}{\sigma_y} = \frac{y(k) - \bar{y}}{\sigma_y}, \tag{7.4}$$

where $(\cdot)^*$, $\mathrm{E}\{\cdot\}$, and $\overline{(\cdot)}$ stand for the normalised, the expected and the mean value, respectively, and σ is the standard deviation of the corresponding signal. The mean value is subtracted to get data centred at the origin. The division by the standard deviation ensures that input dimensions with a large range do not dominate the variance terms. For readability, we do not always differentiate between scaled and non-scaled quantities throughout the book.

7.1.4 Effect of Smoothing, Compression and Quantisation

It is imperative that closed-loop data used for control performance assessment are *non-smoothed and uncompressed*, i.e. the data must be exactly identical to the feedback information utilised by the loop controller. Therefore, it is *not* advisable to use archived data for the purposes of control performance analysis. The reason for caution is that archiving systems often modify the stored data. For example, the measurements may be smoothed to remove noise, and compression may be applied both to keep up with increasing demands for stored information and to extract summary statistics.

Both *smoothing and compression* affect the calculated performance indices. The purpose of smoothing is to reduce the effects of noise by averaging over previous measured values. The smoothed values thereby become correlated, and the sequence is more predictable than the original sequence. Smoothing thus has an impact on the performance index because the index is a measure of predictability. Compression algorithms used in online data historians decide whether or not to archive a point using a variety of rules. The rules are designed to capture the start and end of a trend and any exceptional values, but they are not designed to retain the noise and subtle predictable components that the performance index needs for its assessment. Techniques for data smoothing or compression are reviewed in the paper by Thornhill et al. (2004). The authors also investigated the effect of data compression on different the quantities commonly used in data-driven process analyses. The main conclusion is that minimum-variance and nonlinearity assessment are two procedures that require high-fidelity data; data compression alters these measures significantly.

Thornhill et al. (2004) proposed the following simple methods for automatic detection of compression and quantisation. Suppose that the reconstructed data set is piecewise linear. Its second derivative is zero everywhere apart from the places where the linear segments join. Therefore the presence of the characteristic linear segments can be detected by counting of zero-valued second differences calculated from $(i = 2, \ldots, N-1)$

$$\Delta(\Delta\hat{y})_i = \frac{(\hat{y}_{i+1} - \hat{y}_i)/T_s - (\hat{y}_i - \hat{y}_{i-1})/T_s}{T_s} = \frac{\hat{y}_{i+1} - 2\hat{y}_i + \hat{y}_{i-1}}{T_s^2}, \tag{7.5}$$

where $\hat{y}$ is the reconstructed signal. This gives n_0 zero second derivatives. The compression factor can be estimated from

$$CF_{\text{est}} = \frac{N}{N - n_0}. \tag{7.6}$$

Quantisation estimation is based on computing the non-zero (first) derivatives

$$(\Delta\hat{y})_i = \frac{(\hat{y}_{i+1} - \hat{y}_i)}{T_s} \tag{7.7}$$

giving $\Delta\mathbf{y}_{\neq 0}$, i.e. the vector of elements differing from zero. The quantisation factor can be estimated from

$$QF_{\text{est}} = \frac{\min(\Delta\mathbf{y}_{\neq 0})}{\sigma_{\hat{y}}}. \tag{7.8}$$

The data set should be considered unsuitable for control performance assessment (at least minimum-variance and nonlinearity assessment) if $CF_{\text{est}} > 3 \vee QF_{\text{est}} > 0.4$.

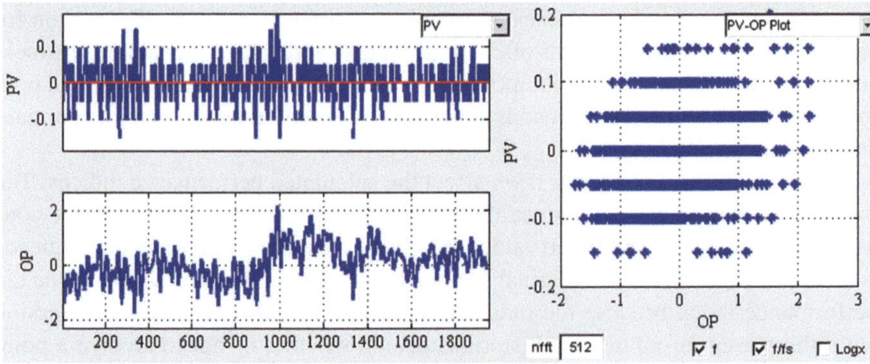

Fig. 7.3 Data from the industrial temperature control loop CHEM3

Example 7.3 Figure 7.3 shows data from a temperature control loop in a chemical plant. One can observe that the loop data are heavily quantised. This is confirmed by the quantisation factor $QF_{\text{est}} = 1.0$. The compression factor $CF_{\text{est}} = 2.0$ indicates that the data are also significantly compressed. To check if the oscillation is due to the quantisation nonlinearity, the bicoherence method combined with ellipse fitting (Sect. 11.7) has been applied to the data. The values of the indices were computed to be $NLI = 0.24$ and $NLI = 0.23$, which signal the presence of a nonlinearity in the loop. An ellipse could be fitted to the PV–OP plot, but with an ellipse fitness of just 10 %. This low value indicates that there is a nonlinearity other than stiction, i.e. in this case a quantisation. Note that a Wiener filter was applied to the data prior to the nonlinearity analysis.

7.2 Prediction Models and Identification Methods

The first step in control-performance assessment is the choice of a (time-series) model structure for describing the net dynamic response associated with the control error. Even though system identification is used as a vehicle, its goal in CPM is *not* the estimation of model parameters themselves, but the estimation of the dynamic of the process and the noise model in impulse-response form. Therefore, any approaches that reduce the modelling burden should be preferred.

7.2.1 Implication of the Use of Routine Operating Data

No matter which model for performance assessment is used, it should always be identified from available operating data without requiring time-consuming active identification experiments. Indeed, safety, product quality or efficiency considerations do not allow the process to run in open loop or to be excited by artificial signals in most practical situations, except in the stage of control system commissioning. Often, the only hope is to have some set-point changes in the data. Nevertheless, closed-loop data are usually sufficient for an accurate closed-loop modelling serving for control performance assessment. Sometimes, it may also be necessary to identify the controller itself if the controller parameters are either not easily available or the controller has been detuned and its actual parameters are not as originally recorded; see Bezergianni and Georgakis (2003) for this topic.

As pointed out by many researchers, e.g. Söderström and Stoica (1989) and Ljung (1999), the fundamental problem with closed-loop data is the correlation between the unmeasurable noise and the input. When feedback is used, the input u and noise will be correlated because u is determined from the process variable, which contains the noise. Under some circumstances, an identification algorithm may lead to the inverse controller transfer function as an estimate of the process model.

The application of closed-loop identification methods for derivation of models that serve as a basis for control design has been established as "identification for control"; see Isermann (1971), Hjalmarsson et al. (1996) and Van den Hof and Schrama (1995). Traditional closed-loop identification approaches fall into the prediction error methods (PEMs) framework. The advantage of PEMs is that the convergence and asymptotic variance results are available (Ljung 1999). The disadvantage of PEMs is that they involve in a complicated parameterisation step, which makes them difficult to apply to MIMO-system identification problems. In general, the estimates from subspace methods are not as accurate as those from PEMs.

It is beyond the scope of this book to discuss these methods in detail. For an overview and more details on closed-loop identification, see e.g. Van den Hof and Schrama (1995) and Qin et al. (2002). A two-step closed-loop identification scheme has been proposed by Huang and Shah (1999: Chap. 15). The motivation of circumventing the complicated parameterisation of PEMs gave birth to subspace identification methods, which are now popular techniques for closed-loop identification in the context of CPM; see Kadali and Huang (2002a, 2002b, 2004) and Huang et al. (2006). These developments are described in a recent book by Huang and Kadali (2008). An overview of subspace identification methods is provided by Qin (2006).

7.2.2 Role of the Estimated Model

In the CPM context, the modelling objective is to create a disturbance model, but not for the traditional purpose of prediction or simulation; rather, only the first τ impulse-response coefficients are used to estimate the minimum variance. The model is only required to be an adequate fit to the (multi-step ahead) predictable component within the data set (Desborough and Harris 1992; Thornhill et al. 1999). The residuals or the innovation sequence are of primary interest. The model parameters themselves are of no interest, and so the model validation does not play a big role in the CPM framework.

Consequently, any models and identification methods can be applied, so long as they deliver a model of sufficient (prediction) quality. Even general nonlinear modelling approaches, such as nonlinear ARMA(X) model structures or artificial neural networks, could be applied, when useful or necessary. In practice, however, it is strongly recommended to keep the model as simple as possible, preferably of the AR(X) type because of its simple and fast estimation.

7.2.3 AR(X)-Type Models

The linear black-box structures frequently used in practice are all variants of the general family, known as Box–Jenkins (BJ) models (Ljung 1999),

$$A(q)y(k) = q^{-\tau}\frac{B(q)}{F(q)}u(k) + \frac{C(q)}{D(q)}e(k), \tag{7.9}$$

using different ways of picking up "poles" of the system and different ways of describing the noise characteristics. $A(q)$, $B(q)$, $C(q)$, $D(q)$ and $F(q)$ are polynomials in q^{-1} of order n, m, p, l and r, respectively:

$$\begin{aligned} A(q) &= 1 + a_1q^{-1} + a_2q^{-2} + \cdots + a_nq^{-n}, \\ B(q) &= b_0 + b_1q^{-1} + b_2q^{-2} + \cdots + b_mq^{-m}, \\ C(q) &= 1 + c_1q^{-1} + c_2q^{-2} + \cdots + c_pq^{-p}, \\ D(q) &= 1 + d_1q^{-1} + d_2q^{-2} + \cdots + d_lq^{-l}, \\ F(q) &= 1 + f_1q^{-1} + f_2q^{-2} + \cdots + f_rq^{-r}. \end{aligned} \tag{7.10}$$

Special forms of the general model are shown in Fig. 7.4.

Usually, a high-order autoregressive (AR) model is used due its simplicity and the ability to be identified by a least-squares estimator without numerical iterations. Too-high-order AR models may, however, be needed to approximate systems that exhibit oscillating behaviour, which might lead to numerical problems or poor estimation. Theoretically, an ARMA model is the better option; the estimation of its parameters needs a nonlinear optimisation routine. ARMA models are now reaching the point of widespread use in control performance analysis owing to their many merits (Kozub 2002): simplicity, easy access and interpretability. An optimal model order can be found by using many methods, such as the Lipschitz quotients method (He and Asada 1993) or the deterministic subspace identification method (Van Overschee and De Moor 1996).

For MV assessment of feedback/feedforward control loops or LQG-based performance assessment, AR(MA)X models may be needed to be identified from input/output data of the process. The assessment of LTV processes requires the estimation of LTV ARMA models using any recursive time-series analysis algorithm; see Huang (2002). The models mentioned can also be formulated in equivalent state-space forms, which can effectively be estimated by means of prediction error methods or subspace-based methods.

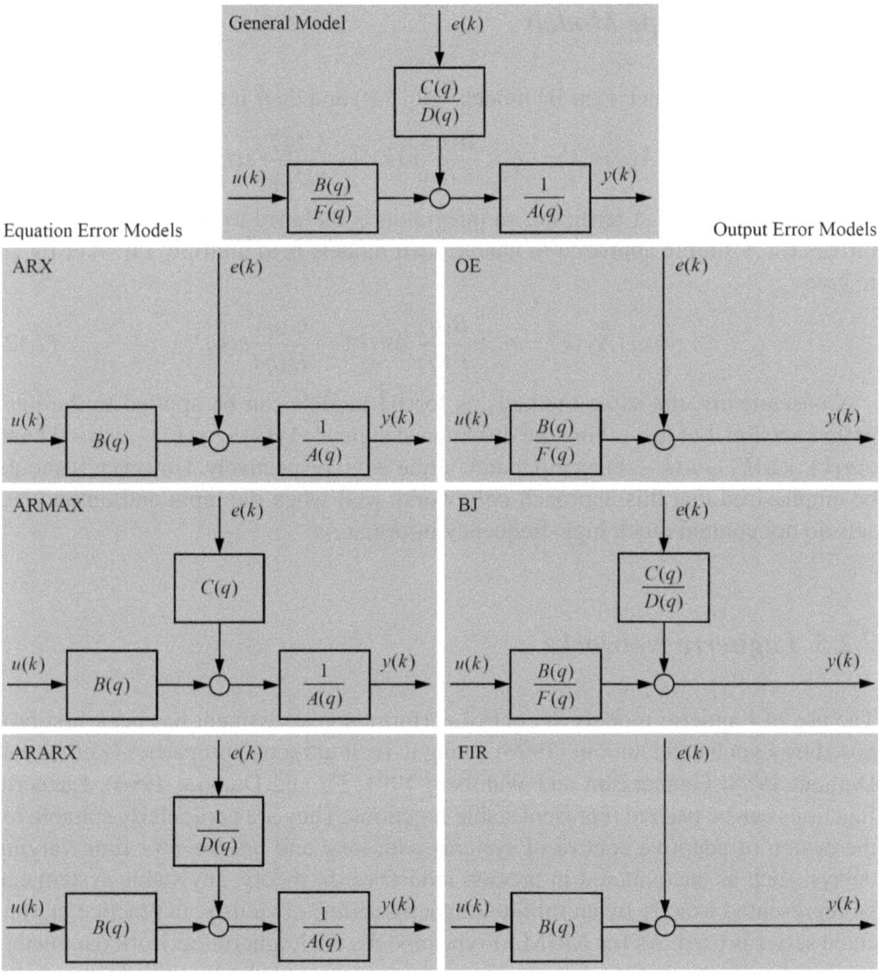

Fig. 7.4 Overview of common linear dynamic models; time-series models, i.e. AR, ARMA and ARAR, result by setting $B = 0$

The topic of identification of linear black-box models using LS or PEM algorithms can be considered mature and will not be treated here. See standard textbooks such as Ljung and Söderström (1987), Söderström and Stoica (1989), Isermann (1992), Johansson (1993) and Ljung (1999). A more recent treatment of the subject, including nonlinear identification methods, can be found in Nelles (2001). Also, many widespread commercial packages contain system identification toolboxes, such as the MATLAB Identification Toolbox and the LabVIEW System Identification Toolkit.

7.2.4 ARI(X)-Type Models

The only difference between BJ models (Eq. 7.9) and their integrated version

$$A(q)y(k) = q^{-\tau}\frac{B(q)}{F(q)}u(k) + \frac{C(q)}{\Delta D(q)}e(k) \tag{7.11}$$

is that in the latter a Δ-term, i.e. an integrator, is included to consider drifting disturbances. A simple approach to handle such models is to multiply Eq. 7.11 by Δ, to give

$$A(q)\Delta y(k) = q^{-\tau}\frac{B(q)}{F(q)}\Delta u(k) + \frac{C(q)}{D(q)}e(k). \tag{7.12}$$

Consequently, the same methods as for BJ models can be applied to the integrated version, but just taking the differenced signals $\Delta y(k) = y(k) - y(k-1)$ and $\Delta u(k) = u(k) - u(k-1)$ as input and output data, respectively. However, it should be emphasised that this approach only works well when the input and output data sets do not contain much high-frequency information.

7.2.5 Laguerre Networks

The use of Laguerre models in control-performance assessment has been first proposed by Lynch and Dumont (1996) owing to their attractive properties (Zervos and Dumont 1988; Gunnarsson and Wahlberg 1991; Fu and Dumont 1993). Laguerre functions can be used to represent stable functions. They are particularly suitable for the design of adaptive control of systems with long and unknown or time-varying delays, such as encountered in process industries. In theory, any stable system can be represented exactly by an infinite Laguerre series; obviously, in practice, a truncated series is used. As for AR(MA)-type models, the Laguerre-network parameters can be identified using least-square or prediction-error methods.

Laguerre-filter models are special versions of the orthonormal basis functions (OBFs)

$$\begin{aligned} y(k) &= g_1L_1(q)u(k) + g_2L_2(q)u(k) + \cdots + g_mL_m(q)u(k) + e(k) \\ &= \sum_{i=1}^{m} g_iL_i(q)u(k) + e(k) \end{aligned} \tag{7.13}$$

as an approximation of the series expansion, i.e. impulse response of the system,

$$y(k) = \sum_{i=1}^{\infty} g_iL_i(q)u(k), \tag{7.14}$$

where g_i is the ith Laguerre gain, and $L_i(q)$ the ith Laguerre filter. OBFs can be viewed as generalised FIR filters, where q^{-i} is replaced by a filter $L_i(q)$. The goal is to find such that filters $L_i(q)$ that yield fast converging coefficients g_i so that the infinite series expansion (Eq. 7.14) is approximated to a required degree of accuracy by Eq. 7.13 with an order m as small as possible.

The choice of the filters $L_i(q)$ in OBF models can be seen as the incorporation of *prior knowledge*, which can stem from many sources like physically based modelling, step responses or correlation analysis. Thus, there are different options for selecting the filters. Perhaps the most popular OBF model is the Laguerre-filter model extensively studied by Wahlberg (1991). It can be expressed by Eq. 7.13, where

$$L_i(q) = \frac{\sqrt{1-\alpha^2}q^{-1}}{1-\alpha q^{-1}}\left(\frac{q^{-1}-\alpha}{1-\alpha q^{-1}}\right)^{i-1}, \tag{7.15}$$

or recursively

$$L_i(q) = \frac{q^{-1}-\alpha}{1-\alpha q^{-1}}L_{i-1}(q), \qquad L_1(q) = \frac{\sqrt{1-\alpha^2}q^{-1}}{1-\alpha q^{-1}}. \tag{7.16}$$

$\alpha \leq 1$ is the pole location of the first filter $L_i(q)$ (a first-order system), known as the filter time scale, or simply filter pole.

Now, the closed-loop transfer function (between ε and y) is approximated by a Laguerre network:

$$y(k) = \left[1 + \sum_{i=1}^{m} g_i L_i(q)\right]\varepsilon(k), \tag{7.17}$$

where the unknown noise sequence $\varepsilon(k)$, as the input of the Laguerre model, has to be simultaneously estimated together with the filter gains g_i. Therefore, the Laguerre network identification, despite sporadic claims to the contrary in literature, is nonlinear in the parameters when used in stochastic identification of disturbance models.

For parameter estimation and calculation of the Harris index, it is convenient to use the discrete Laguerre network in state-space form (Lynch and Dumont 1996):

$$\begin{aligned} \boldsymbol{x}(k+1) &= \boldsymbol{A}\boldsymbol{x}(k) + \boldsymbol{b}\varepsilon(k), \\ y(k) &= \boldsymbol{c}^{\mathrm{T}}\boldsymbol{x}(k) + \varepsilon(k), \end{aligned} \tag{7.18}$$

where

$$\begin{aligned} &\boldsymbol{x}(k) = \left[L_1(q)\varepsilon(k) \quad L_2(q)\varepsilon(k) \quad \ldots \quad L_m(q)\varepsilon(k)\right]^{\mathrm{T}}, \\ &\boldsymbol{A} = [a_{ij}], \qquad a_{ij} = \begin{cases} \alpha, & i = j, \\ 0, & i < j, \\ (-\alpha)^{i-j-1}\sqrt{1-\alpha^2}, & i > j, \end{cases} \\ &\boldsymbol{b} = [b_1 \quad b_2 \quad \ldots \quad b_m]^{\mathrm{T}}, \quad b_i = (-\alpha)^{i-1}\sqrt{1-\alpha^2}, \\ &\boldsymbol{c} = [g_1 \quad g_2 \quad \ldots \quad g_m]^{\mathrm{T}} \equiv \boldsymbol{\Theta}. \end{aligned} \tag{7.19}$$

The minimum variance is determined from

$$\sigma_{\mathrm{MV}}^2 = \left[1 + \sum_{i=0}^{\tau-1}\left(\boldsymbol{c}^{\mathrm{T}}\boldsymbol{A}^i\boldsymbol{b}\right)^2\right]\sigma_\varepsilon^2, \tag{7.20}$$

where $h_i = \boldsymbol{c}^{\mathrm{T}}\boldsymbol{A}^i\boldsymbol{b}$ are the impulse response coefficients or Markov parameters of the system.

The Laguerre model can be effectively estimated by PEM or by applying the extended RLS algorithm (Ljung and Söderström 1987)

$$\begin{aligned}
\boldsymbol{x}(k) &= \boldsymbol{A}\boldsymbol{x}(k-1) + \boldsymbol{b}\eta(k-1),\\
\boldsymbol{P}(k) &= \boldsymbol{P}(k-1) - \frac{\boldsymbol{P}(k-1)\boldsymbol{x}(k)\boldsymbol{x}^{\mathrm{T}}(k)\boldsymbol{P}(k-1)}{1+\boldsymbol{x}^{\mathrm{T}}(k)\boldsymbol{P}(k-1)\boldsymbol{x}(k)},\\
\hat{\boldsymbol{c}}(k) &= \hat{\boldsymbol{c}}(k-1) + \boldsymbol{P}(k)\boldsymbol{x}(k)\big[y(k) - \hat{\boldsymbol{c}}^{\mathrm{T}}(k-1)\boldsymbol{x}(k)\big],\\
\eta(k) &= y(k) - \hat{\boldsymbol{c}}^{\mathrm{T}}(k)\boldsymbol{x}(k).
\end{aligned} \tag{7.21}$$

The residual η gives an estimate of the noise sequence ε and can be used to estimate the noise variance σ_ε^2, which is required in Eq. 7.20. The estimated minimum variance will then be

$$\hat{\sigma}_{\mathrm{MV}}^2 = \left[1 + \sum_{i=0}^{\tau-1}\left(\hat{\boldsymbol{c}}^{\mathrm{T}}\boldsymbol{A}^i\boldsymbol{b}\right)^2\right]\hat{\sigma}_\eta^2. \tag{7.22}$$

The optimum filter pole α depends on the characteristics of the system impulse response, such as its rate of decay, its smoothness and the time delay. An optimum value can be determined using the relationships derived in Fu and Dumont (1993):

$$\alpha_{\mathrm{opt}} = \frac{2M_1 - 1 - M_2}{2M_1 - 1 + \sqrt{4M_1M_2 - M_2^2 - 2M_2}} \tag{7.23}$$

with

$$\begin{aligned}
M_1 &= \frac{\sum_{i=0}^{\infty} i h_i^2}{\sum_{i=0}^{\infty} h_i^2} + (\tau - 1),\\
M_2 &= \frac{\sum_{i=0}^{\infty} i(\Delta h)_i^2}{\sum_{i=0}^{\infty} h_i^2} + 2(\tau-1)\left(1 - \frac{\sum_{i=0}^{\infty} h_i h_{i+1}}{\sum_{i=0}^{\infty} h_i^2}\right).
\end{aligned} \tag{7.24}$$

As this method needs the discrete impulse response of the system, we propose and have good experience with first identifying a higher-order AR model and then using it for generating the impulse response.

A procedure for estimating the Harris index using Laguerre networks can be formulated as follows.

Procedure 7.1 Performance assessment based on Laguerre networks.

1. Select the filter order.
2. Determine/estimate the system time delay τ.
3. Estimate a higher-order AR model from closed-loop data to estimate the impulse response.
4. Determine the optimum filter pole using the impulse response estimated in Step 3 and Eq. 7.23.

5. Identify the Laguerre gains and noise sequence using PEM or RLS from collected output samples.
6. Estimate the minimum variance from Eq. 7.22.
7. Calculate the output variance (Eq. 1.1 or 2.39).
8. Compute the (Harris) performance index (Eq. 2.34).

A thorough treatment of modelling with Laguerre filters can be found by Wahlberg and Hannan (1993). They conclude that—having chosen the Laguerre-filter pole appropriately—the number of parameters needed to obtain useful approximations can be considerably reduced, compared to AR modelling. Another nice feature of Laguerre approximations is that they can well capture the behaviour of time delays (as a part of the model), *without* choosing an explicit delay value τ. Once the Laguerre-filter model is identified, an estimate of τ can be determined from the model zeros; see Sect. 7.3.1.

Even if the Laguerre approach is superior to AR modelling, as advocated by Lynch and Dumont (1996) and Wang and Cluett (2000), it still has some drawbacks: systems with several scattered dominating poles cannot be well described, and resonant poles cause problems in terms of slow convergence since they occur in complex-conjugated pairs. The obvious way to circumvent these difficulties is to use several, possibly complex poles, leading to Kautz filters (Wahlberg 1994), or their generalised versions (Van den Hof et al. 1995). However, the use of these models would require more a priori knowledge about the system in terms of two or more pole locations.

7.2.6 Model-Free (Subspace) Identification

Besides some conceptual novelties, such as re-emphasizing of the state in the field of system identification, subspace methods are characterised by several advantages with respect to PEMs (Favoreel et al. 2000):

- *Parameterisation.* In subspace methods, the model is parameterised by the full state space model, and the model order is decided upon in the identification procedure.
- *Use for MIMO Systems.* There is no basic complication for subspace algorithms in going from SISO to MIMO systems. This is, however, particularly non-trivial for PEMs.
- *Initial Guess.* A non-zero initial state poses no additional problems in terms of parameterisation, which is not the case with input–output-based parameterisations, typically used in PEMs.
- *Numerical Properties.* Subspace methods, when implemented correctly, have better numerical properties than PEMs. For instance, subspace identification methods do not involve nonlinear optimisation techniques, which means that they are fast (since non-iterative) and accurate (since no problems with local minima occur).

- *Use for Unstable Systems.* Stable systems are treated exactly the same way as unstable ones.

The price to be paid for all these nice things is that they are sub-optimal. To demonstrate this trade-off, Favoreel et al. (2000) have compared the two methods on 10 industrial examples.[1] The conclusion of this comparison was that subspace methods represent a valid alternative to the "classical" versions of PEMs. They are fast because no iterative nonlinear optimisation methods are involved, and moreover, they are sufficiently accurate in practical applications. From a theoretical point of view, PEMs are more accurate than subspace methods, as they clearly optimise an objective function. However, if a good initial estimate of the model parameters is not available, the found solution might not be the optimal solution, owing to local minima in the optimisation problem.

Under the framework of subspace identification, a number of approaches and algorithms appeared in the literature. As stated by Favoreel et al. (2000), the difference between the three subspace identification algorithms N4SID/MOESP/CVA is the way the weighting matrices are used in the algorithm. In fact, the MATLAB Identification Toolbox offers a feature called "N4weight" for the "N4SID" command wherein the user can specify MOESP or CVA and the respective weighting matrices will be used, so that it is equivalent to using MOESP/CVA subspace identification algorithms. Other refined algorithms can be found in Wang and Qin (2002) and Huang et al. (2005a, 2005b).

Recent work by Kadali and Huang (2002a) allows the identification of only two of the subspace matrices, namely the deterministic subspace matrix and stochastic subspace matrix, from closed-loop data without requiring any a priori knowledge of' the controller. This method requires, however, set-point excitation and has also been extended to the case of measured disturbances.

A recent paper by Huang et al. (2005a, 2005b) contains a summary and detailed comparison of various closed-loop subspace identification algorithms, including their pros and cons. Two conclusions drawn from this comparative study are:

- For open-loop identification, all algorithms deliver similar performance in both bias and variance aspects.
- The classical MATLAB algorithms (N4SID, MOESP and CVA) yield essentially the same performance, and all are biased in the presence of feedback control.

7.2.7 Estimation of Process Models from Routine Operating Data

Unless some restrictive conditions are satisfied, the open-loop transfer function G_p cannot be identified from normal operating data, i.e. gathered under feedback control, even if the "true" model structure is employed. This will be briefly explained as follows; refer to Söderström et al. (1975) and MacGregor and Fogal (1995) for more details.

[1] The data sets considered can be downloaded freely from www.esat.kuleuven.ac.be/sista/daisy.

7.2.7.1 Basic Identifiability Problem

From block diagram in Fig. 2.1, the two fundamental relationships can be observed:

$$y(k) = G_{\mathrm{p}}(q,\boldsymbol{\Theta}_0)u(k) + G_{\varepsilon}(q,\boldsymbol{\Theta}_0)\varepsilon(k), \tag{7.25}$$

$$u(k) = -G_{\mathrm{c}}(q)y(k) + r(k). \tag{7.26}$$

These equations imply that the input and output data should satisfy

$$y(k) = G_{\mathrm{p}}(q,\boldsymbol{\Theta}_0)u(k) + G_{\varepsilon}(q,\boldsymbol{\Theta}_0)\varepsilon(k), \tag{7.27}$$

$$y(k) = -G_{\mathrm{c}}^{-1}(q)u(k) + r(k), \tag{7.28}$$

where $\boldsymbol{\Theta}_0$ is the real parameter vector. Whereas $\varepsilon(k)$ acts as the disturbance to Eq. 7.27, $r(k)$ plays a similar role as a disturbance to Eq. 7.28. The task of any (process) identification procedure is to find an estimate $\hat{G}_{\mathrm{p}}$ of the process model G_{p}, i.e.

$$y(k) = \hat{G}_{\mathrm{p}}(q,\boldsymbol{\Theta})u(k), \tag{7.29}$$

that fits measured input $u(k)$ and output $y(k)$ data the best.

Comparing Eq. 7.29 with Eqs. 7.27 and 7.28, one can see that, depending on the relative "size" of ε and r, an identification procedure may end up with Eq. 7.27, i.e., $\hat{G}_{\mathrm{p}}(q,\boldsymbol{\Theta}) = G_{\mathrm{p}}(q,\boldsymbol{\Theta}_0)$, Eq. 7.28, i.e., $\hat{G}_{\mathrm{p}}(q,\boldsymbol{\Theta}) = -G_{\mathrm{c}}^{-1}(q)$, or some combination of both. A particular situation occurs when it is desired to undertake completely passive identification, i.e. from routine operating data without any dither signal, and when the reference signal r is constant, i.e., there is no external excitation. This situation is usually the desired scenario in CPM. Then, it is likely that an identification algorithm will pick up $-G_{\mathrm{c}}^{-1}$ as an estimate of G_{p}, provided that G_{p} and G_{c} are assumed to be linear. Note also the fact that under feedback control the input and disturbance sequences are correlated, since the input is calculated from the measured output corrupted by the disturbance. Occasionally, due to nonlinearity or some natural perturbation present in the controller or process, Eq. 7.29 may fit G_{p} of Eq. 7.27 better than $-G_{\mathrm{c}}^{-1}$ of Eq. 7.28, but this is not always the case.

7.2.7.2 Conditions for Identifiability from Pure Closed-Loop Data

Nevertheless, it has been shown by Söderström et al. (1975) that it is possible under *some restrictive necessary and sufficient conditions* on the true process and controller orders, to identify a model of G_{p} only from pure feedback data, without the injection of a dither signal as is usual in closed-loop identification; see also MacGregor and Fogal (1995). Given the system described by an ARMAX model (Fig. 2.1; Eq. 2.1) and the controller by

$$G_{\mathrm{c}}(q) = \frac{\alpha(q)}{\beta(q)}, \tag{7.30}$$

a necessary and sufficient condition for asymptotic identifiability is that (Söderström et al. 1975)

$$\max(n_\beta - n_B, d + n_\alpha - n_A - 1) \geq n_p, \tag{7.31}$$

where n_A, n_B, n_α and n_β are the orders of the polynomials $A(q)$, $B(q)$, $\alpha(q)$ and $\beta(q)$, respectively, n_p is the order of the common factor in the polynomials $C(q)$ and $A(q)\beta(q) - q^{-d}B(q)\alpha(q)$, and $d = \tau - 1$. If the identifiability condition is not satisfied, then *no model can be identified that adequately describes the process, even if the model structure is exactly known.* This condition could be met if the time delay d were sufficiently large and the controller had a sufficiently high order. However, even in that case, the accuracy of the process parameter estimates obtained from routine operating data may be poor (Box and MacGregor 1974; MacGregor and Fogal 1995). A further necessary condition has been formulated by Söderström et al. (1975):

$$\min(\hat{n}_A - n_A, \hat{n}_B + \hat{d} - n_B - d, \hat{n}_C - n_C) \geq 0, \tag{7.32}$$

i.e., the model to be identified has to be over-parameterised. $\hat{n}_A$, $\hat{n}_B$, $\hat{n}_C$ and $\hat{d}$ are the estimated orders/delay of the polynomials. Of course, the condition in Eq. 7.32 cannot be checked a priori, as the exact model is usually unknown.

Example 7.4 To illustrate these points, consider the following system:

$$y(k) = \frac{0.5q^{-\tau}}{1 - 0.8q^{-1}}u(k) + \frac{1}{1 - 0.6q^{-1}}\varepsilon(k) \tag{7.33}$$

with a proportional-only controller $K_c = 0.4$, $r = 0$ and $\sigma_\varepsilon^2 = 1$. No dither signal was applied to the system, and 10 000 observations were collected and used for the identification of an ARMAX(2, 2, 1, τ) model, i.e.

$$y(k) = \frac{(b_0 + b_1q^{-1})q^{-\tau}}{1 + a_1q^{-1} + a_2q^{-2}}u(k) + \frac{1}{1 + c_1q^{-1}}\varepsilon(k). \tag{7.34}$$

The real parameter values corresponding to Eq. 7.33 are: $b_0 = 0.5$, $b_1 = -0.3$, $a_1 = -1.4$, $a_2 = 0.48$ and $c_1 = -0.8$. The simulation and identification tasks have been performed for $\tau = 0$–5.

Figure 7.5 shows the obtained parameter estimates compared to the real values as functions of the time delay. The step responses of the real and identified models are illustrated in Fig. 7.6. It can be clearly seen that the process becomes identifiable only when $\tau \geq 3$. This is in agreement with the identifiability condition in Eq. 7.31, i.e., $\max(0 - 1, d + 0 - 1 - 1) \geq 0 \Rightarrow d \geq 2 \Rightarrow \tau \geq 3$. It is also observed that for $\tau \leq 2$, the estimated step response reflects a weighted combination of the process dynamics and the negative inverse of the controller ($-G_c^{-1}$). For $\tau = 0$, $-G_c^{-1}$ is exactly identified, i.e., $b_0 = -G_c^{-1} = -1/K_c = -2.5$; see also Fig. 7.6, upper subplot/left side.

As Eq. 7.33 has a BJ structure, the aforementioned simulations and identification tasks have been repeated using a BJ(1, 0, 1, 1, τ) model. The obtained results are shown in Fig. 7.7 and Fig. 7.8. As expected, the BJ models captures the process dynamics better than the ARMAX models. Moreover, the identifiability condition seems to be satisfied already for $\tau \geq 2$ when using a BJ model.

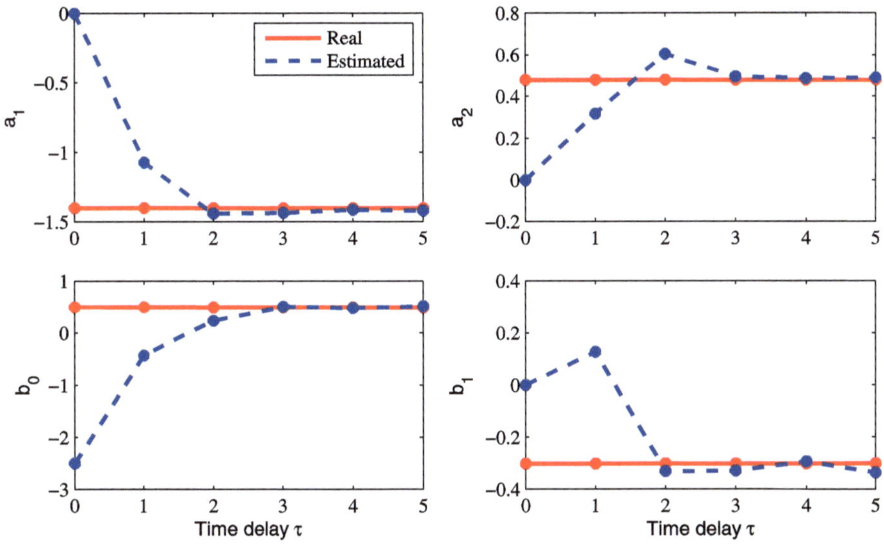

Fig. 7.5 Real and estimated parameters of the process model (ARMAX) in Example 7.4 as function of the time delay

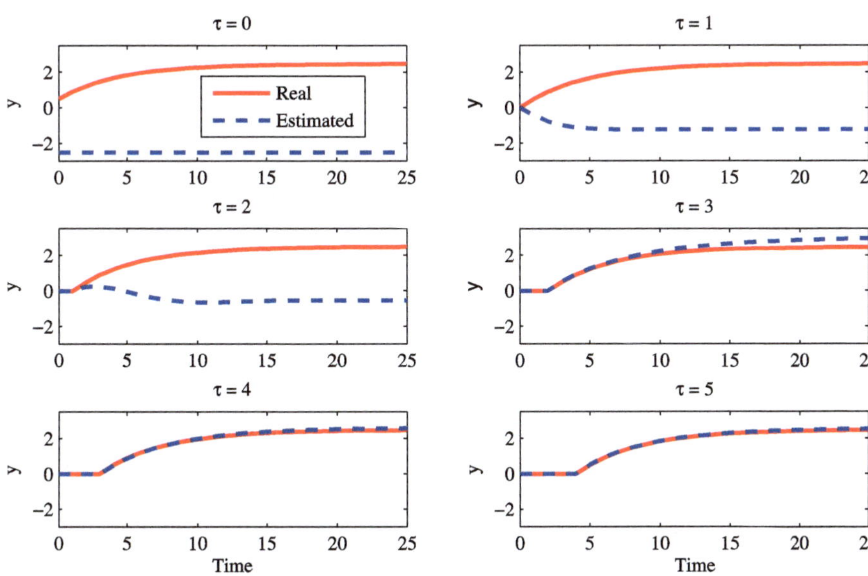

Fig. 7.6 Real and estimated step responses of the process model (ARMAX) in Example 7.4 for different time delay values

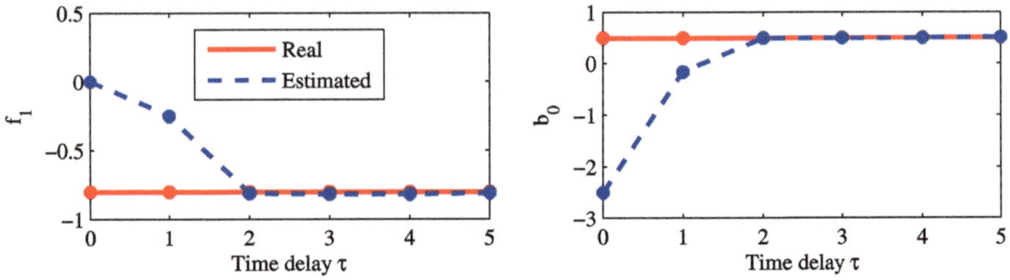

Fig. 7.7 Real and estimated parameters of the process model in Example 7.4 as functions of the time delay

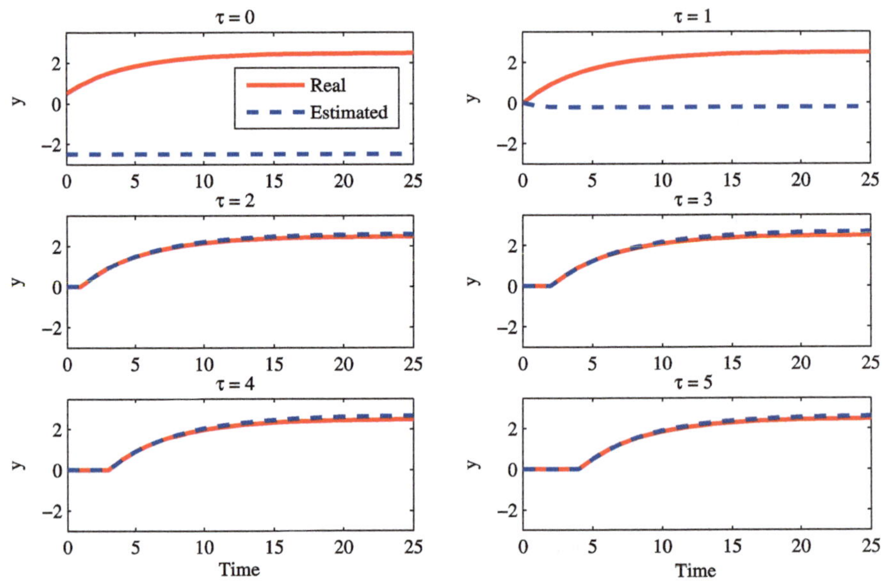

Fig. 7.8 Real and estimated step responses of the process model in Example 7.5 for two time delay values

7.2.7.3 FIR Model Estimation from Pure Closed-Loop Data

Usually, in the identification literature, separately parameterised models for the process G_{p} and disturbance/noise G_{ε} are simultaneously identified. Alternatively, a disturbance model, e.g., of the AR(I)MA type is first estimated from routine data, i.e. using pre-whitening, as carried out for the determination of the Harris index (Sect. 2.4). Then the obtained disturbance model is used as a prefilter for the input and output data, and the process model is identified from the prefiltered data. For this purpose, the system description in Fig. 2.1 can be re-written as

$$\frac{1}{\Delta G_{\varepsilon}(q)}\big(\Delta y(k)\big) = G_{\mathrm{p}}(q)\frac{1}{\Delta G_{\varepsilon}(q)}\big(\Delta u(k)\big) + \varepsilon(k). \tag{7.35}$$

Defining the filtered signals

$$y_{\mathrm{f}} \triangleq \frac{1}{\Delta G_{\varepsilon}}(\Delta y), \qquad u_{\mathrm{f}} \triangleq \frac{1}{\Delta G_{\varepsilon}}(\Delta u), \tag{7.36}$$

leads to

$$y_{\mathrm{f}}(k) = G_{\mathrm{p}}(q)u_{\mathrm{f}}(k) + \varepsilon(k). \tag{7.37}$$

Details about the role of such prefilters in closed-loop identification can be found in MacGregor and Fogal (1995).

Based on the identifiability condition in Eq. 7.31, Julien et al. (2004) have proposed the following procedure for the identification of process and disturbance models from normal operating data and the knowledge of the time delay, but without external excitation. The approach is based on the prefilter Eq. 7.37 and an IIR model in Eq. 2.4, i.e.

$$y_{\mathrm{f}}(k) = G_{\mathrm{p}}(q)u_{\mathrm{f}}(k) + \varepsilon(k) = \left(\sum_{i=\tau}^{\infty} h_i q^{-i}\right)u_{\mathrm{f}}(k) + \varepsilon(k). \tag{7.38}$$

It can be proven that an *unbiased* estimate of the impulse response G_{p} may be determined through linear regression of y_{f} against u_{f} using normal operating data, provided that the time-to-steady-state n_{p} of plant model

$$\hat{G}_{\mathrm{p}}(q) = \sum_{i=\tau}^{n_{\mathrm{p}}} \hat{h}_i q^{-i} \tag{7.39}$$

is chosen sufficiently large. An estimate $\hat{\sigma}_{\varepsilon}^2$ of the variance of the white noise sequence σ_{ε}^2 can be determined from the mean square residual. This result is *asymptotic* but may require sufficiently large data sets to ensure the convergence of the estimates. However, it is not a big concern today, as usually a huge amount of data should be available in plant data bases. For proof of this basic result, the reader should refer to Julien et al. (2004), who pointed out the following cases to be distinguished depending on the values of the time delay τ and the disturbance settling time n_{ε}:

- *Case 1. Disturbance settling time smaller than time delay, i.e.* $n_{\varepsilon} < \tau$. In this case, the coefficients of $\hat{E}(q)$ settle out within $\tau - 1$ control intervals, i.e.

$$\hat{G}_{\varepsilon}(q) \approx \hat{E}(q) + \hat{e}_{\tau-1}\sum_{i=\tau}^{\infty} q^{-i}. \tag{7.40}$$

Using Eq. 2.8 yields

$$\Delta\hat{G}_{\varepsilon}(q) \approx 1 + \sum_{i=1}^{\tau-1} \hat{d}_i q^{-i}, \quad \hat{d}_i = \hat{e}_i - \hat{e}_{i-1}. \tag{7.41}$$

Thus, $\hat{E}(q)$ derived from Eq. 2.37 can be used to generate a FIR estimate of the differenced disturbance dynamics as

$$\Delta G_{\varepsilon}(q) = \frac{C(q)}{A(q)} = \sum_{i=0}^{\infty} d_i q^{-i} \approx 1 + \sum_{i=0}^{n_{\varepsilon}} d_i q^{-i}, \qquad (7.42)$$

with $d_0 = e_0 = 1$ and $d_i = e_i - e_{i-1}$ for $i = 1, \ldots, n_{\varepsilon}$.

Note that random disturbance walks (Eq. 4.29) basically used in MPC fall in this case and the impulse response settles out immediately, i.e.

$$\hat{E}(q) = 1 + 1 \cdot q^{-1} + 1 \cdot q^{-2} + \cdots + 1 \cdot q^{-(\tau-1)} \quad \Rightarrow \quad \Delta G_{\varepsilon}(q) = 1. \quad (7.43)$$

The associated pre-whitening filter is then simply Δ. This implies that G_{p} can always be identified by regression of routine, differenced input–output data when the disturbance is a random walk.

- *Case 2. Disturbance settling time larger than time delay, i.e.* $n_{\varepsilon} \geq \tau$. This is the more general case, where the coefficients of $\hat{E}(q)$ in Eq. 2.37 will not completely describe the disturbance dynamics, and hence, the regression of y_{f} against u_f will produce a biased estimate of the plant dynamics G_{p}. A pragmatic solution in this case is to forecast the remaining coefficients of $\hat{E}(q)$ up to n_{ε} by extrapolation. This approach should reliably work if τ is "sufficiently large" compared to n_{ε}, so that $\hat{E}(q)$ exhibits "most" of the disturbance dynamics. Of course, the quantification of this remains "fuzzy" and up to the user. Julien et al. (2004) suggested that the time delay should be at least 15 % of the disturbance settling time and exceed the duration of any initial wrong-way transient in $\hat{E}(q)$ by at least four intervals, to get satisfactory extrapolations. Otherwise, an open- or closed-loop test, e.g. a step test or the injection of a dither signal, is necessary to get a reliable estimate of the process model G_{p}.

Remark To check that the description in Eq. 7.37 satisfies the identifiability conditions in Eqs. 7.31 and 7.32, one should recognise the analogy between Eqs. 2.1 and 7.37, giving: $A(q) = C(q) = 1$ and $B(q) = q^{-1} \sum_{i=\tau}^{\infty} h_i q^{-i+\tau}$, thus $n_A = 0$, $n_B = \infty$ and $n_p = 0$. Therefore, $\max(n_{\beta} - \infty, \tau + n_{\alpha}) \geq 0$ holds for all plant and controllers. On the other hand, to met the criterion in Eq. 7.32, we must satisfy $\hat{n}_B + \hat{\tau} - n_B - \tau \geq 0 \Rightarrow \hat{n}_B \geq n_B$ when the time delay is known ($\hat{\tau} = \tau$). This implies that the FIR plant model $\hat{G}_{\mathrm{p}}(q) = q^{-\tau+1}\hat{B}(q)$ in Eq. 7.39 can be used to obtain an unbiased estimate of $G_{\mathrm{p}}(q)$ as $N \to \infty$ (Julien et al. 2004).

Remark In stochastic performance assessment, e.g. minimum-variance assessment, it is almost not necessary to know or identify the process model G_{p}. Rather it suffices to get an estimate of the disturbance model G_{ε}, which is always possible using time-series modelling based on normal operating data, as already treated in Sect. 2.4.

7.2.7.4 Extrapolation Model

Julien et al. (2004) proposed fitting a second-order-plus-time-delay (SOPTD) continuous model

$$G(s) = \frac{Ke^{-\tau_d s}}{\frac{s^2}{\omega_0^2} + \frac{2D}{\omega_0}s + 1} \tag{7.44}$$

to the impulse coefficients of $\hat{E}(q)$ before the time delay and then using this model in obtaining the full disturbance model by extrapolation. The SOPTD parameters are estimated so that the objective function

$$J_{\text{extrap}} = \sum_{i=\tau_d+1}^{\tau-1} (\tilde{\psi}_i - \tilde{e}_i)^2, \quad \tilde{e}_i = \hat{e}_i - \hat{e}_{\tau_d}, \tag{7.45}$$

is minimised. $\tilde{\psi}_i$ are the coefficients (up to n_ε) of discrete step response of $G(s)$ parameterised by the current estimates K, ω_0 and D. The integer dead time τ_d has to be specified by the user from inspecting the plot of the coefficients of $\hat{E}(q)$. τ_d is introduced to avoid any undesirable initial transients, e.g. inverse response observed in $\tilde{E}(q)$, from being factored into the extrapolation (Julien et al. 2004). We have found that it may often be sufficient to adopt a FOPTD approximation rather than a SOPTD one. The estimation of the extrapolation model can be easily carried out, e.g. using the fminsearch function of the MATLAB Optimization Toolbox.

The estimated disturbance model is then a composite model containing the first τ terms from $\hat{E}(q)$ and the remaining terms from $\tilde{\psi}(q)$, i.e.

$$\hat{G}_\varepsilon(q) = \hat{E}(q) + \psi_\tau q^{-\tau} + \cdots + \psi_{n_\varepsilon} q^{-n_\varepsilon} + \psi_{n_\varepsilon} \sum_{i=n_\varepsilon+1}^{\infty} q^{-i}, \tag{7.46}$$

where $\psi_i = \tilde{\psi}_i + \hat{e}_{\tau_d}$. From this equation the differenced model can be obtained as

$$\Delta\hat{G}_\varepsilon(q) \approx 1 + \hat{d}_1 q^{-1} + \cdots + \hat{d}_{n_\varepsilon} q^{-n_\varepsilon} \tag{7.47}$$

with $\hat{d}_1 = \hat{e}_1 - 1, \hat{d}_2 = \hat{e}_2 - \hat{e}_1, \ldots, \hat{d}_\tau = \psi_\tau - \hat{e}_{\tau-1}, \hat{d}_{\tau+1} = \psi_{\tau+1} - \psi_\tau$, etc.

The whole procedure for the identification of disturbance and process models from routine operating data is illustrated in Fig. 7.9.

7.2.7.5 Fitting ARIMA Disturbance Model

Another method for estimating the disturbance model involves fitting a model of the ARIMA(n, p, l) type to normal operating data. Ko and Edgar (1998) proposed an iterative procedure based on the nonlinear least-squares minimisation of an objective function similar to Eq. 7.45. However, in the author's experience, it often suffices to fit an ARMA$(2-5, 1, 1)$ model to the differenced data and augment it by an integrator (Sect. 7.2.4). Note that also this disturbance estimation method may be inaccurate for small time delays, i.e. for $\tau \leq 2$.

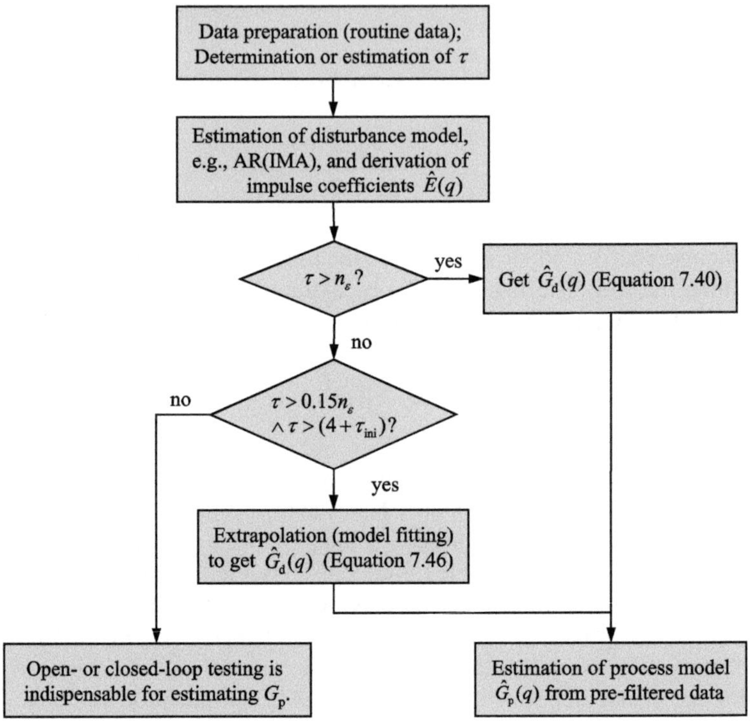

Fig. 7.9 Proposed procedure for estimating disturbance and process models from closed-loop data

Example 7.5 To illustrate the method described above, the following system taken from Julien et al. (2004) is considered:

$$G_{\mathrm{p}} = \frac{0.0059q^{-\tau}}{1-0.9813q^{-1}}, \qquad G_{\varepsilon} = \frac{1-1.1993q^{-1}+0.4910q^{-2}}{(1-0.5q^{-1})(1-0.9032q^{-1})\Delta} \tag{7.48}$$

with an infinite-horizon MPC regulator

$$G_{\mathrm{c}} = \frac{1-0.9813q^{-1}}{1.0470-1.9784q^{-1}+0.9372q^{-2}-0.0059q^{-18}}. \tag{7.49}$$

This control loop has been simulated under measurement noise with the variance $\sigma_{\varepsilon}^2 = 0.002$ and zero set point. 10 000 observations have been utilised to identify a FIR model following the procedure of Julien et al. (2004). Also, a BJ$(1, 2, 2, 1, \tau)$ model augmented by an integrator has been identified for comparison.

Figure 7.10 shows the estimation results in terms of the impulse response of the disturbance and the step response of the process. One can see that the disturbance settling time is higher than 30. Therefore, for small values of time delays, e.g. $\tau = 10$, models identified from closed-loop data are inaccurate whatever the model used. However, the BJ model seems to capture the system dynamics better than the (extrapolation-based) FIR model. For higher time delay values, e.g. $\tau = 20$, both

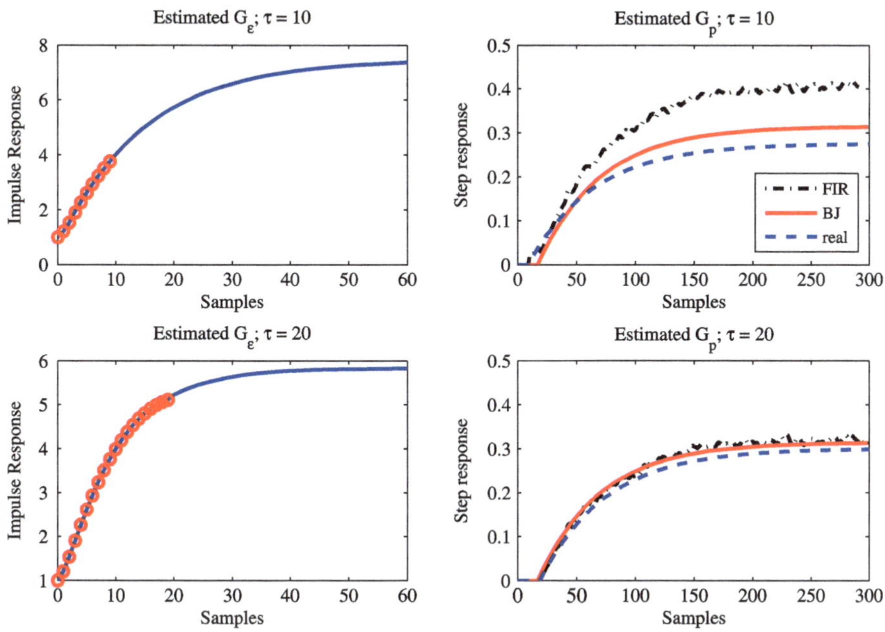

Fig. 7.10 Real and estimated step responses of the process model (BJ) in Example 7.5 for different time delay values

models represent accurate descriptions of the process with an advantage for the FIR model.

7.3 Selection of Model Parameters

Determining the time delay and model orders for the prediction-error methods is typically a trial-and-error procedure. Some useful set of steps that can lead to a suitable model is (National Instruments 2004):

1. Obtain useful information about the model order by observing the number of resonance peaks in the non-parametric frequency response function. Normally, the number of peaks in the magnitude response equals half the order of $A(q)F(q)$.
2. Obtain a reasonable estimate of delay using correlation analysis and/or by testing reasonable values in a medium size ARX model. Choose the delay that provides the best model fit based on prediction errors or other fit criterion.
3. Test various ARX model orders with this delay choosing those that provide the best fit based on e.g. Akaike's Information Criterion (AIC). For this purpose, the MATLAB Identification Toolbox offers the functions arxstruc, ivstruc and selstruc.
4. Since the ARX model describes both the system dynamics and noise properties using the same set of poles, the resulting model may be unnecessarily high in

order. By plotting the zeros and poles (with the uncertainty intervals) and looking for cancellations, the model order can be reduced. The resulting order of the poles and zeros are a good starting point for ARMAX, OE and/or BJ models with these orders used as the $B(q)$ and $F(q)$ model parameters and first- or second-order models for the noise characteristics.

5. If a suitable model is not obtained at this point, try to find additional signals that may influence the output. Measurements of these signals can be incorporated as extra input signals.

However, the application of these recipes to a large number of control loops to be assessed is time consuming and cannot be fully automated. Therefore, more simple rules are needed for selecting the model parameters, as presented below.

From the prediction error viewpoint, the higher the order of the model is, the better the model fits the data because the model has more degrees of freedom or parameters. However, more computation time and memory is needed for higher orders. Moreover, the parsimony principle advocates choosing the model with the smallest number of parameters if all the models fit the data well and pass the verification test.

7.3.1 *Time Delay Estimation*

A reliable estimate for the time delay in the closed-loop is necessary to utilise most of the presented performance assessment methods, particularly the MVC-based techniques. This is problematic when the time delay is unknown or varying. Various time-delay estimation (TDE) approaches have been proposed to address this problem. A classification, comprehensive surveys and comparative (simulation) studies of TDE methods are given in O'Dwyer (1996) and Björklund (2003).

7.3.1.1 Some Time Delay Estimation Methods

In the following, three of the most frequently used approaches for TDE are briefly described:

- *Cross-correlation method.* The classical method for TDE is based on analysing the cross-correlation between u and y as the two signals of interest. Both signals are put close to each other and then time-shifted until they agree the most. This can be formally written as

$$\hat{\tau} = \max_{\tau} \mathrm{E}\{y(k)u(k-\tau)\} \approx \max_{\tau} \sum_{k} y(k)u(k-\tau). \tag{7.50}$$

- *Relational Approximation Method.* The delay term in the continuous-time model is approximated by a low-order rational function, typically a Padé approximation. The time delay is then computed from the pole-zero excess. Isaksson (1997) proposed to estimate first a Laguerre model, followed by a calculation of the (discrete-time) zeros z_i and their conversion into continuous-time zeros s_i. A comparison with a first-order Padé approximation gives an estimate of the (continuous) time delay, assuming that the plant has no non-minimum phase zeros, except those resulting from the time delay:

$$\hat{\tau} = 1 + \frac{\hat{T}}{T_\mathrm{s}} = 1 + \frac{\sum_{i=1}^{r}(2/s_i)}{T_\mathrm{s}}, \qquad s_i \approx \frac{1}{T_\mathrm{s}}\ln(z_i), \tag{7.51}$$

where r is the number of zeros in the right half plane. This method has been modified by Horch (2000) to avoid approximation error sources (i.e. conversion of discrete-time zeros into continuous-time zeros and the Padé approximation itself) by *directly* estimating the time delay from the discrete-time zeros of the Laguerre model

$$\hat{\tau} = 1 - \left.\frac{\varphi(\omega)}{\omega T_\mathrm{s}}\right|_{\omega \ll 1}, \tag{7.52}$$

see also Dumont et al. (2002). Measures, such as *zero guarding*, have to be taken to prevent/remove "false zeros" (close to but outside of the unit circle); see Björklund (2003) for details. Since the intended use here is performance monitoring, it is not at all critical that one actually gets a kind of "apparent time delay" (Swanda and Seborg 1999) when the system (without time delay) is non-minimum phase.

- *Fixed model variable regression estimation (FMVRE).* This is a simple two-step procedure proposed by Elnaggar et al. (1991), consisting of (1) assuming some delay value (interval) and estimating an auxiliary model of fixed order by a least-square method and (2) optimising the least-square error performance index with respect to the delay value. The actual (recursive) algorithm exploits the fact that minimising this index is equivalent to maximising the cross-correlation function between the system input and output increments. This method has been advocated by Lynch and Dumont (1996).

Several time-delay estimation algorithms have been compared in Isaksson et al. (2000) for application in a monitoring tool using Monte Carlo simulations. Most methods make use of the well-known first-order plus time-delay (FOPTD) approximation of the system.

It is very important to realise that *time delay cannot be estimated from routine operating data without external excitations or abrupt changes in the control signals.* This fact often ignored by many researchers has been well proven in the theory of system identification literature; see Ljung (1999). Occasionally, this may be possible due to nonlinearity or some natural perturbation present in the process, but this is not always reliable.

7.3.1.2 Comparative Study

A comparative study of different TDE methods is now presented. Again, the data used come from a strip-thickness control loop. 145 data sets have been considered, each data set corresponds to a rolled steel-strip coil. As the speed is measured for

Table 7.1 Estimation-quality measures for the different time-delay estimation techniques

	TDE	TDEB	TDER	TDOE	TDARX	DELAY-EST	TDMET	TDLAG	
Mean	3.5	4.2	−0.9	2.3	0.7	−1.8	−1.8	1.8	−5.0
RMS	8.8	9.4	1.3	3.1	1.3	1.0	3.1	3.2	7.1

the process, the time delay can be computed accurately. The time delay values are in the range 5–16. The following TDE techniques have been investigated:

- *TDE*. Time-delay estimation using cross-cumulant method (function tde in the MATLAB Higher-order Spectral Analysis Toolbox).
- *TDEB*. Time-delay estimation using conventional bispectrum method (function tdeb in the MATLAB Higher-order Spectral Analysis Toolbox).
- *TDER*. Time-delay estimation using maximum-likelihood windowed cross-correlation (function tder in the MATLAB Higher-order Spectral Analysis Toolbox).
- *TDOE*. Time-delay estimation based on OE model (function oestructd in the MATLAB Identification Toolbox).
- *TDARX*. Time-delay estimation based on ARX model (function arxstructd in the MATLAB Identification Toolbox).
- *TDMET*. Time-delay estimation based on pre-filtered ARX model (function met1structd from Björklund 2003).
- *DELAYEST*. Time-delay estimation based on a comparison of ARX models with different delays (function delayest in the MATLAB Identification Toolbox).
- *TDLAG*. Time-delay estimation based on Laguerre-filter model (Eqs. 7.51 and 7.52).

Table 7.1 contains the mean error ($\tau - \hat{\tau}$) and root mean square (RMS) error values of the time-delay estimates for the different methods. These give estimation quality measures which can be used to see how good the estimates are. It can be concluded that the correlation-based TDER and ARX-model-based (TDARX and DELAYEST) are best. The approximation method based on Laguerre filters is not as good as expected and reported in the literature. A problem with the method based on Laguerre model (Eq. 7.51) is that it can deliver complex-valued time-delay estimates, so only the real part is taken. The higher-order-statistics-based techniques (TDE and TDEB) are worst performing.

7.3.2 Model Order Selection

Care should be taken when selecting proper model orders; it does significantly affect the estimated performance indices. In time-varying closed-loop systems, the proper model order should be determined for each individual data segment. Different disturbances acting on the system will change the model order of the closed-loop

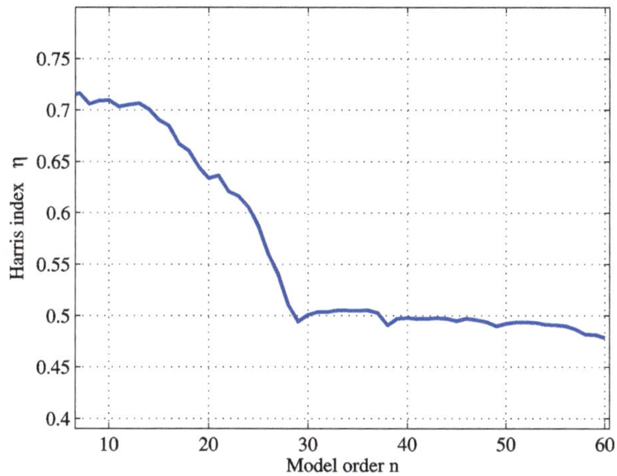

Fig. 7.11 Influence of the model order on the Harris index estimate

system. Different suggestions for selecting model orders have been given in the literature:

- Desborough and Harris (1992) proposed to start with some small model order like $n = 5$ and continuously increase n until the performance index estimate does not change so much; see Fig. 7.11. In this case, a model order of 30 should be sufficient. Note that the time delay is $\tau = 5$ for this control loop.
- Thornhill et al. (1999) used a fixed 30th-order AR model and adjust the sample time such that the closed-loop impulse response is fully captured within 30 samples.
- Horch (2000) found that a suitable model order for AR models is between 15 and 25.
- Haarsma and Nikolaou (2000) recommended the following simple model-order selection procedure: (a) Start with the smallest model possible; (b) Increase the model order until the residuals are statistically white; (c) If a maximum model order is reached without any model producing statistically white residuals, select the model with the least coloured residuals.
- Goradia et al. (2005) suggested to use $n = 20 + \tau$.

In our experience, appropriately selected model orders (typically $n \approx 20 + \tau$), and a minimum length of data, typically $N \geq 100 - 150\tau$, are necessary for obtaining reliable results. When the time delay is unknown, the use of the prediction horizon approach (Sect. 3.3) is highly recommended. However, n should not be too high, as over-parameterisation induce very noisy impulse responses; see Fig. 7.12. In other words, when the disturbance model impulse response is very noisy, it is a clear indication of over-parameterisation, and the model order should be reduced.

When performance assessment is done offline and the residual plots can be inspected, it is recommended to look at the whiteness of the residuals. To check this, whiteness tests (Ljung 1999) or portmanteau lack-of-fit tests (Box and Jenkins 1970) have been developed. MATLAB Identification Toolbox offers the function resid for checking the whiteness of the residuals; see Fig. 7.13.

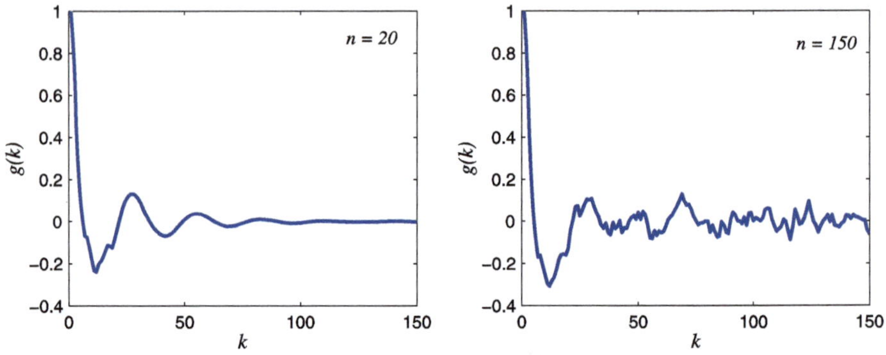

Fig. 7.12 Effect of over-parameterisation, i.e. a too high AR-model order, on the impulse response

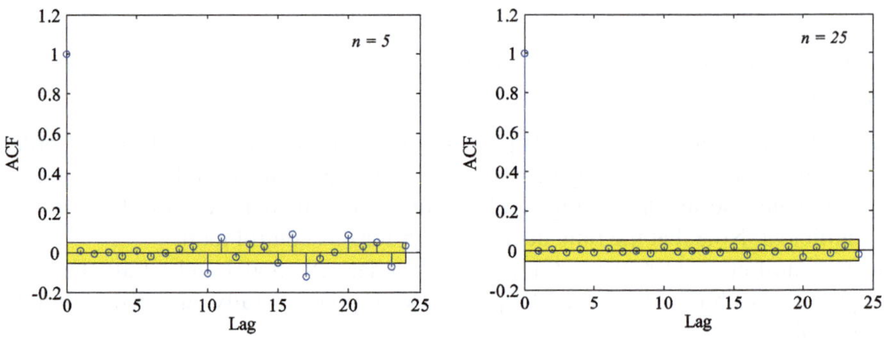

Fig. 7.13 Effect of model order on the auto-correlation of residuals (Loop: thickness control, coil no. 5)

For models with inputs, i.e. of the AR(I)MAX or more general BJ type, a cross-correlation test is also important to decide whether the selected model order set is sufficient. Figure 7.14 shows an example where the model fitting does not pass both residual tests. The model orders have to be increased.

7.4 Comparative Study of Different Identification Techniques

Five different stochastic identification techniques commonly used in the context of control performance assessment will be compared. 144 data sets from a strip-thickness control loop with different features were selected to investigate performance of the different identification methods. A difficulty in comparing the methods and working with real data is that the true model and model structure are unknown. Also cross-validation is difficult to use since the disturbances acting on the system and therefore the disturbance models are constantly changing.

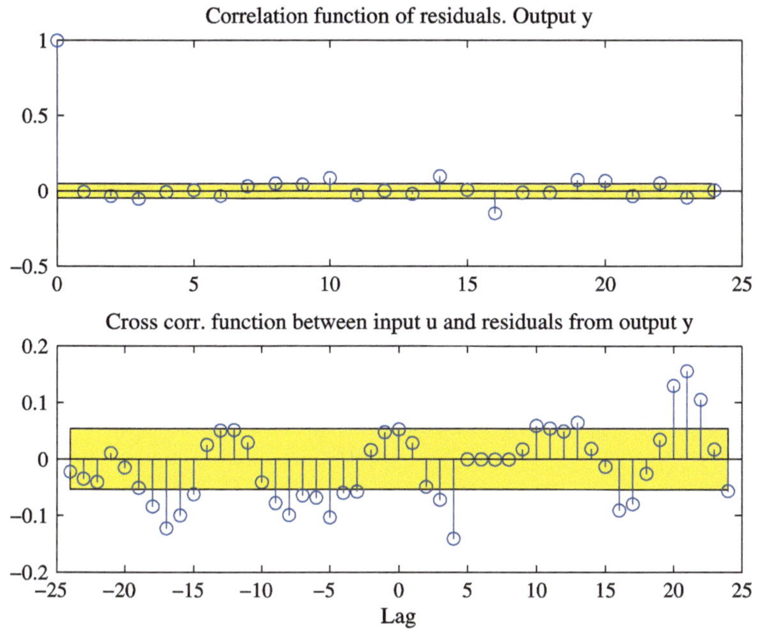

Fig. 7.14 An example of ARMAX fitting that does not pass the residual tests

The data analysed were gathered from a strip-thickness control loop in a rolling mill, described in detail in Sect. 16.3.1. The following assessment algorithms were implemented based on some functions from MATLAB Identification Toolbox:

- *AR LS*. An AR model is identified using [ar] to estimate the noise sequence and calculate the disturbance impulse coefficients, required for computing the Harris index from impulse-response coefficients; see Procedure 2.1.
- *ARMA LS-IV*. An ARMA model is estimated using the MATLAB function armax, by omitting the input and the associated B polynomial order and setting equal orders for the polynomials A and C. The Harris index has been calculated from impulse-response coefficients; see Procedure 2.1.
- *SID*. A state-space model is identified using the subspace algorithm [n4sid], selecting the option "CVA". From this model, the impulse-response coefficients have been generated to compute the Harris index; see Procedure 2.1.
- *FCOR*. The random shocks have been estimated from AR disturbance modelling. The calculation of the cross-correlation function eliminates the need to determine the impulse-response coefficients from the estimated closed-loop transfer function; see Procedure 2.2.
- *Laguerre PEM*. A Laguerre model in state-space form is estimated by [pem], i.e. the batch version of Procedure 7.1.

The loop time delay is accurately known for each coil. The maximum model order was set to 5 for both PEM and subspace identification methods, and 20 for the

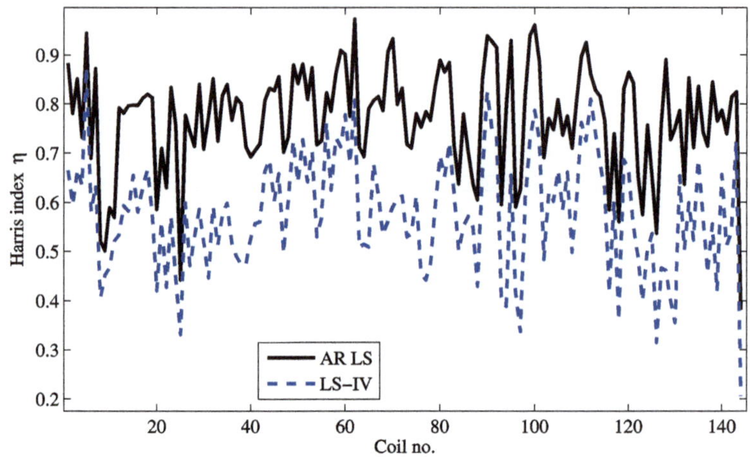

Fig. 7.15 Comparison of Harris indices for a thickness control loop, determined based on AR and ARMA models

AR identification method. Increasing the maximum order did not yield more models with white residuals.

7.4.1 AR vs. ARMA Modelling

AR LS uses only AR model, which is the most restrictive one. It does have the advantage of being simple in computation but poor in capturing all dynamics of the data unless a high-order AR model is used. However, a high-order AR is also problematic in estimation and calculation of the performance index. The AR approach is that expected to have the lowest accuracy. A better alternative is to use ARMA model.

Indeed, it can be learned from the results that calculating the performance index using AR LS clearly over-estimates the performance, compared with the index values resulting from ARMA LS-IV; see Fig. 7.15. This is also due the fact that many data sets analysed show oscillating behaviour. The same conclusion was stated by Haarsma and Nikolaou (2000), who analysed data from a snack food process.

Horch (2000) already discovered this problem from the analysis of measured data from an oscillating flow control loop. Horch thus recommended to not rely on MVC-based index in cases where an oscillation is present, particularly when it caused by nonlinearities. Therefore, oscillation detection is included in a very early stage in the comprehensive procedure proposed in Sect. 13.3.

Therefore, it is always wise to screen out oscillating loops by using specialised indices before evaluating the performance index, as will be described in Chap. 8. If one wants to calculate the Harris index for oscillating signals, ARMA modelling should be used. Note that the estimation of an ARMA model may be difficult, especially since stability cannot be guaranteed.

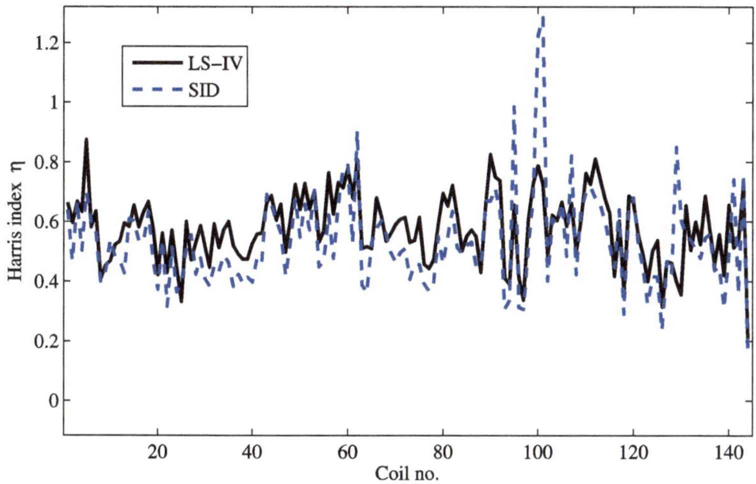

Fig. 7.16 Comparison of Harris indices for a thickness control loop, determined based on ARMA LS-IV and SID methods

7.4.2 Subspace Identification

ARMA LS-IV and SID essentially deal with the same model. Note that ARMA model is equivalent to a state-space model. However, SID uses a subspace algorithm, while ARMA LS-IV uses a prediction error method, which is an optimisation-based algorithm. Thus ARMA LS-IV will be generally better than SID, but ARMA LS-IV may be stuck in local optimal, while SID does not have the same problem.

Non-iterative methods, i.e. subspace and AR LS, are computationally much faster than the iterative PEM methods. The computational cost for the PEM methods is not prohibitive, but it might become a problem for higher-order systems.

The resulting indices from application of the subspace method (with the CVA option) in our study are very similar to those produced by PEM, but with some strange outliers ($\eta > 1$); see Fig. 7.16. Note that per SID identified models were often numerically unstable, i.e. with ill-conditioned covariance matrix. This lowers the practicability of the method.

For data set showing a nice response and perfect performance (without oscillations), the different identification methods work equally well, i.e., they give very similar impulse responses and performance indices. For such cases, AR modelling is obviously to be preferred.

7.4.3 Performance of the FCOR Algorithm

FCOR is a general procedure rather than a specific algorithm. It consists of two steps: filtering/whitening and correlation analysis. The filtering step in FCOR can

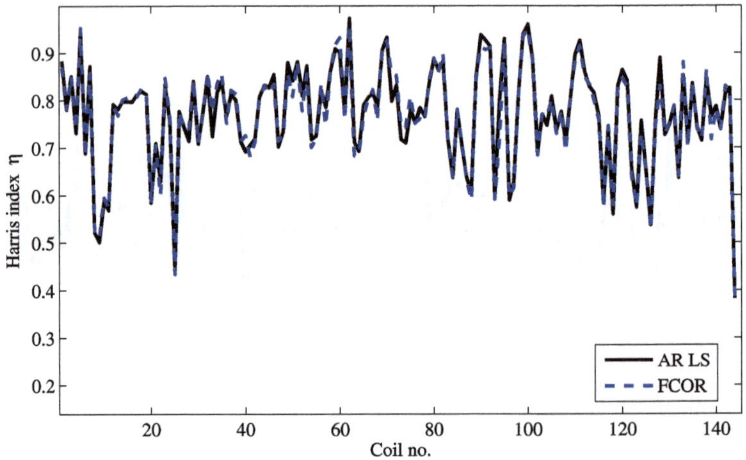

Fig. 7.17 Comparison of Harris indices for a thickness control loop, computed from AR LS and FCOR methods

use any algorithm such as AR, ARMA, state-space or even a nonlinear time series modelling algorithm. It is therefore be of no surprise for FCOR to produce similar performance as AR modelling approach if the AR algorithm is used as its filter; see Fig. 7.17. The advantage of FCOR is its flexibility in the form of the filter as well as it suitability for MIMO systems due to its simplicity in computation, namely computation of the Markov matrices can be avoided.

7.4.4 Use of Laguerre Networks

The assessment indices achieved from the identification of Laguerre networks are even higher than those from AR modelling, so the loop performance is clearly overestimated; see Fig. 7.18. The actual reason cannot be fixed at this time. Possible problems may result from possibly not optimal choice of the filter order or time scale.

Laguerre PEM uses the PEM algorithm. The PEM is optimal, but is a nonlinear optimisation. Thus it may be stuck in a local optimal. However, Laguerre series adds additional selection step for the orthogonal terms. This will unnecessarily complicate the algorithm, so that there is no advantage in comparison to the regular PEM algorithm or ARMA algorithm in this application. This is not in line with what is claimed in some literature.

7.5 Model Estimation for Multivariable Processes

When assessing multivariable control systems, the task of model identification is more crucial than the SISO case. The main problem of using conventional time series analysis is the large number of parameters to be estimated, which grows quickly

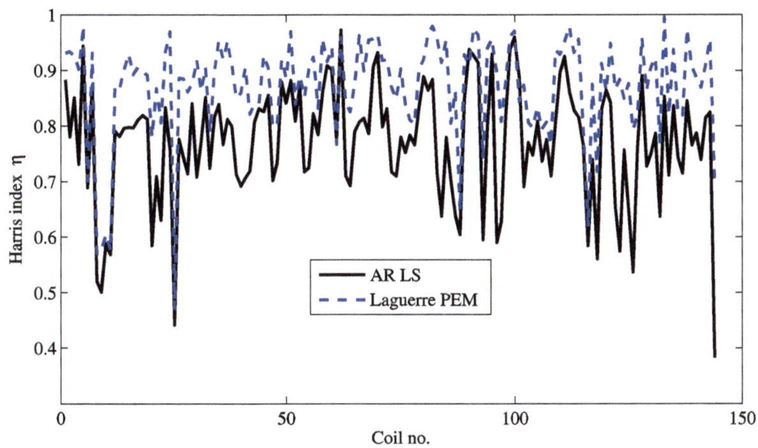

Fig. 7.18 Comparison of Harris indices for a thickness control loop, determined based on AR and Laguerre models

with dimension of the model. Taking a two-inputs/two-outputs process as an example and selecting the model order to be just $n = 2$ means that 16 parameters have to be estimated. If the number of parameters increases, the complexity of the estimation algorithm becomes higher, and the danger of trapping in local minima increases. In this context, it is needed to keep the number of parameters as small as possible. This implies that one should use, for instance, vector ARMA (VARMA) models rather than vector AR (VAR) models, since the former typically requires lower order.

The topic of time series modelling for multivariable processes is discussed, e.g. in Seppala et al. (2002). Subspace methods are also a good option for the identification of multivariable systems, as advocated by Huang et al. (2005b).

7.6 Summary and Conclusions

Different key factors affecting the reliability of the performance assessment results have been discussed in this chapter. Suggestions were given for selecting the right options and parameters. It has been stressed that while data scaling, detrending and eliminating of outliers are recommended, the use of archived data should be strictly avoided. This is because smoothing or compression commonly used in data historians affect the performance index, leading to wrong assessment statements, usually the over-estimation of the control performance. Recommendations were given for selecting a proper sampling time and data length (N) for assessments exercises. Particularly, the data length affects the accuracy of the calculated indices and should lie between 1000 and 2000 (whenever possible). Increasing N may increase the assessment accuracy, but also increases the computational load. Using a lower N is not advisable, as it usually leads to a broader confidence interval for the performance index.

From the variety of models and techniques, which can be used as a basis for performance assessment, AR modelling remains the standard approach, since the associated model estimation is simple and fast by using LS methods. However, there are some situations where other methods such as PEM may be more useful. For instance, it has been concluded that oscillating signals are a potential problem when evaluating the Harris index based on AR modelling, i.e. the performance may be over-estimated. The best way is, therefore, to detect oscillations prior to the computation of the index. If this is not possible or desirable, ARMA or PEM modelling should be used. The only practical reason found for using subspace identification in calculating the MV and GMV benchmark indices is its fast computation compared to PEM. Moreover, this method seems to have more merits when carrying out more advanced assessment such as LGQ or MPC benchmarking.

The knowledge of the time delay is essential to estimating the MV and GMV benchmark indices (and similar ones). It is a real problem and not practical to use routine operating data for such assessments without first having knowledge of the loop delay or trying to estimate it from the data. For the latter task, however, the data must contain clear changes in the control variables, or experimentation with the process in terms of changes in the set point or the addition of a dither signal should be possible. Otherwise an estimation of the time delay will be unreliable. As suggested by Thornhill et al. (1999), the prediction horizon approach can be used to obtain a reasonable estimate of a suitable time delay for use in performance assessment. Note that when any other value for the time delay than the real one is used, the calculated index cannot be interpreted as MV/GMV benchmark, but should be regarded as a kind of user-defined performance index.

The last critical issue discussed in this chapter is the proper selection of the model orders. Different simple rules have been described, which all ensure reliable results. Following these suggestions, the danger of under- or over-estimating the performance index should be minimised. The experience suggests that $n \approx 20 + \tau$ are adequate for most cases to achieve the balance between assessment accuracy and computational load. However, there is no absolute general answer to how large the model order should be, as it depends on the plant-noise model and weighting functions (for GMV).

In practical applications, it is always well spent time to investigate different combinations of data lengths and model orders of defined ranges until the variations in the calculated performance indices are small to achieve accurate assessment results. Of course, this will be only possible when a few control loops are analysed; otherwise, the job would take much more time than can be invested in practice.

Part II
Detection and Diagnosis of Control Performance Problems

Chapter 8
Statistical Process Control

The aim of statistical process control (SPC) techniques, when applied to control-loop operating data, is to identify and track certain variation within the loop and thus highlight situations that show abnormal behaviour, i.e. statistically significant events or abnormalities. Indeed, understanding this variation may be a first step towards the improvement of controller performance. Since variation is present in any process, deciding when the variation is natural and when it needs correction is the key to monitor control performance using SPC. In this context, it should be stressed that SPC is a monitoring rather than control technique because no automatic corrective action is taken after an abnormal process situation is detected.

SPC can be traced back to Walter Shewhart, who introduced the use of SPC *control charts* in the 1920s for adapting management processes to create profitable situations for both businesses and consumers. Today, SPC has become more than control charting alone; it is an umbrella term for the set of activities and methods for data analysis and quality control.

This chapter contains a brief description of selected SPC techniques: Sect. 8.1 presents some control charts typically used in univariate SPC; multivariate SPC methods are given in Sect. 8.2.

8.1 Univariate Statistical Process Control

Univariate SPC techniques are used to signal any violation of control limits, i.e. thresholds defining the boundary region, where the process is on target (or in control or normal). Otherwise, the process is said to be off target (or out-of-control or abnormal). This basic concept of SPC is to utilise process control charts.

The most popular chart is the *Shewhart chart*, in which the data points, the process-variable mean, $\bar{x}$, and the boundaries $T \pm c\sigma$ are plotted. The target T is

M. Jelali, *Control Performance Management in Industrial Automation*,
Advances in Industrial Control, DOI 10.1007/978-1-4471-4546-2_8,

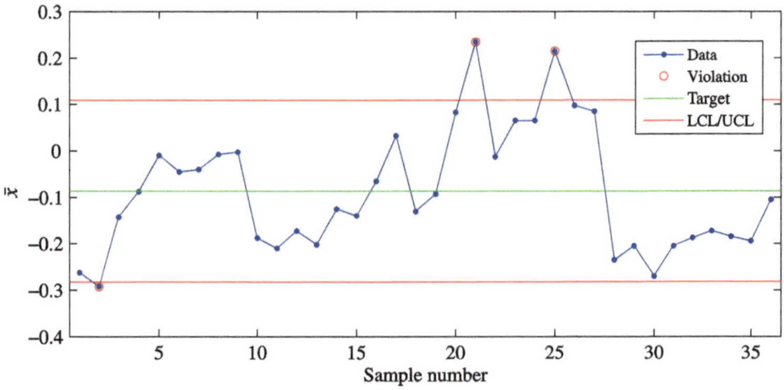

Fig. 8.1 Shewhart control chart

usually specified as the grand mean

$$\bar{\bar{x}} = \frac{1}{N}\sum_{i=1}^{N} \bar{x}_i, \quad \bar{x}_i = \frac{1}{n}\sum_{j=1}^{n} \bar{x}_{ij}, \tag{8.1}$$

or the desired value of $\bar{x}$. N denotes the number of subgroups, each consisting of n selected items, and $\bar{x}_{ij}$ is the jth measurement in the jth group. The limits (LCL: lower control limit and UCL: upper control limit) are normally set at $c = 3\sigma$ (99.73%). Control charts can be generated using the controlchart function in the Statistics Toolbox in MATLAB.

Figure 8.1 shows an example. Points that violates the control limits are marked as red circles on the chart and indicate that the process is misbehaving. Additionally, warning limits commonly set at 2σ can be displayed in the plot. In this case, the process may be off target if the mean is between the warning and control limits. To decide, another sample is taken immediately: if the mean is still outside the warning limits, the process is assumed to be off target. If, however, the second sample mean is within the warning limits, the process works properly.

Enhanced SPS tools include the *cumulative sum (CUSUM)* control chart and the *exponentially weighted moving-average (EWMA)* control chart. The CUSUM chart can be used to determine changes in the process average. The CUMSUM is defined as the running summation of the deviations of the process variable from its target. If the sample mean is plotted, the cumulative sum, $C(k)$, is

$$C(k) = \sum_{j=1}^{k} \big(\bar{x}(j) - T\big). \tag{8.2}$$

During normal process operation, $C(k)$ fluctuates around zero. An abnormal situation would cause a shift in $\bar{x}$, and thus $C(k)$ will drift either upward or downward. An equivalent algebraic version of CUMSUM, which is more convenient for com-

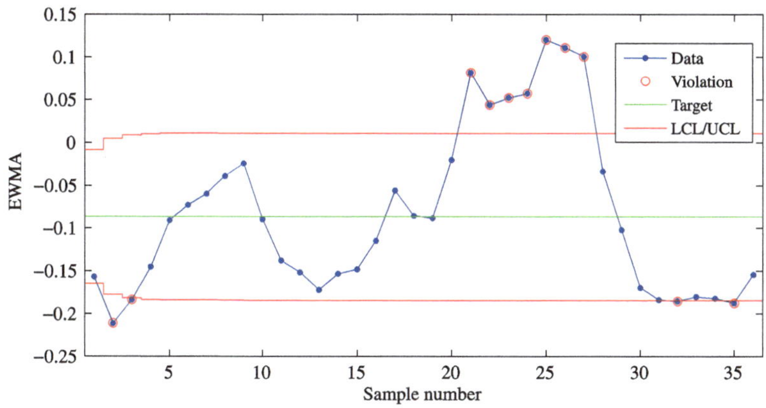

Fig. 8.2 EWMA control chart

puter calculation, consists of the two recursive equations

$$\begin{aligned} C^+(k) &= \max\{0, \bar{x}(k) - (T+K) + C^+(k-1)\}, \quad C^+(0) = 0, \\ C^-(k) &= \max\{0, (T-K) - \bar{x}(k) + C^+(k-1)\}, \quad C^-(0) = 0, \end{aligned} \tag{8.3}$$

where C^+ and C^- denote the sums for the high and low direction, and K is a constant, the slack parameter. An abnormal process situation is assumed when either C^+ or C^- exceeds a specified control limit.

To get a weighted mean of several consecutive observations, the EWMA control chart (Fig. 8.2) can be considered:

$$z(k) = \lambda \bar{x}_k + (1-\lambda) z(k-1), \tag{8.4}$$

where $\lambda \in [0, 1]$ is the forgetting factor, $\bar{x}_k$ is the mean up to the current time k, and z is the EWMA value. A typical value for the forgetting factor is $\lambda = 0.25$. The control limits can determined by

$$T \pm 3\sigma \sqrt{\frac{\lambda}{2-\lambda}}. \tag{8.5}$$

EWMA is also known as geometric moving average, exponential smoothing or first-order pole filter. As for the Shewhart control chart, the process operation is assumed to abnormal if the control limits are violated.

Univariate control chart monitoring, when applied to multivariable processes, does not take into account that the variables may be dependent of each other, and their correlation can be important for understanding the process behaviour. The difficulty of using independent univariate control charts is illustrated in Fig. 8.3. In this figure, the ellipse represents a contour for the in-control process with high confidence limits; circles and triangles represent observations from the process. Individual Shewhart charts are plotted for each quality variable, and it is observed that

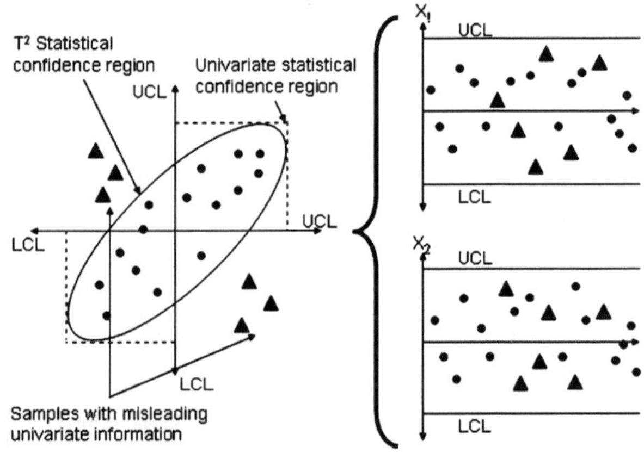

Fig. 8.3 Multivariate vs. univariate statistical analysis and a comparison of the in-control status using T^2 (Ruiz Ordóñez 2008)

each individual Shewhart chart appears in a state of statistical control, and none of the individual observations gives any indications of a problem. This is because the univariate statistical charts do not consider the information contained in the other variables and in the dynamic dependencies of the quality variables (Ruiz Ordóñez 2008). Therefore, multivariate statistical process control (MSPC) has been developed in order to cope with the redundancy problem, extract useful information from process data and utilize it for detecting abnormal system behaviour.

8.2 Multivariate Statistical Process Control

MSPC refers to a collection of algorithms that can be used to extract information from large multivariable data sets, which are more and more frequently recorded in the process industry. Although the algorithms differ considerably, they have one feature in common: they identify several artificial variables, as linear or nonlinear combinations of the original variables, with the aim of obtaining a reduced set of statistically uncorrelated variables. The reduced variable set is then utilised to determine some statistics and bivariate scatter diagrams for online process monitoring. Over the last two decades, a host of MSPC techniques has been developed and successfully applied in many studies. Two prominent MSPC methods are the principal component analysis (PCA) and the partial least squares (PLS). These two algorithms are reviewed briefly in the following sections. More details on PCA and PLS can be find, e.g. by Jackson (1991) and Geladi and Kowalski (1986), respectively.

8.2.1 Principal Component Analysis

PCA is an unsupervised data analysis technique. The method transforms a matrix containing N observations of n process variables, $\mathbf{Z}$, into a matrix of independent artificial variables, or *scores*, $\boldsymbol{t}_k$ $(k = 1, \ldots, n)$. The columns of $\boldsymbol{Z}$ are usually normalised, i.e. they are mean centred and scaled to unit variance (see Sect. 7.1.3). $\boldsymbol{Z}$ is then decomposed into the noise or residual part, $\boldsymbol{E}$, and the systematic part (or predictor), which is conformed by scores $\boldsymbol{T}$ and loadings $\boldsymbol{P}$, according to

$$\boldsymbol{Z} = \boldsymbol{T}\boldsymbol{P}^{\mathrm{T}} + \boldsymbol{E} = \boldsymbol{t}_1\boldsymbol{p}_1^{\mathrm{T}} + \boldsymbol{t}_2\boldsymbol{p}_2^{\mathrm{T}} + \cdots + \boldsymbol{t}_{n_p}\boldsymbol{p}_{n_p}^{\mathrm{T}} = \sum_{k=1}^{n_p} \boldsymbol{t}_k\boldsymbol{p}_k^{\mathrm{T}} + \boldsymbol{E}, \tag{8.6}$$

where the $\boldsymbol{p}_k$ vectors of length n are the eigenvalues of the data covariance matrix $\boldsymbol{Z}^{\mathrm{T}}\boldsymbol{Z}$; $\boldsymbol{p}_k$ are also known as the loadings. The $\boldsymbol{t}_k$ and $\boldsymbol{p}_k$ pairs are ordered so that the first pair captures the largest amount of variation in the data and the last pair captures the least. In the case where significant co-linearity exists, it is generally found that a small number of principal components (n_p) can account for much of the power, and hence information, in the covariance matrix. When Eq. 8.6 is applied to a single vector of new observations, $\boldsymbol{z}^{\mathrm{T}}$, the resulting term $\boldsymbol{e}$ is called the prediction error. A suitable value for n_p can be determined, for instance, as suggested by Wire and Gallagher (1996). This approach applies cross-validation, with due consideration of the variance explained by the scores, to select an appropriate value for n_p.

Having identified the PCA model, the score matrix and residual matrix can be used to detect abnormal states in a process (or control system). For this purpose, two monitoring statistics, T^2 and Q, are introduced as

$$T_k^2 = \sum_{i=1}^{n_p} \frac{\boldsymbol{t}_i^{\mathrm{T}}\boldsymbol{t}_i}{\sigma_i^2} = \sum_{i=1}^{n_p} \frac{t_{ki}^2}{\sigma_i^2}, \tag{8.7}$$

$$Q_k = \|\boldsymbol{E}\|^2 = \sum_{j=n_p+1}^{n} t_{kj}^2, \tag{8.8}$$

where T_k^2 and Q_k are samples of the T^2 and Q statistics at time instance k; t_{ki} and t_{kj} are the values of the ith and jth score variables, respectively, at time instance k, and σ_i^2 is the estimated variance of the ith score variable.

The T^2 statistic is a measure of the variation within the PCA model, and the Q statistic is a measure of the amount of variation not captured by the PCA model (Q is the sum of squared prediction errors). Both statistics are independent of each other, since the residuals of the PCA model are orthogonal to the model plane. Usually, Q-statistics is much more sensitive than T^2. This is because Q is very small, and therefore any minor change in the system characteristics will be observable. T^2 has great variance and therefore requires a great change in the system characteristics for it to be detectable.

For on-line monitoring, the PCA model is applied to project a new data point onto the model plane, spanned by the retained loading vectors, and estimate the residual of the PCA model. Then, the associated values of the T^2 and Q statistics are calculated, and the monitoring charts are updated. Elevated values of these statistics provide an indication of abnormal process behaviour, and an alarm is raised to the process operator. Jackson (1991) and Goulding et al. (2000) provide further details regarding the use of these statistics to detect and isolate abnormal conditions. Miletic et al. (2004) and AlGhazzawi and Lennox (2008) provide further information on data processing for MSPC applications. PCA models can be calculated using the princomp function in the Statistics Toolbox in MATLAB.

8.2.2 Partial Least Squares

PLS is a regression tool that can be applied whenever process variables can be partitioned into cause ($\boldsymbol{X}$) and effect ($\boldsymbol{Y}$) values. The method is commonly used in preference to alternative identification algorithms, such as multiple linear regression (MLR), when developing data driven models. It has the advantage to produce accurate and robust models in situations, where high levels of correlations exist between the cause variables (Geladi and Kowalski 1986).

The PLS algorithm is similar to PCA and selects factors of cause variables in a sequence, which successively maximises the explained covariance between the cause and effect variables. Given a matrix of cause data, $\boldsymbol{X}$, and effect data, $\boldsymbol{Y}$, a factor of the cause data, $\boldsymbol{t}_1$, and effect data, $\boldsymbol{u}_1$, is evaluated, such that

$$\boldsymbol{X} = \hat{\boldsymbol{X}} + \boldsymbol{E} = \sum_{k=1}^{n_p} \boldsymbol{t}_k \boldsymbol{p}_k^{\mathrm{T}} + \boldsymbol{E}, \tag{8.9}$$

$$\boldsymbol{Y} = \hat{\boldsymbol{Y}} + \boldsymbol{F} = \sum_{k=1}^{n_p} \boldsymbol{t}_k \boldsymbol{q}_k^{\mathrm{T}} + \boldsymbol{F}, \tag{8.10}$$

where $\boldsymbol{q}_k$ is the loading vectors of the effect matrix, $\boldsymbol{t}_k$ and $\boldsymbol{p}_k$ are the score and loading vectors of the predictor matrix, and $\boldsymbol{E}$ and $\boldsymbol{F}$ are the residual matrices of the predictor and effect matrices. These equations are referred to as the outer relationships. The score vectors $\boldsymbol{t}_k$, which are different to those obtained using PCA, are mutually orthogonal.

The PLS method provides the potential for a regularised model through selecting an appropriate number of scores, or latent variables in the model. Furthermore, it is often found that a relatively small number of the low-index latent variables can explain the greater part of the variation in both the cause and effect variables. Again, cross-validation can be used to select the necessary number of latent variables.

As with PCA, three statistics associated with the PLS algorithm can be used to identify when new data are inconsistent with the data that was used to identify the PLS model: T^2, $Q^{(X)}$ and $Q^{(Y)}$, which are defined as

$$T_k^2 = \sum_{i=1}^{n_p} \frac{t_{ki}^2}{\sigma_i^2}, \tag{8.11}$$

$$Q_k^{(X)} = \sum_{j=n_p+1}^{M} t_{kj}^2, \tag{8.12}$$

$$Q_k^{(Y)} = \sum_{i=1}^{N} (y_{ki} - \hat{y}_{ki})^2. \tag{8.13}$$

The key differences between PLS and PCA are that PLS considers the explicit relationship between cause and effect variables, which is particularly important when considering a multivariable control system. The statistics identified for PLS relate specifically to either the cause or effect variables, which can help isolate the cause of any abnormality. As in PCA, the three statistics, T^2, $Q^{(X)}$ and $Q^{(Y)}$, are independent of each other due to the orthogonality of the PLS-model residuals to the model planes.

For process monitoring, it is sufficient to use only the T^2 and $Q^{(Y)}$ statistics. This is because any changes of the correlation structure between the predictor variables imply the change in $Q^{(X)}$ statistic. If these changes are due to abnormal process behaviour, then the $Q^{(Y)}$ statistic would also respond to this behaviour, which makes the $Q^{(X)}$ and $Q^{(Y)}$ statistics provide redundant information for condition monitoring (Chen et al. 2004).

8.2.3 Advanced Methods for Multivariate Statistical Process Control

The conventional MSPC methods described above may not always function well, since they cannot detect any changes in the operating condition if the statistics are inside the control limits. In order to avoid this problem, improved techniques MSPC were described and compared by Kano et al. (2000). One method, termed moving principal component analysis (MPCA), is based on the idea that changes in the operating condition, i.e. changes in the correlation among process variables, can be detected by monitoring directions of principal components (PCs). An index was defined for evaluating the changes in the direction of each PC and the changes in the subspace spanned by selected PCs. The second method, called DISSIM, is based on the idea that changes in the operating condition can be detected by monitoring a distribution of time-series data, which reflects the corresponding operating condition. In order to evaluate quantitatively the difference between two data sets, a dissimilarity index was defined. Another extension of MSPC, termed as multiscale PCA (MS-PCA), is to combine PCA with wavelet analysis: PCA has the ability to extract the relationship between variables and decorrelate the cross-correlation, and wavelet analysis helps extract features in the measurements and approximately decorrelate the autocorrelation. The procedures for these techniques and a comparative study can be found by Kano et al. (2000) and the references cited therein.

Another limitation of conventional PCA-SPC and similar methods is that the monitored variables are assumed to be normally distributed. Kano et al. (2002) proposed to use independent component analysis (ICA) rather than PCA. ICA (Jutten and Herault 1991) is a signal processing technique for transforming observed multivariate data into statistically independent components, which are expressed as linear combinations of observed variables. ICA not only decorrelates the signals (2nd-order statistics), but also reduces higher-order statistics dependencies, attempting to make the signals as independent as possible. Kano et al. (2002) suggested to use ICA-SPC functions for generating and monitoring non-Gaussian independent variables, while PCA-SPC for monitoring Gaussian variables.

8.3 Use for Control Performance Assessment

The techniques of statistical process control (initiated by Shewhart) and those of control performance assessment (initiated by Harris) are examples of two different groups of data-driven techniques for investigating process performance issues. Both groups of methods have been developed and are growing within different research communities. Only recently, a few works appear, which try to bridge both technologies; see Loquasto III and Seborg (2003), AlGhazzawi and Lennox (2009) and Chen and Wang (2010).

Statistical methods are often combined with other methods in order to produce the final performance assessment statistic. For example, PCA and distance similarity factors were used by Loquasto III and Seborg (2003) to monitor MPC performance, where several PCA pattern classifiers were developed to monitor the control system and to identify abnormal behaviour. This technique was used to monitor the performance of an MPC controlled Wood–Berry distillation column model.

MSPC techniques are also used by AlGhazzawi and Lennox (2009) to diagnosis and monitor MPC systems. The proposed method uses recursive PCA and PLS regression to develop charts that allow the user to easily identify and troubleshoot abnormal behaviour of MPC systems.

A combination of PCA with an ARMA model was recently proposed by Chen and Wang (2010) to perform performance assessment of MIMO systems. Using PCA, a minimum-variance performance bound for the MIMO feedback control system is established by projecting the difference between the process outputs and the desired set points to determine the scores and residuals. An ARMA filter is used to eliminate any autocorrelation between the scores. The filtered scores and residuals are then used to compute two indices, which can serve as the performance monitoring statistics for the given process. Finally, since it may not be feasible to use the minimum-variance benchmark in all cases, the integrated PCA and ARMA method can be used to design an achievable minimum-variance benchmark and hence tune controllers so as to minimise the weighted sum of variances of the scores.

8.4 Summary and Conclusions

Techniques for statistic process control have been reviewed, including univariate control charts, multivariate principal component analysis and partial least squares and some extensions of them. The key advantages of these approaches are simple, long-standing, widely known and routinely applied for the detection of faults and abnormal situations in the process industry since decades—noticeable by the large number of publications in this area. However, MSPC methods are not yet widely used in relation with control performance assessment and, hence, offer great potential for developing easy-to-use control-systems condition-monitoring tools. Furthermore, there is a need to apply the available MSPC/CPA techniques to large data sets from industrial processes and to prove their suitability.

Chapter 9
Detection of Oscillating Control Loops

Since a control loop may exhibit poor performance for various reasons, it is not only important to detect poor performance, but the challenge is to trace the bad performance to its root cause. Not only controller design and tuning but also other elements in the control systems, such as sensors and actuators, are often responsible for the poor performance. There are many reasons for poor control performance, which can be detected using specialised methods and indices, without requiring the knowledge of time delays or model identification.

9.1 Root Causes of Poor Performance

Attempt of this Part II of the book is to review and suggest procedures for semi- or fully automatic diagnosis of poor performance of control loops. This covers:

- *Detection of Process Nonlinearity*. Nonlinearities in actuators (such as saturation, deadband or hysteresis in control valves), sensors or in the process itself may cause limit cycle oscillations. This chapter describes two of the prominent methods for detecting nonlinearities in control loops. Section 10.4 describes two simple saturation indices for the automatic detection of saturated actuator actions.
- *Oscillation Detection*. Oscillations in process control loops are a very common problem. Oscillations often indicate a more severe problem than irregular variability increase and hence require more advanced maintenance than simple controller re-tuning. There are several reasons for oscillations in control loops. They may be caused by excessively high controller gains, oscillating disturbances or interactions, but the most common reason for oscillations is friction in control valves. In Chap. 8, the most important methods for non-invasive oscillation detection are presented and compared.
- *Oscillation Diagnosis*. Particular attention will be paid to the diagnosis of stiction (i.e. static friction resulting in "stick-slip" motion) in valves, which are often used as the final control elements in plants of the process industries. Stiction can be easily detected using invasive methods, such as the valve travel or bump test;

M. Jelali, *Control Performance Management in Industrial Automation*,
Advances in Industrial Control, DOI 10.1007/978-1-4471-4546-2_9,

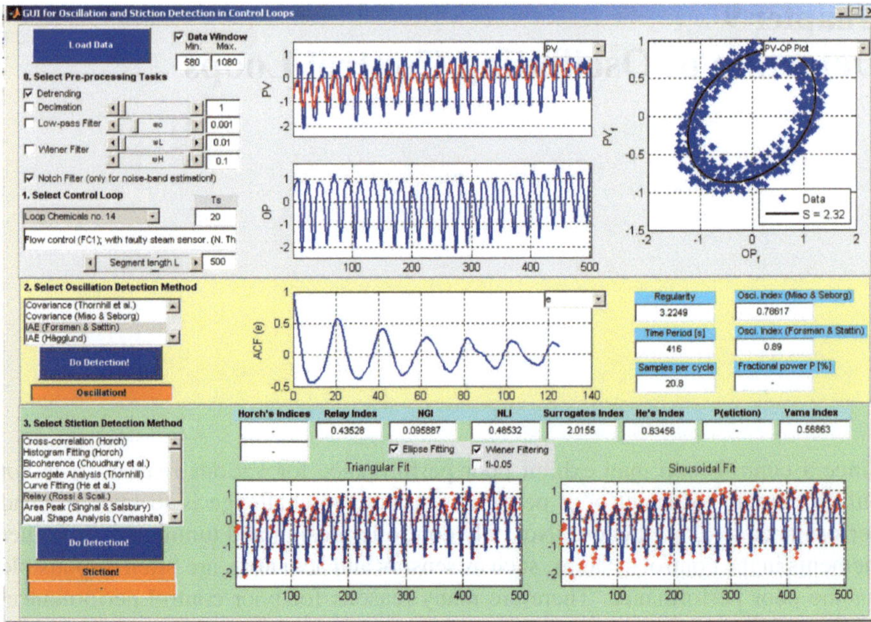

Fig. 9.1 Oscillation and stiction detection GUI showing the analysis of industrial data from a sensor fault in a refinery

see Sect. 11.8.2. However, it is neither feasible nor effective to apply an invasive method across an entire plant site due to the requirement of significant manpower, cost and plant downtime. It is more convenient to first investigate gathered process data using a non-invasive method. Chapter 11 presents an overview of the methods for diagnosing the source of control valve problems. These methods enable one to discriminate between oscillation due to valve nonlinearities, aggressive controller tuning or disturbances affecting the loop.

For each method described, the basic assumptions, limitations, strengths and weaknesses will be clearly stated. The parameterisation of the methods is thoroughly discussed to give default settings for real applications. Industrial examples from different industrial fields (chemicals, refining, petro-chemicals, pulp & paper, mining, mineral and metal processing) are presented throughout the chapters to demonstrate the applicability of the methods.

Throughout the Chaps. 9 and 10, the techniques presented for oscillation, nonlinearity and stiction detection will be demonstrated using a MATLAB GUI (Fig. 9.1) developed by the author, where all techniques are implemented, including some preprocessing methods. The GUI gives users the possibility to compare the method on a data set side-by-side, in a fast and user-friendly way. The upper part of Fig. 9.1 contains functions for data pre-processing (detrending, filtering, decimation, etc.) and for generating plots (time trends, PV-OP shape, power spectrum, etc.). In the middle part of the figure, methods for oscillation detection can be selected and applied to the data. The lower part of the figure provides the user with techniques for stiction detection.

9.2 Characterisation and Sources of Oscillations in Control Loops

There is no clear mathematical definition of "oscillation" that could be applied. Therefore, oscillation detection is usually done somewhat heuristically: one speaks of oscillations as *periodic variations that are not completely hidden in noise* and hence visible to human eyes (Horch 2007). According to Choudhury et al. (2008: Chap. 18), an oscillatory signal is a *periodic signal with well-defined amplitude and frequency*, e.g. a sinusoidal signal. However, the amplitude specification remains an ambiguous question since small oscillations are usually not a serious problem. There are other applications, e.g. signal processing for communication applications, where periodic signals with very low signal-to-noise ratio are very common. These are, however, of no interest in the process industries.

Therefore, the knowledge of the quantifiable properties of oscillatory signals is necessary to make an expert assessment whether the fault is significantly large and requires a corrective action to be taken. The strength of oscillations can be quantified using period, regularity and power (Thornhill et al. 2003a, 2003b, 2003c):

- *Period*. The reciprocal of the oscillation frequency is termed as the period of oscillation. In other words, it is twice the time lapse between two zero crossings of oscillatory signal. In reality, this period may vary around a mean value due to the presence of measurement noise and other stochastic components of the process. In some cases, multiple oscillations with different periods of oscillation may exist in the process variables resulting from multiple fault sources, and hence the period may vary along the time.
- *Regularity*. Regularity of oscillatory signal translates into a quantity that represents the non-randomness behaviour. If the variation in the signal is due to random disturbances, the period of oscillation will hold a wider distribution compared to that of a true oscillatory nature. Regularity of oscillations can be defined as

$$r = f\left(\frac{\bar{T}_\mathrm{p}}{\sigma_{T\mathrm{p}}}\right), \tag{9.1}$$

  where $\bar{T}_\mathrm{p}$ is the mean value, and $\sigma_{T\mathrm{p}}$ the standard deviation of the periods $T_{\mathrm{p}i}$ at adjacent signal intervals; see Sect. 9.7.
- *Power*. Power of oscillations is a means to quantify the amplitude of the oscillatory signal. It is the sum of the spectral power in the selected frequency channels as a fraction of the total power; see Sect. 9.3.

Oscillations (or vibrations) are a very drastic form of plant performance degradation. The most important sources of oscillations in control loops are:

- *Aggressive Tuning.* Too high controller gains may lead to unacceptable oscillations in the process variable. If the controller is tuned such that the loop is (nearly) unstable, i.e., the controller gains are selected too high, there will be an oscillation due to saturation nonlinearity (as control-signal constraints always exist in real systems).
- *Nonlinearities.* Perhaps the most likely reason for control-loop oscillations is the presence of static nonlinearities in the system, such as *static friction* (leading to *stick-slip effect*), *dead zone*, *backlash*, *saturation* and *quantisation*. Refer to the discussion of these phenomena by Horch (2000).
- *Disturbances.* These are a challenge for an automatic CPM system. When having detected an oscillation, it is important to distinguish between internally and externally generated oscillations. External disturbances usually come from upstream processes with the material transferred, or from other control loops due to interactions.
- *Loop Interactions.* Control loops are often mutually interacting. Therefore, if one loop is oscillating, it will likely affect other loops too. In many cases, the oscillations are in a frequency range such that the controller cannot remove them. Then, an oscillation is present even though the controller is well tuned (it might have been tuned for some other control task) (Horch 2000).

For reasons of safety and profitability, it is important to detect and diagnose oscillations. A number of researchers have suggested methods for detecting oscillating control loops. The methods fall into the following categories:

1. Detecting spectral peaks (classical approach); see Sect. 9.3.
2. Methods based on time-domain criteria like the integral of absolute error (IAE) (Hägglund 1995; Thornhill and Hägglund 1997; Forsman and Stattin 1999; Singhal and Salsbury 2005; Salsbury 2006). The method by Hägglund is described in Sect. 9.4, and that by Forsman and Stattin in Sect. 9.5.
3. Methods based on the auto-covariance function (Miao and Seborg 1999; Thornhill et al. 2003c). Both techniques are presented in Sects. 9.6 and 9.7.
4. Use of wavelet plots; see Sect. 9.9.

9.3 Detection of Peaks in the Power Spectrum

Detecting oscillations by looking for peaks in the power spectrum is an obvious and classical approach. The amplitude of the highest peak outside the low-frequency area has to be compared to the total energy in this frequency area. Visual inspection of spectra is therefore helpful because strong peaks can be easily seen, but determination of period and regularity from the spectrum is not recommended.

The ratio between the position of a peak and its bandwidth gives a measure of the regularity of the oscillation, but the presence of noise in the same frequency channels causes difficulties with the determination of bandwidth (Thornhill et al. 2003a, 2003b, 2003c). Also, automating the use of spectra for several hundreds or even

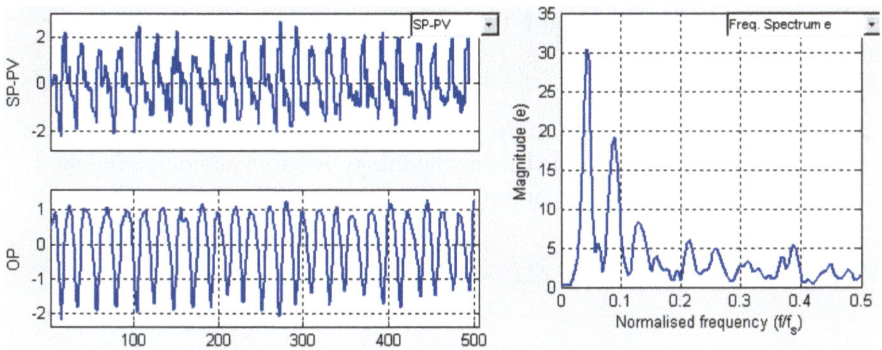

Fig. 9.2 Spectral analysis of the data from the industrial Loop CHEM14

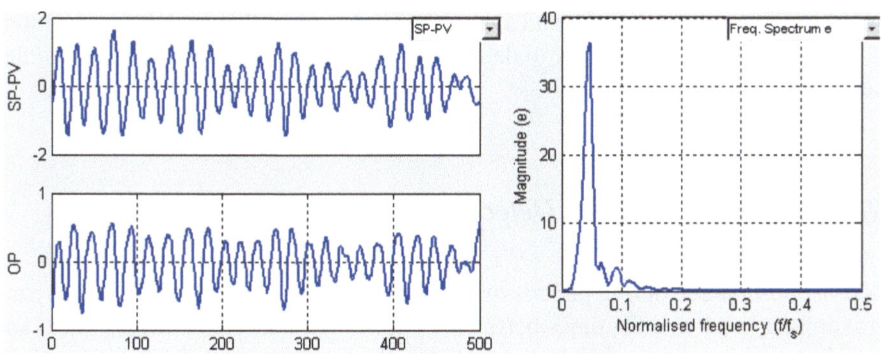

Fig. 9.3 Spectral analysis of the data from the industrial Loop CHEM17

thousand of loops is a difficult task, as visual inspection is generally necessary and the tuning parameters are manually specified. Moreover, the application of spectral analysis becomes difficult if the oscillation is intermittent and periods vary every cycle. There are many methods for peak detection in a spectrum. For instance, a peak is defined as a point that is more than three times greater than the average of the surrounding (e.g. 40) samples.

Example 9.1 Figures 9.2 and 9.3 show the data and power spectrum for the control error of two industrial loops CHEM14 and CHEM17, respectively. The power spectrum of CHEM14 has a non-sinusoidal nature of oscillation characterised by the presence of a harmonic in the spectrum. This is a clear indication of nonlinearity in the loop, which has indeed a faulty steam flow sensor. By contrast, the power spectrum of the loop CHEM17 has a distinct spectral peak due to oscillation, which can be classified as a disturbance.

9.4 Regularity of "Large Enough" Integral of Absolute Error (IAE)

Perhaps the first automatic procedure for detecting oscillations in control loops was presented by Hägglund (1995). This methodology is based on computing the IAE between zero-crossings of the control error e, i.e.

$$IAE = \int_{t_{i-1}}^{t_i} \left| e(t) \right| \mathrm{d}t, \tag{9.2}$$

where t_{i-1} and t_i are two consecutive instances of zero crossings. It is assumed that the controller has integral action, so that the average control error is zero. If no integral action is present, it is suggested to study the difference between the measurement signal and its average value obtained from a low-pass filter. It might be advantageous to use the second approach also for controllers with integral action, since it provides the possibility to detect oscillations that are not centred around the set point.

9.4.1 Load-Disturbance Detection

A basic observation during periods of good control is that the magnitude of the control error is small and the times between the zero crossings are relatively short; see Fig. 9.4. In this case, the IAE values calculated according to Eq. 9.2 become small. When a load disturbance occurs, the magnitude of $e(t)$ increases, and a relatively long period without zero crossings, and thus a large IAE value, results. When the IAE exceeds a certain limit, denoted IAE_{lim}, it is therefore likely that a load disturbance has occurred. The choice of this limit is a trade-off between the demand for a high probability of detection and the requirement for a small probability of getting false detections.

The load-disturbance detection procedure can be used to detect oscillations. Suppose that the control error is a pure sinusoidal wave with the amplitude A and the frequency ω, and that this signal is to be detected as a sequence of load disturbances. This means that the integral of each half period of the oscillation must be greater than IAE_{lim}. The following upper limit of IAE_{lim} is then obtained:

$$IAE_{\mathrm{lim}} \le \int_0^{\pi/\omega} \left| A \sin(\omega t) \right| \mathrm{d}t = \frac{2A}{\omega}. \tag{9.3}$$

The procedure should be able to detect oscillations in the low- and middle-range area. A requirement is therefore that frequencies up to the ultimate frequency ω_u should be detected. A reasonable choice of A is 1 %, which means that 2 % peak-to-peak oscillations are acceptable but that oscillations with higher amplitude should

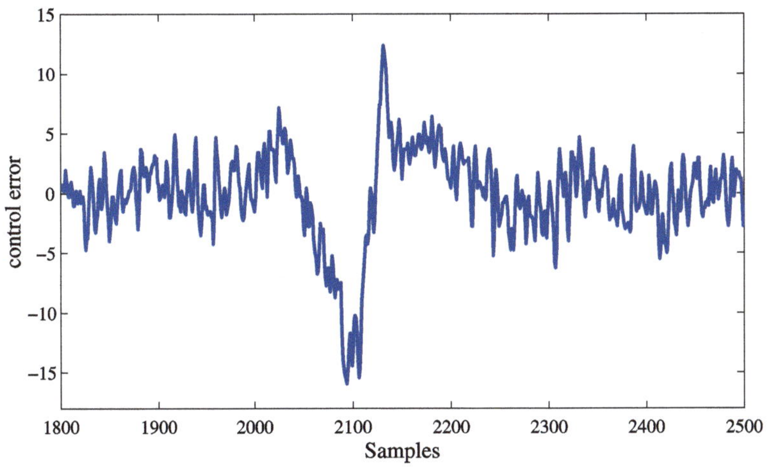

Fig. 9.4 Control error e during a period of good control affected by a load disturbance (data from strip thickness control loop in a tandem rolling mill)

be detected (Hägglund 1995). Such a parameter choice gives

$$IAE_{\mathrm{lim}} = \frac{2}{\omega_{\mathrm{u}}}. \tag{9.4}$$

This quantifies what "large enough" IAE means. The ultimate frequency ω_{u} may be known when the controller is tuned with a relay auto-tuner, but normally it is unknown. With a properly tuned PI(D) controller, the integral time T_{I} is of the same magnitude as the ultimate period $T_{\mathrm{u}} = 1/\omega_{\mathrm{u}}$. If ω_{u} is unknown, it can therefore be replaced by the integral frequency $\omega_{\mathrm{I}} = 2\pi/T_{\mathrm{I}}$ in Eq. 9.4, i.e.

$$IAE_{\mathrm{lim}} = \frac{T_{\mathrm{I}}}{\pi}. \tag{9.5}$$

9.4.2 Basic Appraoch

The basic idea of the oscillation-detection procedure is to conclude the presence of an oscillation if the rate of load-disturbance detections becomes high. For this purpose, the behaviour of the control performance is monitored over a supervision time T_{sup}: if the number of detected load disturbances exceeds a certain limit, n_{lim}, during this time, it can be concluded that an oscillation is present.

The oscillation detection procedure has three parameters that must be selected: IAE_{lim}, T_{sup} and n_{lim} (Hägglund 1995):

- Set $n_{\mathrm{lim}} = 10$.
- Select $T_{\mathrm{sup}} = 5 n_{\mathrm{lim}} T_{\mathrm{u}}$, where T_{u} is the ultimate period obtained from a relay auto-tuning experiment. If such an experiment is not performed, one should replace T_{u} with the integral time T_{I}.

- Use $IAE_{\mathrm{lim}} = 2A/\omega_{\mathrm{u}}$, where A as the lower limit of the acceptable oscillation amplitude at the ultimate frequency ω_{u}: a suggested value of A is 1 % of the signal range. Again, if ω_{u} is not available, it is replaced by $\omega_{\mathrm{I}} = 2\pi/T_{\mathrm{I}}$.

9.4.3 Detection Procedure

The whole procedure for load-disturbance and oscillation detection can be summarised as follows.

Procedure 9.1 load-disturbance and oscillation detection (Hägglund 1995).

1. Choose a suitable acceptable oscillation amplitude A, say $A = 1$ %.
2. Calculate $IAE_{\mathrm{lim}} = 2A/\omega_{\mathrm{u}}$ if ω_{u} is available; otherwise as $IAE_{\mathrm{lim}} = 2A/\omega_{\mathrm{I}} = AT_{\mathrm{I}}/\pi$.
3. Monitor the IAE, where the integration is restarted every time the control error changes sign.
4. If the IAE exceeds IAE_{lim}, conclude that a load disturbance has occurred.
5. Monitor the number of detected load disturbances (n_l).
6. If n_{l} exceeds n_{lim}, conclude that an oscillation is present.

Note that, for on-line applications, it is more convenient to perform the calculations recursively. The number of detected load disturbances is determined by

$$n_{\mathrm{l}}(k) = \gamma n_{\mathrm{l}}(k-1) + l, \quad l = \begin{cases} 1 & \text{if a load disturbance is detected,} \\ 0 & \text{else,} \end{cases} \tag{9.6}$$

where γ is a weighting factor, to be set as $\gamma = 1 - T_{\mathrm{s}}/T_{\mathrm{sup}}$.

9.4.4 Method Enhancement for Real-Time Oscillation Detection

If the signal range of the process output is known and if an estimate of the time constant of the loop is available, at least as T_{I}, then it is possible to run the oscillation detection procedure automatically without any further process information. This property is of vital importance in process control applications. A modified version of the method was suggested in Thornhill and Hägglund (1997), where no knowledge of the signal ranges is required, or no integral term is used in the controller, as is often the case for integrating processes, such as in level-control loops. In these situations, the use of Δt_i, the time between zero crossings, in the criterion for detection of oscillations provides an alternative means to generate a threshold for real-time detection of deviations. The criterion for a 1 % deviation (Eq. 9.5) becomes

$$IAE_{\mathrm{lim}} = \frac{2\Delta t_i}{\pi}. \tag{9.7}$$

That is, the local period of oscillation is taken to be $2\Delta t_i$ instead of T_I. Note that in this case the calculation of T_p from zero crossings is sensitive to noise, and countermeasures (filtering, noise band) have to be taken to avoid spurious zero crossings.

The use of $2\Delta t_i$ in place of T_I also has benefits when the data are sub-sampled. Such a case might arise if the real-time oscillation detection were to reside in the plant information system layer of a DCS rather than in the PID layer, because the data may be sub-sampled to reduce traffic across the communications link. In sub-sampling, the value captured at the sample instant is held constant until the next sampling instant, and the IAE value calculated by integration from sub-sampled data can therefore be larger than expected. The effect gets worse as the sub-sampling period becomes longer. If the sub-sampling interval is close to the controller integration time, then a deviation may be detected after just one sample. Therefore, the use of the alternative time-scale parameter Δt_i is again helpful. When the controller output range is unknown, the detection of oscillations is enhanced by use of the RMS value of the noise as a scaling parameter. Of course, the range is recorded with other loop parameters in the DCS, but assessing a scaling factor from the data reduces the dependence on extraneous information. For online use, the noise assessment would need to be assessed over, say, the past 24 hours using a recursive method of filtering (Thornhill and Hägglund 1997).

All these observations and enhancements show that Hägglungs' online detection method can be applied even in cases where neither the controller-tuning settings nor the range of the process variables are known, and that it can be used for sub-sampled data. Its applicability has therefore been extended to wider range of cases often met in the industrial practice.

Example 9.2 Hägglund's oscillation-detection procedure is now demonstrated on the data of the flow control loop CHEM35. The flow controller was a PI controller with gain $K_\mathrm{c} = 0.6$ and integral time $T_\mathrm{I} = 42$ s. Figure 9.5 (upper panel) shows a window of data, i.e. control error, from the loop (Samples 1:800; Wiener filter [0.05, 1.0]). Because of stiction, the process oscillates with an amplitude of a few percent. The second panel shows the IAE value calculated between successive zero crossings of the control error, as well as IAE_lim: since the ultimate period was not available, IAE_lim was calculated from the integral time T_I as $IAE_\mathrm{lim} = 2/\omega_\mathrm{I} = T_\mathrm{I}/\pi \approx 13.3$. The IAE values are significantly larger than IAE_lim, as can be expected because of the high oscillation amplitude. Finally, the third panel shows the rate of load detections n_l and the rate limit n_lim. The rate n_l exceeds the rate limit n_lim after about 18 min, and the detection procedure gives an alert. The rate n_l converges to about 25, 2.5 times larger than the rate limit. However, in the implemented version, n_l is reinitialized to zero every time n_l exceeds n_lim.

To summarise, Hägglund's oscillation-detection method is very appealing but has two disadvantages: (i) it is assumed that the loop oscillates at its ultimate frequency which may not be true, e.g. in the case of stiction; (ii) the ultimate frequency is seldom available, and the integral time (also not always available) may be a bad indicator for the ultimate period. A strength of the method is that it can be applied for online detection of oscillations.

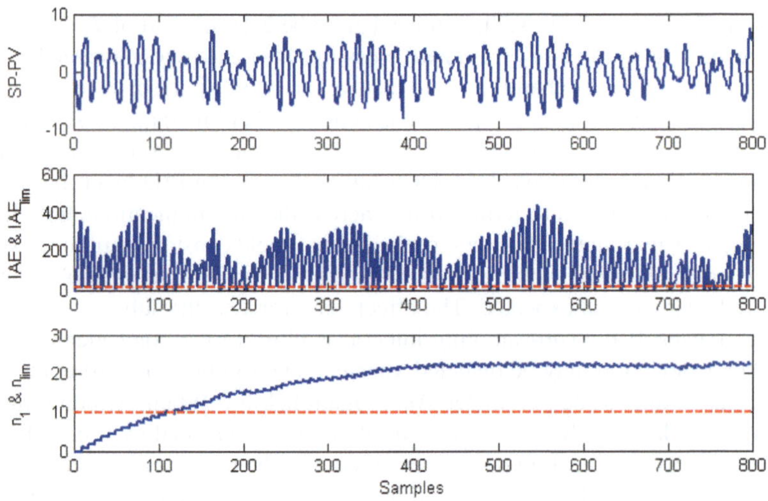

Fig. 9.5 Hägglund's oscillation detection procedure applied to the control loop CHEM35

9.5 Regularity of Upper and Lower Integral of Absolute Errors and Zero Crossings

9.5.1 Basic Methodology

The underlying idea of this method introduced by Forsman and Stattin (1999) is that if e is near periodic, then the time between successive zero crossings and the successive IAEs should not vary so much over time. The IAEs are separated for positive and negative errors (Fig. 9.6)

$$A_i = \int_{t_{2i}}^{t_{2i+1}} \left|e(t)\right| \mathrm{d}t, \qquad B_i = \int_{t_{2i+1}}^{t_{2i+2}} \left|e(t)\right| \mathrm{d}t \tag{9.8}$$

to generate the oscillation index

$$h := \frac{h_A + h_B}{N_{\mathrm{zc}}}, \tag{9.9}$$

$$h_A = \#\left\{i < \frac{N_{\mathrm{zc}}}{2}; \alpha < \frac{A_{i+1}}{A_i} < \frac{1}{\alpha} \wedge \gamma < \frac{\delta_{i+1}}{\delta_i} < \frac{1}{\gamma}\right\}, \tag{9.10}$$

$$h_B = \#\left\{i < \frac{N_{\mathrm{zc}}}{2}; \alpha < \frac{B_{i+1}}{B_i} < \frac{1}{\alpha} \wedge \gamma < \frac{\varepsilon_{i+1}}{\varepsilon_i} < \frac{1}{\gamma}\right\}, \tag{9.11}$$

where N_{zc} is the number of zero crossings, and $\#S$ denotes the number of elements in the set S.

The oscillation index can be interpreted in the following way (Forsman and Stattin 1999):

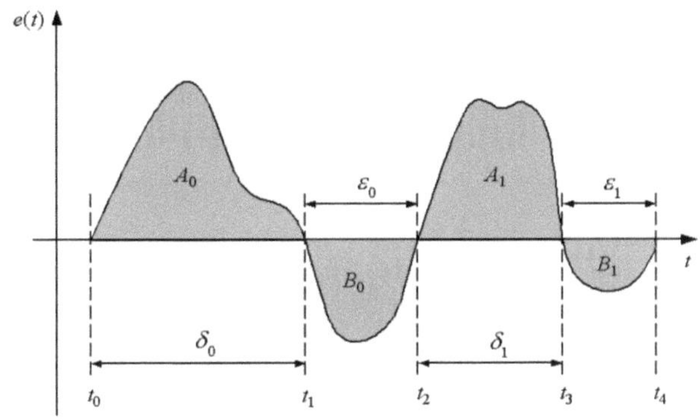

Fig. 9.6 Parameters for the calculation of Forsman and Stattin's oscillation index

- Loops having $h > 0.4$ are oscillatory, i.e. candidates for closer examination.
- If $h > 0.8$, a very distinct oscillatory pattern in the signal is expected.
- White noise has $h \approx 0.1$.

9.5.2 Practical Conditions and Parameter Selection

Forsman and Stattin suggested to select $\alpha = 0.5$–0.7, $\gamma = 0.7$–0.8 and stated that the criterion is fairly robust to variations in these tuning parameters. One reason for this is that there is a coupling between the condition on the IAE and the condition of the time between zero crossings. In practice, the error e should be pre-filtered prior to the index calculation, so that high-frequency noise is attenuated. A simple low-pass filter (or exponential filter) can be used:

$$e_f(k) = \alpha e(k) + (1 - \alpha) e_f(k), \qquad e_f(1) = e(1), \tag{9.12}$$

where e_f is the filtered control error. Equation 9.12 is the time-discrete version of the first-order filter

$$T_f \dot{e}_f(t) + e_f(t) = e(t), \tag{9.13}$$

where T_f denotes the filter time constant. α and T_f are related by $\alpha = T_s/(T_f + T_s)$. The choice of filter constant $0 < \alpha \leq 1$ represents a trade-off between detecting fast, small oscillations ($\alpha = 1$) and attenuating high-frequency noise ($\alpha \to 0$), typically $\alpha = 0.1$. That is, a smaller value α provides more filtering. For offline analysis, a non-causal filter [filtfilt] from MATLAB Signal Processing Toolbox can also be used.

Example 9.3 The method by Forsman and Stattin (1999) is applied to the data of the room temperature control loop BAS6 in a commercial building. This loop has also been analysed by Singhal and Salsbury (2005). Figure 9.7 shows a window of data, i.e. control error and controller output, from the loop. Because of stiction and tight tuning, the process clearly oscillates. With the parameters $\alpha = 0.5$–0.7, $\gamma = 0.7$–0.8, we get $h_A = 31$, $h_B = 28$, $N_{zc} = 76$ and thus the oscillation index value $h = 0.78$. This value is higher than the threshold 0.4, indicating a clear oscillatory pattern.

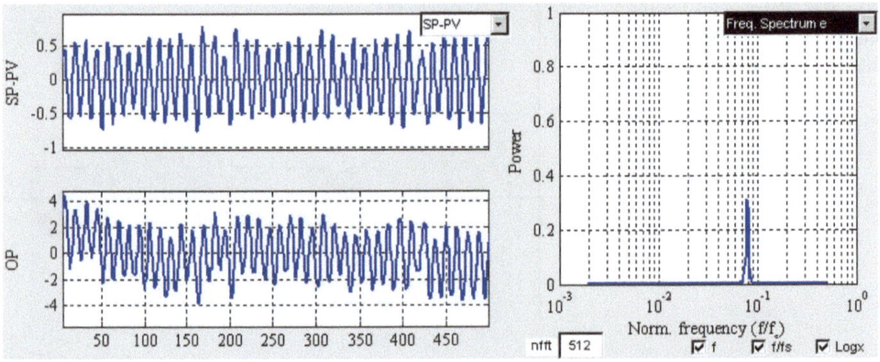

Fig. 9.7 Data from Loop BAS6

9.6 Decay Ratio Approach of the Auto-covariance Function

The ACF of an oscillating signal is itself oscillatory with the *same* period as the oscillation in the time trend. The advantage of using the auto-covariance function (ACF) for oscillation detection over time-trend-based methods is that the ACF in a sense provides a kind of filtering. The impact of noise is reduced because white noise has an ACF that is theoretically zero for lags greater than zero. Figure 9.8 clearly shows an example of the filtering effect of the auto-correlation function. Although the OP signal is noisy, its ACF is very clean.

9.6.1 Methodology

The patented[1] method of Miao and Seborg (1999) is based on the analysis of the auto-covariance function of normal operating data of the controlled variable or the control error. The approach utilises the decay ratio R_{acf} of the auto-covariance function, which provides a measure of oscillation in the time trend is. Figure 9.9 illustrates the definition of R_{acf}, described by the equation

$$R_{\mathrm{acf}} = \frac{a}{b}, \tag{9.14}$$

where a is the distance from the first maximum to the straight line connecting the first two minima, and b is the distance from the first minimum to the straight line that connects the zero-lag auto-covariance coefficient and the first maximum.

As the decay ratio of the auto-covariance function is directly related to the decay ratio of the signal itself, it is a convenient oscillation index. A value R_{acf} smaller than 0.5, corresponding to a decay ratio in the time domain smaller than 0.35,

---

[1] US Patent #5,719,788 (1998).

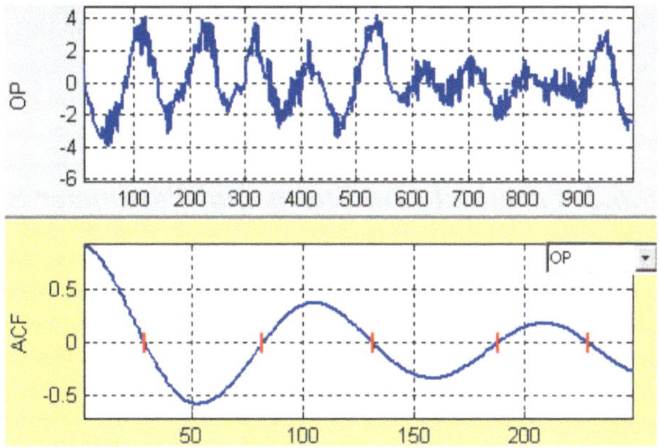

Fig. 9.8 OP time trend (*top*) and its auto-covariance function (*bottom*) for loop POW3

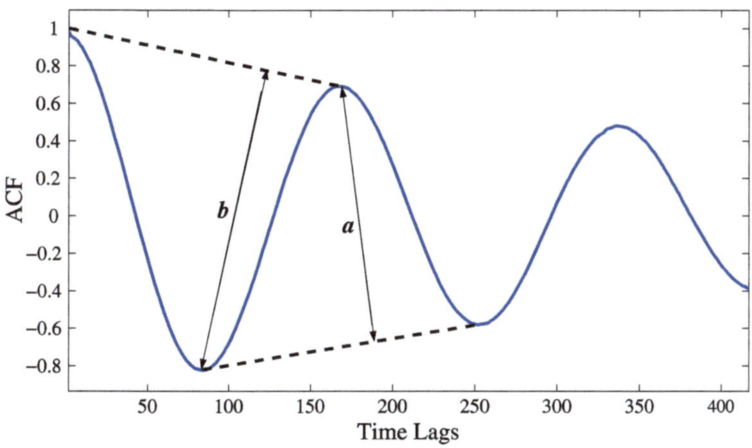

Fig. 9.9 Determination of the decay ratio of the auto-covariance function

may be considered as acceptable for many control problems. On the other hand, if $R_{acf} \geq 0.5$, then the signal is considered to exhibit an excessive degree of oscillation. Therefore, R_{acf} can be used to detect excessive oscillations in control loops according to the following simple procedure.

Procedure 9.2 Oscillation detection using the decay ratio of the auto-covariance function.

1. Calculate the auto-covariance function of the measured y or e and determine the ratio R_{acf} (Eq. 9.14). For the case where there are less than two minima, set the index value to zero.
2. If R_{acf} is greater than a specified threshold, say 0.5, it is concluded that the considered signal is excessively oscillatory.

Note that, as in every method, the selection of the threshold is somewhat subjective and application-dependent.

9.6.2 Practical Conditions and Parameter Selection

To calculate the oscillation index R_{acf} from auto-correlation coefficients, it is necessary to have at least two minima and one maximum, i.e. 1.25 cycles, in the correlogram. Because the maximum lag is selected to be one quarter of the number of data points (Box et al. 1994), 1.25 cycles in the correlogram corresponds to five cycles in the signal data. Hence, an oscillation can be detected by the decay-ratio method if the data exhibit at least five cycles of a damped oscillation during the data-collection period T_{c}. This means that T_{c} must be at least five times the period of the lowest frequency of interest. This frequency range can be specified as suggested by Hägglund (1995), i.e. depending on the ultimate frequency ω_{u} (when known),

$$\frac{\omega_{\mathrm{u}}}{10} \leq \omega \leq \omega_{\mathrm{u}}, \tag{9.15}$$

or based on the controller integration time T_{I},

$$\frac{2\pi}{10T_{\mathrm{I}}} \leq \omega \leq \frac{2\pi}{T_{\mathrm{I}}}. \tag{9.16}$$

This also suggests selecting

$$T_{\mathrm{c}} > 50T_{\mathrm{I}}. \tag{9.17}$$

However, when the controller is poorly tuned, T_{I} may be too small, and the nominal data collection period $T_{\mathrm{c}} = 50T_{\mathrm{I}}$ would then be too short to detect low-frequency oscillations. To avoid this potential problem, Miao and Seborg (1999) recommended repeating the auto-covariance analysis for $T_{\mathrm{c}} = 250T_{\mathrm{I}}$ whenever no oscillation is detected using $T_{\mathrm{c}} = 50T_{\mathrm{I}}$.

When the signal-to-noise ratio is small, a large number of local maxima and minima may occur in the correlogram. This would cause problems in the determination of the R_{acf} index. Therefore, it is desirable to remove high-frequency noise by filtering. Again, the selection of the filter time constant is a compromise between detecting fast, small oscillations and attenuating high-frequency noise. Excessive filtering attenuates the considered signal over the frequency range of interest and thus adversely affects the calculated R_{acf} value.

Example 9.4 The data from the loop BAS6 analysed in Example 9.3 are considered again. The application of the decay-ratio oscillation detection technique leads to $a = 1.74$, $b = 1.85$ and thus the oscillation index value $R_{\mathrm{acf}} = 0.94$, confirming that the loop has very distinct oscillation.

9.7 Regularity of Zero Crossings of the Auto-covariance Function

The strength of oscillations can be quantified using three characteristics: period T_p, regularity r and power P of oscillation. A regular oscillation will cross the signal mean at regular intervals. Therefore, the intervals between zero crossings of an oscillatory time trend can be exploited for off-line detection of oscillations: the deviation of the intervals between zero crossings is compared to the mean interval length; a small deviation indicates an oscillation. The threshold selection is signal independent, i.e., there is no need for scaling the individual signals. However, noise can cause "false" crossings, and drift and transients will destroy the notion of a signal mean.

Instead of looking at the zero crossings of the time trend, Thornhill et al. (2003c) suggested to use the zero crossings of the ACF. By looking at the regularity of the period, an oscillation can be detected. Regularity is assessed by the use of a statistic, r, termed the regularity factor. It is derived from the sequence of ratios between adjacent intervals Δt_i at which deviations cross the threshold. Thus, the mean period of the oscillation $\bar{T}_{\mathrm{p}}$ can be determined from (Fig. 9.8)

$$\bar{T}_{\mathrm{p}} = 2\frac{\sum_{i=1}^{n} \Delta t_i}{n} = \frac{2}{n}\sum_{i=1}^{n}(t_i - t_{i-1}), \tag{9.18}$$

and the dimensionless *regularity factor* r is (Thornhill et al. 2003c)

$$r = \frac{1}{3}\frac{\bar{T}_{\mathrm{p}}}{\sigma_{T\mathrm{p}}}, \tag{9.19}$$

where $\sigma_{T\mathrm{p}}$ the standard deviation of $T_{\mathrm{p}i}$. An oscillation is considered to be regular with a well-defined period if r is greater than unity, i.e.

$$r > 1. \tag{9.20}$$

The regularity factor r can thus be regarded as an oscillation index.

It is recommended to exclude the interval from zero lag to the first zero crossing because it corresponds to only one half of a completed deviation. Also, one should not use the last zero crossings as they can be spurious in the case of very persistent oscillations. Thornhill et al. (2003c) suggested taking 10 intervals between the first 11 zero crossings for calculating the oscillation period.

Again, the benefit of the ACF for oscillation detection is that the impact of noise is much reduced. The pattern of zero crossings of the ACF therefore reveals the presence of an oscillation more clearly than the zero crossings of the time trend. Practical considerations require that only signals with significant activity in the chosen frequency band be considered. The regularity test (Eq. 9.20) should thus be only applied if the filtered signal has sufficient *fractional power* defined as

$$P = \frac{\sum_{\omega=\omega_{\mathrm{n}1}}^{\omega_{\mathrm{n}2}} \Phi(j\omega)}{\sum_{\omega=0}^{\pi} \Phi(j\omega)}, \tag{9.21}$$

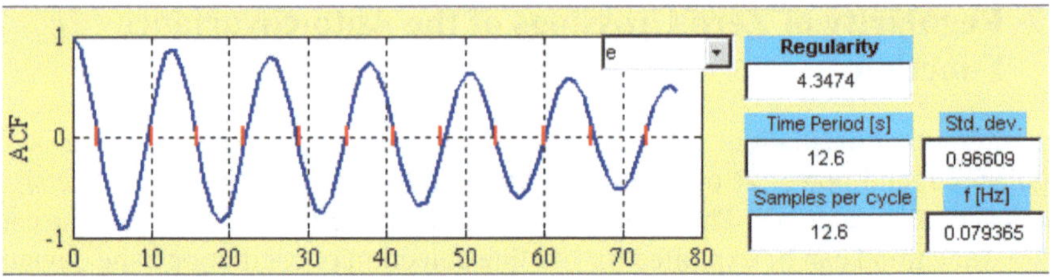

Fig. 9.10 Oscillation detection results for Loop BAS6

where Φ is the power spectrum, and ω_{n1} and ω_{n2} denote the lower and upper boundaries of the filter, respectively (Sect. 9.8). A low value of P indicates that the signal does not have significant activity in the selected frequency, i.e., the behaviour of the signal is dominated by other frequencies. Thornhill et al. (2003c) suggested a threshold of 1 % for P, but higher values, e.g. 5 %, can be used to avoid detection of insignificant oscillations.

Example 9.5 Thornhill's oscillation detection technique is applied on the data from the loop BAS6 analysed in the previous examples, to yield the results shown in Fig. 9.10. The regularity factor of $r = 4.3$ indicates a clear oscillation of the loop. The fractional power can be determined from the spectrum shown in Fig. 9.7 (right side) to be $P = 31$ %.

9.8 Pitfalls of Multiple Oscillations—Need for Band-Pass Filtering

In many data sets obtained from industrial control loops, *slowly varying trends*, *high-frequency noise* and *multiple oscillations* are observed, as can be seen in Fig. 9.11. These effects often destroy the regularity of oscillations, which are then difficult to analyse. In this case, an automated algorithm may detect none or only one oscillation despite the spectrum shows multiple peaks.

Example 9.6 Figure 9.11 shows the oscillation detection results for the unfiltered data from the industrial loop CHEM33 having a slowly varying trend induced by set point changes and two superimposed oscillations of different periods. This can clearly be seen in the power spectrum. The lower panel in the figure marks the positions of the zero crossings: the intervals between zero crossings of the auto-covariance function reflect neither oscillation accurately because the zero crossings of the fast and slow oscillations each destroy the regularity of each other's pattern. Therefore, no oscillation detection method indicates clear oscillation, i.e. all oscillation indices have values below the alert thresholds.

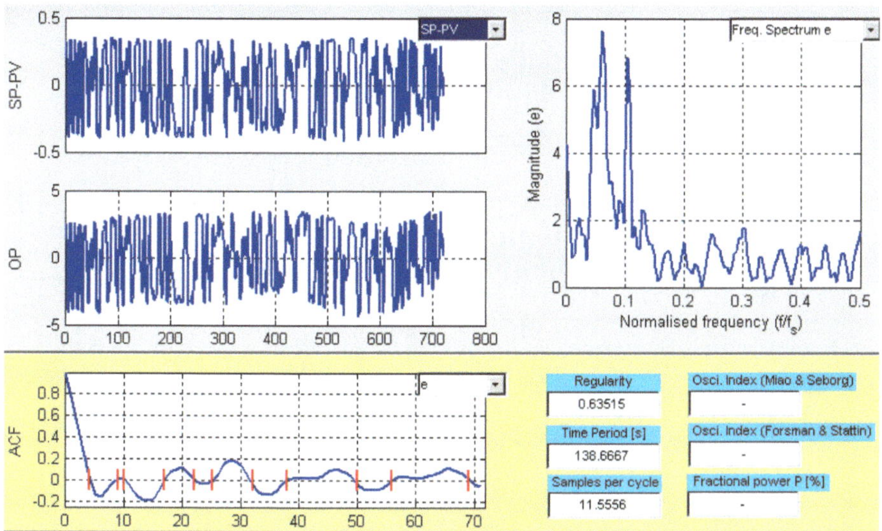

Fig. 9.11 Oscillation detection results for the unfiltered data from Loop CHEM33

This difficulty can be overcome by means of band-pass filtering, where the filter boundaries are selected from the inspection of the peaks present in the spectrum or the peaks in the bicoherence plot. Frequency domain-based filtering, e.g. Wiener filter, which sets the power in unwanted frequency channels to zero, is one good option. The approximate realisation of a Wiener filter (Press et al. 1986) should be used because a true Wiener filter requires an estimate of the noise power within the wanted frequency channels, which would then be subtracted from those channels. The detailed design algorithm is given in Thornhill et al. (2003c), who explain how to deal with aliased frequencies above the Nyquist frequency and constraints on the filter width and also discuss the automation of the frequency-domain filter. It has been suggested to select the filter width Δf centred at $\pm f_0$ so that

$$\frac{2}{\Delta f} \leq \frac{5}{f_0} \quad \text{or} \quad \Delta f \geq \frac{f_0}{2.5}. \tag{9.22}$$

Example 9.7 When the data from loop CHEM33 are filtered using a Wiener filter with the boundaries [0.050, 0.080], we get the oscillation detection results shown in Fig. 9.12. Now a distinctively regular oscillation with ca. 15 samples per cycle is detected by Thornhill's method. One should particularly notice the marked regular zero crossings of the control error. Also the other oscillation detection methods signal distinctive oscillation, as all oscillation indices have values higher than the alert thresholds. The second oscillation having an oscillation frequency of ca. 10 samples per cycle is also detected by all methods when the filter boundaries are placed on [0.095, 0.110]; see Fig. 9.13. The filter boundaries have been selected by inspecting the power spectrum in Fig. 9.11.

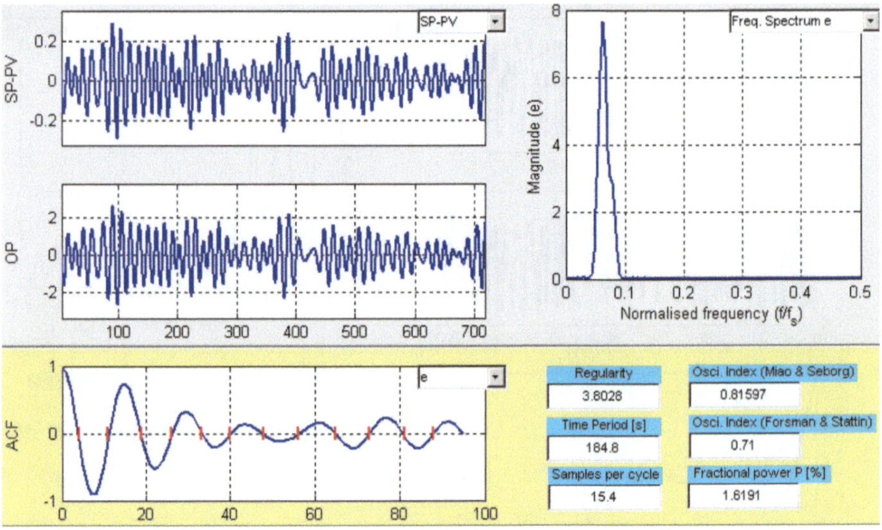

Fig. 9.12 Oscillation detection results for the filtered [0.050, 0.080] data from Loop CHEM33

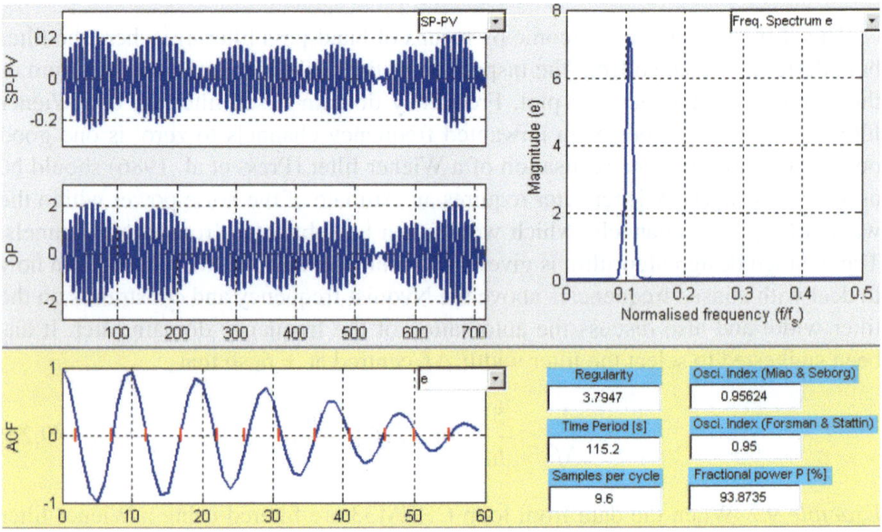

Fig. 9.13 Oscillation detection results for the filtered [0.095, 0.110] data from Loop CHEM33

Our experience revealed that any oscillation-detection method should be combined with a calculation of the regularity factor to avoid possibly "false" detection or "false" determination of the oscillation period. This can occur when more than one oscillation is present in the signal. The accurate value of the oscillation period is needed in the stage of the root-cause diagnosis of the oscillation. Also the inspection of the power spectrum is highly recommended to roughly set the filter boundaries.

Indeed, this hybrid approach suggested here is a semi-automated method for oscillation detection. Although the method can be fully automated (Thornhill et al. 2003c), care must be taken to not get misleading results due to non-proper selection of the band-pass filter. Therefore, it is recommended to use the method in a semi-automated fashion.

9.9 Detection of Intermittent Oscillations

In the case of persistent oscillation in the time trend, the power spectrum gives a clear signature for the oscillation since it has a sharp peak of large magnitude at the frequency of oscillation. However, there are some cases where the oscillation is intermittent, i.e. non-persistent. In a set of such time trends, where the nature of the signal changes over time, the Fourier transform has to be used on subsets of the data to observe the time-varying frequency content. Alternatively, wavelet analysis may be used.

The technique of wavelet transform and algorithms for its computation (Kaiser 1994) can treat time and frequency simultaneously in time-frequency domain. This provides the signal amplitude as a function of frequency of oscillation (the resolution) and time of occurrence. One method of presentation shows times and resolution plotted on the horizontal and vertical axis, and amplitudes represented by hues in the contour lines corresponding to them on the time-frequency plane. It is then possible to analyse the relation between the timing of frequency emerging and disappearing in the process, thus providing more precise and deeper insights into the process behaviour. Wavelet analysis has been successfully applied to plant-wide disturbance (oscillation) detection or diagnosis by Matsuo et al. (2004). However, the evaluation of wavelet plots is difficult to automate, so a visual inspection by a human expert is required.

9.10 Summary and Conclusions

The detection of oscillations in control loops can be regarded as a largely solved problem. Many methods exist for this purpose; some of them have been reviewed in this chapter. Emphasis was placed on discussing possible problems that can occur when the techniques are applied to real-world data. These can be noisy, subject to abrupt changes, and may contain slowly varying trends and different superposed oscillations, i.e. with different frequencies. Particularly, the latter problem is still a challenge for automatic detection, without human interaction. Moreover, the detection of plant-wide oscillations and finding their sources and propagation routes is an active research topic. Thornhill and co-workers provided some methods and strategies to deal with plant-wide oscillations. An overview of the topic was written by Thornhill and Horch (2007).

Chapter 10
Detection of Loop Nonlinearities

Most of control performance methods assume that the system is (at least locally) linear. However, the presence of certain type of nonlinearity may cause severe performance problems. For instance, stiction, hysteresis and dead-band in actuators, or faulty sensors can induce unwanted oscillations; see Chap. 11 for a thorough discussion. Thus, it is recommended to evaluate "how linear (or nonlinear)" the closed loop under consideration actually is in an early step of the assessment procedure.

To understand the nonlinearity detection techniques presented in this chapter, we recall the following important observations:

- A control loop containing a nonlinearity such as a sticking valve often exhibits self-generated and self-sustained limit cycles. The waveform in a limit cycle is periodic but non-sinusoidal and therefore has *harmonics*.
- It is observed that the first- and second-order statistic (mean, variance, autocorrelation, power spectrum, etc.) are only sufficient to describe linear systems. Nonlinear behaviour must be detected using *higher-order* statistics. For instance, the bicoherence, based on the Fourier transform of the data, is a measure of interaction of frequencies: its plot shows large peaks for frequencies that are interacting, indicating a strong nonlinearity.
- A distinctive characteristic of a nonlinear time series is the presence of *phase coupling* which creates *coherence* between the frequency bands occupied by the harmonics such that the phases are *non-random* and form a *regular pattern* (Thornhill 2005).

These features of nonlinear behaviour have been used as a basis for the development of some nonlinearity detection methods. The exploitation of the bicoherence property led to the bicoherence technique described in Sect. 10.2. Section 10.3 presents the surrogates analysis method, which is based on the regularity of phase patterns in nonlinear time series. If an actuator hits a constraint, then the discontinuity in its movement indicates a transient response, which can take the control loop back into saturation. Such a repeating pattern is also a limit cycle, which can be detected by specialised indices, as shown in Sect. 10.4. Section 10.5 contains a comparative study of both nonlinearity detection techniques on some industrial data sets.

M. Jelali, *Control Performance Management in Industrial Automation*, Advances in Industrial Control, DOI 10.1007/978-1-4471-4546-2_10,

10.1 Methods Review

In the classical identification literature, many nonlinearity test methods have been proposed; see the survey by Haber and Keviczky (1999: Chap. 4) and the references therein. Almost all nonlinearity detection methods assume that the system under investigation can be excited with certain input signals, which is not always possible or allowable in practice, and that both the controlled variables and the manipulated variables are available. Due to their invasiveness, these methods for nonlinearity detection are of no interest at all within the CPM framework.

More interesting is the use of higher-order statistics-based methods (Appendix B) to detect certain types of nonlinearities in time series. Such a test determines whether a time series could plausibly be the output of a linear system driven by Gaussian white noise or whether its properties can only be explained as the output of nonlinearity. For a control loop, this test is applied on the control error signal (SP–PV) to the controller because the error signal is more stationary than PV or OP signal. Moreover, If any disturbance is measurable, the test can be applied to check the linearity of the disturbance.

The assumption of Gaussianity allows implementation of statistically efficient parameter estimators, such as maximum likelihood estimators. A stationary Gaussian process is completely characterised by its second-order statistics (auto-correlation function or, equivalently, its power spectral density, PSD) and it can always be represented by a linear process. Since PSD depends only on the magnitude of the underlying transfer function, it does not yield information about the phase of the transfer function. Determination of the true phase characteristic is crucial in several applications, such as seismic deconvolution, blind equalisation of digital communications channels and analysis of the nonlinearity of a system operating under random inputs or disturbances. Use of higher-order statistics allows one to uniquely identify non-minimum-phase parametric models. Higher-order cumulants of Gaussian processes vanish; hence, if the data are stationary Gaussian, a minimum-phase (or maximum-phase) model is the "best" that one can estimate. Given these facts, it has been of some interest to investigate the nature of the given signal: whether it is a Gaussian process and if it is non-Gaussian, whether it is a linear process.

Several cumulant-based methods have been designed to detect certain types of nonlinearities in time series. The earliest tests of this type can be tracked back to Subba Rao and Gabr (1980) and then Hinich (1982). One of the earliest tests based upon testing of a sample estimate of the bispectrum has been presented by Subba Rao and Gabr (1980). Hinich (1982) has simplified the test of Subba Rao and Gabr (1980) by using the known expression for the asymptotic covariance of the bispectrum estimators. Modifications of Hinich's linearity test has been presented by Fackrell (1996) and Yuan (1999). The next logical step would be to test for vanishing trispectrum of the record. This has been done in Molle and Hinich (1995) using the approach of Hinich (1982); extensions of the approach by Subba Rao and Gabr (1980) are too complicated. A computationally simpler test using "integrated polyspectrum" of the data have been proposed in Tugnait (1987). The integrated polyspectrum (bispectrum or trispectrum) is computed as cross-power spectrum, and it is zero for Gaussian processes. Alternatively, one may test higher-order cumulant functions of the data in time domain. This has been done in Giannakis and Tstatsanis (1994). More recently, Choudhury et al. (2004) proposed an easy-to-use bicoherence-based technique for nonlinearity detection, which is described in Sect. 10.2.

Other methods of nonlinearity detection use surrogate data (Theiler et al. 1992; Kantz and Schreiber 1997), which have been found in many applications ranging from analysis of EEG recording of people with epilepsy (Casdagli et al. 1996), over the analysis of X-rays emitted from a suspected astrophysical black hole (Timmer et al. 2000), to the finding of nonlinearity sources in chemical plants (Thornhill 2005). Surrogate data are time series superficially constructed by randomisation of phases to remove the phase coupling, but under preservation of the same power spectrum and, hence, the same auto-correlation function as the test data.

Nonlinearity detection methods based on surrogate data consider a key statistic of the time series under test compared to that of a sufficiently large number of surrogates: nonlinearity is diagnosed if the statistic significantly differs in the test data; otherwise, the null hypothesis that a linear model fully explains the data is expected. This will be described in-depth in Sect. 10.3.

10.2 Bicoherence Technique

Common to majority of the cumulant-based methods is to check whether the (squared) bicoherence (Kim and Powers 1979)—a measure of quadratic phase coupling and thus an indicator of nonlinear signal generation mechanisms—is constant or not by performing two tests. One is for testing the zero squared bicoherence (10.2.1),

$$bic^2(f_1, f_2) := \frac{|B(f_1, f_2)|^2}{\mathrm{E}\{|X(f_1)X(f_2)|^2\}\mathrm{E}\{|X(f_1+f_2)|^2\}}, \tag{10.1}$$

which shows that the signal is Gaussian and thereby the signal generating process is linear. The other is to test for a non-zero constant (squared) bicoherence, which shows that the signal is non-Gaussian but the signal generating process is linear. Recall that the bispectrum is defined as (see Appendix B)

$$B(f_1, f_2) = \mathrm{E}\big\{X(f_1)X(f_2)X^*(f_1+f_2)\big\}, \tag{10.2}$$

where $X(f_1)$ is the discrete Fourier transform of the test data $x(k)$ at the frequency f_1, $X^*(f_1)$ the complex conjugate, and E the expectation operator. All frequencies are normalised so that the sampling frequency is 1.

A simple way to check the constancy of the squared bicoherence (i.e., the *linearity*) is to have a look at the 3D squared bicoherence plot and observe the "*flatness*" of the plot; see Fig. 10.1. This method is, however, tedious and cumbersome when a large number of signals have to be analysed. Based on these approaches, Choudhury et al. (2004) have proposed practical automated tests, which are described in the following.

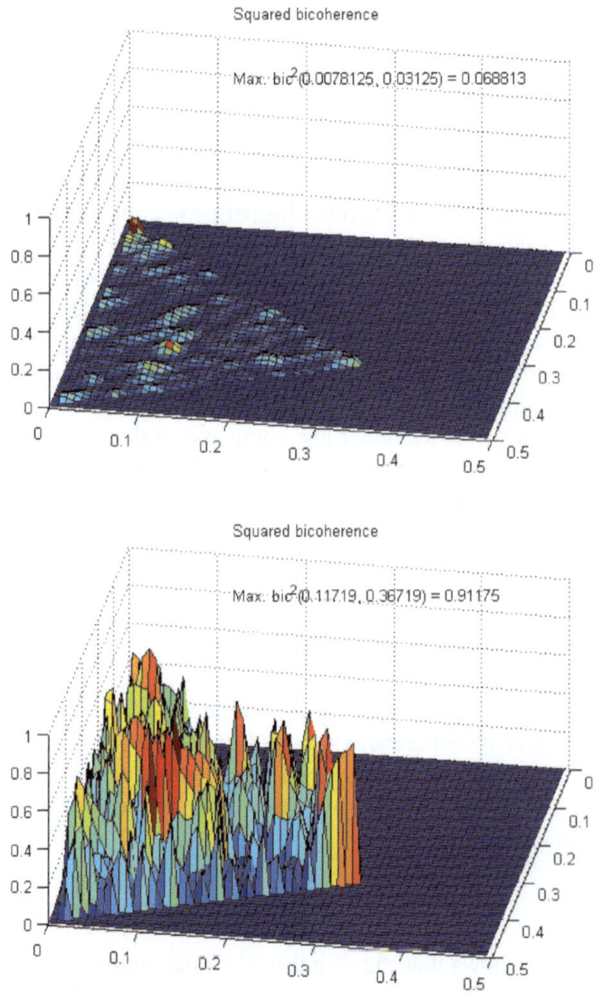

Fig. 10.1 Plot of the bicoherence function in the principal domain for a linear system (*top*) and a nonlinear system (*bottom*)

10.2.1 Non-Gaussianity Index

A modified test with better statistical properties but no frequency resolution is formulated by averaging the squared bicoherence over the triangle of the principal domain (PD: $0 < f_1 < 0.5 \wedge 2f_1 + f_2 < 1.0$) containing L bifrequencies. The test can be stated as follows:

- Null hypothesis (H0): The signal is *Gaussian*.
- Alternate hypothesis (H1): The signal is *not Gaussian*.

Under H0, the test for the squared bicoherence average can be based on the following equation (Fackrell 1996; Choudhury et al. 2004):

$$P\left(\overline{bic^2} > \overline{bic^2}_{\text{crit}}\right) = \alpha, \quad \overline{bic^2}_{\text{crit}} := \frac{1}{4KL}\left[c_\alpha^z + \sqrt{4L - 1}\right], \tag{10.3}$$

where L is the number of bifrequencies inside the principal domain of the bispectrum, K the number of data segments used for DFT, and c_α^z the one-sided critical value from the standard normal distribution for a significance level of α. Typically, one selects $c_\alpha^z = 5.99$ for the central $z = \chi^2$ distribution with $\alpha = 0.05$. Equation 10.3 can be rewritten as (Choudhury et al. 2004)

$$P(NGI > 0) = \alpha, \quad NGI := \overline{bic^2} - \overline{bic^2}_{\text{crit}}, \tag{10.4}$$

where *NGI* stands for the *non-Gaussianity index*. Therefore, at a confidence level α the following rule-based decision can be formulated:

- If $NGI \leq 0$, the signal is *Gaussian*.
- If $NGI > 0$, the signal is *non-Gaussian*.

Therefore, a signal is Gaussian (non-skewed) at a confidence level of α if the NGI is less than or equal to zero. This index has been defined to automate the decision instead of checking the flatness of the bicoherence plots for all bifrequencies in the principal domain.

If the signal is found to be Gaussian, the signal generating process is assumed to be linear. In the case of non-Gaussianity, the signal generating process should be tested for its linearity. If the signal is non-Gaussian and linear, the magnitude of the squared bicoherence should be a non-zero constant at all bifrequencies in the principal domain.

10.2.2 Nonlinearity Index

If the squared bicoherence is of a constant magnitude at all bifrequencies in the principal domain, the variance of the estimated bicoherence should be zero. To practically check the flatness of the plot or the constancy of the squared bicoherence (Fig. 10.1), the maximum squared bicoherence can be compared with the average squared bicoherence plus two or three times the standard deviation of the estimated squared bicoherence (depending on the confidence level desired). The automatic detection of this can be performed using the following *nonlinearity index* (NLI), which is defined for a 95 % confidence level as (Choudhury et al. 2004)

$$NLI = \left| bic^2_{\max} - \left(\overline{bic^2} + 2\sigma_{bic^2} \right) \right|, \tag{10.5}$$

where σ_{bic^2} is the standard deviation of the estimated squared bicoherence, and $\overline{bic^2}$ is the average of the estimated squared bicoherence. Ideally, the NLI should be 0 for a linear process, i.e. the magnitudes of squared bicoherence are assumed to be constant or the surface is flat. This is because if the squared bicoherence is a constant at all frequencies, the variance will be zero, and both the maximum and the mean will be same. Therefore, it can be concluded that

- if $NLI = 0$, the signal generating process is *linear*,
- if $NLI > 0$, the signal generating process is *nonlinear*.

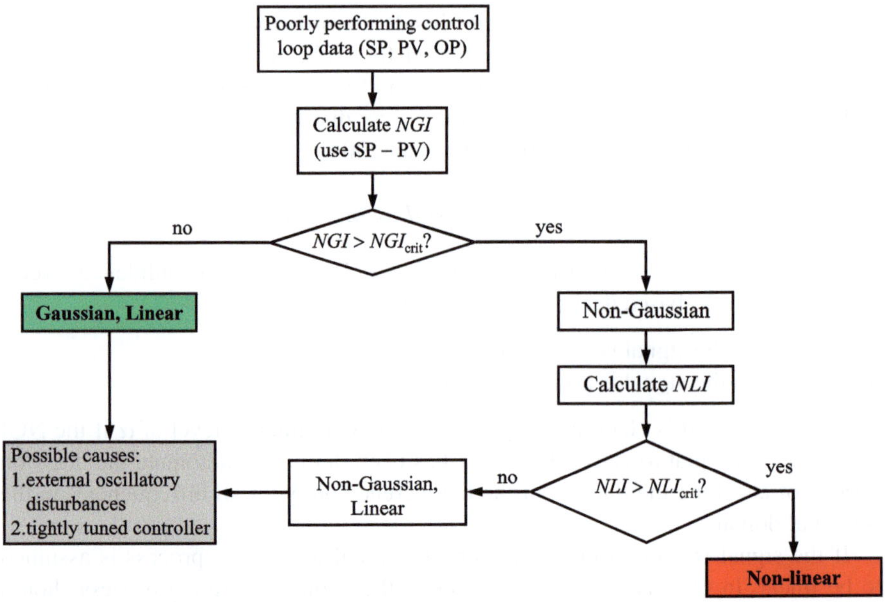

Fig. 10.2 Decision flow diagram of the bicoherence-based methodology for the detection and diagnosis of loop nonlinearity

Since the squared bicoherence is bounded between 0 and 1, the NLI is also bounded between 0 and 1.

10.2.3 Procedure and Practical Conditions

Figure 10.2 shows the flow diagram of the bicoherence-based nonlinearity detection method. Once a control loop is identified as nonlinear based on the analysis of the control error time trend, the cause of nonlinearity should be diagnosed. This may be due to a nonlinear process (component), the presence of stiction in the actuator(s), faulty sensors, etc. If the diagnosis signals that the loop is linear, other causes should be considered as a possible source for poor performance, such as an external oscillatory disturbance or an aggressively tuned controller. As for any data analysis, it is useful in practice to spent time in properly pre-processing the data. Some aspects to be considered when using the bicoherence technique are described in the following.

10.2.3.1 Default Parameters

Whenever possible, a large number of data points (e.g. 4096 samples) have to be used for the nonlinearity detection algorithm. Standard choices of the parameters

Table 10.1 Threshold values for NGI and NLI (Choudhury et al. 2006a, 2006b)

Data length	NGI_{crit}	NLI_{crit}
1024	0.004	0.04
2048	0.002	0.02
4096	0.001	0.01

are: data length (N) of 4096, a segment length of 64, a 50 % overlap, Hanning window with a length of 64 and a discrete Fourier transform (DFT) with a length of 128.

10.2.3.2 Selection of Critical Values

In practice, it is difficult to obtain an exact zero value for NGI for Gaussian signals. Therefore, a threshold value, NGI_{crit}, of NGI such that $NGI < NGI_{\text{crit}}$ implies a Gaussian signal. An NGI value of less than NGI_{crit} should be assumed to be zero. Consequently, if $NGI \leq NGI_{\text{crit}}$, the signal can be assumed to be Gaussian at a 95 % confidence level. Similarly, an NLI value less than NLI_{crit} is assumed to be zero, and consequently, the process is considered to be linear at a 95 % confidence level. The larger the NLI, the higher is the extent of nonlinearity. Recommended values for the thresholds NGI_{crit} and NLI_{crit} are given in Table 10.1.

10.2.3.3 Bicoherence Estimation

The bispectrum can be normalised in various ways to obtain bicoherence. There are more than one normalization in the literature. Particularly, some normalisation may not deliver bicoherence magnitudes bounded between 0 and 1. For example, the bicoherence function included in the freely available Higher-order Spectral Analysis Toolbox in MATLAB does not provide bounded values. Therefore, users are suggested to use the normalisation provided in Eq. 10.1. More details on bicoherence estimation can be found in Appendix B and the references cited therein.

10.2.3.4 Non-stationarity/Drift of the Data

Most of the statistical analyses including bicoherence estimation assume that the signal is stationary. Therefore, slowly drifting trends should be eliminated (Sect. 9.8). The contribution of non-stationarity of the signal in NGI and NLI indices has been reduced by excluding the outer triangle and using the inner triangle of the principal domain of the bispectrum only during the calculation of the average squared bicoherence. To exclude any peak(s) obtained from the non-stationary of the data, the outer triangle of the principal domain was excluded during the calculation of the average squared bicoherence (Nikias and Petropulu 1993).

10.2.3.5 Problem of Outliers and Abrupt Changes

Bicoherence estimation is very susceptible to outliers or abrupt changes in the signal; see the thorough discussion by Fackrell (1996). Outliers should be removed and replaced by a suitable statistical method. Also, the portion of the signal used for bicoherence calculation should not have any step change or abrupt change.

10.2.3.6 Dealing with Short Length Data

Though in recent time, it is easy to obtain a longer length data set (e.g. $N = 4096$), sometime there is no alternative to a shorter length data set. In those cases, depending on the length of the data, certain amount of overlap can be used during the calculation of bicoherence using a direct method similar to Welch periodogram method (Choudhury et al. 2004). Also, the threshold values used for NGI and NLI should also be changed for obtaining reliable results with a minimum number of false positives; see Table 10.1.

10.2.4 Modified Indices

In the estimation of the bicoherence, many spurious peaks arise due to the occurrence of small magnitudes of the denominator used to normalise the bispectrum (Eq. 10.1). Recently, Choudhury et al. (2006a, 2006b) suggested an addition of a small and dynamically adapted constant ε to the denominator to remove these spurious peaks:

$$bic^2(f_1, f_2) := \frac{|B(f_1, f_2)|^2}{\mathrm{E}\{|X(f_1)X(f_2)|^2\}\mathrm{E}\{|X(f_1 + f_2)|^2\} + \varepsilon}. \tag{10.6}$$

The selection of ε depends on the noise level. To obtain the value of ε automatically, it can be chosen as the maximum of the Pth percentiles of the columns of D (the denominator of Eq. 10.1). If it is assumed that there will be a maximum of 25 % peaks in each column of D, and the value of P can be chosen as 75 (Choudhury et al. 2006a, 2006b).

The previously described NGI and NLI have then been modified to overcome their limitations:

$$NGI_{\mathrm{mod}} := \frac{\sum bic^2_{\mathrm{significant}}}{L} - \frac{c^z_\alpha}{2K}, \tag{10.7}$$

$$NLI_{\mathrm{mod}} := bic^2_{\max} - \left(\overline{bic^2_{\mathrm{robust}}} + 2\sigma_{bic^2,\mathrm{robust}}\right), \tag{10.8}$$

where $bic^2_{\mathrm{significant}}$ are those bicoherences which fail the hypothesis test in Equation 10.3, i.e. $bic^2(f_1, f_2) > \frac{c^z_\alpha}{2K}$, L is the number of $bic^2_{\mathrm{significant}}$, and $\overline{bic^2_{\mathrm{robust}}}$ and

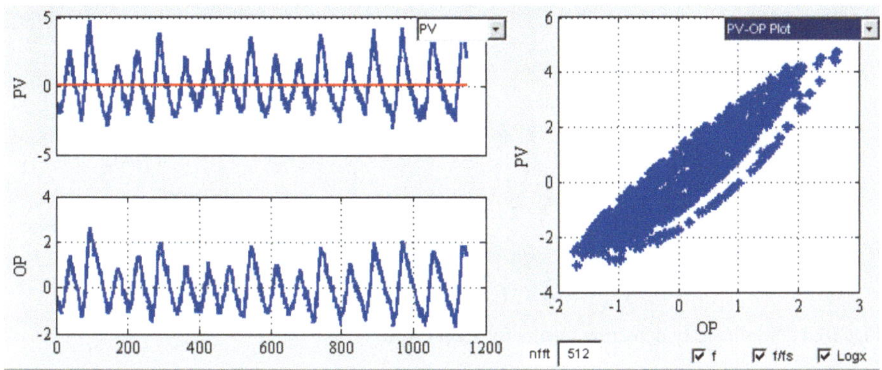

Fig. 10.3 Data and PV–OP plot for loop PAP3

$\sigma_{bic^2,\text{robust}}$ are the robust mean and the robust standard deviation of the estimated squared bicoherence, respectively. They are calculated by excluding the largest and smallest $Q\%$ of the bicoherences. A good value of Q may be chosen as 10.

Using NGI_{mod}, the following modified rule-based decision is suggested:

- If $NGI_{\text{mod}} \leq \alpha$, the signal is *Gaussian.*
- If $NGI_{\text{mod}} > \alpha$, the signal is *non-Gaussian.*

Similarly, the nonlinearity can be checked by

- if $NLI_{\text{mod}} \leq 0$, the signal generating process is *linear*,
- if $NLI_{\text{mod}} > 0$, the signal generating process is *nonlinear*.

Moreover, a total nonlinearity index (*TNLI*) has been introduced by Choudhury et al. (2006a, 2006b) as a metric or measure quantify nonlinearities:

$$TNLI := \sum bic^2_{\text{significant}}. \tag{10.9}$$

TNLI is bounded between 0 and L. *TNLI* quantifies the total nonlinearity present in a time series if it is detected as nonlinear by NGI_{mod} and NLI_{mod}. This is particularly important when comparing the extent of nonlinearities in various time series to detect the source of nonlinearity; see Sect. 10.5.

Example 10.1 Figure 10.3 shows the time trends and the PV–OP plot from a level control loop in a paper plant. Applying the bicoherence technique to 1024 data points with a segment length of 128, a 50 % overlap, Hanning window and a DFT length of 128 yields the results given in Fig. 10.4. The value of the non-Gaussianity Index, $NGI_{\text{mod}} = 0.36 > \alpha = 0.05$, implies that the process is non-Gaussian. The nonlinearity index value $NLI_{\text{mod}} = 0.76 > 0$ obtained reveals the presence of a nonlinearity in the process data. The total nonlinearity index was $TNLI = 2.57$. Indeed, it was known that this loop suffers from valve stiction.

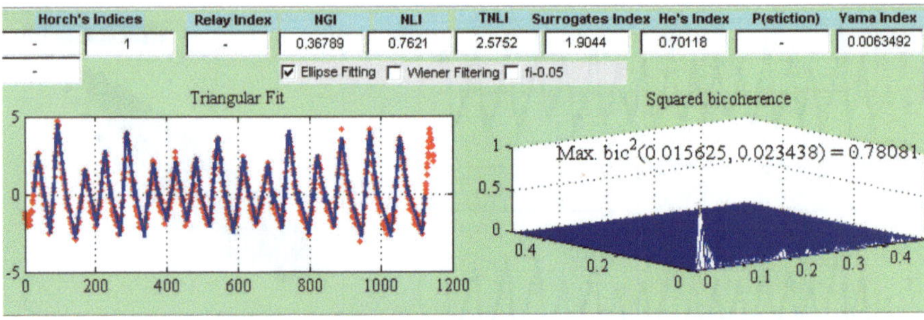

Fig. 10.4 Nonlinearity detection results for loop PAP3

10.3 Surrogate Data Analysis

Nonlinearity detection using surrogates analysis is based on the assumption that a time series with phase coupling (as symptom of nonlinearity) has more regular pattern and, hence, is more predictable than a surrogate having the same power spectrum but with randomised phases (Theiler et al. 1992). The phase randomisation process aims at destroying the phase coupling. The first step of the analysis is to define some discriminating statistic that quantifies the predictability of the time trend compared to that of an ensemble of surrogates, which mimics "linear properties" of the data under study. The decision whether dynamic nonlinearities are present in the data or not is made as follows (Kaplan 1997):

- If the value of the statistic for the test data falls into the distribution for the surrogate data, then the statistic does not allow us to distinguish the test data from the surrogates, and there is no evidence of dynamic nonlinearities.
- If the test data has a value for the statistic that is outside of the distribution for the surrogate data, then it can be concluded that the test data is somehow different from the surrogates. If the dynamics underlying the test data are assumed to be stationary, then some dynamic nonlinearity is evident in the data.

In this context, there are many algorithms for generating surrogate data and many statistics suitable for the discrimination task.

10.3.1 Generation of Surrogate Data

As already mentioned, surrogate data are time series superficially constructed to have the same power spectrum and, hence, the same auto-correlation function as the original test data. For the generation of surrogate data, the following three-step procedure is usually used.

Procedure 10.1 Generation of surrogate data.

1. Calculate the FFT of the test data, which gives amplitude and phase at each frequency:

$$\mathbf{z} = FFT(\text{test series}). \tag{10.10}$$

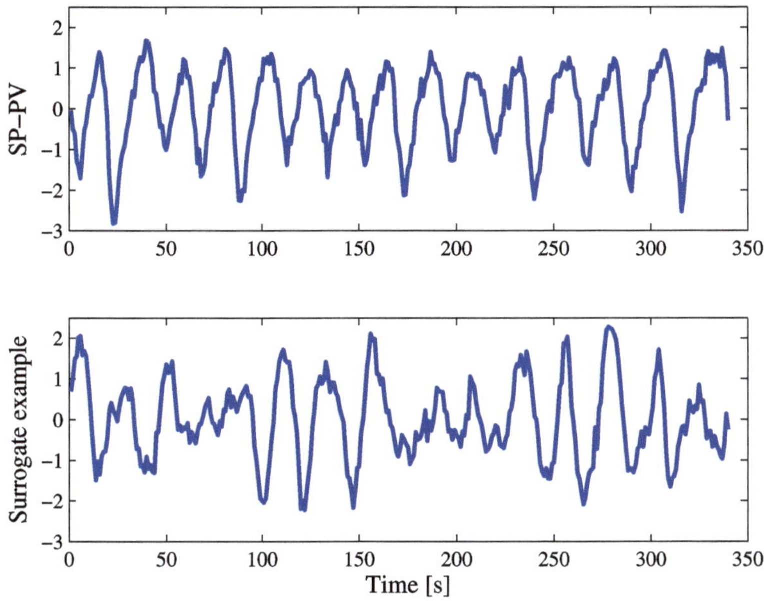

Fig. 10.5 Time trend of an industrial control loop (PAP3) and an example of its surrogate data

2. Randomise the phase at each frequency to be uniformly distributed in $[0, 2\pi]$ but preserve the asymmetry around frequency 0:

$$z_{\text{surr}} = \begin{cases} z(i), & i = 1, \\ z(i)e^{j\phi_{i-1}}, & i = 2, \ldots, N/2, \\ z(i), & i = N/2+1, \\ z(i)e^{-j\phi_{N-i+1}}, & i = (N/2+2), \ldots, N. \end{cases} \tag{10.11}$$

3. Take the inverse Fourier transformation (IFFT):

$$\text{surrogate data} = IFFT(z_{\text{surr}}). \tag{10.12}$$

As test data have often a nonnormal (non-Gaussian) distribution in practice, it may be necessary to transform the test data's distribution into a normal one before taking the DFT and to transform the generated Gaussian surrogate data back to the distribution of the test data. These transformations can be effectively done through sorting algorithms (Theiler et al. 1992) and would avoid problems with comparing non-normal test data to normal surrogate data. Different algorithms for the generation of surrogate data and their properties are described in detail by Theiler et al. (1992), Kaplan and Glass (1995) and Small and Tse (2002).

Example 10.2 Figure 10.5 (upper panel) illustrates the time trend of the control error from Loop PAP3. It has a clearly defined pattern and thus a good prediction of where the trend will go after reaching a given position. The lower panel of the figure shows an example of a surrogate of the time trend. By contrast to the original time trend, the surrogate lacks structure even though it has the same power spectrum. The removal of phase coherence has upset the regular pattern of peaks, i.e., it is not easy to forecast where the trajectory will go next from region to another. This signals nonlinearity of the time trend.

10.3.2 Discriminating Statistics—Nonlinear Predictability Index

Statistics used for nonlinearity detection based on surrogate data usually determine the difference between a property obtained for the test data and the mean value of this property for a set of surrogates. The result is considered significant if the difference is clearly larger than some standard deviations. The property can be a *correlation dimension*, *Lyapunov exponents*, *entropy*, a *nonlinear predictability* measure (Stam et al. 1998), or *redundancy*, known as *mutual information* in the case of two variables (Paluš 1995).

A statistic based on a three-sigma test can be formulated using nonlinear predictability as (Thornhill 2005)

$$NPI = \frac{\bar{\Gamma}_{\text{surr}} - \Gamma_{\text{test}}}{3\sigma_{\text{surr}}}, \tag{10.13}$$

where Γ_{test} is the mean squared prediction error of the test data, $\bar{\Gamma}_{\text{surr}}$ the mean or the reference distribution, and σ_{surr} its standard deviation. This nonlinearity test can be interpreted as follows:

- If $NPI > 1$, then nonlinearity is inferred in the data. The higher *NPI*, the more nonlinear is the underlying process.
- If $NPI \leq 1$, the process is considered to be linear.
- Negative values in the range $-1 \leq NPI < 0$ are not statistically significant and arise from the stochastic nature of the test.
- Results giving $NPI < -1$ do not arise at all because the surrogate sequences, which have no phase coherence are always less predictable than nonlinear time series with phase coherence.

The basis of this test proposed by Thornhill (2005) is to generate predictions from near neighbours under exclusion of near-in-time neighbours so that the neighbours are only selected from other cycles in the oscillation, following a slightly modified version of the algorithm described by Sugihara and May (1990). When n_{n} nearest neighbours have been identified, then those near neighbours are used to make an H-step-ahead prediction. A sequence of prediction errors can thus be created by subtracting the average of the predictions of the n_n nearest neighbours from the observed values. The root of mean square (RMS) value of the prediction error sequence is built to give the overall prediction error; see Sect. 10.3.3.

Note that the uncertainty of the aforementioned statistical test is minimal, since the index is a three-sigma test and was set up like that to err on the side of caution, i.e. it gives false negatives rather than false positives. Thus, one can generally believe that if $NPI > 1$, then there really is some nonlinearity present in the process considered.

10.3.3 Nonlinearity Detection Procedure

The procedure for detecting nonlinearities based on surrogate data analysis is summarised as follows.

Procedure 10.2 Nonlinearity detection using surrogate data analysis (Thornhill 2005).

1. Determine the period of oscillation T_p and thus the number S of samples per cycle; pre-process the data: mean-centring, possibly scaling and particularly end-matching (Sect. 10.3.4). As for the bicoherence technique, the elimination of slowly varying trends and high-frequency noise by means of frequency filtering is highly recommended (Sect. 9.8).
2. Form the embedded matrix from the pre-processed data subset of the test data $x(1), \ldots, x(N)$ as

$$\boldsymbol{X} = \begin{bmatrix} x(1) & x(2) & \cdots & x(E) \\ x(2) & x(3) & \cdots & x(E+1) \\ x(3) & x(4) & \cdots & x(E+2) \\ \vdots & \vdots & \vdots & \vdots \\ x(N-E+1) & x(N-E+2) & \cdots & x(N) \end{bmatrix}. \tag{10.14}$$

1. For each row $\boldsymbol{x}_i$ of $\boldsymbol{X}$, find the indices j_p $(p = 1, \ldots, n_n)$ of n_n nearest neighbour rows $\boldsymbol{x}_{j_p}$ having the n_n smallest values of the (Euclidean) norm $\|\boldsymbol{x}_{j_p} - \boldsymbol{x}_i\|$ subject to a near-in-time neighbour exclusion constraint $|j_p - i| > E/2$. E is the number of columns in the embedded matrix.
2. Calculate the sum of squared prediction errors for the test data

$$\Gamma_{\text{test}} = \sum_{i=1}^{N-H} \left(x(i+H) - \frac{1}{n_n} \sum_{p=1}^{n_n} x(j_p + H) \right)^2, \tag{10.15}$$

where H is the prediction horizon.
3. Create M surrogate prediction errors Γ_{surr} by applying the above Steps 2–4 to M surrogate data sets.
4. Calculate the nonlinearity index according to Eq. 10.13.

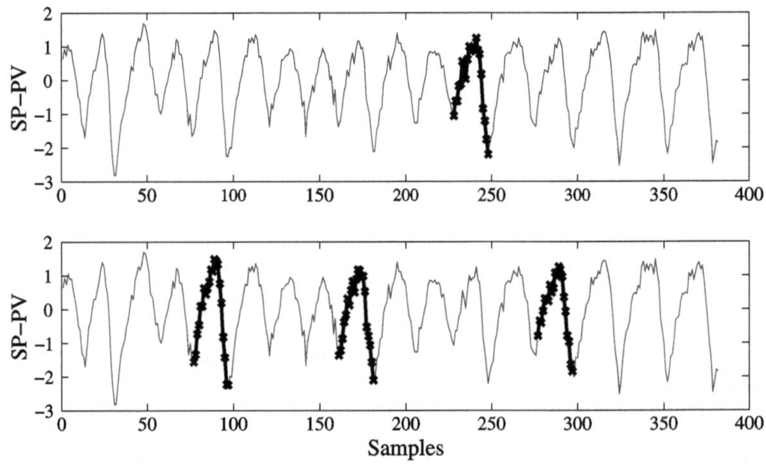

Fig. 10.6 Illustration of the nearest neighbour concept: the highlighted (*bold* and *star*) cycles in the *lower panel* are the three nearest neighbours of the cycle in the *upper panel* (Loop PAP3)

Example 10.3 Figure 10.6 illustrates the prediction principle using the data (decimated by the factor 3) from the industrial Loop PAP3, where the embedding dimension E is 21, and thus the prediction is made 21 steps ahead. The upper panel shows the 228th row of the data matrix $\boldsymbol{X}$, which is a full cycle starting at sample 228, marked with a heavy line. Rows of $\boldsymbol{X}$ that are nearest neighbours of that cycle begin at samples 77, 161 and 277 and are also shown as a heavy lines in the lower panel. Note that the analysis is non-causal and any element in the time series may be predicted from both earlier and later values.

10.3.4 Spurious Nonlinearity—Pitfalls in the Surrogate Data

There are some situations that can cause surrogate data to be a misleading poor match to the test data, even though the power spectrum and histogram are the same in the test and surrogate data.

10.3.4.1 Spikes in the Data

Sharp spikes in the test data are transformed into white noise within the surrogate data generation. Therefore, one should carefully remove such spikes from the data set: even a single spike can have a statistically significant effect on the surrogate data. In the data set already shown in Fig. 10.5, a superficial spike is inserted in the data at $t = 150$ s; see Fig. 10.7. If we compare the plots in Fig. 10.5 and Fig. 10.7, the surrogate data in the latter figure has additional noise turned by the spike. Even

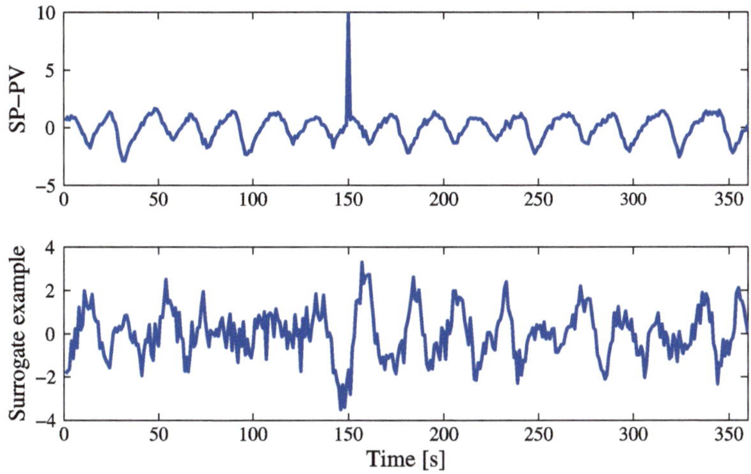

Fig. 10.7 Time trend of an industrial control loop (PAP3) and an example of its surrogate data when introducing a sharp spike in the data at $t = 150$ s

though the power spectrum of both data sets and their surrogates are identical, the spike is highly localised in time in the test data but spread throughout the surrogate data.

10.3.4.2 Strongly Periodic Time Series

In the case of strongly cyclic data, the length of the data set will almost never be an exact integer multiple of the dominant period. Unless care is taken with data end-matching, a surrogate-data-based nonlinearity test may give false positive results, i.e. indicate nonlinearity for a linear time series. The reason for this is the phenomenon of spectral leakage in the DFT caused by the use of a finite length data sequence. A phase-randomised surrogate derived from the DFT of non-properly end-matched test data would contain frequencies that are not present in the original signal and will, thus, be less predictable than the original signal giving a false indication of nonlinearity. It is therefore essential to take special precautions before generating surrogate data. The simplest measure is to adjust the length of the test data. Some algorithms are briefly described as follows. Note, however, that industrial process data do not often suffer from the strong periodicity problem because the cyclic behaviour is not normally strong, as found out by Thornhill (2005).

10.3.4.3 Length Adjustment Based on Zero Crossings

Since the determination of the zero crossings is a part of many oscillation detection algorithms (Chap. 9), it is not difficult at all to take a window of the data containing exactly an even number of zero crossings. Care should be taken when noise is present in the data. A noise band should be defined to exclude spurious zero crossings.

10.3.4.4 Data End-Matching by Minimising Discontinuities

Hegger et al. (2004) recommend finding a subset of the non-matched data with N samples starting at x_i and ending with x_{i+N-1} which minimises the sum of the normalised discontinuities (d_0+d_1) between the initial and end values and the initial and end gradients, where

$$d_0 = \frac{(x_i - x_{i+N-1})^2}{\sum_{j=i}^{i+N-1}(x_j - \bar{x})^2}, \qquad d_1 = \frac{((x_{i+1} - x_i) - (x_{i+N-1} - x_{i+N-2}))^2}{\sum_{j=i}^{i+N-1}(x_j - \bar{x})^2}, \tag{10.16}$$

where $\bar{x}$ is the mean of the sequence $x_i, \ldots, x_{i+N-1}$. However, the procedure should be modified to avoid the artifacts due to spectral leakage in oscillating signals, as proposed by Barnard et al. (2001). The aim is to creates a time trend where the last value is the first sample of another cycle. An end-matched sequence which contains an exact number of cycles is $x_i, \ldots, x_{i+N-2}$ derived from the $x_i, \ldots, x_{i+N-1}$ sequence by omitting the last sample (Thornhill 2005).

10.3.4.5 Data End-Matching by Minimising the Frequency Mismatch Error

Stam et al. (1998) proposed an algorithm to adjust the length of the data set so that it becomes an exact integer multiple of the length of the fundamental period (and its higher harmonics). For this purpose, a frequency mismatch error E_{fmm} is defined as follows:

$$E_{\mathrm{fmm}} = \sum_{i=1}^{m}(x_i - x_{k+i})^2, \tag{10.17}$$

where $\{x_i, i = 1, \ldots, N\}$ is the set of non-matched data. One starts with $k = N - m$, then decreases k in steps of 1 and calculates E_{fmm} for each value of k. The new end-point of the time series N' is the value of k for which E_{fmm} reaches its first minimum. In this way, the length of the time series has been adjusted so that the beginning and the end will fit very smoothly. The goodness of fit can be adjusted by setting m. Stam et al. (1998) found out that choosing $m = 10$ is sufficient even in the case of periodic time series with very complex waveforms. A nice by-product of this algorithm is that the problem of the "jump phenomenon" between the beginning and the end of the time series is solved. Because N' will typically not be a power of 2, the DFT, instead of the faster FFT, must be used for generating phase-randomised surrogate data.

Example 10.4 An example showing the extreme need for data end-matching is illustrated in Fig. 10.8: when the data are not properly end-matched, then the resulting

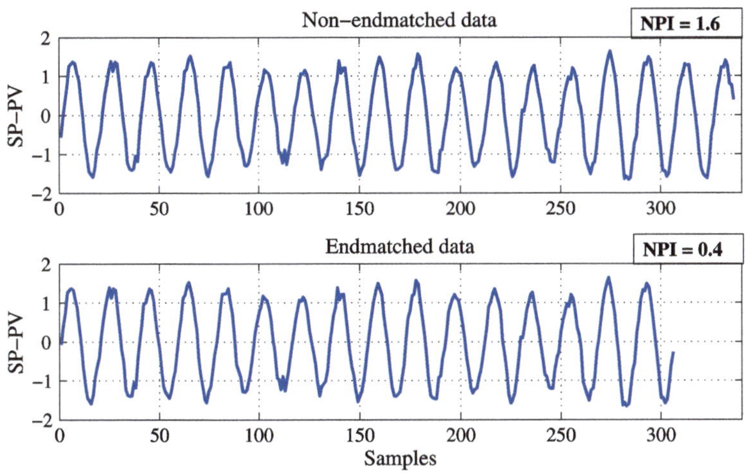

Fig. 10.8 Illustration of the effect of data end-matching on the nonlinearity index for the industrial control loop (PAP13), which does not have any nonlinearity

NPI value of 1.6 signals nonlinearity (upper panel), which is however a wrong indication. One gets the right decision when end-matching (here based on zero crossings) is applied (lower panel).

10.3.5 Default Parameter Values and Practical Issues

Empirical studies carried out by Thornhill (2005) have shown that reliable results of surrogates testing can be achieved using the default values (Table 10.2) given for the parameters involved in the algorithm. In this context, some important remarks should be mentioned:

- The maximum number of samples per cycle ($S = T_p/T_s$) is limited to have a trade-off between the S value needed to properly define the shape of a non-sinusoidal oscillation on the one hand and the speed of the computation on the other. Also, it would be infeasible to operate with fewer than seven samples per cycle because harmonics would not be satisfactory captured. Using a minimum $S = 7$ ensures capturing the third harmonic, as the most prominent harmonic in symmetrical oscillations having square or triangular wave, just to meet the Nyquist criterion of two samples per cycle.
- The recommendation $E = \text{floor}(S)$ is easy to implement since S is already known as a by-product of oscillation detection algorithms (Chap. 9). E should not be too small to avoid the phenomenon of false near neighbours (Rhodes and Morari 1997), especially when time trends have high-frequency features or noise.

Table 10.2 Suggested default values for the parameters involved in surrogates-analysis-based nonlinearity detection algorithm (Thornhill 2005)

Description	Value
Number of samples per cycle (S)	$7 \leq S \leq 25$
Number of columns in the embedded matrix (E)	$E = \text{floor}(S)$
Prediction horizon (H)	$H = E$
Number of cycles of oscillation (C)	$C \geq 12$
Number of near neighbours (k)	$n_\text{n} = 8$
Number of surrogates (M)	$M = 50$

- It is not advisable to use less than 12 cycles of oscillation to avoid inconsistency in the nonlinearity index. Though in recent time, it is easy to obtain longer length data sets, sometimes there is no alternative to a shorter length data set. This condition is therefore something restrictive in practice.
- From common sense of reasoning, n_n should be smaller than C, as some cycles may be lost during the end-matching process. This implies to select n_n so that $n_\text{n} \leq C - 4$. Thornhill (2005) found that the choice $n_\text{n} = 8$ is quite satisfactory in practice.
- Thornhill also concluded from studying different values for the number M of surrogates to be used in the statistical test that $M = 50$ should be sufficient.
- Without further caution, the surrogates testing procedure will lead to nonlinearity indices, which vary from one run to the other, due to the randomised surrogate generation. This is, however, very undesirable in practice. A pragmatic solution to the variability of NPI is to always use the same random number seed for the first surrogate, which forces all the surrogates to be the same each time the nonlinearity test is applied to the same data set. Alternatively, one can repeat the surrogates testing some times and display the averaged index value.

Example 10.5 The same loop data considered in Example 10.1 are now analysed using the surrogates method. The default parameters in Table 10.2 have been used, i.e., $S = E = H = 21$, $C = 25$, $n_\text{n} = 8$ and $M = 50$. Note that the data have been decimated by factor 3, as S is too high without decimation. End-matching was achieved based zero crossing. The analysis leads to a nonlinearity index $NPI = 2.13 > 1.0$, indicating the presence of nonlinearity in the loop.

10.4 Detection of Saturated Actuators

The range of values taken by OP in control loops is generally scaled to be between 0 and 100 %. Saturation is easily recognised by visual inspection because the time trend becomes flat and constrained either at the value 0 or at the value 100. Saturation cannot only result from poor controller tuning or missing anti-windup, but is often an indication of an inadequate actuator, e.g. control valve, sizing in practice.

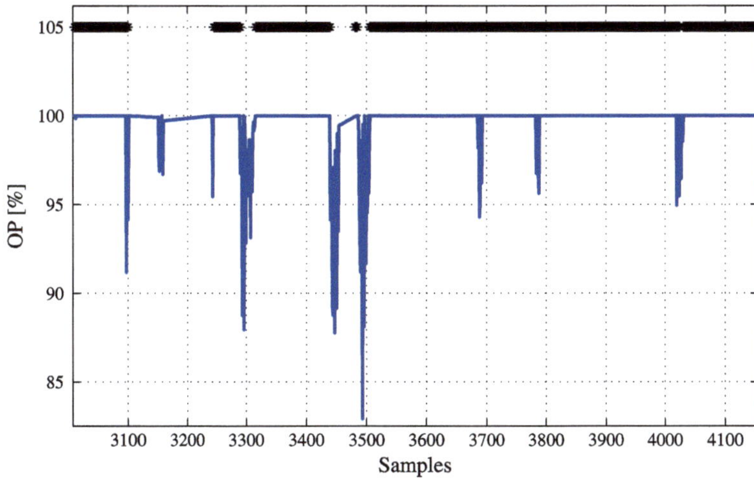

Fig. 10.9 Saturation analysis for loop POW5: the heavy black lines at 105 show episodes where OP became saturated at 100 %, detected by the Kolmogorov–Smirnov test method

10.4.1 Saturation Test Based on Statistical Distribution

A saturation test proposed by Matsuo et al. (2004) evaluates the statistical distribution of samples in short sequence of the time trend and tests whether the sequence matches test sequences of the same length that are all 0 or 100. Such a statistical test can be applied to data having any arbitrary statistical distribution and requires little tuning. The method uses a two-sample Kolmogorov–Smirnov goodness-of-fit hypothesis test [kstest2]: a Kolmogorov–Smirnov test is performed to determine if independent random samples are drawn from the same underlying continuous population with specified significance level α (default: $\alpha = 0.05$). The result is interpreted as follows:

- $H = 0 \Rightarrow$ The null hypothesis is supported at significance level α, i.e. both considered sequences are sampled from the same underlying distribution, and thus saturation occurs.
- $H = 1 \Rightarrow$ The null hypothesis is rejected at significance level α, i.e. both considered sequences are not sampled from the same underlying distribution, and thus no saturation occurs.

Therefore a saturation index can be defined as

$$\eta_{\text{sat}} = 1 - H. \tag{10.18}$$

A saturation index of 1 indicates saturation, and $\eta_{\text{sat}} = 0$ implies no saturation.

Example 10.6 The heavy black lines over the OP time trend in Fig. 10.9 show episodes of saturated operation detected in the data of loop POW5 by applying the

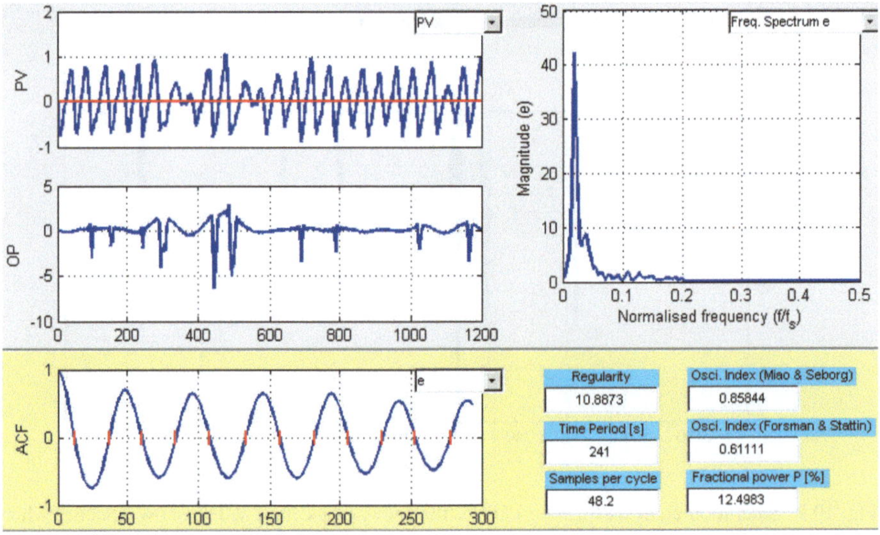

Fig. 10.10 Oscillation detection results for loop POW5; the data shown in the figure are pre-processed using a Wiener filter $[0.01, 0.20]$

saturation index (Eq. 10.18) based on the Kolmogorov–Smirnov test. The saturation index confirms the observation that the loop has a periodically saturated actuator. This saturation problem causes a limit cycle, as shown by the oscillation analysis results in Fig. 10.10.

10.4.2 Saturation Index for Valve Monitoring

Another saturation index proposed by Jämsa-Jounela et al. (2003) for valve monitoring is defined as the ratio of the time that a valve opening (MV) is greater than 90 % or smaller than 10 % to the time needed to carry out a set-point change:

$$\eta_{\text{sat}} = \frac{\int_0^\tau t_{\text{vc}}\,\mathrm{d}t}{\tau}, \qquad t_{\text{vc}} = \begin{cases} 0 & \text{if } (u_{\text{v}} \geq 10\,\%) \wedge (u_{\text{v}} \leq 90\,\%), \\ 1 & \text{if } (u_{\text{v}} < 10\,\%) \vee (u_{\text{v}} > 90\,\%), \end{cases} \tag{10.19}$$

where τ is an estimate of the time constant of the process. Values of η_{sat} close to 0 indicate a correct actuator sizing; values close to 1 are a sign of deficient actuator sizing. In the latter case, one should check if the valve has to be resized.

10.5 Comparative Studies

The nonlinearity techniques presented above are now demonstrated and compared with two industrial case studies. The main task in both studies is to find out the

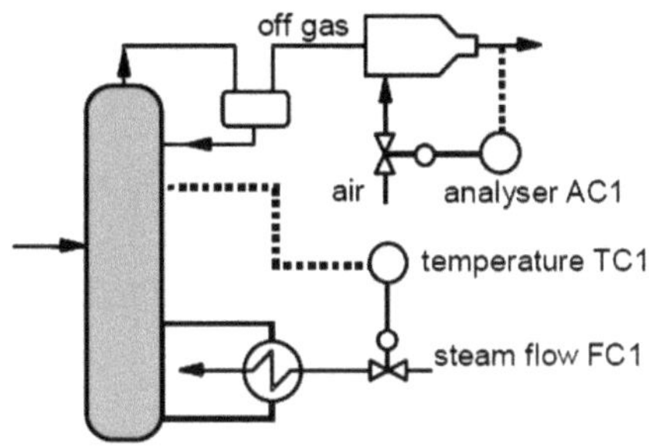

Fig. 10.11 Refinery separation unit showing the locations of TC, FC and AC loops; upstream and downstream pressure control loops PC1 and PC2 are not shown (Thornhill 2007)

source of oscillations propagating through whole plants. It is thus important to realise the following propagation mechanism (Thornhill et al. 2002, 2003a, 2003b, 2003c): a common source of oscillation is a limit cycle caused by a nonlinearity, e.g. control valve with a deadband or excessive static friction. A process variable oscillating for that reason can readily propagate the oscillation to other variables and disturb other control loops, hence causing a plant-wide disturbance. A candidate for the root cause is the time series with the strongest nonlinearity, i.e. showing the largest nonlinearity index. This is due to the fact that the dynamic behaviour of physical processes gives low-pass filtering and therefore removes nonlinearity from the time series. Consequently, harmonics of a limit cycle are expected to become smaller further from the root cause, and the time trends become more sinusoidal.

10.5.1 Unit-Wide Oscillation Caused by a Sensor Fault

The data in this study are from a refinery separation unit; see Fig. 10.11. The sampling interval was 20 s. The control errors in Fig. 10.12 show the presence of a unit-wide oscillation in the loops FC1, TC1 and AC1 with a period of 21 sampling intervals or 7 min. Measurements from upstream and downstream pressure controllers PC1 and PC2 are also available and show evidence of the same oscillation along with other disturbances and noise.

It is known that there was a faulty steam sensor in the steam flow loop FC1. It was an orifice plate flow meter, but there was no sweep-hole in the plate which had the effect that condensate collected on the upstream side until it reached a critical level, and the accumulated liquid would then periodically clear itself by siphoning through the orifice. The challenge for the analysis of this unit is to verify that the faulty steam flow loop is the root cause of the disturbance. A fully worked analysis and diagnosis of the sources of disturbances in this data set (spectral and oscillation analysis; root-cause analysis using the surrogates method) is published by Thornhill (2007).

In this study, we analyse the data using both nonlinearity detection techniques, the bicoherence method and the surrogates method and compare their performance. The numerical values on the right-hand side of Figs. 10.12 and 10.13 show the results from nonlinearity detection using the bicoherence method (NLI/TNLI) and

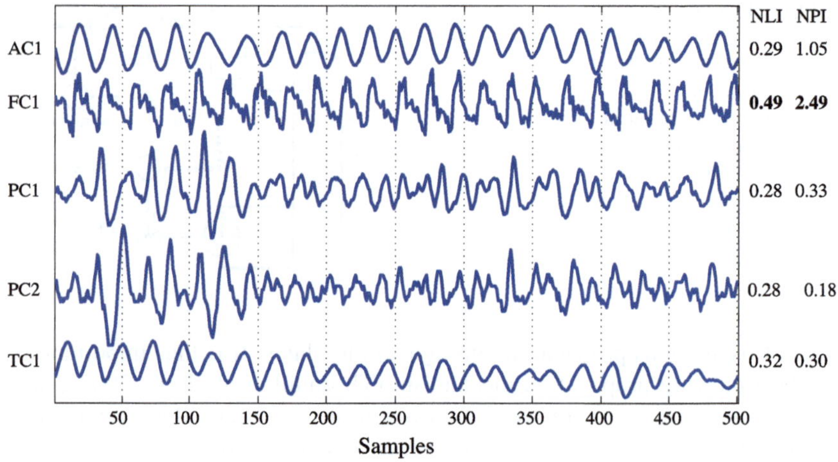

Fig. 10.12 Time trends and nonlinearity analysis results (NLI and NPI) for the refinery-wide oscillation data

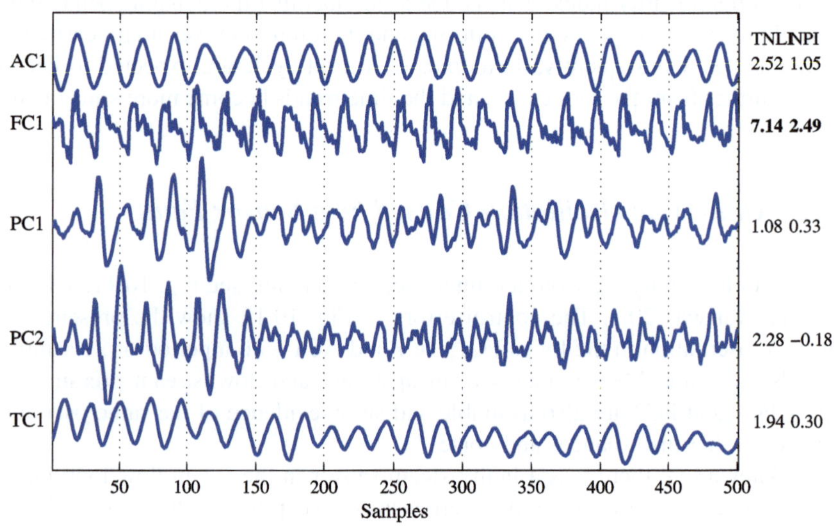

Fig. 10.13 Time trends and nonlinearity analysis results (TNLI and NPI) for the refinery-wide oscillation data

the surrogate data analysis (NPI). Both methods clearly indicate that the FC1 control loop contains the source of the oscillation, since the highest index values have been determined for its control error signal. Nevertheless, whereas the surrogates method unambiguously points to FC1 as the sole nonlinear signal, the bicoherence method detects also all other signals as nonlinear.

The same analysis was also carried out for the OP signals and led to similar conclusions, with the exception that also the OP signal of AC1 was also detected as

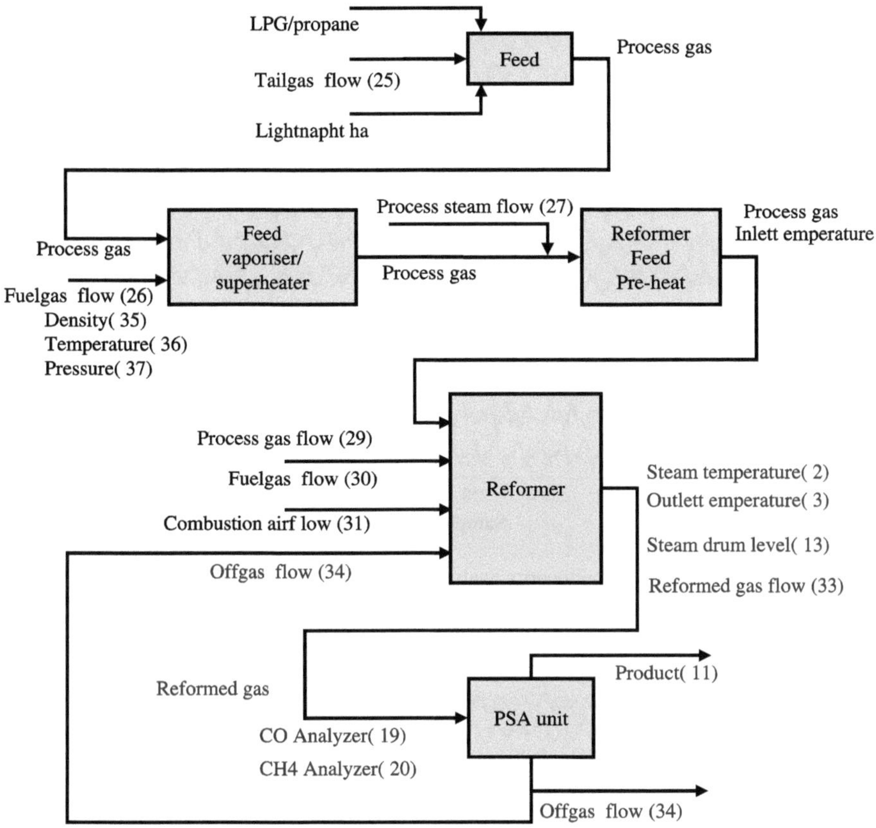

Fig. 10.14 Process schematic of a hydrogen reformer from a SE Asian refinery (Thornhill 2005)

nonlinear. This is due to the fact that the TC1 controller output is the set point of the FC1 loop because of the cascade configuration. Thornhill (2007) pointed out that the reason for the oscillation in PC1 and PC2 is that (a) the tuning might be rather tight and (b) that these two pressure loops are interacting with one another.

10.5.2 Plant-Wide Oscillation Caused by a Valve Fault

A set of refinery data (courtesy of an SE Asian refinery; Fig. 10.14) are examined in this section. This data set was previously used as benchmark for oscillation detection and diagnosis methods. Thornhill et al. (2001) had performed spectral PCA on the 34 loops recorded in the plant data and found 12 loops that were associated with a plant-wide oscillation; see Fig. 10.15.

Thornhill et al. (2001) and Thornhill (2005) applied nonlinearity tests and concluded that the source of nonlinearity was one of Tags 13, 33 or 34. The measure-

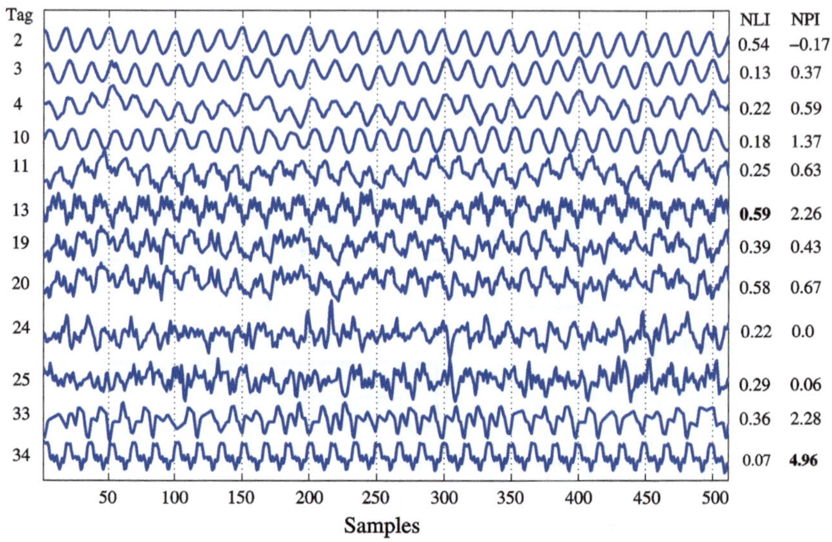

Fig. 10.15 Time trends and nonlinearity analysis results (NLI and NPI) for the SE Asian refinery data

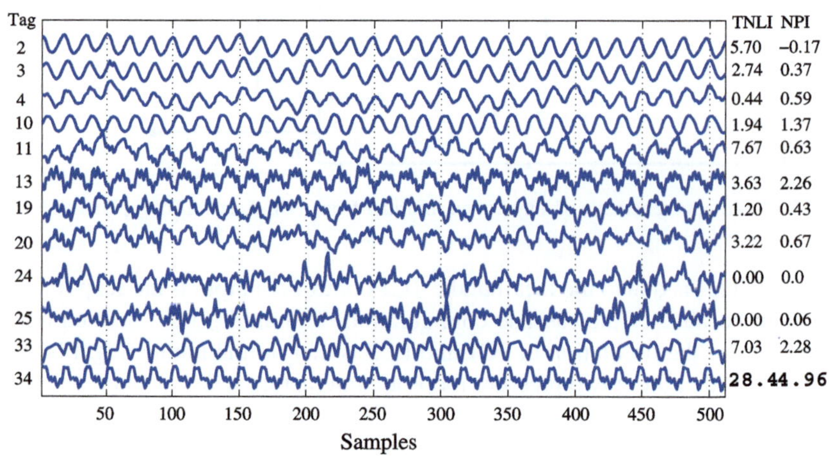

Fig. 10.16 Time trends and nonlinearity analysis results (TNLI and NPI) for the SE Asian refinery data

ments from this plant have been discussed by Thornhill et al. (2002). Moreover, Zang and Howell (2004) compared some oscillation diagnosis techniques on this data set. Note that the real root cause for the 16.7-min oscillation was not exactly known, but it has been emphasised that it is a valve fault.

The purpose of this study is the application of the bicoherence and surrogates analysis methods to the data set and the comparison of both techniques. The results are shown on the right-hand side of Fig. 10.15. Surprisingly, both methods disagreed about the source of nonlinearity. Whereas the surrogates testing method clearly indicates that Tag 34 has the largest nonlinearity index, the bicoherence technique identifies Tag 13 as the signal with the strongest nonlinearity. So far both tags belong to that group where the root cause is likely to be found. However, there is a complete disagreement of the nonlinearity indices for Tag 34. Considering the results of the earlier studies mentioned above, it must be argued that the surrogates analysis method points out the right root cause and thus provides a better measure. This result is however confirmed when using the total nonlinearity index (Sect. 10.2.4); see Fig. 10.16.

10.6 Summary and Conclusions

The very common problem of oscillating control loops can result from process nonlinearities often present in sensors and actuators. Two nonlinearity testing methods have been studied that help detect the root cause of such problems. The bicoherence technique based on higher-order statistics defines two indices, the non-Gaussianity index (NGI) and the nonlinearity index (NLI), to determine whether a time series could plausibly be the output of a linear system driven by Gaussian white noise, or whether its properties could only be explained as the output of a nonlinear system. The surrogates testing method is based on the relative predictability of test data and surrogate data and also provides a nonlinearity index that can be calculated from given routine operating data. The key issues to be addressed when applying both methods to real-world data from different industrial control loops have been discussed. It was pointed out that the bicoherence method is sensitive to nonstationary trends and abrupt changes. The surrogates method should be applied with extreme care concerning end-matching of the data.

Both techniques are useful to diagnose the root cause of limit cycles not only in single control loops, but also in whole plants that usually contain a large of number loops. The root cause of a limit cycle is to be found in the part of the plant where the nonlinearity index is largest. The methods presented have been examined and compared on two data sets from industrial plants showing unit-wide oscillations. The results revealed that both techniques do not always agree about the root cause location, and it seems that surrogates testing is more reliable. Anyway, the total nonlinearity index has to be applied when comparing the nonlinearity extent of different time series.

So far data-based nonlinearity detection and diagnosis provide quantitative and rapid information about the source of nonlinearity-induced limit cycles in processing units. However, process understanding and know-how and/or active testing are still needed in the final stage of performance monitoring to confirm the root cause and explain the interaction routes or mechanisms of propagation.

Chapter 11
Diagnosis of Stiction-Related Actuator Problems

Control valves are the most commonly used actuators or final control elements in the process industries. They are mechanical devices subject to wear and tear with respect to time. Therefore, valves may develop serious problems and should be regularly maintained. Surveys by Bialkowski (1993), Ender (1993), Desborough and Miller (2002) and Paulonis and Cox (2003) indicate that about 20–30 % of all control loops oscillate due to valve problems caused by valve nonlinearities, such as stiction, hysteresis, dead-band or dead-zone. Many control loops in process plants perform poorly due to valve static friction (*stiction*), as one of the most common equipment problems. It is well known that valve stiction in control loops causes oscillations in a form of periodic finite-amplitude instabilities, known as *limit cycles*. This phenomenon increases variability in product quality, accelerates equipment wear or leads to control system instability.

The literature contains several non-invasive methods to detect stiction in control loops by only using OP and PV signals. Among others, the following approaches are mentioned: the cross-correlation method of Horch (1999), the area-peak method of Singhal and Salsbury (2005) and Salsbury (2006), the relay method of Rossi and Scali (2005), the curve-fitting technique of He et al. (2007), the pattern recognition technique of Srinivasan et al. (2005a) and the bicoherence and ellipse fitting method of Choudhury et al. (2006a). Some other techniques available are based on additional knowledge about the characteristic curve of the valve or values of MV, i.e. valve position, e.g. Kano et al. (2004) and Yamashita (2006). Fairly complicated methods for detecting stiction were proposed by Horch and Isaksson (1998) and Stenman et al. (2003). Some stiction detection techniques have been reviewed and compared by Rossi and Scali (2005) and Horch (2007). Note that only a few methods have been published about the quantification of stiction, i.e. Choudhury et al. (2006a) and Srinivasan et al. (2005b).

This chapter is devoted to the illustration of the actuator stiction effect on control-loop performance and to a review of the most important techniques for automatic stiction detection, to be incorporated in performance monitoring. In Sect. 11.1, a typical control loop with control valve is explained. Section 11.2 gives a qualitative illustration of the stiction phenomenon and related effects in actuators. Sec-

M. Jelali, *Control Performance Management in Industrial Automation*,
Advances in Industrial Control, DOI 10.1007/978-1-4471-4546-2_11,

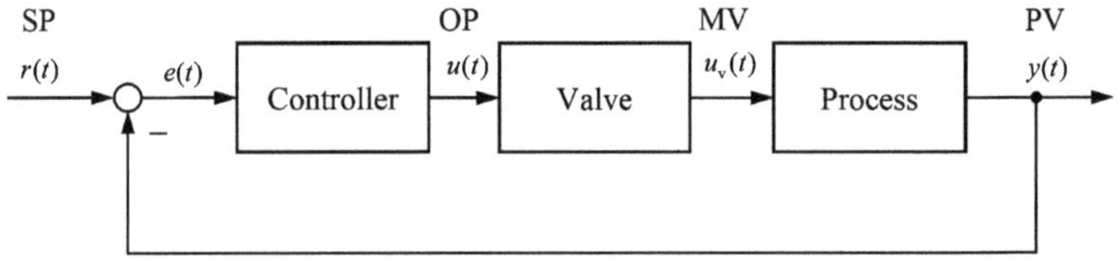

Fig. 11.1 Simple feedback scheme for a valve-controlled process with definition of the variables (SP: set point; OP: controller output; MV: manipulating variable: valve position; PV: process variable) used in this section

tion 11.3 contains a brief description of models and an analysis of how closed-loop variables change with stiction and process parameters.

A review of some popular methods for automatic stiction detection will be given in the rest of this chapter. Section 11.4 describes methods that are based on MV–OP shape analysis, i.e. require the measurement of the valve position or an equivalent variable. Among many techniques, which are completely automatic and require only values of SP, OP and PV, the cross correlation method (Sect. 11.5), the curve-fitting technique and similar methods (Sect. 11.6) and nonlinearity techniques combined with ellipse fitting to PV–OP maps (Sect. 11.7) are presented. In Sect. 11.9, a basic oscillation diagnosis procedure is proposed.

11.1 Typical Valve-Controlled Loop

Figure 11.1 shows a simple configuration of control loops actuated with a control valve. A typical example of such a configuration, i.e. a level control loop, is illustrated in Fig. 11.2. In many applications in the process industry, pneumatic control valves are used. The diagram of a typical pneumatic valve is shown in Fig. 11.3. The valve aims to restrict the flow of process fluid through the pipe that can be seen at the very bottom of the figure. The valve plug is rigidly attached to a stem that is attached to a diaphragm in an air pressure chamber in the actuator section at the top of the valve. When compressed air is applied, the diaphragm moves up, and the valve closes. At the same time, the spring is compressed.

Control valves should be maintained to have acceptable values for the parameters given in Table 11.1. Out of these, stiction is the most severe problem that can occur in a control valve. In many processes, stiction of 0.5 % is considered too much, as stiction guarantees cycling and variability and is thus more harmful than other valve problems. For instance, hysteresis is also undesirable but usually not really a problem up to 5 %. Nonlinear valve characteristic is another example, which can be handled using a nonlinearity compensation technique (Ruel 2000). As stiction is the most severe problem, it is important to detect it early on so that appropriate action can be taken and major disruptions to the operation can be avoided. Owing to the large number of loops in an industrial plant, this analysis should be performed automatically, limiting the possibility of false alerts and performing quantitative evaluation of performance loss.

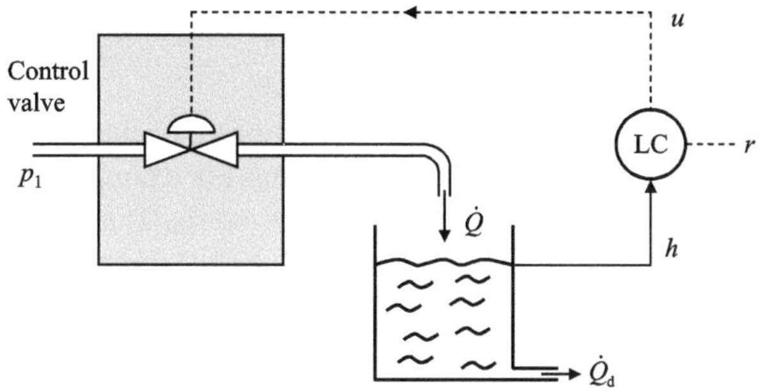

Fig. 11.2 Level control loop

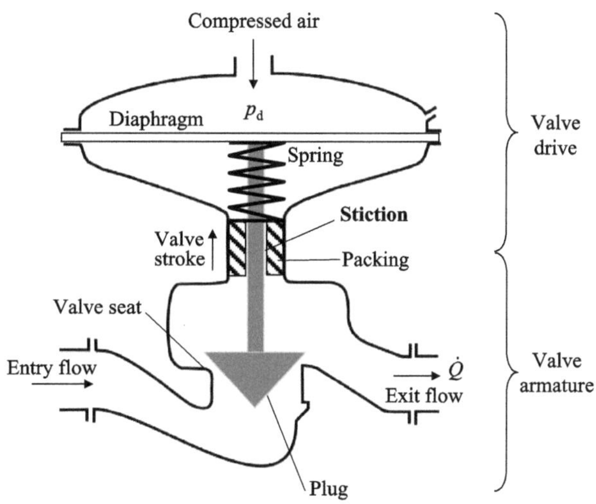

Fig. 11.3 Diagram of a pneumatic control valve (Lunze 2007)

Table 11.1 Ideal values and acceptable ranges of valve parameters (Ruel 2000)

Parameter	Ideal	Practical
Process gain	1	<0.5: too small; 0.5–3.0: acceptable; >3.0: too high.
Noise band	0	<0.5 %: acceptable.
Hysteresis	0	<3 %: acceptable; >3 %: to be checked.
Stiction	0	≪1 %: desirable; >1 %: to be checked.

Note that all stiction methods described in this chapter are based on utilising available industrial data for OP and PV. This reflects industrial practice, where MV is usually not known, except for flow control loops, where PV and MV are considered to be coincident.

11.2 Effects Relating to Valve Nonlinearity

There are some terms such as dead-band, backlash and hysteresis, which are often misused in describing valve problems. For example, quite commonly a dead-band in a valve is referred to backlash or hysteresis. The following items review the American National Standard Institution's (ANSI) formal definition of terms related to stiction. The aim is to differentiate clearly between the key concepts that underlie the ensuing discussion of friction in control valves. The following definitions can also be found in (EnTech 1998; Fisher-Rosemount 1999), which also make reference to ISA (1979):

- *Backlash* is a relative movement between interacting mechanical parts, resulting from looseness, when the motion is reversed.
- *Hysteresis* is the property of the element evidenced by the dependence of the value of the output, for a given excursion of the input, upon the history of prior excursions and the direction of the current traverse. Hysteresis is usually determined by subtracting the value of dead-band from the maximum measured separation between upscale-going and downscale-going indications of the measured variable during a full-range traverse after transients have decayed. Figure 11.4a and Fig. 11.4c illustrate the concept.
- *Dead-band* is the range through which an input signal may be varied, upon reversal of direction, without initiating an observable change in output signal. Dead-band produces phase lag between input and output and is usually expressed in percent of span; see Fig. 11.4b.
- *Dead-zone* is a predetermined range of input through which the output remains unchanged, irrespective of the direction of change of the input signal. Dead-zone produces no phase lag between input and output; see Fig. 11.4d

The aforementioned definitions show that the term "backlash" specifically applies to the slack or looseness of the mechanical part when the motion changes its direction. Therefore, in control valves it may only add dead-band effects if there is some slack in rack-and-pinion type actuators (Fisher-Rosemount 1999) or loose connections in rotary valve shaft. ISA (1979) definitions and Fig. 11.4 show that hysteresis and dead-band are distinct effects. Dead-band is quantified in terms of input signal span (i.e. on the x-axis), while hysteresis refers to a separation in the measured (output) response (i.e. on the y-axis).

Also for the term "stiction", there exist numerous definitions in the literature. See Choudhury et al. (2005a, 2005b), who proposed a formal and general definition of stiction and its causing mechanism: "*stiction is a property of an element such that its smooth movement in response to a varying input is preceded by a sudden abrupt jump called the slip jump. Slip jump is expressed as a percentage of the output span. Its origin in a mechanical system is static friction which exceeds the friction during smooth movement.*"

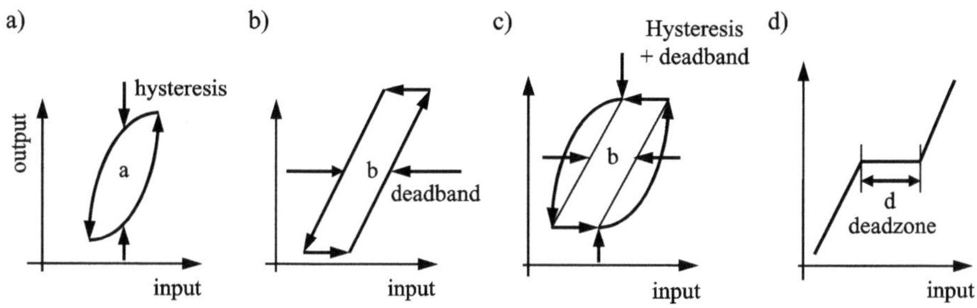

Fig. 11.4 Hysteresis, dead-band and dead-zone (redrawn from ISA 1979)

11.3 Stiction Analysis

The basis for most detection techniques is the qualitative illustration of the phenomenon of stiction and how closed-loop variables change with stiction and process parameters, the topic of this section.

11.3.1 Effect of Stiction in Control Loops

Figure 11.5 shows the typical input–output behaviour of a sticky valve. Without stiction, the valve would move along the dash-dotted line crossing the origin: any amount of OP adjustment would result in the same amount of valve-position (VP) change. However, for a sticky valve, static and kinetic/dynamic friction components have to be taken into account. The input–output behaviour then consists of four components, *dead-band*, *stick band*, *slip jump* and the *moving phase*, and is characterised by the three phases (Rossi and Scali 2005):

1. *Sticking*. MV is constant with the time, as the valve is stuck by the presence of the static friction force F_{S} (dead-band plus stick band). Valve dead-band is due to the presence of Coulomb friction F_{C}, a constant friction which acts in the opposite direction to the velocity (see Fig. 11.7).
2. *Jump*. MV changes abruptly, as the active force F_{a} unblocks the valve;
3. *Motion*. MV varies gradually; F_{a} is opposed only by the dynamic friction force F_{d}.

In Fig. 11.5, S and J denote dead-band plus stick band and slip jump, respectively. Because stiction is generally measured as percentage of the valve travel range, for simplicity, all variables, i.e. S, J, u(OP), y(PV) and u_{v} (MV), are translated into percentage of the valve range so that algebra can be performed among them directly. To illustrate how OP adjustments drive VP change in a sticky valve in Fig. 11.5, suppose that the valve rests at a neutral position A at the beginning. If the OP adjustment is between $A'B'$, the valve will not be able to overcome the static friction band, and so the VP will not change. However, if the OP moves outside of $A'B'$,

Fig. 11.5 Relation between controller output and valve position under valve stiction

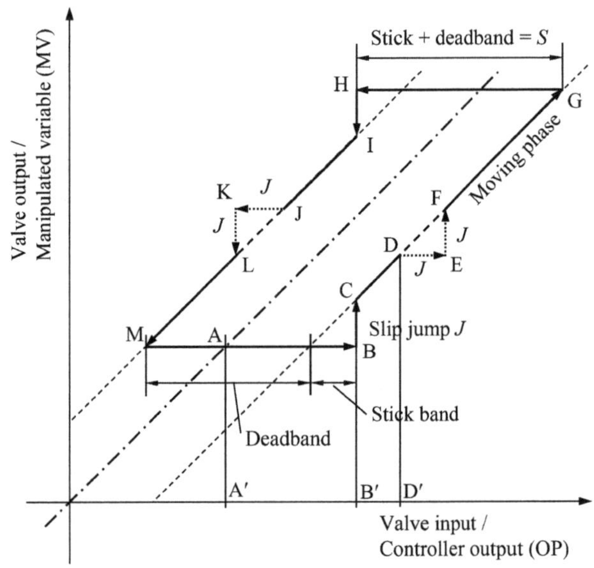

say D', then the valve is able to overcome the static friction band at point B and jumps to point C. After that, the valve moves from C to D, overcoming the kinetic friction band only.

Due to very low or zero velocity, the valve may stick again in between points C and G in Fig. 11.5 while travelling in the same direction (EnTech 1998). In such a case, the magnitude of dead-band is zero, and only stick-band (DE/EF) is present. This can be overcome if OP is larger than the stick-band only but is uncommon in industrial practice. The dead-band and stick band represent the behaviour of the valve when it is not moving, though the input to the valve keeps changing. Slip-jump represents the abrupt release of potential energy stored in the actuator chambers due to high static friction in the form of kinetic energy as the valve starts to move. The magnitude of the slip-jump is very crucial in determining the limit cyclic behaviour introduced by stiction (McMillan 1995; Piipponen 1996). Once the valve slips, it continues to move until it sticks again (point G in Fig. 11.5). In this moving phase, dynamic friction is present which may be much lower than the static friction. The sequence motion/stop of the valve due to stiction is called *stick-slip motion*.

In industrial practice, MV is usually not known, except for flow control loops, where PV and MV are considered to be coincident. The PV–OP plots for some industrial flow control loops with sticky valves are shown in Fig. 11.6. In contrast to the idealised plot in Fig. 11.5, real PV–OP plots have sometimes *destroyed* patterns produced due to the effect of process and controller dynamics and external disturbances. This fact makes it difficult to detect stiction based on PV–OP plots and to estimate stiction model parameters from usually measured PV and OP signals.

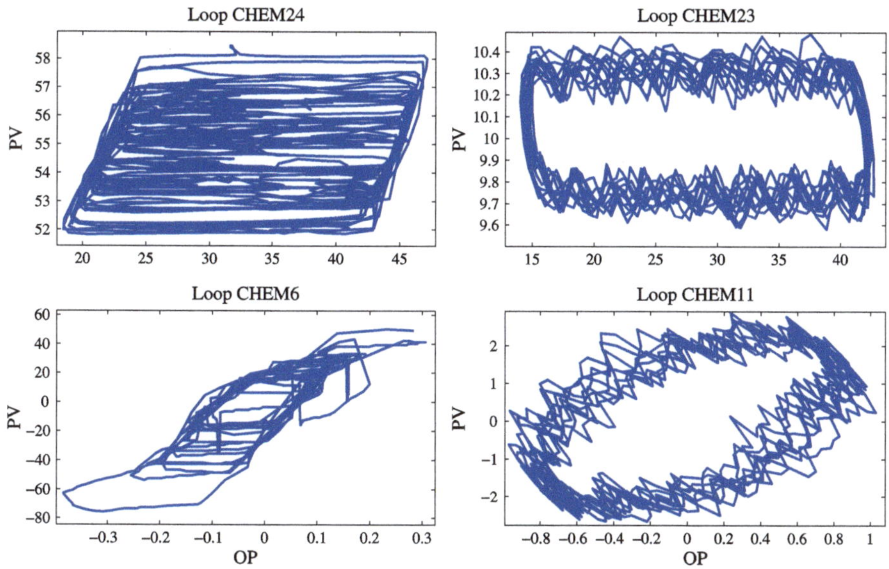

Fig. 11.6 PV–OP plots of four industrial flow control loops with stiction

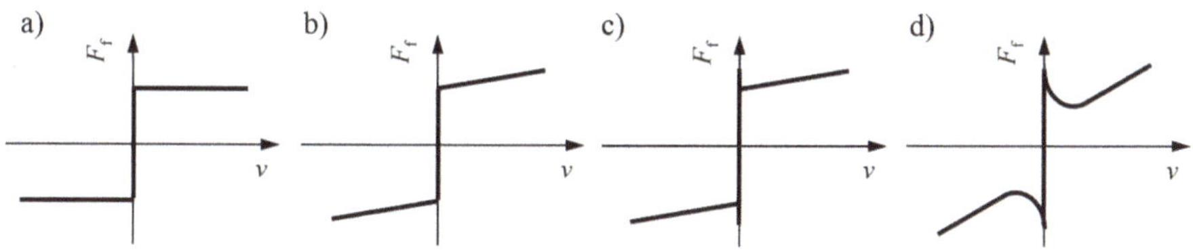

Fig. 11.7 Examples of static friction models: (**a**) Coulomb friction; (**b**) Coulomb plus viscous friction; (**c**) stiction plus Coulomb and viscous friction; (**d**) Stribeck friction (friction force decreasing continuously from the static friction level) (Olsson et al. 1998)

11.3.2 Physically Based Stiction Modelling

Many models have been proposed in literature to describe the presence of friction in the actuators. Surveys are reported in Armstrong-Hélouvry et al. (1994) and Olsson (1996). Different static friction models are shown in Fig. 11.7. One standard method is to model friction as a function of velocity, which is referred to as the Stribeck friction curve (after Stribeck 1902); see Fig. 11.7d. For more details, see Jelali and Kroll (2003). Particularly, the modelling of static friction is treated in Karnopp (1985).

Even though the number of parameters is not large, compared with other models, the lack of knowledge on values of critical variables is a major problem to describe the situation in a real plant. This problem is simplified with alternative approaches presented recently by Kano et al. (2004), Choudhury et al. (2005a, 2005b) and He et al. (2007). Data-driven models are adopted to describe the relationship MV = f(OP) illustrated in Fig. 11.5. In particular, only the two parameters S and J are used. Depending on these parameters, various characteristic curves of the control valve result. The models will be briefly discussed in the next section.

11.3.3 Data-Driven Stiction Modelling

Data-driven models have parameters that can be directly related to plant data and can produce the same behaviour as physical models. Such a model needs only an input signal and the specification of dead-band plus stick band (S) and slip jump (J). It avoids the main disadvantages of physical modelling of a control valve, namely that it requires the knowledge of the mass of the moving parts of the actuator, spring constant and the friction forces. Using data-driven models, the effect of the change of these parameters on the control loop can be determined in an easy way, and stiction detection methods can be tested in simulation.

A first data-driven stiction model has been proposed by Choudhury et al. (2005a, 2005b).[1] Kano et al. (2004) presented an improved stiction model, derived for pneumatic control valves. Recently, He et al. (2007) claimed that both models have complicated logic and some deficiencies. A model that is fairly simple and overcomes these problems was also suggested. However, He's model does not consider all stiction cases, i.e. undershooting stiction ($J < S$), no-offset stiction ($J = S$) and overshooting stiction ($J > S$).

It is not within the scope of this section to describe all these models, so the readers are referred to the mentioned papers. We only give the expression of the describing function for the stiction nonlinearity derived by Choudhury et al. (2005a, 2005b):

$$N(A) = -\frac{1}{\pi A}(P_{\text{real}} - jP_{\text{im}}), \tag{11.1}$$

where

$$\begin{aligned} P_{\text{real}} &= \frac{A}{2}\sin 2\varphi - 2A\cos\varphi - A\left(\frac{\pi}{2} + \varphi\right) + 2(S-J)\cos\varphi, \\ P_{\text{im}} &= -3\frac{A}{2} + \frac{A}{2}\cos 2\varphi + 2A\sin\varphi - 2(S-J)\sin\varphi, \\ \varphi &= \sin^{-1}\left(\frac{A-S}{A}\right). \end{aligned} \tag{11.2}$$

A is the amplitude of the input sinusoid. Describing function analysis is useful to study the stability of closed loops with stiction and to find out how to influence the loop behaviour.

---

[1]A Simulink model of this stiction model is available for download at http://www.ualberta.ca/slshah/valvestictionform.htm.

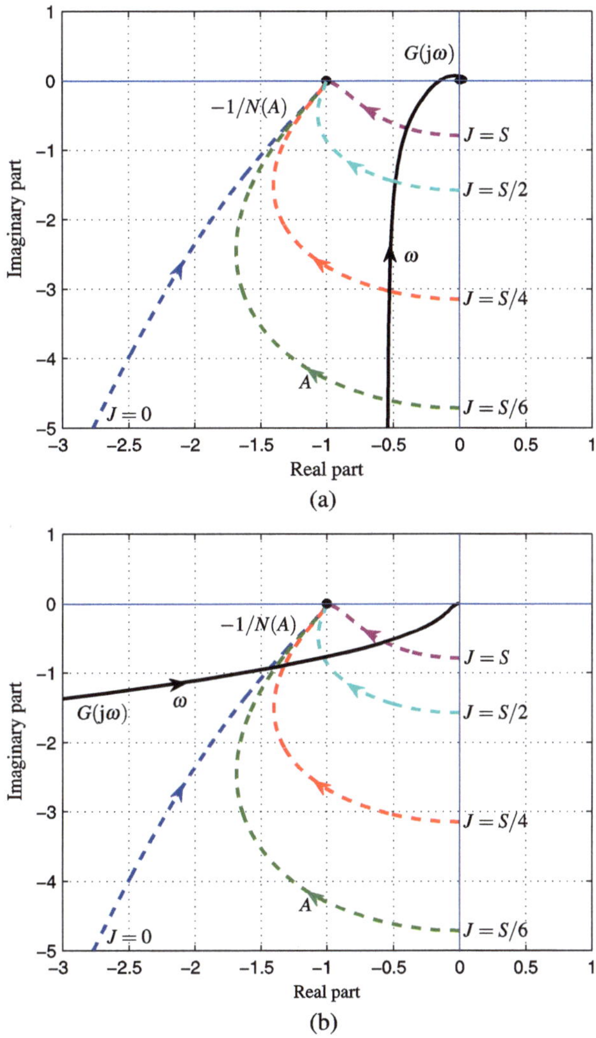

Fig. 11.8 Nyquist plots for closed loops with PI-controllers: (**a**) self-regulating process; (**b**) integrating process

Typical Nyquist plots for a self-regulating process (left panel) and an integrating process (right panel) with PI controllers are shown in Fig. 11.8. The describing function is parameterised by A, and the open-loop frequency response function of the controller and controlled system by ω. Both systems are closed-loop stable and thus intersect the negative real axis between 0 and 1. One can see in the left-hand panel of Fig. 11.8 that there will be a limit cycle for the self-regulating control loop if a slip-jump (J) is present. J forces the $-1/N$ curve onto the negative imaginary axis in the $A = S/2$ limit. Thus, the frequency response curve of the self-regulating loop and its PI controller is guaranteed to intersect with the describing function because the integral action means that the open-loop phase is always below $-\pi/2$, i.e., it is in the third quadrant of the complex plane at low frequency.

The figure also shows the $-1/N$ curve for the dead-band limit cycle. In the $A = S/2$ limit, the curve becomes large, negative and imaginary. The self-regulating loop does not have a limit cycle if the nonlinearity is a pure dead-band, because the frequency response curve does not intersect the $-1/N$ curve; consult also McMillan (1995) and Piipponen (1996).

The integrating loop with PI controller has a frequency response for which the phase becomes $-\pi$ at low frequency. The right-hand panel of Fig. 11.8 shows that it will intersect the $-1/N$ curves for the slip-jump cases and also for the pure dead-band case. Therefore, a valve with a dead-band and no slip-jump can cause a limit cycle oscillation for an integrating process with a PI controller. The frequency of oscillation is higher, and the period of oscillation shorter when the slip-jump is present because the $-1/N$ curves with the slip-jump intersect the frequency response curve at higher frequencies than the $-1/N$ curve for the dead-band.

11.3.4 Typical Trends of Variables and Input–Output Shape Analysis

In this section, an analysis of main features of the stiction phenomenon is carried out by observing trends of the loop variables (OP, MV and PV) and the resulting input–output shapes, as functions of some parameters of process, stiction and controller. For this purpose, a FOPTD process (with different values of the ratio between the time delay τ and the time constant T) controlled by a PI controller is considered. In Fig. 11.9, the effect of stiction in term of the ratio $S/(2J)$ on MV, OP, PV, as well as MV–OP plot and PV–OP plot are illustrated for three values of the ratio T_d/T. Similar analysis can be carried out for integrating processes.

The following main features can be observed:

- The controller output (OP) shows *always triangular* wave.
- The manipulated variable (MV) maintains always typical square wave elements; the almost perfect square wave shape, shown for $T_d/T \gg 1$, can be slightly modified to saw-tooth shape, but the discontinuity on the derivative is maintained.
- The controlled variable (PV) presents also the limit cycle, but the effect of the process modifies the typical square wave showed in MV. For decreasing values of $S/(2J)$ and T_d/T, the shape changes from square wave to triangular, much closer to a sinusoidal form.

The effect of stiction can be distinguished from other oscillation root causes by the following observations:

- In general, nonlinearity-induced oscillations, which are observed both on controller outputs and process variables, contain harmonics.
- *In the case of poor performance or external disturbance, both controller outputs and process variables follow sinusoidal waves*; the induced oscillations have low harmonic content.

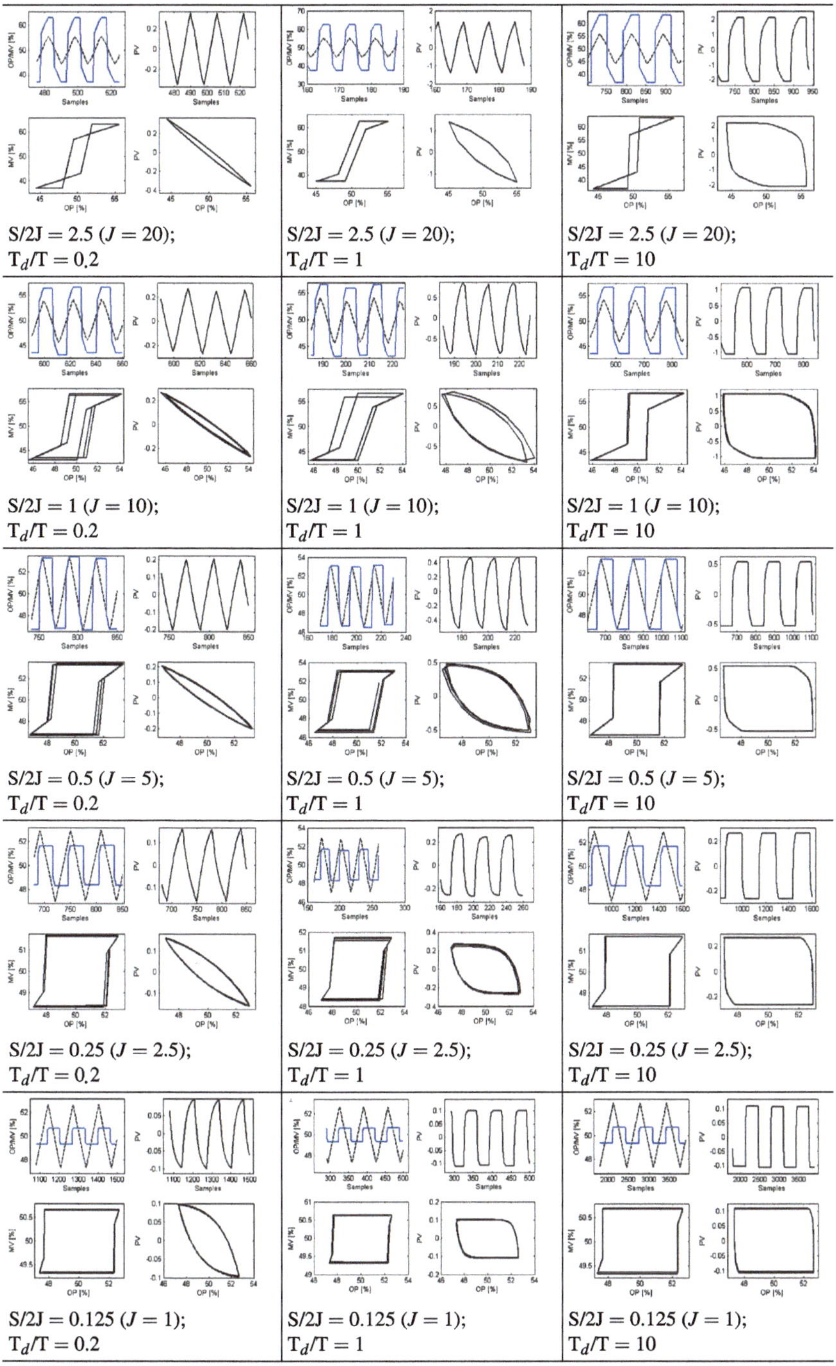

Fig. 11.9 Wave shape of OP, MV and PV for a self-regulating process in case of stiction varying the system parameter: (**a**) variations of T_d/T; (**b**) variations of $S/2J$ ($T_s = 2$; $S = 5$)

Table 11.2 Typical stiction-pattern shapes for different process types (Forsman 2000; Srinivasan et al. 2005a)

Process type	Fast processes (flow)		Slow processes (pressure & temperature)	Integrating processes (Level)	Level with PI control
	Dominant I action	Dominant P action			
OP	Triangular (Sharp)	Rectangular	Triangular (Smooth)	Triangular (Sharp)	Triangular (Sharp)
PV	Square	Rectangular	Sinusoidal	Triangular (Sharp)	Parabolic

Table 11.3 Occurrence of limit cycles in control loops (Choudhury et al. 2008)

Process	Controller	Dead-band ($J = 0$, $S \neq 0$)	Stick slip ($J \neq 0$, $S \neq 0$)
Self-regulating	P only	No limit cycles	No limit cycles
Self-regulating	PI	No limit cycles	Limit cycles
Integrating	P only	No limit cycles	Limit cycles
Integrating	PI	Limit cycles	Limit cycles

- *In the case of stiction, the controller output usually follows a triangular wave for self-regulating processes*; *for integrating processes, the process variable shows a triangular wave*. Triangular (symmetric) waveforms only contain odd harmonics. Triangular waves that are observed in controller outputs are usually asymmetric and contain both even and odd harmonics.

The reason for this behaviour is that while the plant input is continuous for aggressive control (except when the controller output is saturated), valve stiction results in a discontinuous plant input that closely resembles a rectangular pulse signal. Moreover, stiction-pattern shapes differ for different processes, as given in Table 11.2.

Note that flow loops, e.g. steam flow loops, can be integrating. The same applies for pressure loops: gas pressure is self-regulating when the vessel (or pipeline) admits more feed when the pressure is low, and reduces the intake when the pressure becomes high. Integrating processes occur when there is a pump for the exit stream (Seborg et al. 2004: 326).

However, it is important to realise that valve stiction does not always lead to limit cycles in a control loop. Rather, the occurrence of limit cycles depends on the type of the process and controller and of the presence of dead-band and stick slip; see Table 11.3. From this one can conclude:

- Dead-band only cannot produce limit cycles in self-regulating processes.
- A short-term solution that may solve the stiction problem is to change the controller to P-only. In a self-regulating process, the limit cycle should then disappear. This is not the case for integrating processes, but the amplitude of the limit cycle will probably decrease.

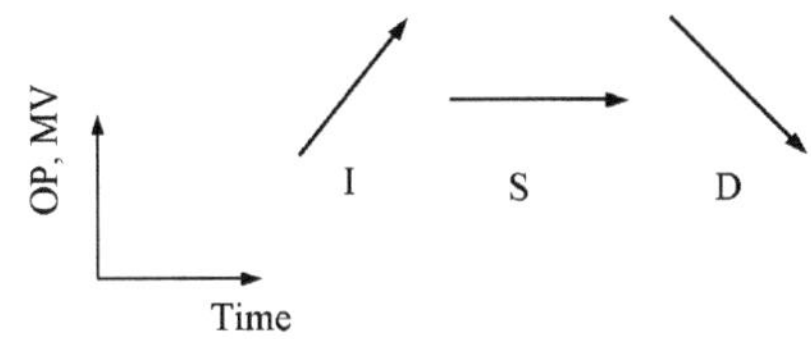

Fig. 11.10 Symbolic representations of a time series: Increasing (I), Steady (S) and Decreasing (D)

11.4 Stiction Diagnosis Based on Shape Analysis of MV–OP Plots

Qualitative trends or shapes of a time series signal can be represented as a sequence of symbolic values. A stiction detection method based on the analysis of MV–OP patterns is presented in this section. The description is largely adopted from Yamashita (2006) and Manum (2006).

The simplest representation would use three symbols: increasing (I), steady (S) and decreasing (D). Figure 11.10 shows these three primitives in a time value plane. These three symbols are sometimes called plus (+), zero (0) and minus (−). They correspond to the signs of their respective derivatives. The identification of the symbols for each sampling point is based on the time derivatives of the signal. In the absence of noise, these symbols can be identified using finite differences with an appropriate window size. Thresholds must be defined to distinguish steady and increasing/decreasing states. In a noisy signal, some form of filter or a neural network can be used to identify the primitives (Rengaswamy and Venkatasubramanian 1995). However, this makes the method much more complicated.

For a given time series signal, the simplest way to describe the signal by symbols is to use the following three primers: increasing (I), decreasing (D) and steady (S). The primers can be identified using standard deviation of the differentials of the recorded signals as a threshold for identification. In words, the identification works like this:

1. Calculate the differentials of the given signals.
2. Normalise the differentials with the mean and standard deviation.
3. Quantise each variable in three symbols using the following scheme (x is the recorded signal, and $\dot{x}$ is the normalised differential):

   - If $\dot{x} > 1$, x is increasing (I).
   - If $\dot{x} < -1$, x is decreasing (D).
   - If $-1 \leq \dot{x} \leq 1$, x is steady (S).

By combining the symbols for the OP and MV signals we get a symbolic representation of the development in an (OP, MV) plot with time. The primers for the combined plot are shown in Table 11.4. The sticky motions, IS and DS, are framed. These are the two primers when the controller is either increasing (I) or decreasing (D) its output, while the valve position is steady (S).

Table 11.4 Symbolic representation of behaviour of a time series in OP–MV plots

OP/MV	D	S	I
I	ID	IS	II
S	SD	SS	SI
D	DD	DS	DI

11.4.1 Stuck Indices

The simplest idea for detecting stiction is counting the periods of sticky movement by finding IS and DS patterns in the input–output plots of the valve. Based on this idea, an index ρ_1 to detect the loop with stiction can be defined in an appropriate time window:

$$\rho_1 = \frac{\tau_{\mathrm{IS}} + \tau_{\mathrm{DS}}}{\tau_{\mathrm{total}} - \tau_{\mathrm{DS}}}, \tag{11.3}$$

where τ_{total} is the width of the time window, and τ_{IS} and τ_{DS} are time periods for patterns IS and DS, respectively. This index will become large if the valve has severe stiction ($0 \leq \rho_1 \leq 1$). As an extreme case, the index ρ_1 becomes unity if the valve does not move at all for changes of controller output. If the signals are random, the value of ρ_1 is likely to become 0.25 because ρ_1 represents two out of eight patterns, i.e. $0.25 = 2/8$. These two patterns can occur by various causes other than stiction: disturbances, time delay and noise. Improvement of the accuracy of detection is attainable by reducing these irrelevant causes in the movement sequence. Therefore, one can infer that the loop is likely to have valve stiction if the index value is greater than 0.25.

A fragment of the movement sequence can be represented by a sequence of two successive patterns. For example, if pattern II follows the pattern IS, the movement is represented as (IS II). Using this representation, typical movement for valve stiction can be represented as four fragments (IS II), (DS DD), (IS SI) and (DS SD), as shown in Fig. 11.11. All sticky motions of valve stiction, IS and DS, should be a part of these patterns. The degree of stiction can be evaluated by counting the time period of IS and DS in these four fragments of patterns. Subsequently, an improved index ρ_2 can be defined as

$$\rho_2 = \frac{\tau_{\mathrm{IS\ II}} + \tau_{\mathrm{IS\ SI}} + \tau_{\mathrm{DS\ DD}} + \tau_{\mathrm{DS\ SD}}}{\tau_{\mathrm{total}} - \tau_{\mathrm{SS}}}, \tag{11.4}$$

where $\tau_{\mathrm{IS\ II}}$ is the total number of IS samples in all the found (IS II) movements in the observation window, $\tau_{\mathrm{DS\ DD}}$ is the number of DS samples in the found (DS DD) movements, and so on. This index includes only the sticky movements matched with the four typical fragments of a sequence of movement patterns. If the entire pattern is a typical stiction pattern, the value of ρ_2 should be identical to ρ_1. One may consider that a loop has stiction if ρ_2 is greater than 0.25, which is the same criterion for ρ_1.

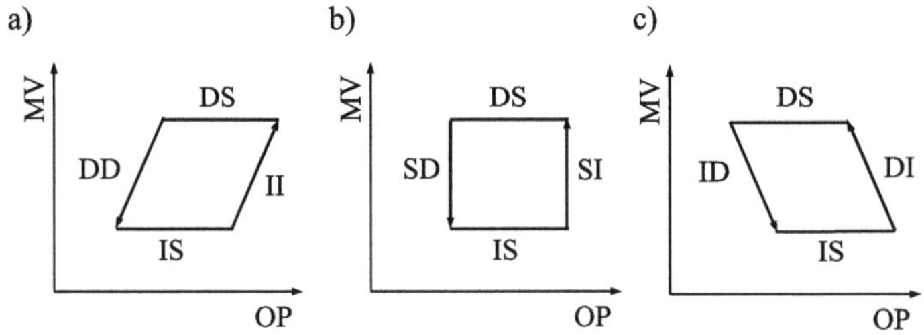

Fig. 11.11 Qualitative shapes typically found in sticky valves

Only patterns IS, SS and DS are found for the extreme case in which the valve does not move; thereby, the index ρ_2 becomes 0, not 1. A more convenient index, ρ_3, is introduced to avoid this inconvenience.

Table 11.4 shows eight symbols for qualitative patterns. In general, to represent a sequence of two successive symbols, seven symbols can be used followed by the symbol, except for the symbol used in the first segment. Typical patterns for stiction are shown in Fig. 11.11a and b; they each include two patterns. Therefore, patterns that have nothing to do with stiction can be represented by one of five possible symbols following the symbol IS or DS. By removing these five patterns each from the index ρ_1, a new index, ρ_3, can be defined as

$$\rho_3 = \rho_1 - \frac{\sum_{x \in W} \tau_x}{\tau_{\text{total}} - \tau_{\text{SS}}}, \tag{11.5}$$

where W is the set of all patterns that have nothing to do with stiction, i.e. in symbols, $W = \{$IS DD, IS DI, IS SD, IS ID, IS DS, DS DI, DS SI, DS ID, DS II, DS IS$\}$. For example, if the movement was (IS IS IS DD IS IS II), we should subtract 3/7 from the original, because the three first IS primes could not be a part of a stiction pattern since they were followed by a DD. Except for the special case that the valve does not move, $\rho_3 = \rho_2$. The experience with Yamashita's method shows that ρ_1 is always too high, particularly for cases without stiction, whereas ρ_3 correctly rejects stiction in these cases. This implies that ρ_3 should be used as the determining stiction index, not ρ_1.

11.4.1.1 Detection Procedure and Practical Conditions

The automatic stiction detection procedure can be summarised as follows.

Procedure 11.1 Stiction detection based on the shape analysis of OP–MV plots (Yamashita 2006).

1. Obtain data of the controller output and the valve position (or the corresponding flow rate).
2. Calculate the time difference for each measured variable.
3. Normalise the difference values using the mean and standard deviation.

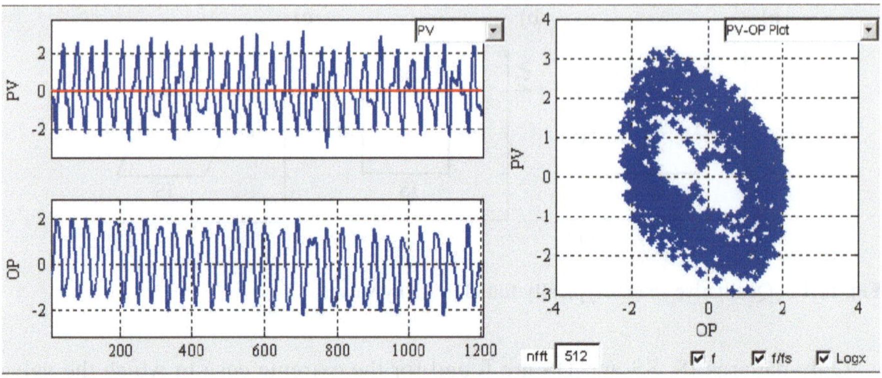

Fig. 11.12 Data and PV–OP plot for loop PAP2 (flow control)

4. Quantise each variable in three symbols I, S and D. Use the standard deviation of the differentials of the recorded signals as a threshold.
5. Describe qualitative movements in OP–MV plots by combining symbolic values of each variable.
6. Skip SS patterns for the symbolic sequence.
7. Evaluate the index ρ_1 by counting IS and DS periods in the patterns found (Eq. 11.3).
8. Find specific patterns and count stuck periods. Then evaluate the index ρ_3 (Eq. 11.5). Conclude stiction if $\rho_3 \geq 0.25$.

This procedure is very easy to implement. The main practical problem of the method is that it requires measurement of the valve position or the flow rate, which are only available for smart valves. The method is therefore straightforward for flow control loops, of which there are a vast number in the chemical process industries. Connell (1996) notes that about half of the control loops in oil refineries are used for flow control. Also, the method is sensitive to noise as we use the derivative for finding the symbolic representations. The sampling time may affect the performance of the method. Lowering the sampling time makes the method inefficient, as the calculation of the differentials will be too dominated by the noise. Setting the sampling time very high is also disadvantageous, so there must be an optimum where we avoid sampling too much. A good default setting for the sampling time is the dominant time constant of the process. Moreover, it has been reported that the method does not detect stiction for loops showing patterns like those illustrated in Fig. 11.11c, found in many industrial data sets. Yamashita's method has been deeply examined by Manum (2006). The main results of this study can also be found by Manum and Scali (2006).

Example 11.1 Figure 11.12 shows the time trends and the PV–OP plot from a flow control loop in a paper plant. The application of Yamashita's stiction detection method gives the indices $\rho_1 = 0.48 > 0.25$ and $\rho_3 = 0.41 > 0.25$. The indices have different values, but both indicate valve stiction, which is the correct conclusion.

Note that two other methods based on qualitative shape analysis have also been suggested by Kano et al. (2004) for detecting valve stiction. The methods are based on the following observations (Fig. 11.5) when stiction occurs:

- *Method A*. There are sections where the valve position does not change even though the controller output changes. Stiction is stronger as such sections are longer.
- *Method B*. The relationship between the controller output and the valve position takes the shape of a parallelogram if the slip jump J is neglected. Stiction is stronger as the distance between l_1 and l_2 is longer.

Detection algorithms are described in Kano et al. (2004). The main drawback of the methods of Kano et al. (2004) is that they require the valve position or the flow rate to be measured, which may not always available. Furthermore, many parameters have to be selected and seem to be process specific and sensitive to noise. Examples, which confirm problems with these methods, have been reported by Kano et al. (2004) and He et al. (2007).

11.5 Cross-Correlation-Based Stiction Detection

The simplest method for stiction detection is that proposed by Horch (1999). It is based on the following idea (Fig. 11.13):

- If the cross-correlation function $\Phi_{uy}(\tau)$ between controller output u and process output y is an odd function (i.e. asymmetric w.r.t. the vertical axis), the likely cause of the oscillation is stiction.
- If the cross-correlation function $\Phi_{uy}(\tau)$ is even (i.e. symmetric w.r.t. the vertical axis), then stiction is not likely to have caused the oscillation. In this case, the oscillation may be due to external disturbances, interaction or aggressive tuning of the controller.

The following assumptions are needed to apply this stiction detection method:

- The process does not have an integral action.
- The process is controlled by a PI controller.
- The oscillating loop has been detected as being oscillatory with a significantly large amplitude.

For automatic distinction between odd and even $\Phi_{uy}(\tau)$, the following measures are used (Fig. 11.14):

$$\Delta\tau = \frac{|\tau_l - \tau_r|}{|\tau_l + \tau_r|}, \tag{11.6}$$

$$\Delta\rho = \frac{|\Phi_0 - \Phi_{\max}|}{|\Phi_0 + \Phi_{\max}|}, \tag{11.7}$$

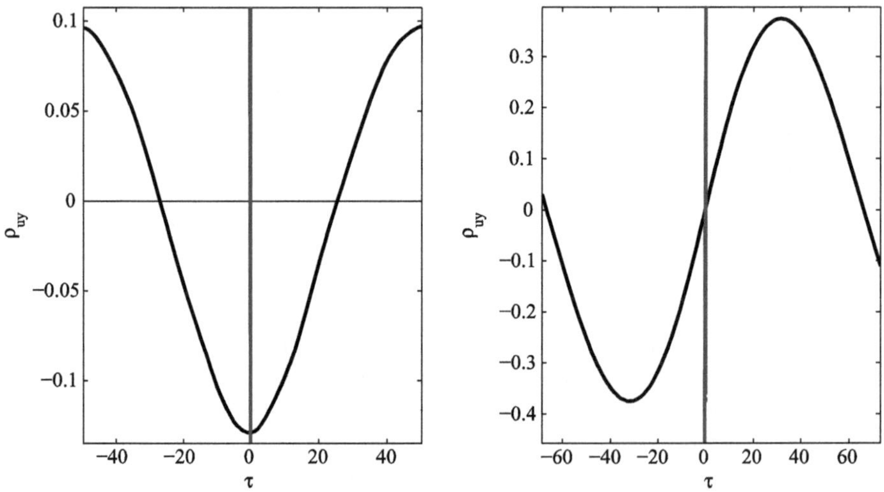

Fig. 11.13 Cross-correlation between control signal and process output for the case of no stiction (*left*) and stiction (*right*)

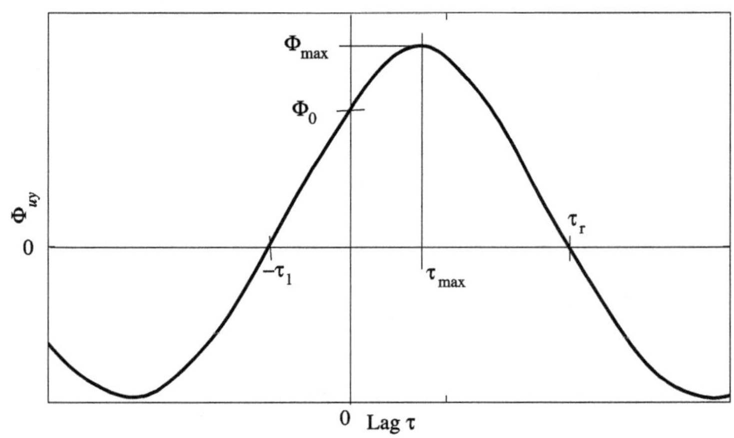

Fig. 11.14 Definition of key variables for the cross-correlation function

where τ_r is the zero crossing for positive lags, $-\tau_l$ is the zero crossing for negative lags, Φ_0 is the cross-correlation value at lag 0, i.e. $\Phi_{uy}(w)$, and

$$\Phi_{\max} = \operatorname{sign}(\Phi_0) \max_{\tau \in [-\tau_l, \tau_r]} \left|\Phi_{uy}(\tau)\right|. \tag{11.8}$$

The approach is intended to distinguish the phase shift $\Delta\varphi$ by $\pi/2$ (odd CCF) and π (even CCF). Introducing $\pi/6$ margins (deviations of the CCF from "ideal"

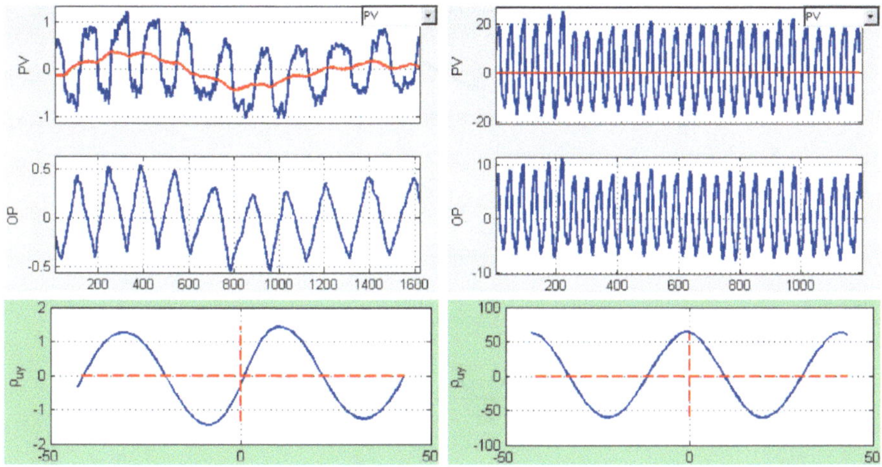

Fig. 11.15 Cross-correlation for loop CHEM1 ($\Delta\tau = 0.73$; $\Delta\rho = 0.4$) (*left*); and loop PAP4 ($\Delta\tau = 0.13$; $\Delta\rho = 0.0$) (*right*)

positions as shown in Fig. 11.13), the diagnostic method can be written as follows:

$$\left.\begin{array}{l} 0.0 < \Delta\rho \le \frac{2-\sqrt{3}}{2+\sqrt{3}} \approx 0.072 \\ 0.0 < \Delta\tau \le \frac{1}{3} \end{array}\right\} \Rightarrow \Delta\varphi = \pi \Rightarrow \text{no stiction,}$$
$$\left.\begin{array}{l} \frac{2-\sqrt{3}}{2+\sqrt{3}} \approx 0.072 < \Delta\rho < \frac{1}{3} \\ \frac{1}{3} < \Delta\tau < \frac{2}{3} \end{array}\right\} \Rightarrow \text{no decision,} \qquad (11.9)$$
$$\left.\begin{array}{l} \frac{1}{3} \le \Delta\rho \le 1.0 \\ \frac{2}{3} \le \Delta\tau \le 1.0 \end{array}\right\} \Rightarrow \Delta\varphi = \pi/2 \Rightarrow \text{stiction.}$$

The interval, where no decision is taken, corresponds to a CCF which is neither odd nor even. This will typically occur when the oscillation is strongly asymmetric. It may be an indication of a more unusual problem such as sensor or other equipment faults (Horch 1999).

Example 11.2 The results of the application of Horch's stiction detection method to measured data from loop CHEM1 and PAP4 are illustrated. The cross-correlation plots and indices given in Fig. 11.15 confirm the presence of stiction in the first loop and its absence in the second loop. These are the right conclusions, as it is observed that the signals have typical stiction patterns (triangular) in the OP signals.

To summarise, the stiction detection method by Horch (1999) is indeed simple and thus easy to use. Besides normal operating data, neither detailed process knowledge nor user interaction is needed. The drawbacks/pitfalls are that (i) the method

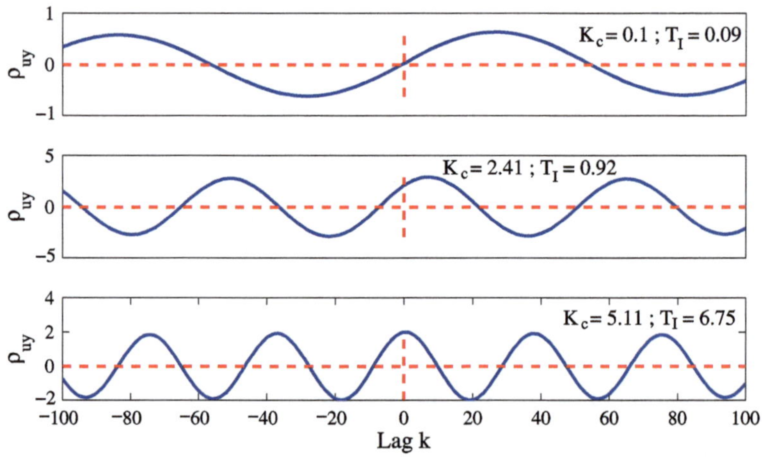

Fig. 11.16 Different controller tunings result in different types of cross-correlation function between OP and PV

cannot be applied to integrating systems and (ii) the phase shift depends on controller tuning: the phase lag is π for an aggressive controller when the loop cycles due to controller output saturation; however, when stiction is present and the controller output is not saturated, the phase lag can lie between $\pi/2$ and π for a PI controller; see below. Examples, which confirm these problems, have been given by Yamashita (2006) and He et al. (2007). The latter researcher has also theoretically analysed Horch's first method and demonstrated its general inconsistency.

Example 11.3 To illustrate this point, a FOPTD system $e^{-s}/(3s+1)$ controlled by a PI controller is considered. Figure 11.16 shows the cross-correlation functions for three different controller settings, corresponding to phase shifts of $-\pi$, $-3\pi/4$ and $-\pi/2$. For these cases, Horch's method would conclude that there is stiction for the first case, undetermined for the second case, and no stiction for the third case, although there is no stiction in all three cases.

11.6 Diagnosis Based on Curve Fitting

A common characteristic of the techniques described in this section is that they perform a fitting of PV, OP or SP–PV data, to detect the typical signature of stiction and distinguish it from other causes. Based on the observations described in Sect. 11.3.4, He et al. (2007) proposed a valve-stiction-detection technique, in which the controller output or process variable are fitted piece-wise (every *significant half cycle*) to both triangular (Fig. 11.17a) and sinusoidal (Fig. 11.17b) curve segments using an LS method. If the fit for the triangular wave is better, then stiction is concluded;

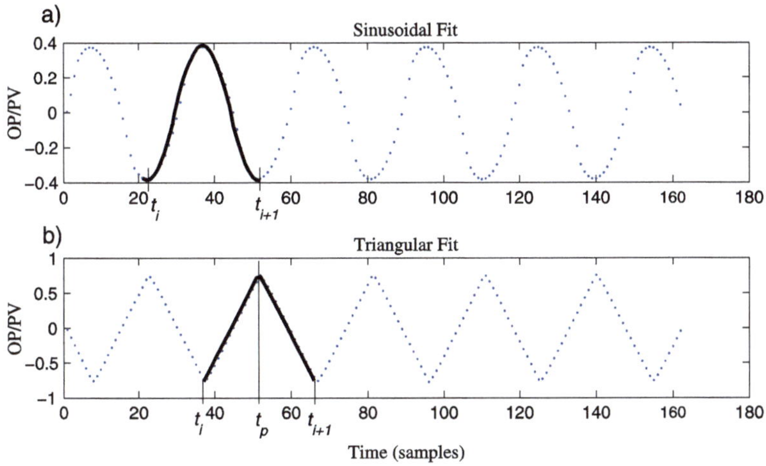

Fig. 11.17 Schematic of the curve fittings: (**a**) sinusoidal fitting; (**b**) triangular fitting

otherwise no stiction occurs. By comparing the error between real and fitted data, an evaluation of the accuracy of approximation and then a stiction index can be obtained.

11.6.1 Sinusoidal Fitting

The OP or PV signal is fitted piece-wisely for each half-period of oscillation (see Fig. 11.17a), which means that each fitting piece may have different amplitude and/or frequency. This consideration is reasonable, since real processes of noise and disturbances are present in the signals, thus the oscillation magnitude and frequency may change from time to time, and also unsymmetrical signals may result.

Denoting the signal to be fitted as $y(t)$, the objective function for the sinusoidal fitting is

$$J_{\sin} = \underbrace{\min}_{A,\omega,\phi} \left\| y(t) - A\sin\left[\omega(t_i : t_{i+1} - t_i) + \phi\right] \right\|_2, \tag{11.10}$$

where A is the amplitude, ω the frequency, and φ the phase shift of the sinusoid. $(t_i : t_{i+1})$ is the time range of fitting as in Fig. 11.17a. Because the curve is fitted piece-wisely, we can set $\varphi = 0$. We use numerical iterative method, i.e. nonlinear least-square method (MATLAB's lsqnonlin/lsqcurvefit) to find the best fitting. The initial values for the oscillation period T_p, and thus for ω, are determined from the oscillation-detection method. A can be initially set to the half peak value of the signal. However, ω and A are fitted for each half-period of the signal, as mentioned above. To get smooth transition of the approximating curve between both half-periods, the use of some overlapping data may be useful. The overall mean squared error for sinusoidal fitting MSE_{sin} is the average of MSEs over all considered time periods.

11.6.2 Triangular Fitting

Triangular fitting as shown in Fig. 11.17b is more difficult because it is a piece-wise curve fitting with two degrees of freedom: the location and the magnitude of the maxima. In our implementation, we do not use the iterative algorithm proposed by He et al. (2007), but fit two linear pieces to data in each half-period:

$$J_{\text{tri}} = \underbrace{\min}_{a,b} \left\| y(t) - a\Delta t + b \right\|_2. \tag{11.11}$$

Δt is the time difference (or time-index difference) between each t_i between t_{p} and for the first linear LS fitting, and that difference between t_{p} and t_{i+1} for the second linear LS fitting. The fitting is performed by the MATLAB function polyfit with a polynomial degree of unity. Also here, the use some overlapping data may be useful to achieve smooth transition of the approximating curve between the half-periods. The overall mean squared error for triangular fitting MSE_{tri} is the average of MSEs over all considered time periods.

11.6.3 Stiction Index and Detection Procedure

The algorithm of stiction detection based on piecewise curve fitting can be summarised as follows.

Procedure 11.2 Stiction detection based on curve fitting.

1. Check if the loop is oscillating and determine the zero crossings and the oscillation period (use one of the methods described in Chap. 9).
2. Fit a sinusoidal wave and a triangular wave to the measurements of controller output for a self-regulating process or to the measurements of process variable for an integrating process as well as possible (i.e. solve the optimisation problems in Eqs. 11.10 and 11.11).
3. If the fit for the triangular wave is significantly better than the fit for sinusoidal wave, conclude *stiction*. In the opposite case, conclude *no stiction*. If both are approximately equal (e.g. the difference is smaller than 10 %), no decision is made.

Step 3 of the algorithm can be based on the *stiction index* defined by

$$\eta_{\text{stic}} = \frac{MSE_{\text{sin}}}{MSE_{\text{sin}} + MSE_{\text{tri}}}, \tag{11.12}$$

that is

$$\begin{aligned} \eta_{\text{stic}} &\geq 0.6 \Rightarrow \text{Stiction}, \\ 0.4 < \eta_{\text{stic}} &< 0.6 \Rightarrow \text{No decision}, \\ \eta_{\text{stic}} &\leq 0.4 \Rightarrow \text{No stiction}. \end{aligned} \tag{11.13}$$

This stiction-detection method has many advantages:

- Only measurements of the controller output for a self-regulating process or the process variable for an integrating process are required, which are usually available.
- The method is robust in handling noise owing to the formulation as least-squares problems.
- It works consistently for both non-integral and integral processes.

The following remarks discuss some limitations and application related issues of this method (He et al. 2007):

- There is a grey area, where neither sinusoid nor triangle fits the signal well. In this case, He's method cannot provide meaningful detection result.
- External disturbances are assumed to be sinusoidal. This is true for most of the cases because disturbances will eventually become more sinusoidal as they propagate away from the source due to low-pass plant dynamics. However, if the disturbance source that leads to oscillation is close to the valve being diagnosed, such as limited cycle due to process and/or controller nonlinearity, the method can fail or lead to wrong diagnosis.
- An exact triangular wave will be obtained only if there is a pure integrator in the controller or process. However, a clear triangle is not required in order for the method to work. Processes with no or extremely weak integration should be treated as self-regulating processes, and OP fitting should be applied.
- For the case of varying load, a carefully designed high-pass or band-pass filter (see also Sect. 9.8) might be a better approach to eliminate the low-frequency drift in the process. However, because moving window approach is applied and piecewise fitting is utilised, the impact of the slightly biased crossing point determined by the algorithm is small.
- The low-pass filter associated with a PI controller has no significant impact on He's method. For valve stiction in integrating processes, although the triangular PV signal can be smoothed by the filter, because PV is fitted, the smoothing effect does not matter. For valve stiction in self-regulating processes, the rectangular wave will be smoothed by the filter, but after the integration action of the PI controller, the OP fitting still favours triangle if the stiction is not too weak, i.e. in the grey area.

Example 11.4 We consider two industrial data sets; one is from a flow control loop in a refinery and the other a level control loop in a paper mill. The results of the curve fitting are shown in Fig. 11.18. For the (self-regulating) flow loop, a triangular shape is the better fitting to the OP data, giving a stiction index $\eta_{\text{stic}} = 0.83$, indicating the

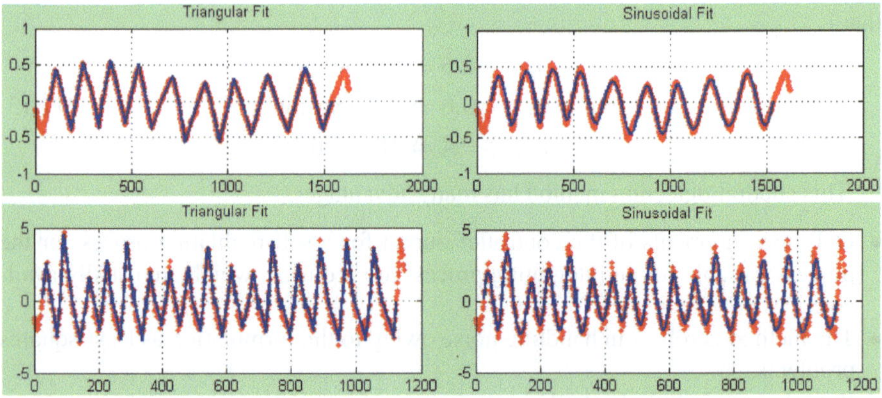

Fig. 11.18 Curve fitting for loop CHEM1 (OP data) (*top*) and loop PAP3 (PV data) (*right*)

presence of stiction. The PV signal of the (integrating) level loop can be better fitted by triangular shape. The corresponding stiction index has the value $\eta_{\mathrm{stic}} = 0.70$, also indicating that the loop suffers from a stiction problem.

11.6.4 Similar Techniques

Rossi and Scali (2005) proposed a method, called *relay technique*, to fit the PV using three different models: the output response of a first-order plus time delay under relay control (relay wave), triangular wave and sinusoidal wave. Relay and triangular waves are associated with the presence of stiction, while sinusoidal shape with the presence of external perturbations. A stiction index has been defined as

$$\eta_{\mathrm{stic}} = \frac{MSE_{\mathrm{sin}} - MSE_{\mathrm{relay}}}{MSE_{\mathrm{sin}} + MSE_{\mathrm{relay}}}. \tag{11.14}$$

This index takes values in the range $[-1, 1]$: negative values indicate a better approximation by means of sinusoids, positive values by means of relay or triangular approximations. Values close to zero indicate that the two approximations have similar errors and the procedure gives an uncertain answer; by considering that noise can change the shape of the curve, the uncertainty zone is defined by $|\eta_{\mathrm{stic}}| < 0.21$. This corresponds to a ratio $\mathrm{MSE}_{\mathrm{sin}}/\mathrm{MSE}_{\mathrm{relay}} = 0.66$ (Rossi and Scali 2005), in analogy to the limit value $\Delta\tau \le 2/3$ in Horch's method (Sect. 11.5). Although this method is very similar to He's method described above, the relay-wave approximation is very complex and thus requires high computation effort.

Note that the extension of the curve-fitting method for other signal patterns, such as rectangular and trapezoidal waves, is straightforward. For this purpose, Srinivasan et al. (2005a) proposed a technique based on qualitative pattern recognition. The algorithm aims to distinguish square, triangular and saw-tooth-like signal shapes in both OP and PV. The technique used to classify the signals is dynamic time warping (DTW), and this is applied for each oscillation cycle individually rather than for a complete data set at once. DTW is a classical technique from speech recognition used to compare signals with stored patterns (Table 11.2). DTW takes into account that different parts of the signal under investigation may be shorter or longer than the reference and also enables the detection of similarities in such cases.

11.7 Nonlinearity Detection and PV–OP Pattern Analysis

In this method proposed by Choudhury et al. (2006a), the detection of valve or process nonlinearity is first carried out using higher-order statistical method-based NGI and NLI indices (Sect. 10.2). Once a nonlinearity is detected, then the data are treated by a Wiener filter (PV_f, OP_f), and the PV_f–OP_f plot, generated from a segment of the data that has regular oscillations, is used to isolate its cause. A signature of valve stiction is when PV_f–OP_f plot shows cyclic or elliptic patterns. If no such patterns are observed, it is concluded that there are valve problems but these are not due to the stiction.

For the purpose of this section, the following assumptions are important:

- *The process nonlinearity is negligible in the vicinity of the operating point* where the data has been collected. This is a standard and reasonable assumption because the forthcoming diagnosis methods work with routine operating data of a control loop *under regulatory control*. In general, when processes are fairly well regulated at standard operating conditions, the loop can be assumed to behave linearly since a linear controller is capable of satisfactory regulation of the plant.
- *The valve movement or change in input signal to the valve is kept within a range*, so that the control loop exhibits linear behaviour under steady-state regulatory control. Choudhury et al. (2004) found out that this is the case when the operating range is about 25 % of the full span (0–100 %) of the valve, even when it is a square-root valve. This assumption is necessary to exclude that the nonlinearity may come from the possibly nonlinear valve characteristics, which is definitively not a fault. Therefore, when implementing any detection procedure, a check on the range of OP signal should be performed. If the range of OP is larger than 25 %, a corresponding message should be issued to the user and let him consult valve characteristic information.

11.7.1 Stiction Detection and Estimation Procedure

The whole procedure is illustrated in Fig. 11.19 and summarised as follows. It is largely adopted from Choudhury et al. (2006a) but extended with the possibility to use surrogates analysis instead of the bicoherence technique.

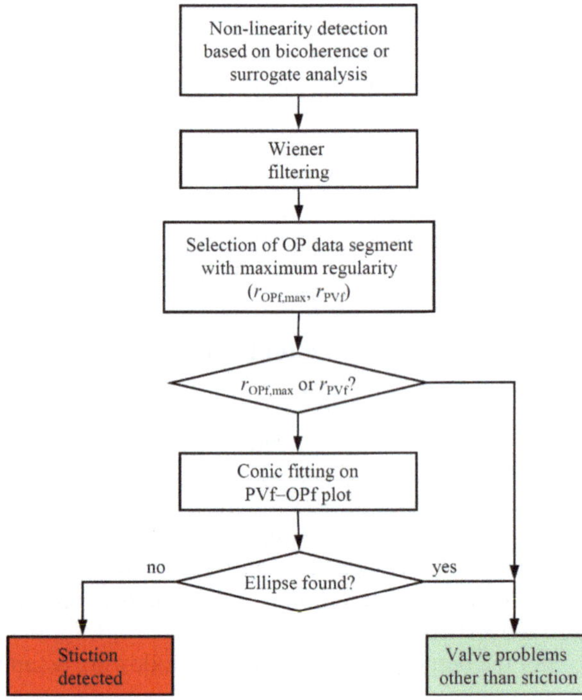

Fig. 11.19 Flow chart of the methodology for stiction detection based on nonlinearity analysis and conic fitting (Choudhury et al. 2006a)

Procedure 11.3 Stiction diagnosis based on nonlinearity and ellipse fitting.

1. *Detection of Nonlinearity*. Calculate NGI and NLI using the bicoherence method (Sect. 10.2) or NPI using the surrogate technique (Sect. 10.3) for the control error signal (SP–PV). If the indices do not indicate loop nonlinearity, the poor performance is probably caused by a poorly tuned controller or an external oscillatory disturbance (refer to Fig. 10.2), and the procedure is stopped.
2. *Pre process Data*.
   (a) Once the nonlinearity is detected, select appropriate filter boundaries f_{min} and f_{max}.
   (b) Filter PV and OP data using the Wiener filter to obtain $\mathrm{PV_f}$ and $\mathrm{OP_f}$. Ideally, the filter should remove the effect of noise and the set point changes (if any), leaving only the clear stiction pattern in the filtered variables. When surrogate analysis is considered, elimination of spikes and data end-matching are essential.
3. *Determination of the Segment of Data with most Regular Oscillations*.
   (a) Choose a segment length L, say $L = 1000$ (if data length permits). When surrogate analysis is used, the default parameter values (Table 10.2) are important.
   (b) Divide the OPf data into segments of length L. OPf is chosen instead of $\mathrm{PV_f}$ because often the OP signal is less noisy than the PV signal.
   (c) Calculate the regularity factor r_i and oscillation period $T_{\mathrm{p},i}$ for each segment of OPf data.

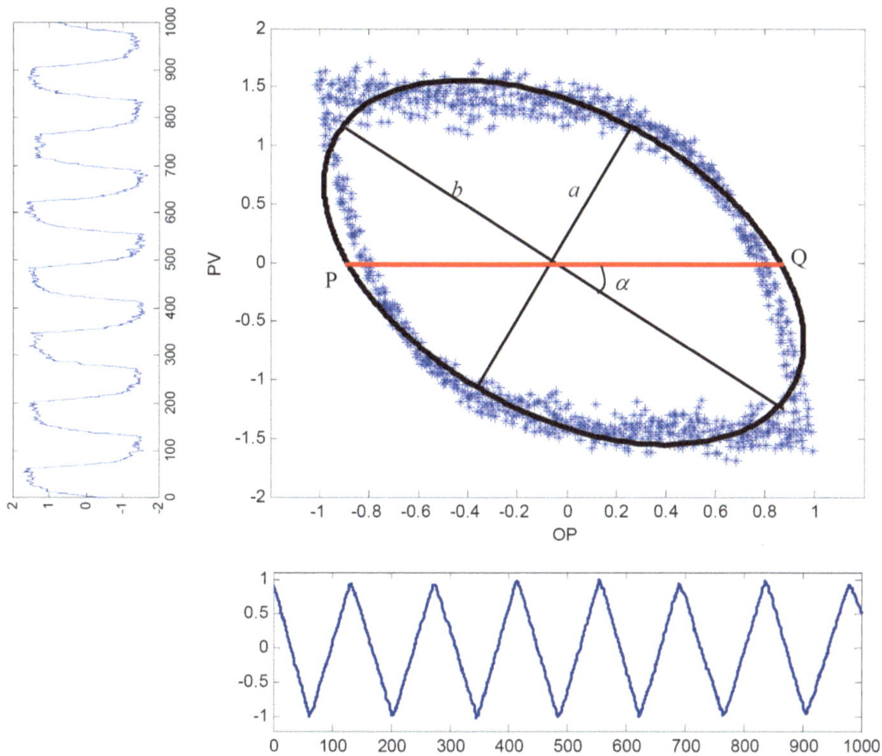

Fig. 11.20 Ellipse fitted to the PV_f–OP_f plot

(d) Obtain the maximum regularity factor $r = \max(r_i)$.
(e) Take T_p which is equal to the $T_{p,i}$ of the segment of OP with r.
(f) If $L > T_p$, then choose $L = 4T_p$ and go to step 3b.
(g) Now, OPf is the segment of the OPf data that corresponds to r, and PV_f is the part of the corresponding PV_f data.

4. *Fitting an Ellipse*. Fit a conic to the mapping of PV_f–OP_f (Fig. 11.20). If an ellipse can be fitted to the PV_f–OP_f plot, it can be concluded that the valve suffers from stiction problem. Algorithms for fitting an ellipse to a set of data are given by Choudhury et al. (2006a) and Manum (2006).

As a by-product of this method, apparent stiction can be quantified from the maximum width of the ellipse fitted in the PV–OP plot measured in the direction of OP quantifies stiction, i.e. (Choudhury et al. 2006a):

$$\text{Stiction plus dead-band } S\,[\%] \approx \frac{2ab}{\sqrt{a^2\sin^2\alpha + b^2\cos^2\alpha}}, \tag{11.15}$$

where a and b are the lengths of the minor and major axes of the fitted ellipse, respectively, and α is the angle of rotation of the ellipse from positive x-axis; see

Fig. 11.20. The quantified stiction is termed as "apparent stiction", because it may be equal to or different from the actual amount of stiction due to the influence of loop dynamics on PV and OP, in particular due to the effect of the controller to regulate PV and thus smooth the stiction effect (Choudhury et al. 2006a).

11.7.2 Practical Issues

As for any method, some issues have to considered, when dealing with industrial data that are usually subject to noise, drifts, etc.

11.7.2.1 Choosing an Appropriate Segment of the Data

In reality, the valve may stick for sometime and may not stick for some other time. Also the oscillation regularity factor values may differ from one data segment to another. Therefore, a data segment that show maximum regularity should be picked up.

11.7.2.2 Selection of Filter Boundaries

The selection of filter frequencies is crucial. Choudhury et al. (2006a) suggested to obtain the frequency band (f_1, f_2) corresponding to the maximum bicoherence peak in step 1, i.e. $[\omega_L = \max(0.004, f_{\min} - 0.05), \omega_H = \min(0.5, f_{\max} + 0.05)]$ with $f_{\min} = \min(f_1, f_2)$ and $f_{\max} = \max(f_1, f_2)$. Recall that all frequencies are normalised so that the sampling frequency is 1 and that 0.05 is subtracted or added from the frequencies to ensure that the exact location of the significant peak does not fall on the filter boundaries. The minimum possible value for the lower boundary is 0.004 or 250 samples/cycle. For oscillations with periods longer than this, the data can be down-sampled in such a way that the FFT length used in the bicoherence calculation consists of at least 3 or 4 cycles.

However, the experience with this automatic determination of filter boundaries revealed that even for simple loops, this initial "automatic" setting often does not work and "tuning" was needed to get good results. This initial investigation of the method lowered the hopes of implementing the method automatically, and research on other schemes is needed (Manum 2006). Also, a look at the frequency spectrum is always recommended to adjust the filter boundaries. Even in the original work by Choudhury et al. (2006a), the aforementioned rule has not been strictly followed.

Example 11.5 The necessity of filtering to remove non-stationary trends is illustrated in this example. The row data shown in Fig. 11.21 are from an industrial flow control loop in a refinery. Without filtering, it is clear that no distinct ellipse can be fitted to the PV–OP plot. On the contrary, if a Wiener filter with the boundaries

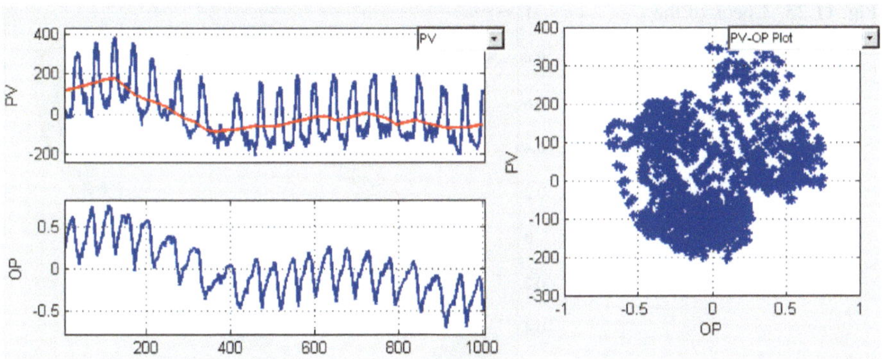

Fig. 11.21 Nonlinearity detection and PV–OP plot for loop CHEM30 without Wiener filtering (NGI = 0.03; NLI = 0.58)

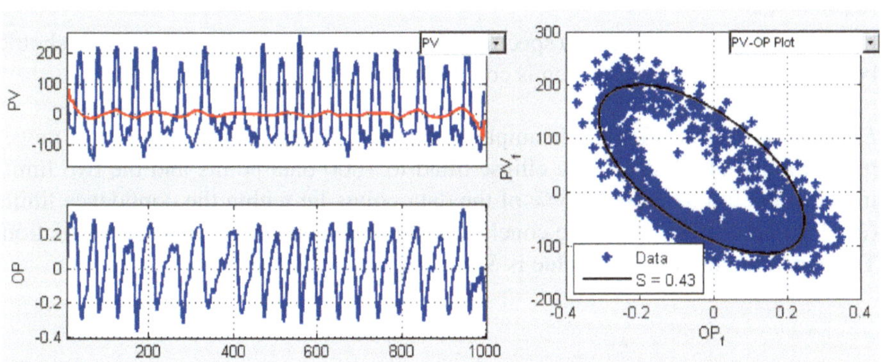

Fig. 11.22 Nonlinearity detection and ellipse fitting results for loop CHEM30 when the data are pre-processed using a Wiener filter [0.01, 1.0] (NGI = 0.04; NLI = 0.57)

[0.01, 1.0] is applied to the data, the PV–OP mapping has a clear and distinct elliptical pattern, as shown in Fig. 11.22. It is also observed that the filtering has removed the slowly varying mean-shift and high-frequency noise from the PV and OP signal. Equation 11.15 yields an estimate of the apparent stiction $S = 0.43$ %.

11.7.2.3 Ellipse Fitness

When dealing with real data, the ellipse fitting algorithm will tend to produce an ellipse even in the case where it is not clearly distinctive. Therefore, it is useful to introduce a measure of fitness for the ellipse. This draws two confidence limit ellipses around the fitted ellipse and checks how many data points are within the limits; see Fig. 11.23. The percentage of theses data points is defined as the ellipse

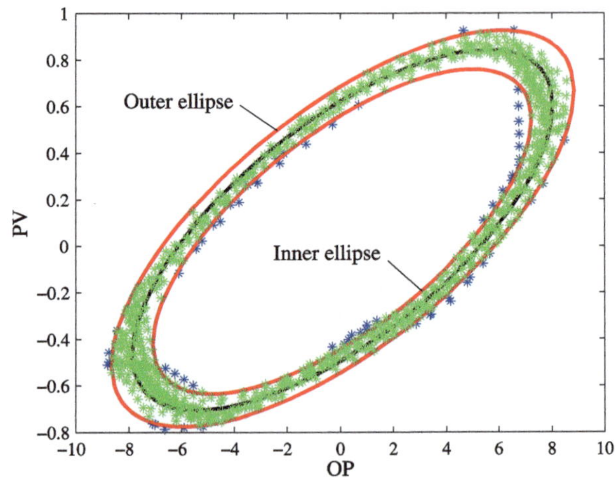

Fig. 11.23 Check of the validity of fitted ellipse for loop POW2

fitness. If the fitness is below a specified threshold, say 60 %, the fitted ellipse should be rejected, and thus no stiction is concluded.

Example 11.6 Data for this example come from a level control loop in a power plant. Figure 11.23 shows the ellipse fitted to 1000 data points and the two limiting ellipses. In this case, 93.5 % of the data points lie within the confidence limits (± 10 %). Therefore, it can be concluded that this loop suffers from valve stiction. The apparent stiction band value is $S \approx 11.4$.

11.8 Tests to Confirm Stiction

All stiction-detection methods presented above have an uncertainty zone where no decision is taken. In this case, the proposed methods cannot provide meaningful detection results. Even in the case a control loop is identified to have an oscillation problem probably caused by valve stiction, additional tests on the valve should be performed to pinpoint and confirm the problem.

Moreover, the non-invasive detection methods presented above are capable of detecting stiction in control valves. However, all of three methods work with single control loops and do not take into account the propagation of oscillation. Stiction in one valve may generate limit cycle oscillations that can easily propagate to other loops of the connected adjacent units. That is why all non-invasive methods when applied to an entire plant site may falsely signal stiction in a large number of control valves. To avoid this, plant tests are needed at the last stage to confirm and isolate the valves detected to be sticky. Using the stiction detection techniques presented below, the number of valves to be tested will be kept at minimum.

A well-known test to confirm stiction in industrial practice is the *valve travel test* (Sect. 11.8.2). Such a test should be done with the valve in service, as in-service checks are more accurate than testing the valve out of service, i.e. at test-beds.

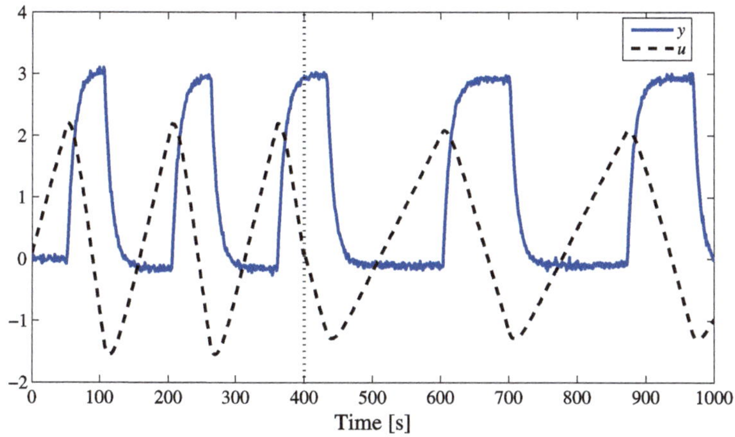

Fig. 11.24 Effect of changing controller gains on oscillating signals in case of stiction: a frequency decrease is clearly seen

Prior to these tests, we strongly advise to carry out a *controller-gain change test* (Sect. 11.8.1) to keep the other tests as the last mean of stiction confirmation.

11.8.1 Controller Gain Change Test

The presence of stiction in a control loop produces limit cycle oscillations in the controller variable and the controller output. *Changes in controller gain* (K_c) *have the peculiarity of affecting the oscillation frequency, without influencing its shape.* Decreasing K_c causes a decrease of the oscillation frequency. An intuitive explanation of this behaviour can be given by considering that a lower value of K_c causes a lower rate of increase of the active force F_a. Therefore, longer times occur to overcome the static friction force F_s, and, consequently, a lower frequency of oscillation results. A theoretical justification of this behaviour using describing function analysis is provided by Choudhury et al. (2005a, 2005b).

Once stiction is detected in a loop using any non-invasive method, changes in oscillation frequency due to variation in controller gain can help confirm the presence of stiction in the loop; see Fig. 11.24 and Fig. 11.25. This method is a simple alternative test that can be applied online without significant disruption of the plant production before applying an invasive stiction detection method.

The technique of changing the controller gain is also very useful for confirmation of stiction in (connected) multi-loop systems. In such an environment, all (non-invasive) stiction-detection methods may detect stiction in all connected loops, irrespective of the nature of the acting disturbances, i.e. whether the loop oscillation is really caused by stiction or due to an external oscillatory disturbance from a neighbouring loop. Thornhill et al. (2003a, 2003b, 2003c) showed an impressive example of this phenomenon, where 26 variables were identified to oscillate with a similar

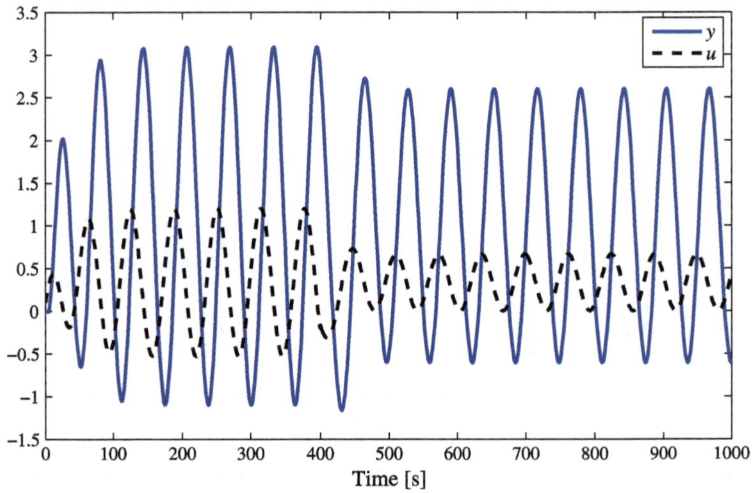

Fig. 11.25 Effect of changing controller gains on oscillating signals for the no-stiction case (*bottom*): the oscillation frequency does not change

time period as a condenser level. In such situations, the controller-gain-change can help distinguish between the loops suffering from stiction and those oscillating due to propagated oscillatory disturbance: *if a limit oscillation enters in a loop as an external disturbance, a change in controller gain will NOT change the frequency of oscillation.*

11.8.2 Valve Travel or Bump Test

Stiction in control valves is usually confirmed by putting the valve in manual and increasing the control signal in small increments until there is an observable change in the process output. More specifically, measuring stiction online can be performed by the following steps (Gerry and Ruel 2001):

Procedure 11.4 Valve travel test for stiction quantification.

1. Put the controller in manual with the output near the normal operating range.
2. Change the controller output by 5 to 10 % to overcome the hysteresis on the loop. If the process variable does not move from this change, repeat it until the process variable moves.
3. Wait for the process variable to settle.
4. Make a small change in the controller output, say about 0.2 %, in the same direction as the last step. Wait for the same amount of time as the previous step to see if the process variable moves.
5. Repeat Step 4 until the process variable moves, i.e. until the stiction band is overcome.

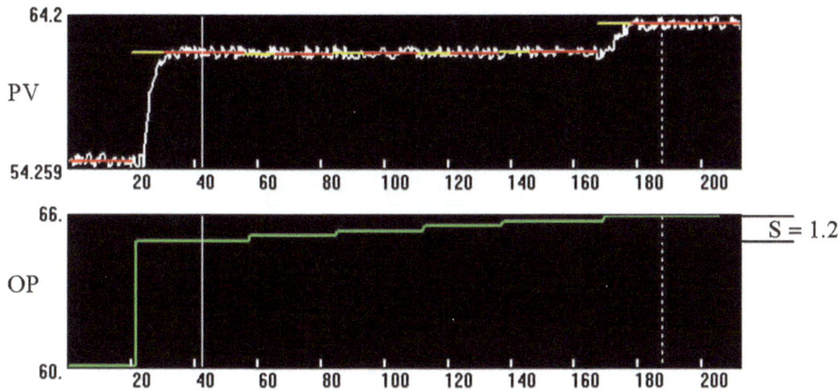

Fig. 11.26 Data collected for stiction check (Gerry and Ruel 2001)

The stiction in the loop is the total amount of OP change required to detect a change in the process variable. Figure 11.26 shows this series of tests performed on a flow loop. The vertical lines on the plot mark the "stiction band" or the difference in OP indicating the amount of stiction present in the valve, ca. 1.2 %. The test shows that stiction is a problem for this valve, since stiction of more than 0.2 % will usually cause cycling.

Again, this method of confirming stiction by putting the loop in manual is not convenient and cost-effective due to the risk of plant upset and production of more "off-spec" products. It should therefore be the last stage of any oscillation diagnosis procedure.

11.9 Stiction Diagnosis Procedure

The diagnosis procedure we propose in this section combines techniques individually described before; see Fig. 11.27. It starts with oscillation detection (Chap. 9): when the control loop is found to be oscillating, the most probable origin is assumed to be valve stiction. However, nonlinearity and saturation detection (Chap. 10) are also useful as a possible source of oscillation. We then apply a stiction detection method (Sects. 11.5 to 11.7) and estimate its level (Chaps. 9 and 11). If stiction is detected and has a significant level, then a gain-change test (Sect. 11.8.1) should be performed following a cautious "belt and braces" approach before the costs of downtime and maintenance can be justified. When stiction is confirmed, valve travel test should be carried out (Sect. 11.8.2), preferably with the valve in service. If the level of friction is high, the best solution is to undertake valve maintenance. If the valve is oversized, the suggestion is to resize it, otherwise the impact of friction will stay high, and thus improvement in the control performance will not be possible. The negative effects of stiction cannot be totally eliminated without repairing the valve.

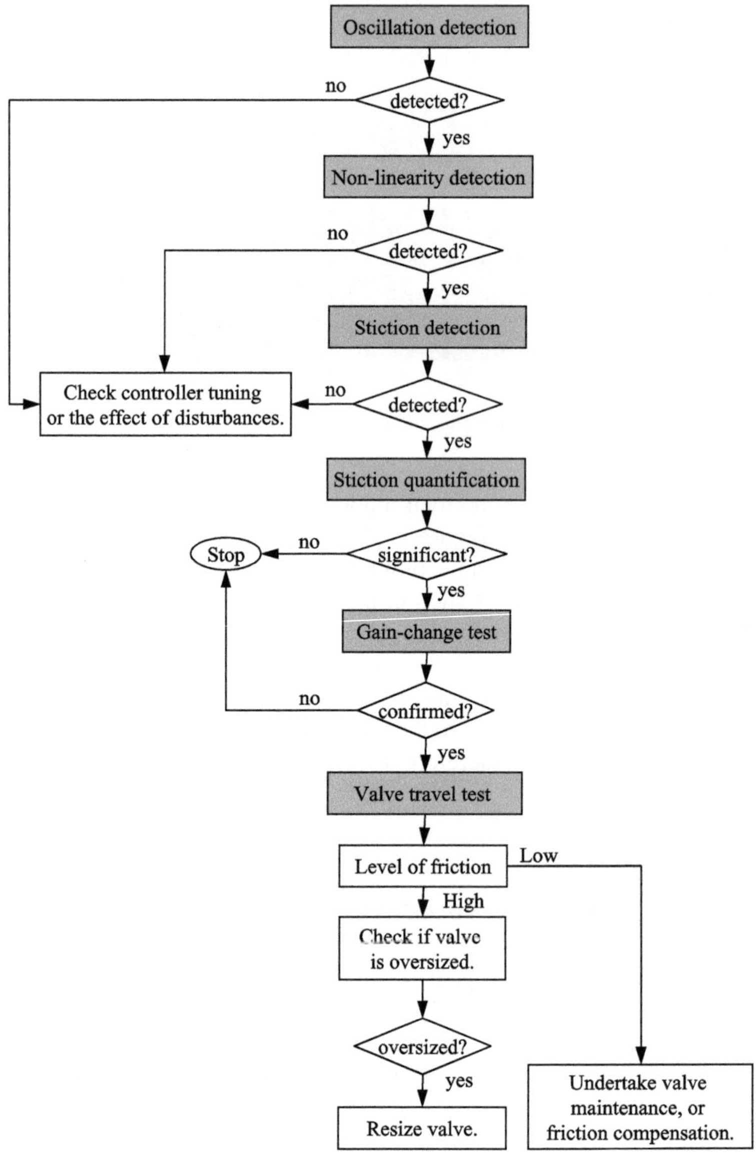

Fig. 11.27 Diagnosis procedure to discover the cause of oscillations and recommended actions to eliminate them

In many cases, however, this is not possible because of several reasons (Gerry and Ruel 2001):

- It is economically not feasible to stop the production.
- The valve/actuator type is the problem, and it is necessary to use this type of valve/actuator for fail-safe considerations.

- Replacing the valve/actuator could be too expensive.
- Tune the positioner using a large proportional gain and no integral action. If derivative action is available, use some to make the valve continuously move. With integral action in the positioner, it may wind up, causing the valve to seemingly have a mind of its own. After some period of time, the stem will jump, after the positioner has wound up enough. By removing integral action from the positioner, this windup problem is eliminated.
- If a smart positioner is used, adjust the parameters. Some positioners do not use PID but rather special algorithms to send a burst of pressure each time a new position is requested. The positioner action is to stop the valve at the requested position.
- Use a PID controller (for the control loop) where the integral action has a variable strength: if the absolute error is smaller than some value, then take out the integral action, otherwise use it. Using this method, when the valve is within the stiction band, the integral action is missing from the controller, the controller output will not integrate, having the end effect of removing the stiction cycle from the loop.
- Use a PID with gap: if the absolute error is smaller than a given threshold, the controller output is frozen; if not, the amount of error from the gap is used as the controller input.

In these cases, the negative effects of stiction cannot be totally eliminated, but methods for combating the stiction to reduce these effects are beneficial. Such techniques include (Gerry and Ruel 2001):

Another way to reduce the effect of friction is to compensate for it. A very simple way to eliminate some effects of friction is to use a dither (high-frequency) signal, which is added to the control signal. The effect of the dither is that it introduces extra forces that make the system move before the stiction level is reached. The effect is thus similar to removing the stiction. The effects of dither in systems with dynamic friction were studied by Panteley et al. (1998).

Systems for motion control typically have a cascade structure with a current loop, a velocity loop and a position loop. Since friction appears in the inner loop, it would be advantageous to introduce friction compensation in that loop. The friction force is estimated using some model, and a signal that compensates this force is added to the control signal (Olsson et al. 1998). Some friction compensation techniques for servo-hydraulic control systems have also been discussed by Jelali and Kroll (2003).

11.10 Summary and Conclusions

The main problems that can occur in actuators have been described with a focus on the analysis of stiction in control valves. The most important stiction detection techniques have been reviewed, including their assumptions, strengths and weaknesses. These are essential when applying any method to real-world data. The cross-correlation technique is simple and easy to implement but may have problems with phase shift induced by controller tuning and is limited to self-regulating processes.

Curve fitting techniques are available in different versions for stiction detection. These methods rely on the signal pattern characterising stiction, which, however, may be modified if other malfunctions, e.g. aggressive tuning, are simultaneously present in the loop. Nonlinearity detection followed by ellipse fitting has also been proven to be very efficient in detecting stiction, but the complexity of this technique and the difficulty of automatically selecting proper filter boundaries are clear weaknesses of the method. Therefore, we recommend to apply more than one technique to have redundancy.

A systematic oscillation diagnosis procedure has been proposed, combining some oscillation and nonlinearity detection techniques, as well as additional tests and methods for checking, elimination or compensation of valve stiction.

Chapter 12
Complete Oscillation Diagnosis Based on Hammerstein Modelling

As should be learnt from the previous chapters, when a control loop is detected to be oscillating, the root cause of that oscillation may be aggressive controller tuning, external disturbance, or valve nonlinearity and particularly stiction. Even in the case where stiction is detected by applying any of the techniques described in Chap. 11, it is desirable to quantify the extent of stiction. This is because we usually attempt to evaluate a large number of control loops, with valves being more or less sticky. To our best knowledge, only two methods have been published about the quantification of stiction, i.e. Srinivasan et al. (2005a, 2005b) and Choudhury et al. (2006a, 2006b). The next useful feature missing up to now in the literature is to distinctly detect multiple faults in an oscillating control loop.

This chapter presents a novel technique for detection and estimation of valve stiction in control loops from normal closed-loop operating data based on a two-stage identification algorithm. The control system is represented by a Hammerstein model including a two-parameter stiction model and a linear model for describing the remaining system part. Only OP and PV data are required for the proposed technique. This not only detects the presence of stiction but also provides estimates of the stiction parameters. Therefore, the method is useful in short-listing a large number of control valves more or less suffering from stiction in chemical or other plants, containing hundreds or thousands control loops. This helps reduce the plant maintenance cost and increase the overall profitability of the plant. A unique feature of the proposed technique is also its capability to discriminate between the different oscillation sources or detect the situation when the loop suffers from *two or more* oscillation root causes *simultaneously*. Note that the method does not need any further experimentation with plant and, hence, uses only closed-loop operating data.

The chapter is organised as follows. Section 12.1 presents the features of the framework proposed for stiction quantification. The considered model structure is introduced in Sect. 12.2. The new stiction estimation algorithm is presented in Sect. 12.3. Some practical issues when analysing industrial data with the presented method are discussed in Sect. 12.4. Section 12.5 contains simulation and industrial case studies to demonstrate the practicality and applicability of the proposed technique. The extension of the technique for complete oscillation diagnosis, i.e. detect-

M. Jelali, *Control Performance Management in Industrial Automation*,
Advances in Industrial Control, DOI 10.1007/978-1-4471-4546-2_12,

ing and distinguishing multiple faults occurring simultaneously in control loops is given in Sect. 12.6. The complete oscillation diagnosis approach is illustrated with simulation and industrial examples.

12.1 Features of the Proposed Framework

The proposed procedure is partly an extension of similar approaches, e.g. Srinivasan et al. (2005b), having in common the fact that the nonlinear part is represented by Hammerstein models, but the stiction model and the identification techniques are different. Global search techniques, i.e. pattern search (PS) methods or genetic algorithms (GA), are used here to estimate the nonlinear model parameters, subordinated with a least-squares (LS) identification of the linear model parameters. The author is not aware of any reports on the application of such algorithms in the valve-stiction estimation. Moreover, estimates of both stiction-model parameters, dead-band plus stick band (S) and slip jump (J), are provided in this work. Only when both parameters S and J are known, it is possible to estimate the inner signal MV. In contrast, the ellipse-fitting method by Choudhury et al. (2006a, 2006b) estimates only the parameter S. The approach of Srinivasan et al. (2005a, 2005b) is based on a separable LS identification algorithm proposed by Bai (2002) and applicable only to nonlinearities with one single unknown parameter. Thus, Srinivasan et al. (2005a, 2005b) uses a simple relay model for describing the valve stiction. This model is, however, physically unrealistic and fails to capture the behaviour of loops with non-integrating processes. Note also that the slip jump J is very difficult to observe in PV–OP plots because the process dynamics destroys the pattern. This makes the estimation of a two-parameter model much more challenging than those considered so far in the literature.

The framework for oscillation diagnosis proposed consists of the following features:

1. Two-parameter stiction models, which are more accurate and suitable for both self-regulating and integrating processes, particularly when time delay is present, are considered.
2. The linear dynamics is represented through simple, i.e. low-order, models. Only partly automated structure selection is suggested. This choice keeps complexity and thus the required computational burden limited.
3. Global search techniques, i.e. pattern search (PS) methods or genetic algorithms (GA), are used to estimate the nonlinear model parameters, subordinated with an LS identification of the linear model parameters.
4. The proposed parameter estimation is recommended as a second diagnosis stage, i.e. for stiction quantification after detecting stiction using other non-invasive methods.
5. Since both the linear and the nonlinear parts will be estimated, closed-loop simulations without the stiction model help identify possibly bad controller tuning

or external disturbances, affecting simultaneously the loop performance (in addition to stiction). This ability of detecting multiple loop faults occurring simultaneously is a unique feature of this technique.
6. The method is not limited to sticky loops but can also be applied when other nonlinearities, such as hysteresis or backlash, are present.
7. This identification-based technique is robust against noise and drifting trends, usually corrupting real-word data.

12.2 Identification Model Structure

In 2003, Jelali and Kroll have shown that hydraulic valves can be represented by models combining linear dynamic blocks and one or more static nonlinear block(s), such as the Hammerstein model, Wiener model or even more general model structures. This depends on the valve type/complexity and the nonlinearity under consideration. Note also that the estimation of friction-model parameters was proposed by Jelali and Kroll (2003) within the framework of grey-box identification for hydraulic control systems, but not in the context of control performance monitoring.

Restricting attention to stiction nonlinearities, it can easily be seen that a simple control valve model is of the Hammerstein type, as also recently considered in Srinivasan et al. (2005b) and illustrated in Fig. 12.1. In our approach, we use one of the two-parameter models $M_{\mathrm{stic}}(J, S)$ mentioned in Sect. 11.3.3. Such a model can realistically capture the closed-loop behaviour, as shown by Choudhury et al. (2005a, 2005b) in contrast to the simple model used in Srinivasan et al. (2005b). A comparison of the relay and Choudhury's models can be found in Singhal and Salsbury (2005). It is pointed out that the relay model is a good approximation of Choudhury's model only when the ratio time delay to time constant (θ/T) is small and the system order is low. The discrepancy between the results using both models increases with increasing θ/T and system order. This is due to the missing consideration of the sliding part in the valve characteristic as shown in Fig. 12.2. As the delay increases, the sliding part becomes larger compared to the slip jump, and valve output moves with the controller output for a longer period of time. As the processes we mainly consider are those with significant time delays, a two-parameter stiction model is needed.

The process dynamics, including the valve dynamics, is represented by an ARMAX(n, m, p, τ) model, as in Eq. 2.1. For modelling the sticky valve, we use one of the data-driven two-parameter models $NL_{\mathrm{stic}}(J, S)$, proposed by Kano et al. (2004), Choudhury et al. (2005a, 2005b) and He et al. (2007).

The stiction nonlinearity can be written in the general form

$$u_{\mathrm{v}}(k) = NL_{\mathrm{stic}}\big(u(k), \ldots, u(0), u_{\mathrm{v}}(k-1), \ldots, u_{\mathrm{v}}(0), J, S\big), \qquad (12.1)$$

parameterised by the parameter pair J and S, which are assumed to be constant. Note that the internal signal $u_{\mathrm{v}}(k)$ is not measurable. Stiction nonlinearity is a discontinuous function or "hard" nonlinearity, which does not belong to the class

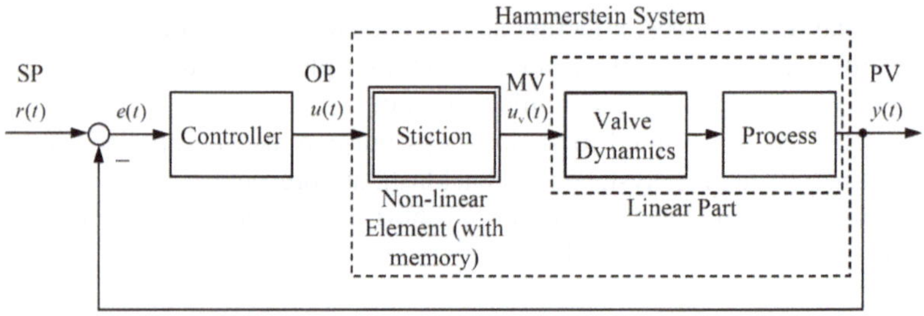

Fig. 12.1 Process-control loop with valve stiction within an identification framework

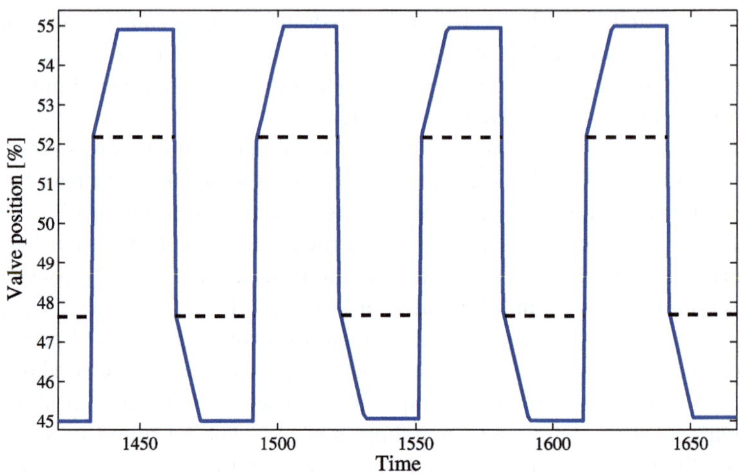

Fig. 12.2 Typical discrepancy between a relay model (*dashed lines*) and a two-parameter stiction model (*solid lines*)

of memoryless nonlinearities, usually assumed in the estimation of block-oriented models. Stiction nonlinearity thus results in non-smooth and non-convex objective maps. These facts have two important implications:

1. Most of known techniques, e.g., correlation analysis, prediction-error methods, for the estimation of Hammerstein models cannot be applied for the stiction estimation and diagnosis problem.
2. Gradient-based algorithms, known as *local* optimisation methods, would always stuck in a local minimum near the starting point. In contrast, global search, usually gradient-free, algorithms are not affected by random noise in the objective functions, as they require only function values and not the derivatives.

We do not intend to discuss all approaches for the identification of Hammerstein models, but refer the reader to the excellent overview by Srinivasan et al. (2005b), where the inherent limitations of "classical" methods in the estimation and diagnosis of stiction are pointed out. In this work, the use of a separable least-squares estimator combined with a global search algorithm is proposed.

12.3 Identification Algorithm

To simplify the Hammerstein model identification, a decoupling between the linear and nonlinear parts is useful, as recommended by Seborg and Viberg (1997) and Bai (2002) in the context of open-loop identification of input nonlinearities other than stiction. This technique is extended in this section for the application to the identification of stiction models parameterised by J and S.

12.3.1 Linear Model Estimation

To simplify the derivation, we first consider an ARX model (with $m = n$) to describe the linear system part in the time domain as

$$y(k) = \boldsymbol{\Theta}^{\mathrm{T}}\big[y(k-1), \ldots, y(k-n), u_{\mathrm{v}}(k-\tau-1), \ldots, u_{\mathrm{v}}(k-\tau-n)\big] + \varepsilon(k) \quad (12.2)$$

with the unknown parameter vector

$$\boldsymbol{\Theta} = [-a_1, \ldots, -a_n, b_1, \ldots, b_n]^{\mathrm{T}} \quad (12.3)$$

as well as J and S because of Eq. 12.1. Let

$$\hat{u}_{\mathrm{v}}(k) = NL_{\mathrm{stic}}\big(u(k), \ldots, u(0), \hat{u}_{\mathrm{v}}(k-1), \ldots, \hat{u}_{\mathrm{v}}(0), \hat{J}, \hat{S}\big), \quad (12.4)$$

which provides an estimate of MV, i.e. the valve position $u_v(k)$, using $\hat{J}$ and $\hat{S}$. Define the prediction error

$$\varepsilon_{\hat{\boldsymbol{\Theta}},\hat{J},\hat{S}}(k) = y(k) - \hat{y}(k) \quad (12.5)$$

$$= \hat{\boldsymbol{\Theta}}^{\mathrm{T}}\big[y(k-1), \ldots, y(k-n), u_{\mathrm{v}}(k-\tau-1), \ldots, u_{\mathrm{v}}(k-\tau-n)\big] \quad (12.6)$$

for $k = 1, 2, \ldots, N$ and the objective function (MSE: mean squared error)

$$V_N(\hat{J}, \hat{S}, \hat{\boldsymbol{\Theta}}) = \frac{1}{N}\sum_{k=1}^{N} \varepsilon^2_{\hat{\boldsymbol{\Theta}},\hat{J},\hat{S}}(k). \quad (12.7)$$

The associated estimates are

$$\big[\hat{J}, \hat{S}, \hat{\boldsymbol{\Theta}}^{\mathrm{T}}\big]^{\mathrm{T}} = \arg\min V_N(\hat{J}, \hat{S}, \hat{\boldsymbol{\Theta}}), \quad (12.8)$$

where N is the number of data samples $y(k)$, $\hat{y}(k)$ is the estimate of $y(k)$ or Hammerstein-model output, and $\hat{\boldsymbol{\Theta}}$ is the estimated parameter vector of the linear model part.

With

$$\boldsymbol{y} = \begin{bmatrix} y(n+\tau) \\ y(n+\tau+1) \\ \vdots \\ y(N) \end{bmatrix},$$

$$\boldsymbol{\Phi}(\hat{J},\hat{S}) = \begin{bmatrix} y(n+\tau-1) & \dots & y(\tau) & \hat{u}_{\mathrm{V}}(n-1) & \dots & \hat{u}_{\mathrm{V}}(0) \\ y(n+\tau) & \dots & y(\tau+1) & \hat{u}_{\mathrm{V}}(n) & \dots & \hat{u}_{\mathrm{V}}(1) \\ \vdots & \vdots & \vdots & \vdots & \vdots & \vdots \\ y(N-1) & \dots & y(N-n) & \hat{u}_{\mathrm{V}}(N-\tau-1) & \dots & \hat{u}_{\mathrm{V}}(N-n-\tau) \end{bmatrix} \tag{12.9}$$

the objective function V_N can be rewritten as

$$V_N = \frac{1}{N}\left\| \boldsymbol{y} - \boldsymbol{\Phi}(\hat{J},\hat{S})\hat{\boldsymbol{\Theta}} \right\|_2. \tag{12.10}$$

For a given data set $\{y(k), u(k)\}$, i.e. $\{\mathrm{PV}(k), \mathrm{OP}(k)\}$, this minimisation involves three variables $\hat{J}$, $\hat{S}$ and $\hat{\boldsymbol{\Theta}}$. V_N is non-smooth in $\hat{J}$ and $\hat{S}$, but smooth in $\hat{\boldsymbol{\Theta}}$. Moreover,

$$0 = \frac{1}{2}\frac{\partial V_N}{\partial \hat{\boldsymbol{\Theta}}} = -\boldsymbol{\Phi}(\hat{J},\hat{S})\boldsymbol{y} + \boldsymbol{\Phi}^{\mathrm{T}}(\hat{J},\hat{S})\boldsymbol{\Phi}(\hat{J},\hat{S})\hat{\boldsymbol{\Theta}}. \tag{12.11}$$

Hence, if $\boldsymbol{\Phi}^{\mathrm{T}}(\hat{J},\hat{S})\boldsymbol{\Phi}(\hat{J},\hat{S})$ is invertible, a necessary and sufficient condition for $\hat{\boldsymbol{\Theta}}$ to be optimal is

$$\hat{\boldsymbol{\Theta}} = \left[\boldsymbol{\Phi}^{\mathrm{T}}(\hat{J},\hat{S})\boldsymbol{\Phi}(\hat{J},\hat{S})\right]^{-1}\boldsymbol{\Phi}^{\mathrm{T}}(\hat{J},\hat{S})\boldsymbol{y}, \tag{12.12}$$

provided that $\hat{J}$ and $\hat{S}$ are optimal. Therefore, by substituting $\hat{\boldsymbol{\Theta}}$ in terms of $\hat{J}$ and $\hat{S}$ back into V_N, it follows that

$$V_N = \frac{1}{N}\left\| \boldsymbol{I} - \boldsymbol{\Phi}(\hat{J},\hat{S})\left[\boldsymbol{\Phi}^{\mathrm{T}}(\hat{J},\hat{S})\boldsymbol{\Phi}(\hat{J},\hat{S})\right]^{-1}\boldsymbol{\Phi}^{\mathrm{T}}(\hat{J},\hat{S})\boldsymbol{y} \right\|_2. \tag{12.13}$$

When the two-parameter stiction model is used to calculate $\hat{u}_{\mathrm{V}}(k)$, the dimension of the search space is reduced from $2n+2$ to 2, independently of the linear part. However, note that, in contrast to the nonlinearities considered in Bai (2002), the stiction nonlinearity cannot be expressed in closed form.

The identification algorithm proposed for systems with stiction nonlinearities parameterised by J and S can now be summarised as follows.

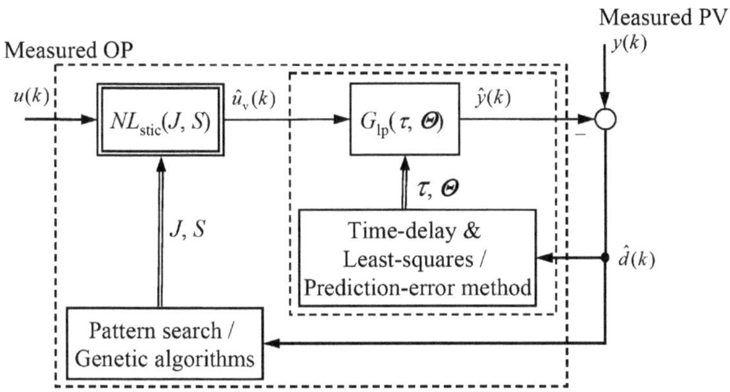

Fig. 12.3 Two-stage identification of the system parameters

Procedure 12.1 Hammerstein identification for stiction estimation.

1. Determine the time delay and initial values for J and S.
2. Consider the system Eq. 12.2, collect a data set $\{y(k), u(k)\}$ and construct $\boldsymbol{y}$ and $\boldsymbol{\Phi}(\hat{J}, \hat{S})$, as in Eq. 12.9.
3. Solve Eq. 12.13 for the optimal $\hat{J}$ and $\hat{S}$.
4. Calculate the optimal $\hat{\boldsymbol{\Theta}}$ as in Eq. 12.12.

Thus, when an ARX is used, iterative optimisation needs only be performed with respect to the nonlinear parameters (J and S).

To generalise the method, one may use ARMAX instead of ARX, which now implies the need for iterative (nonlinear) estimation also for the linear model. Once the optimal values $\hat{J}$ and $\hat{S}$ are obtained by a global search method, as described below, the stiction model is computed to generate $\hat{u}_v(k)$. Based on $\hat{u}_v(k)$ and $y(k)$, the parameters $\hat{\boldsymbol{\Theta}}$ of the linear model are identified using an LS-IV (instrumental variables) algorithm or a prediction error method (PEM) (Ljung 1999), combined with model-structure selection, i.e. for estimating the model orders and time delay (when not known or specified); see Sect. 12.4. This two-stage identification method is illustrated in Fig. 12.3. The key elements of our approach are as follows:

- The linear dynamics is modelled using a low-order model, the polynomial parameters of which are estimated using LS or PEM in a subordinated identification task. The number of unit delays τ, when not known, is determined by applying an appropriate time-delay estimation method.
- A pattern search algorithm or a genetic algorithm is employed to estimate the values of the parameters S and J of the stiction model so that the MSE is minimised (Eq. 12.10). The time trend of MV can thus be estimated.

The successful use of this approach requires the system to be sufficiently excited. As the loops of interest are those detected to be oscillating, this condition should be satisfied.

Remark Although the proposed stiction estimation method is principally able to detect stiction, it is more recommended as a second diagnosis stage, i.e. for stiction quantification after valve stiction has been detected by another simpler method. The presence of stiction nonlinearity also ensures the identifiability of the model from normal closed-loop operating data. Also, the identifiability is ensured when another nonlinearity, such as dead-band or dead-zone, is present instead of stiction. However, when the loop oscillates due "linear" causes, such as external disturbances, the closed-loop identification can become difficult when sufficient excitation is not present. In such cases, the stiction diagnosis may yield wrong results.

12.3.2 Nonlinear Model Estimation

Traditional derivative-based optimisation methods, like those found in the MATLAB Optimization Toolbox, are fast and accurate for many types of similar optimisation problems. These methods are designed to solve "smooth", i.e. continuous and differentiable, minimisation problems, as they use derivatives to determine the direction of descent. While using derivatives makes these methods fast and accurate, they are not suitable when problems lack smoothness, as is the case in the valve-stiction estimation problem. When faced with solving such non-smooth problems, methods like genetic algorithms or pattern search methods are the effective alternatives. Both methods are briefly discussed in the following.

12.3.2.1 Genetic Algorithms

Genetic algorithms (GAs) are search techniques which imitate the concepts of natural selection and genetics. They were formally introduced by Holland (1975). GAs search the solution space of a function through the use of simulated evolution, i.e. the survival-of-the-fittest strategy. This provides an implicit as well as explicit parallelism that allows for the exploitation of several promising areas of the solution space at the same time. Instead of looking at one point at a time and stepping to a new point for each iteration, a whole population of solutions is iterated towards the optimum at the same time. Using a population allows us to explore multiple "buckets" (local minima) simultaneously, increasing the likelihood of finding the global optimum. GAs are thus well suited for solving difficult optimisation problems with objective functions that possess "bad" properties such as discontinuity and non-differentiability, as is the case for the stiction estimation problem.

In the GA-based optimisation approach (Fig. 12.4), the (unknown) parameters are represented as *genes*, hence the name "genetic", on a *chromosome*, representing an individual. Similar to the simplex search, a GA features a group of candidate solutions, the *population* or *gene pool*, on the response surface. Applying *natural selection* and using the *genetic operators*, recombination and mutation, chromosomes

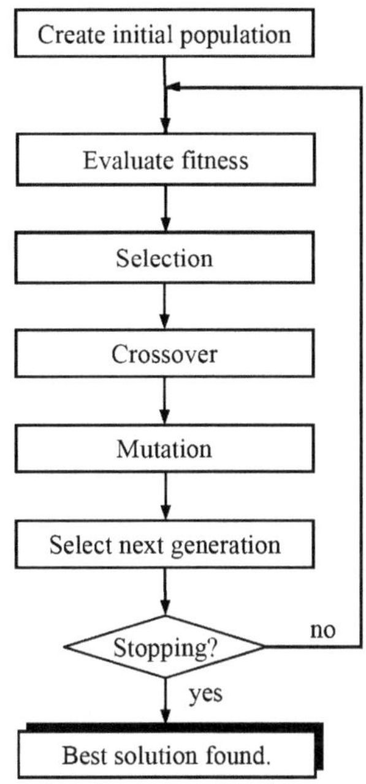

Fig. 12.4 Basic procedure of genetic algorithms

with better *fitness*, i.e. degree of "*goodness*", are determined. Natural selection guarantees that chromosomes with the best fitness will propagate in future populations. Using the *recombination operator*, the GA combines genes from two parent chromosomes to form two new chromosomes (children) that have a high probability of having better fitness than their parents. *Mutation* allows new areas of the response surface to be explored. One of the reasons GAs work so well is that they offer a combination of hill-climbing ability through natural selection and a stochastic method through recombination and mutation.

The main drawback of using GAs is the high computational burden. This is due their prohibitively slow convergence to the optimum, when compared to gradient-based methods, especially for problems with a large number of design variables. Nevertheless, the continual progress of computer technologies has greatly reduced the effort required to implement such methods. Therefore, GAs are becoming increasingly popular in the last years. Complete discussions of genetic algorithms can be found in the books by Goldberg (1989) and Michalewicz (1994).

12.3.2.2 Pattern Search Methods

Pattern search (PS) is an attractive alternative to GAs, as it is often computationally less expensive. Pattern search operates by searching a set of points called a pattern, which expands or shrinks depending on whether any point within the pattern has a lower objective function value than the current point. The search stops after a minimum pattern size is reached. Like the genetic algorithm, the pattern search algorithm does not use derivatives to determine descent and so works well on non-differentiable, stochastic and discontinuous objective functions. Also, similar to the genetic algorithm, pattern search is often very effective at finding the global minimum because of the nature of its search.

A pattern search algorithm can be generally described as follows.

Procedure 12.2 General pattern search procedure.

1. Initialise direction and mesh size. At each iteration k, the mesh is defined by the set $M_k = \bigcup_{x \in S_k} \{x + \Delta_k D z : z \in \mathrm{N}^{n_D}\}$, where $S_k \in \mathrm{R}^n$ is the set of points where the objective function f had been evaluated by the start of iteration k, and $\Delta_k > 0$ is the mesh size parameter that controls the fineness of the mesh. D is a set of positive spanning directions in R^n.
2. Perform the following steps until convergence:

   - *SEARCH step.* Employ some finite strategy seeking an improved mesh point, i.e. where the value of the objective function is lower than that at the current point. The SEARCH step usually includes a few iterations using a heuristic, such as GA, random sampling, or the approximate optimisation on the mesh of surrogate function.
   - *POLL step.* If the SEARCH step was unsuccessful, evaluate the objective function at points in the poll set $P_k = \{x_k + \Delta_k d : d \in D_k \subseteq D\} \subset M_k$, i.e. at points neighbouring the current one on the mesh, until an improved mesh point is found.
   - *Parameter UPDATE.*
     - *Success*, i.e. when SEARCH or POLL finds an improved mesh point: accept new iterate and coarsen the mesh.
     - *Failure*: refine the mesh.

     The mesh size is updated according to the rule $\Delta_{k+1} = \tau^{w_k} \Delta_k$, where $\tau > 1$ is a fixed rational number, $w_k \in \{0, 1, \ldots, w^+\}$ for mesh coarsening and $w_k \in \{w^-, w^- + 1, \ldots, -1\}$ for mesh refining, and $w^- \leq -1$ and $w+ \geq 0$ are two fixed integers. Typically, $\Delta_{k+1} = 2.0\Delta_k$ is used for mesh coarsening, and $\Delta_{k+1} = 0.5\Delta_k$ for mesh refining.

Depending on the mesh-forming method, the search heuristic and the polling strategy, different pattern-search algorithms result. For detailed discussions of these algorithms, the reader should consult, for instance, Lewis and Torczon (1999, 2000).

In fact, our experience with this method has early led to the conclusion that PS is very fast in finding a point somewhere in the region of the global minimum. Therefore, (GA-based) pattern search is the recommended approach for solving the stiction identification problem.

12.4 Key Issues

12.4.1 Model Structure Selection

As a good engineering rule says that one "should not try to estimate what is already known" (Ljung 1993), it is highly recommended to use any a priori knowledge available, e.g., from physical insight or from experimentation with the process during the controller commissioning stage. In CPM, it is usually assumed that the time delay is known or can be estimated accurately. At least an interval, where the time delay may lie, should be given. All this information, when available, should be included in the model structure. This significantly accelerates the convergence rate of the parameter estimation algorithm.

Moreover, simulation studies carried out by Srinivasan et al. (2005b) indicate that a decoupling in the accuracy of the estimate between the nonlinear and the linear component of Hammerstein models with stiction nonlinearity may exist. Particularly, the accuracy of the estimate of the nonlinear component may be not affected by the complexity of the model structure used to describe the linear model part. This point also would suggest using a simple model for the linear valve and process dynamics.

However, this nice feature seems to be valid only for one-parameter nonlinearities. In fact, we observed an interaction between the two parameters slip jump J and time delay τ. No change is shown in MV up to the time when the controller output becomes larger than S; PV will change after a further delay since that. This may result in an identifiability problem with the consequence that a good estimate of τ is required to accurately estimate J. To ensure this, a time-delay estimation (TDE) algorithm is included in our stiction identification approach.

The following approaches for selecting the model structure of the linear model part have been investigated:

1. *Automatic Structure Selection.* One may use the functions arxstruc/ivstruc and selstruc of MATLAB's Optimization Toolbox, minimising Akaike's information criterion (AIC) (Ljung 1999). This estimation is quick since the ARX model can be written as a linear regression and can be estimated by solving a linear equation system. Our experience reveals that this method often fails to provide a good estimate of time delay, particularly when substantial noise is present. The same approach but using an ARMAX model, as suggested by Srinivasan et al. (2005b), can be followed. To simplify and thus accelerate the model identification process, we advice choosing the model order to be equal, i.e., $n_A = n_B = n_C = n$, so that it is only necessary to search for two parameters. A search is performed over the range of possible orders $[n_{\min} : n_{\max}]$ and numbers of unit delays $[\tau_{\min} : \tau_{\max}]$ to find the minimum AIC for the ARMAX model. This approach, which requires relatively high computation burden, was not always successful, due to existence of many local minima.
2. *Time-Delay Estimation.* As mentioned above, we had good experience and thus recommend using a fixed-order model, e.g. ARMAX$(2, 1, 2, \hat{\tau})$, for which the

optimal time delay is determined (when not known). For this purpose, numerous methods exist; see Björklund (2003) for an intensive discussion and comparison. We investigated many of them, e.g. the *pre-filtered* arxstruc method proposed by Björklund (2003) (and called met1struc) and some methods included in the MATLAB Higher-Order Spectral Analysis Toolbox, e.g. tder (windowed cross-correlation). None of these techniques was successful in *all* simulation and practical cases we studied. Therefore, we implemented a simple search over a range $[\tau_{\min} : \tau_{\max}]$ for ARMAX$(2, 1, 2, \hat{\tau})$ and picked up the one with the minimum AIC. This approach led to good results, at the expense of higher computation burden. See Sect. 12.5.1.

12.4.2 Determination of Initial Parameters and Incorporation of Constraints

As for every nonlinear optimisation technique, pattern search needs some initial values for the parameters at the first iteration. It is obvious that a good initial guess speeds up the convergence of the algorithm and increases the probability of finding the global optimum. A good initialisation for the stiction estimation approach proposed can be determined as follows:

1. Use the ellipse-fitting method by Choudhury et al. (2006a, 2006b) to yield an initial guess $\hat{S}_0$ of the dead-band plus stick band S.
2. Assuming $S = S_0$, use a simple grid search with a rough step size to get a complete picture $V_N(J)$, whose minimum represents an initial estimate $\hat{J}_0$ for the slip jump J.

This procedure provides a region where the global optimum should lie. Therefore, it is useful to constrain the search of the algorithm within this region, i.e. to specify a lower and upper bound for the parameters J and S, say $J \in [0.8J_0, 1.2J_0]$ and $S \in [0.8S_0, 1.2S_0]$. This helps avoid problems with falling in local minima. The specified regions should be not too narrow, because the curve fitting method estimates "apparent stiction" which may differ from real stiction. However, it should be stressed that the proposed method can also be applied on its own without using the ellipse fitting method or any other technique.

Remark It may occur that a loop with a sticky valve has two (or more) distinct behaviours, e.g. stiction undershoot for one part of the data and stiction overshoot for the other part. An example of such behaviour is illustrated in Choudhury et al. (2005a, 2005b: Fig. 4). It is obvious that the stiction-estimation algorithm will have problems in this case. Therefore, the careful inspection of the PV–OP plot should reveal this situation, and the data parts should be separated.

Remark Decisive for successful optimisation is adding constraints to discriminate parameter combinations (S and J) that drive the stiction model output to zero for all time, which is useless as input for the inner identification process. If such a parameter combination is found, the estimation error is assigned an extremely large value to make sure that the algorithm does not reselect this candidate solution.

12.5 Application and Results

The utility of the proposed stiction-estimation technique is now illustrated in some simulation and industrial studies. Over a dozen industrial case studies (not all presented here) from different industrial fields have demonstrated the wide applicability and accuracy of this method as a useful stiction quantification technique. In all case studies, the number of cycles taken for the analysis lies in the range 10–15.

All computations reported in this study were carried out using MATLAB and Simulink (Release 14). All open-loop and closed-loop simulations were accomplished using Simulink. To perform the optimisation tasks, we employed MATLAB in conjunction with the Genetic Algorithms and Direct Search (GADS) Toolbox, i.e. the function ga and the function patternsearch (with the option "@searchga"). The code for the algorithms used in this work can be provided up to a request from the author. Most reliable results in the simulation and practical cases we studied were achieved by searching the time delay value τ that minimises AIC for an ARMAX model with fixed orders $n = 3$, $m = p = 2$. Computations were performed on a Pentium M 1.70 GHz personal computer.

12.5.1 Simulation Studies

Two of the investigated simulation studies are discussed below in a separate section for each of the loops. The transfer functions and (PI) controllers are shown in Table 12.1. The magnitudes of S and J are specified as a percentage (%) of valve input span and process output span, respectively. Kano's stiction model was used, but the same results can be found by considering Choudhury's model. Results for the ideal case, where no noise is present and the time delay is specified, are not given, as they are not spectacular: the algorithm yields very accurate parameter estimates. Below, a few simulation results are given; however, a broad range of other test conditions, i.e. low-order/high-order, self-regulating/integrating ($\tau/T = 0.1$–10) and different stiction strengths ($J/S = 0$–5), have also been successfully tested. (The results are not shown here due to brevity.) The proposed algorithm produces good estimates of the stiction model parameters with deviations less than 10 % of the actual values.

12.5.1.1 First-Order-Plus-Time Delay Process

This example, which models a concentration loop with slow dynamics and a large time delay, is taken from Choudhury et al. (2005a, 2005b). A zero-mean Gaussian

Table 12.1 Process models and controllers used in the simulation studies

Process type	Process model	Controller
FOPTD	$G_{\mathrm{lp}}(s)=\frac{3\mathrm{e}^{-\tau s}}{1+Ts}$, $\tau=10$, $T=10$	$G_{\mathrm{c}}(s)=0.2(1+\frac{1}{10s})$
ITPTD	$G_{\mathrm{lp}}(s)=\frac{\mathrm{e}^{-\tau s}}{(1+Ts)s}$, $\tau=1$, $T=0.5$	$G_{\mathrm{c}}(s)=0.4(1+\frac{1}{0.2s})$

Table 12.2 Results for the process simulations

FOPTD				ITPTD			
Test conditions ($T_{\mathrm{s}}=1$ s)		Estimated stiction [%]		Test conditions ($T_{\mathrm{s}}=1$ s)		Estimated stiction [%]	
J	S	$\hat{J}$	$\hat{S}$	J	S	$\hat{J}$	$\hat{S}$
2.0	5.00	2.02	5.00	4.0	6.0	4.30	5.66
5.0	5.00	4.97	4.89	3.0	3.0	3.12	3.27
7.0	5.00	6.34	4.86	5.0	3.0	4.90	3.01

noise signal was added to process output. The stiction estimation algorithm with time-delay identification was applied to the "data" obtained for different stiction cases (undershoot, no offset and overshoot). Table 12.2 lists the test conditions and results for each scenario. (T_{s} denotes the sampling time.) It can be concluded that the presented technique accurately quantifies stiction in all scenarios considered. The estimates of the recovered stiction models are very close to the true values.

12.5.1.2 Integrating Process with Time Delay

The stiction estimation for a closed loop with an integrating process with time delay is tried next. As before, several scenarios were considered. The conditions tested and the results achieved are given in Table 12.2. They prove the reliability of the presented method for stiction quantification also for integrating processes. The estimates of the recovered stiction models are close to the true values.

12.5.2 Industrial Case Studies

The objective of this section is to evaluate the proposed framework and techniques on different industrial control loops, including flow control (FC), pressure control (PC), level control (LC) and temperature control (TC). For each loop, the set point (SP), controlled output (PV) and controller output (OP) data were available. The procedure suggested in Sect. 12.4.2 was used for finding good initial stiction parameters. Table 12.3 gives the summary of the results achieved by the application of the methodology described in Sect. 12.3. The results are commented below for each

Table 12.3 Summary of results for the industrial control loops

Loop no.	Loop type	Oscillation and nonlinearity detection results			Initial guess for S	Estimated stiction [%]		CPU time [min]
		r	T_{lp} [s]	*NLI*	S_0	$\hat{J}$	$\hat{S}$	
CHEM25	PC	5.6	192	0.56	1.80	0.59	1.80	21
PAP2	FC	11.2	42.4	0.16	3.00	0.84	3.00	26
CHEM24	FC	2.87	136	0.17	23.00	0.81	22.90	29
POW2	LC	21.4	288	0.55	11.40	1.10	11.47	24
POW4	LC	16.2	237	0.36	4.80	2.49	4.49	34
MIN1	TC	4.0	6940	0.15	1.10	0.96	1.02	18
CHEM70	FC	54.6	3135	–	–	0.04	0.14	20

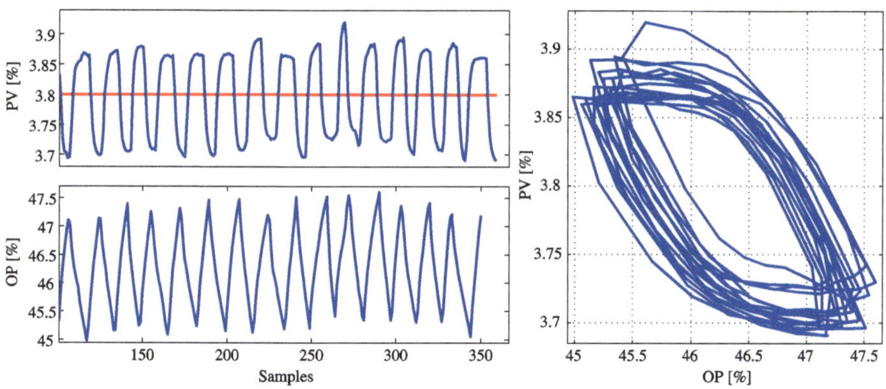

Fig. 12.5 Data from loop CHEM25 (pressure control)

loop. In all examples, the linear part was approximated by an ARMAX$(3, 2, 2, \hat{\tau})$ model with $\hat{\tau}$ determined for minimum AIC. Table 12.3 also contains values of the oscillation regularity factor r and period T_{lp} (Sect. 9.7) and the nonlinearity index (*NLI*) (Sect. 10.2.2) in relation with ellipse fitting of the PV–OP plot (Sect. 11.7).

It is important to see that, when the time delay is known, the computation time reduces significantly, by up to 90 %. This means that a major portion of the computation burden results from time delay estimation. Therefore, an efficient algorithm for this task is worth consideration in future research.

12.5.2.1 Pressure Control Loop

This is a pressure control loop in a refinery. Data from this loop were analysed by the ellipse-fitting method to give a stiction band $S_0 = 1.8$; see Fig. 12.5. The presence of stiction is confirmed by the index value $NLI = 0.56$. Setting this valve as S_0 and

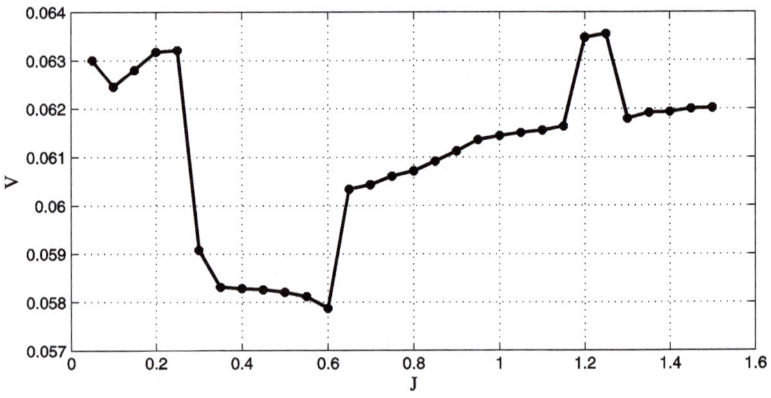

Fig. 12.6 Objective function V vs. J for loop 1 setting $S = 1.8$ for loop CHEM25 (pressure control)

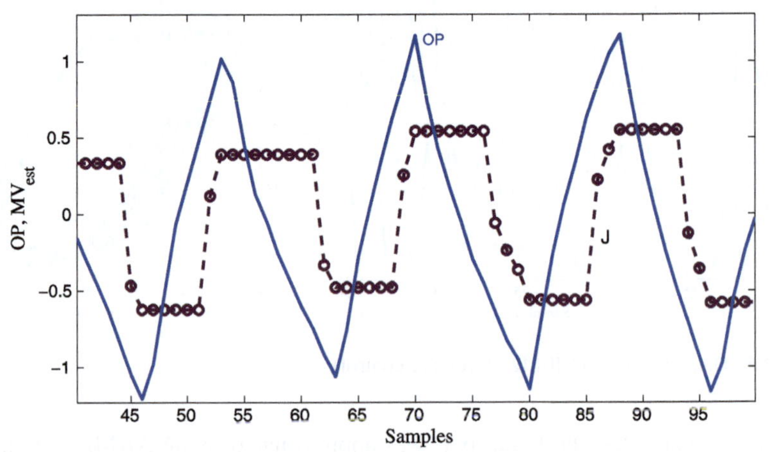

Fig. 12.7 Time trends (window) of controller output (OP) and estimated valve position (MV) for loop CHEM25 (pressure control)

varying J yields $V(J)$ shown in Fig. 12.6. It can be seen that the minimum lies in the neighbourhood of $J_0 = 0.60$. Using these initial parameter values, the proposed algorithm was run with the constraints $J \in [0.25, 1.0]$ and $S \in [1.5, 2.2]$ to give $J = 0.59$ and $S = 1.80$.

Specifically for this loop, the estimated inner signal MV is illustrated in Fig. 12.7. This figure clearly indicates both dead-band plus stick band and slip jump effects. The latter is large and visible, especially when the valve is moving in downward and upward directions (J is marked in the figure). Figure 12.8 shows how well the estimated model fits the measured data. The linear model identified has the polynomials

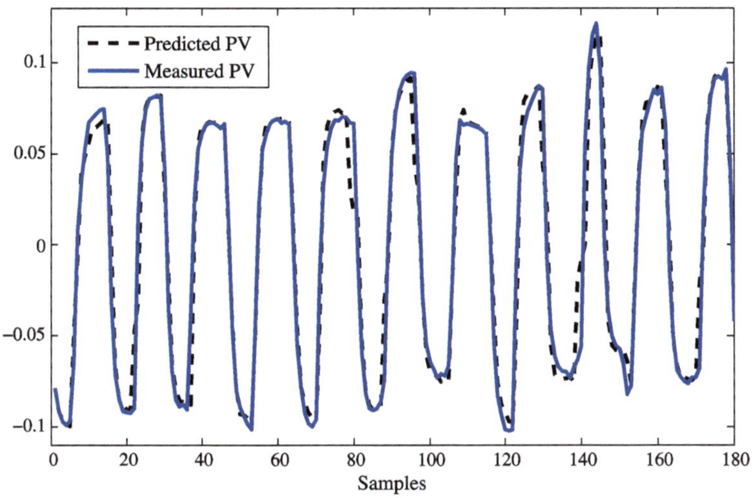

Fig. 12.8 Measured vs. predicted PV data for loop CHEM25 (pressure control)

(Eq. 2.1, $\tau = 2$)

$$A(q) = 1 - 1.718q^{-1} + 0.9234q^{-2} - 0.1861q^{-3},$$
$$B(q) = 0.07152 - 0.0682q^{-1},$$
$$C(q) = 1 - 0.9129q^{-1} + 0.1184q^{-2}.$$

If we use a specified linear model structure, i.e. ARMAX(2, 2, 1, 1), and run the algorithm again, we get the estimates $J = 0.60$ and $S = 1.87$, which are close to the ones obtained above. The linear model has now been estimated to be (Eq. 2.1, $\tau = 1$)

$$A(q) = 1 - 0.7181q^{-1} + 0.1256q^{-2}, \qquad B(q) = 0.00659 + 0.0658q^{-1},$$
$$C(q) = 1 + 0.2376q^{-1}.$$

The total model shows similar prediction quality, but the computation time reduces to 1.7 min. This is to show the price we pay for not a priori knowing the time delay.

12.5.2.2 Flow Control Loop

This loop with valve stiction is taken from Horch (2007: FC525). Figure 12.9 illustrates the data considered for stiction quantification. The application of the approach proposed in this chapter led to the estimates $J = 0.84$ and $S = 3.0$.

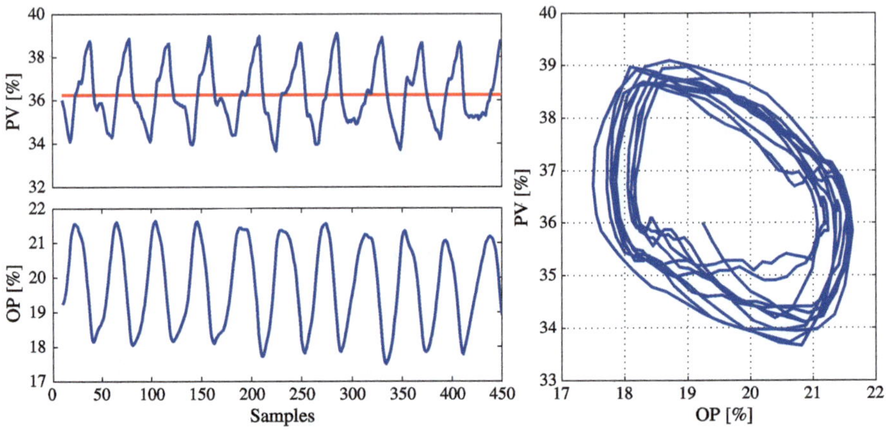

Fig. 12.9 Data from loop PAP2 (flow control)

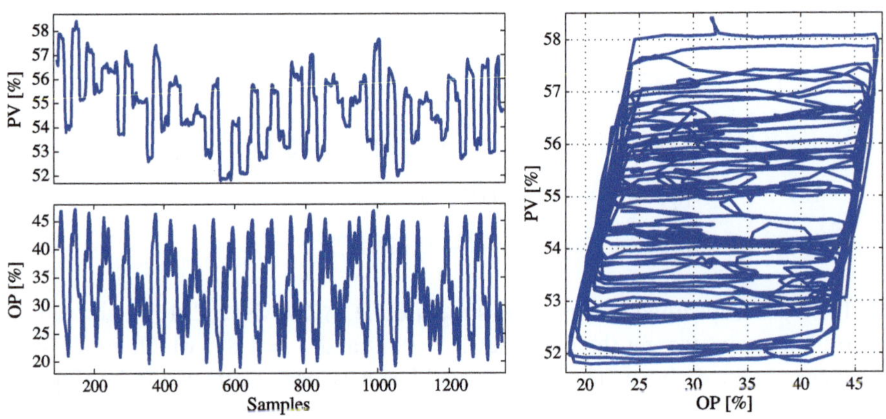

Fig. 12.10 Data from loop CHEM24 (flow control)

12.5.2.3 Flow Control Loop with Set-Point Changes

This flow control loop has excessive stiction. It is an inner loop of a cascade control system and thus is subject to rapid set point changes. The plant data are shown in Fig. 12.10. The PV–OP plot shows a shape (parallelogram) very similar to Fig. 10.5. The fact that the oscillations in OP and PV are varying in amplitude and time period makes stiction detection and quantification challenging. The stiction index clearly signals the presence of stiction in the loop. The stiction parameter estimates were found to be $J = 0.81$ and $S = 22.9$, which can be confirmed by a look at the PV–OP plot in the right side of Fig. 12.10.

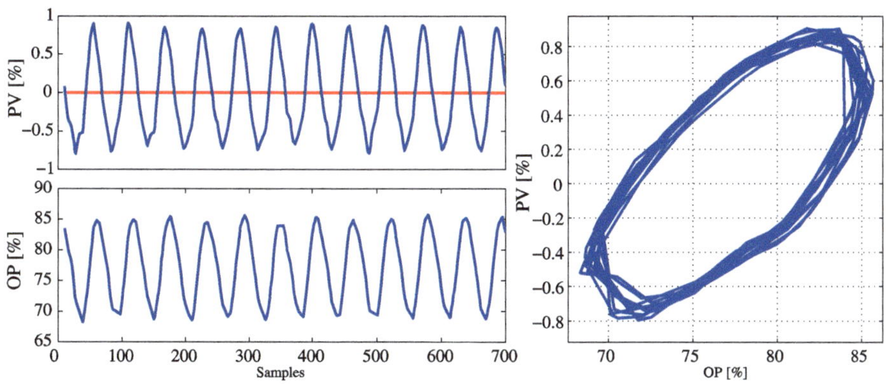

Fig. 12.11 Data from loop POW2 (level control)

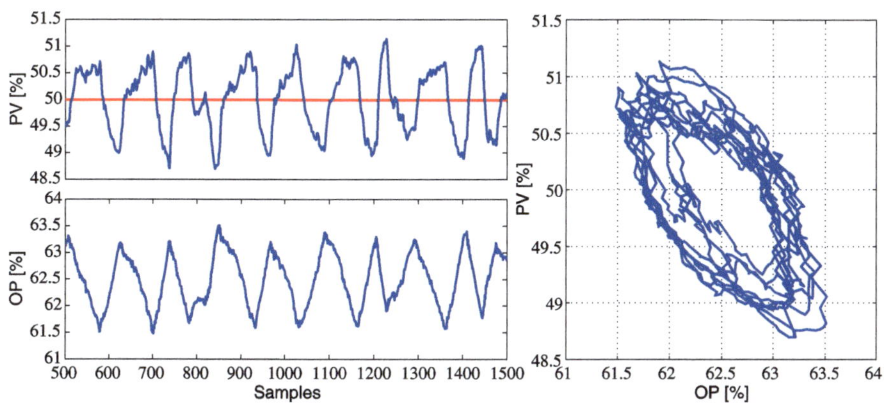

Fig. 12.12 Data from loop POW4 (level control)

12.5.2.4 Level Control Loop

This example represents a level control loop in a power plant; see Fig. 12.11. Data from this loop were already analysed in Choudhury et al. (2005a, 2005b: Fig. 3) and Choudhury et al. (2006a, 2006b: Fig. 4), and the ellipse-fitting method was applied to give a stiction band $S_0 = 11.4$. The estimated stiction parameters using the technique proposed here were $J = 1.10$ and $S = 11.47$. These results are in good agreement with the data plots given in the aforementioned literature, where the MV trend and MV–OP plot are also shown.

12.5.2.5 Level Control Loop

The data for this level control loop was also obtained from a power plant and are illustrated in Fig. 12.12. The dead-band plus stick band was estimated to be $S_0 = 4.8$

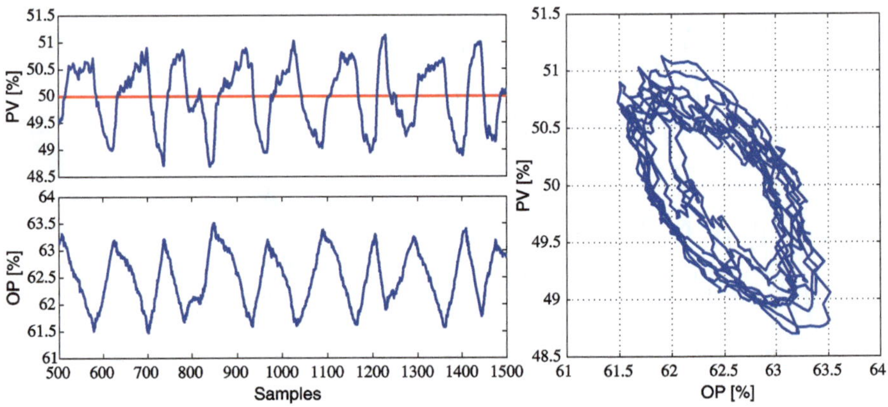

Fig. 12.13 Data from loop MIN1 (temperature control)

by applying the ellipse-fitting method. The stiction parameters in this loop were estimated to be $J = 2.49$ and $S = 4.49$.

12.5.2.6 Temperature Control Loop

This loop on a furnace feed dryer system was also considered in Choudhury et al. (2005a, 2005b: Fig. 6) and Choudhury et al. (2006a, 2006b: Fig. 10). Using the data shown in Fig. 12.13, the proposed estimation algorithm leads to the parameter estimates $J = 0.96$ and $S = 1.02$ (very small undershoot), which are in good agreement with the data plots given in Choudhury et al. (2005a, 2005b) and Choudhury et al. (2006a, 2006b), where the MV trend and MV–OP plot are also shown. The initial value for S was $S_0 = 1.1$, determined using the ellipse-fitting method.

12.5.2.7 Flow Control Loop with External Disturbances

The purpose of this example is to show that the presented approach can be principally used to detect stiction. A flow control loop is considered, for which the stiction indices indicate no stiction; see Fig. 12.14. It is also known that this loop suffers from external disturbances. This is confirmed by the stiction estimation algorithm. The latter yields negligible values of J and S. Recall that dead-band, i.e. $J = 0$ and $S > 0$, cannot induce oscillation for *self-regulating* processes.

12.6 Detection of Multiple Loop Faults

Several different causes can have the responsibility of poor performance in a control loop, more common being incorrect tuning of controllers, presence of stiction in actuators and external disturbances. It is of crucial importance to be able to distinguish

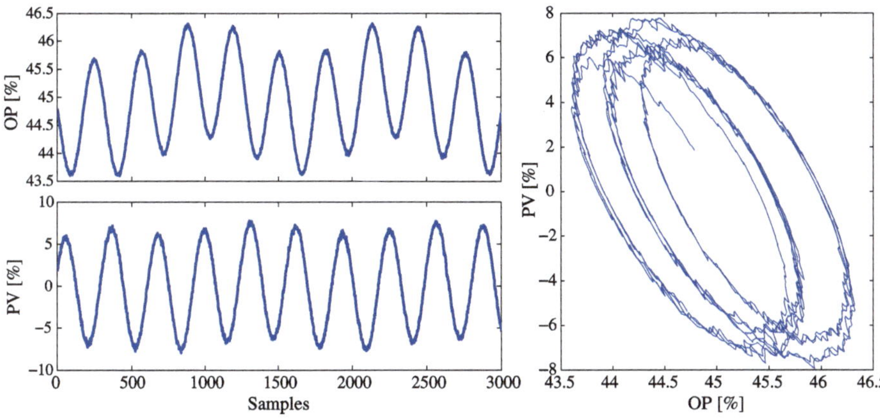

Fig. 12.14 Data from loop CHEM70 (flow control)

correctly the cause in order to take the appropriate action. However, most of the monitoring techniques target only one specific problem in a control loop with the assumption that other problems do not occur simultaneously. Obviously, different problems can produce similar symptoms, thus triggering the performance monitoring algorithm to alarm.

In this section, we focus on control loops that have been detected to oscillate. In the preceding part of this chapter, emphasis was on detecting and discrimination the main (distinct) source of oscillation: stiction on the one hand and aggressive controller or external disturbance on the other hand. If the loop suffers from two or more faults simultaneously, the techniques presented so far will fail in most cases. At best, some techniques may indicate that the one root cause is more probable than the other.

In this section, a new comprehensive oscillation diagnosis is proposed to detect possibly occurring multiple faults in valve-controlled loops; see Procedure 12.3 and Fig. 12.15. The diagnostic procedure can be described as follows. It is a straightforward extension of the stiction estimation technique based on Hammerstein modelling (Sect. 12.3).

Procedure 12.3 Complete oscillation diagnosis based on Hammerstein modelling (Fig. 12.15).

1. Detect the presence of oscillations using one of the techniques in Chap. 9.
2. Detect the presence of stiction using one of the techniques in Chap. 11 (optional).
3. Identify a Hammerstein model and quantify stiction using the technique proposed in Sect. 12.3.
4. Use the identified Hammerstein model to estimate the PV trend $\hat{y}(k)$. The signal $\hat{d}(k) = y(k) - \hat{y}(k)$ gives an estimate of the external disturbances. If $\hat{d}(k)$ is oscillatory, which can be quantified by an oscillation index (Chap. 9), the oscillation is due to external disturbances.

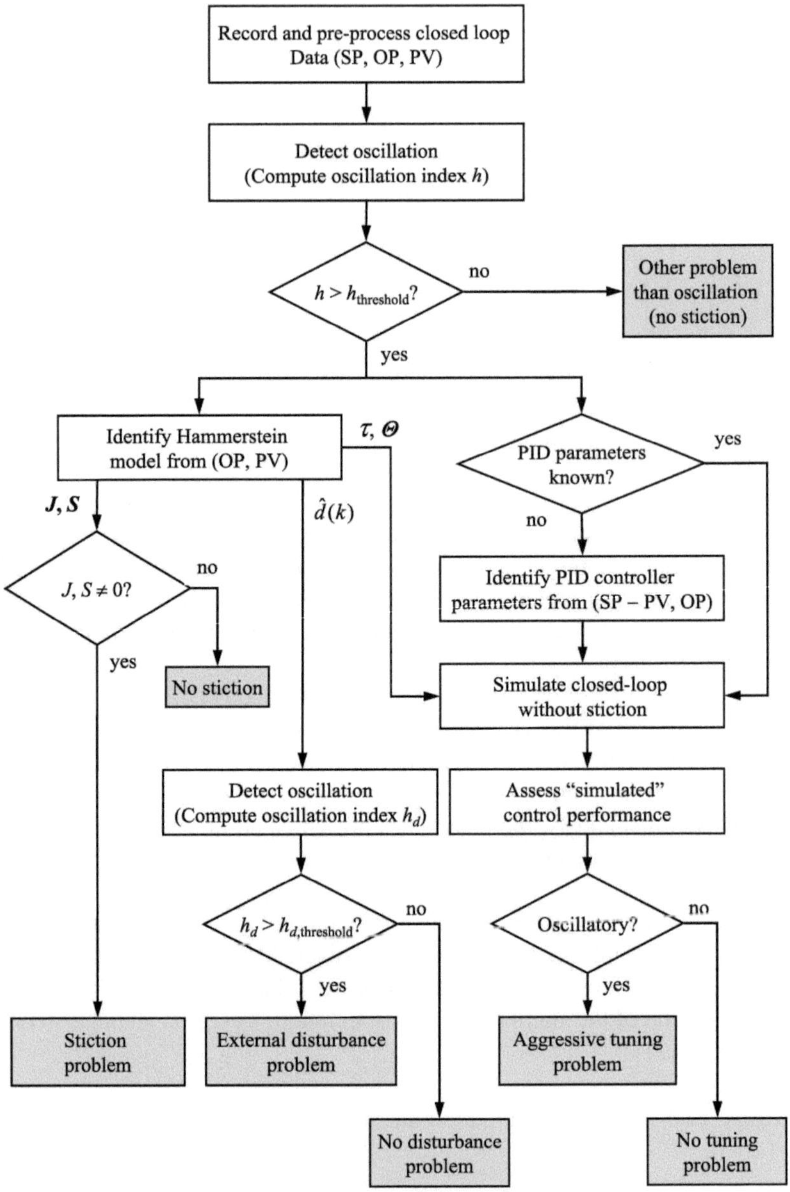

Fig. 12.15 Flow chart of the oscillation detection and diagnosis procedure

5. Estimate the controller model (when not known) based on measured data SP–PV and OP, e.g., using the arx function from MATLAB identification Toolbox. Calculate the controller parameters depending on the controller type and representation.

6. Use the identified linear model and the (estimated) controller to simulate the closed loop without stiction. Apply, for instance, step changes on the loop and assess the controller, e.g. using the step-response-based assessment method in Sect. 5.2. If the assessment indicates oscillatory/aggressive behaviour, aggressive controller tuning contributes to the loop oscillation.

Step 5 is introduced to account for the fact that sometimes either the settings of the installed controller are not known, or some components have been added that affect the controller characteristics. In the latter case, it might be useful to use operating data for identifying a controller model with higher order than the assumed structure. If, for instance, a PI controller of the discrete form

$$G_{\mathrm{PI}}(q^{-1}) = \frac{K_1 + K_2 q^{-1}}{1 - q^{-1}} \tag{12.14}$$

is adopted, the controller model can be identified using the MATLAB command arx($[OP, \mathrm{SP} - -\mathrm{PV}, T_s], [1, 2, 0]$). From Eq. 12.14, the controller parameters are obtained as (based on the forward difference approximation)

$$K_c = -K_2, \qquad T_I = -\frac{K_2 T_s}{K_1 + K_2}. \tag{12.15}$$

Even in cases where the controller type is unknown, a controller model can be identified from measured OP/SP–PV data. In such a situation, one can consider subspace identification [N4SID] for the controller-model estimation, since it provides an automatic selection of the model order. Readers should refer to Bezergianni and Georgakis (2003) for a collection of controller discrete transfer functions and more details about controller model estimation based on subspace identification.

In the following sections, the oscillation diagnosis technique presented above is applied on simulated and real data. The performance of the method is checked in terms of the detection of multiple faults: valve stiction, aggressive tuning and oscillatory external disturbances. The estimates of stiction parameters J and S, of the PI controller model K_c and T_I and of the external disturbance are compared with the real values, when known.

12.6.1 Simulation Examples

We illustrate the proposed technique on the FOPTD process with stiction and PI controller considered in Sect. 12.5.1.1. However, we additionally superpose PV with an external sinusoidal disturbance, i.e. $d = 0.1\sin(0.02t)$.

If we now apply the proposed technique for oscillation diagnosis, the results in Fig. 12.16 are achieved. The first subplot (top/left) shows the simulated and predicted PV for the estimated stiction parameters $J = 2.02$ and $S = 5.0$ (nearly identical with the real ones). In the second subplot (top/right), the simulated and estimated

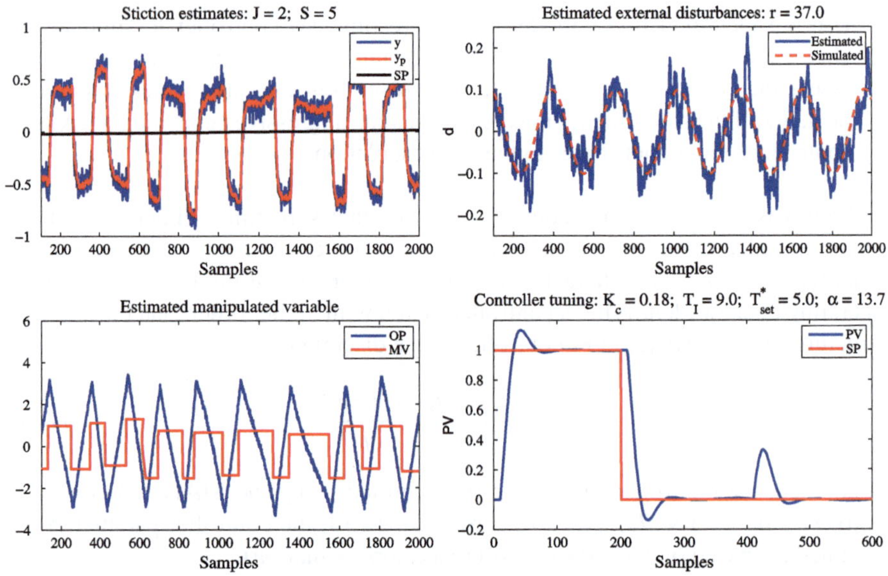

Fig. 12.16 Results of the oscillation diagnosis for the FOPTD example with stiction and external disturbance

external disturbance trends are given. The oscillation indices (regularity factor r) clearly indicate this. The third subplot (bottom/left) illustrates the OP and estimated MV signals, indicating stiction patterns. The last subplot shows the responses of the closed loop without stiction to set-point and input steps. The controller parameters have been estimated as $K_c = 0.18$ and $T_I = 9.0$ s, which are close to the used settings. The resulting step response and its assessment are also shown in the figure. The obtained estimated external disturbances and step responses are close to those used in the simulation. The value $T^*_{set} = 5.0$ indicates acceptable (but not optimal) controller (deterministic) performance; see Sect. 5.2. Overall, both faults and the controller settings have been correctly detected by the proposed diagnosis method.

In a second scenario, we change the proportional gain to $K_c = 0.45$ to get aggressive controller behaviour. Also in this case, the achieved estimates of the controller parameters and of the external disturbance are close to the real ones; see Fig. 12.17. Moreover, the aggressive controller tuning has been recognised by considering the values of the normalised settling time $T^*_{set} = 18.7$ and the overshoot $\alpha = 63.4$. In summary, all three faults introduced have been correctly detected by the proposed oscillation diagnosis technique.

12.6.2 Industrial Examples

The objective of this section is to demonstrate the application of the proposed method on some selected industrial control loop data. Date sets, containing the set

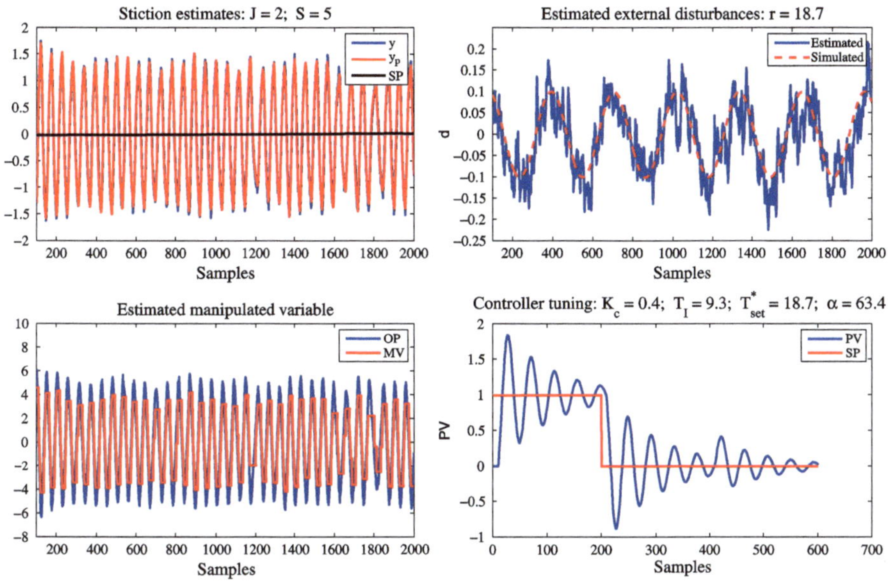

Fig. 12.17 Results of the oscillation diagnosis for the FOPTD example with stiction, external disturbance and aggressive tuning

point (SP), controller output (OP) and process variable (PV), are used for the diagnosis. Such data sets are usually available in the industrial practice. Controller tuning parameters and particularly time delays are, however, often unknown. From the industrial loops listed in Appendix C, the controller settings for the loops CHEM18–28 and CHEM32–39 were provided. For the loops investigated in the following, we now assume a Box–Jenkins model BJ$(2, 2, 2, 2, \hat{\tau})$ with automatically estimated time delay $\hat{\tau}$.

12.6.2.1 Flow Control Loop with Stiction

The flow control loop data considered show clear stiction pattern; see Fig. 12.18 (left side). The stiction index value $\eta_{\text{stic}} = 0.95$ confirms the presence of stiction in the loop. The stiction parameter estimates are $\hat{J} = 3.9$ and $\hat{S} = 26.8$. No regular external oscillation is detected ($r = 0.7$). The identification of the controller model gives the values $\hat{K}_c = 1.2$ and $\hat{T}_I = 7.3$ s. The simulation of closed loop using the estimated linear model and controller parameters lead to step responses shown in the figure (fourth subplot). The corresponding performance belongs to the acceptable performance class, as indicated by the normalised settling time value $T^*_{\text{set}} = 9.3$. Hence, this loop suffers mainly from excessive stiction in the control valve, which should be repaired at the next shutdown of the plant.

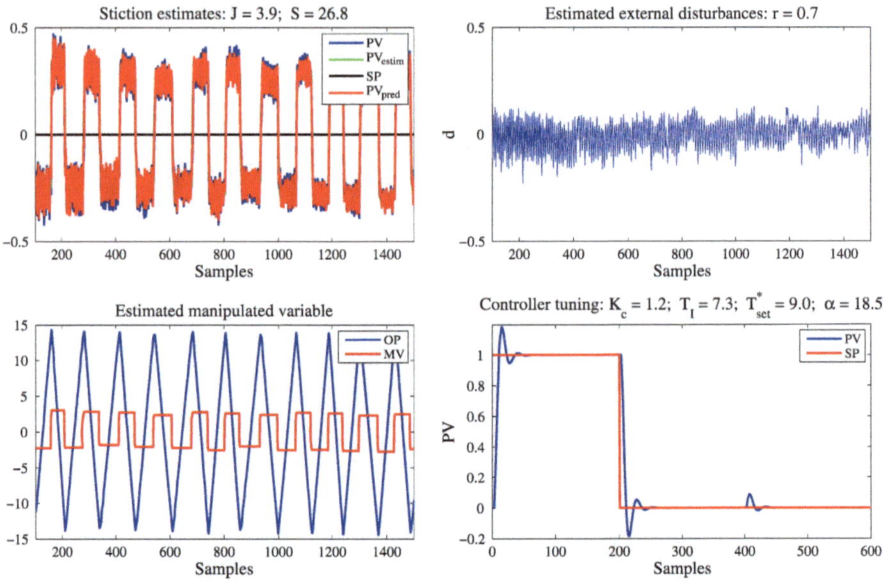

Fig. 12.18 Results of the oscillation diagnosis for Loop CHEM23 (flow control)

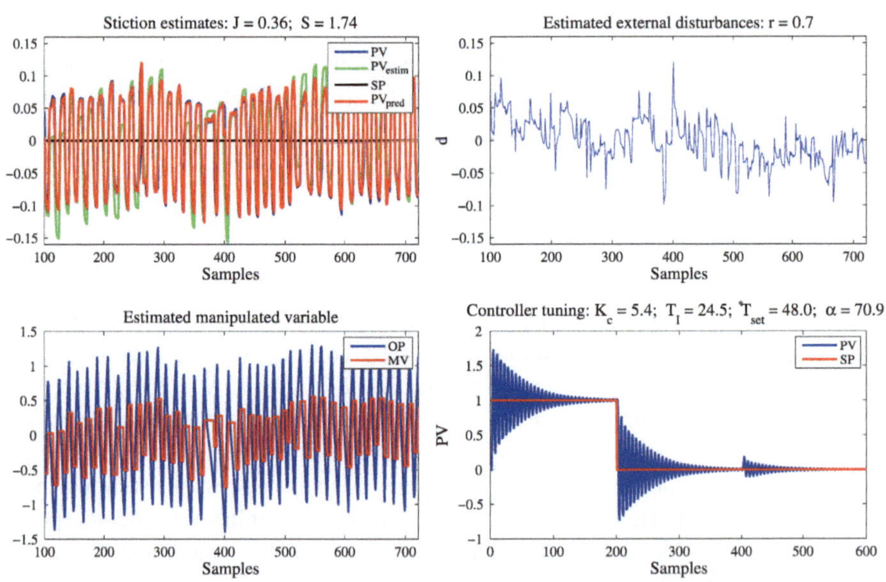

Fig. 12.19 Results of the oscillation diagnosis for Loop CHEM25 (pressure control)

12.6.2.2 Pressure Control Loop with Stiction and Aggressive Controller Tuning

The diagnosis is performed on the data obtained from the pressure control loop CHEM25. It was known a priori that the control valve in this control loop contained stiction. The diagnosis results for this control loop are shown in Fig. 12.19.

The oscillation detection algorithm applied to the tracking error signal d leads to a regularity value of 0.45, indicating no significant external oscillations. The stiction parameter estimates $\hat{J} = 0.36$ and $\hat{S} = 1.74$ indicate the presence of quantifiable stiction, which was causing the sustained oscillations in OP and PV signals. The step responses obtained from closed-loop simulations of identified process and controller models are excessively oscillatory. The estimated controller parameters are $\hat{K}_c = 5.4$ and $\hat{T}_I = 24.5$ s, which are close to the known real values $K_c = 6.6$ and $T_I = 30.0$ s. Hence, it can be concluded that stiction in the control valve and aggressive controller tuning are responsible for the oscillatory trend in this control loop. Therefore, repairing the valve and changing the PI tuning parameters would solve the oscillation problem.

12.6.2.3 Level Control Loop with Stiction, External Disturbance and Aggressive Controller Tuning

Data from the level control loop in Sect. 12.5.2.4 are investigated again using the proposed discrimination procedure. This leads to slightly different estimates of stiction parameters $\hat{J} = 3.75$ and $\hat{S} = 10.7$. Note that these values (particularly J) are different from those estimated using an ARMAX model in Sect. 12.5.2.4. The diagnosis results are shown in Fig. 12.20. It can be seen that loop does not only suffer from stiction but also from aggressive controller tuning, leading to oscillatory step responses, as indicated by the values of the normalised settling time $T^*_{set} = 97.5$ and the overshoot $\alpha = 85.2$. There is also an external oscillating disturbance acting on the loop, as indicated by the oscillation index $r = 1.9$. It can be concluded that this loop is oscillating due to three faults: stiction, oscillatory disturbance and aggressive controller tuning. To completely solve the oscillation problem of this loop, it is not only necessary to repair the valve and change the controller tuning, but also to find out the source of the external disturbance and eliminate it.

For this loop, the MV (or valve position) data were available and can now be compared with the estimated MV values from the identified stiction model; see Fig. 12.21. One the one hand, the figure shows that the estimated stiction model behaviour is close to the measured MV. One the other hand, it can be seen that both two-parameter stiction models can successfully reproduce the stiction behaviour of the valve.

However, it is interesting to note that the actual industrial MV moves around, while the model predicted MV says it should be stuck. This is especially true in the valleys of the MV oscillation in the lower signal part. This means that the data-driven stiction model is failing to capture some physics. One possible reason is a

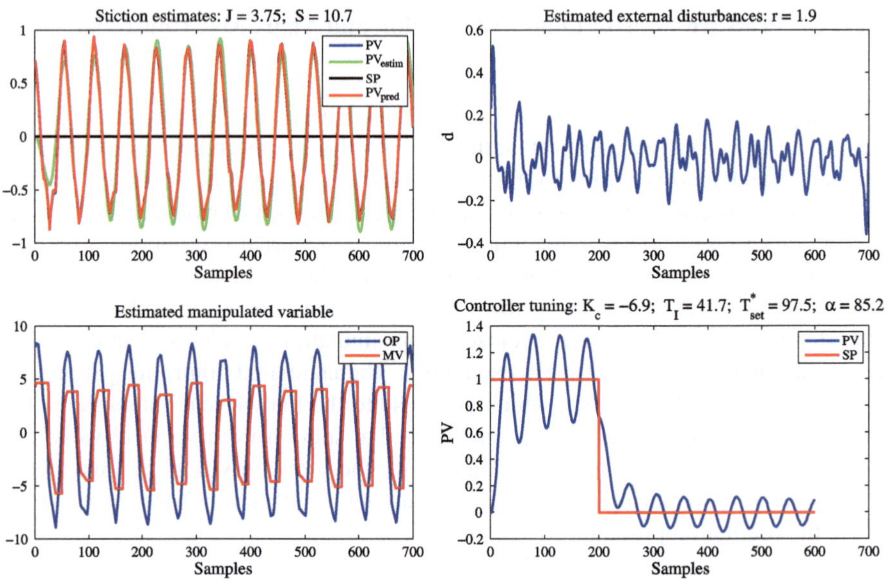

Fig. 12.20 Results of the oscillation diagnosis for Loop POW2 (level control)

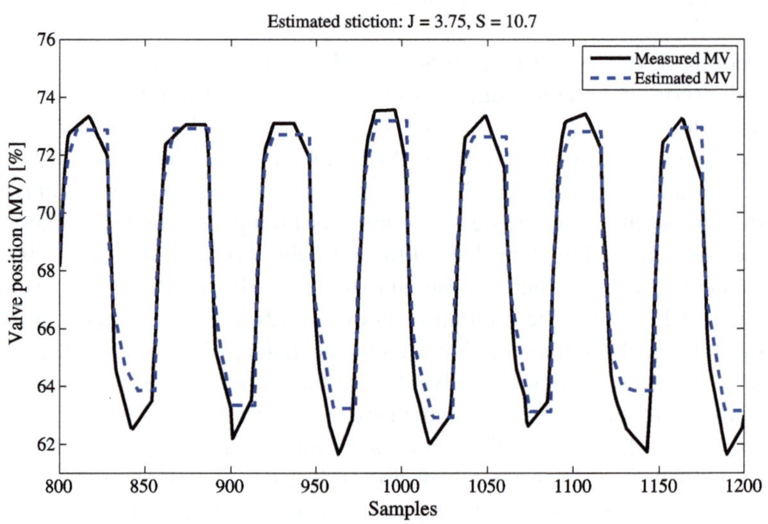

Fig. 12.21 Measured vs. predicted manipulated variable for loop POW2 (level control)

more complicated friction behaviour. For instance, in addition to the static and dynamic friction, the friction in the moving phase may have regimes with different gradients in the velocity versus friction force characteristic. Moreover, it seems that the valve shows some asymmetrical behaviour since it sometimes sticks in one direction but not in the other (Choudhury et al. 2009).

12.7 Summary and Conclusions

A novel procedure for quantifying valve stiction in control loops based on two-stage identification has been presented in this chapter. The proposed approach uses PV and OP signals to estimate the parameters of a Hammerstein system, consisting of a connection of a two-parameter stiction model and a linear low-order process model. A pattern search or genetic algorithm subordinated by a least-squares estimator was proposed for the parameter identification. This particularly yields a quantification of the stiction, i.e. estimating the parameters dead-band plus stick band (S) and slip jump (J), thus enabling one to estimate time trends of the valve position (MV). Needless to say that the method can also be applied in the case of one-parameter stiction models.

The results on different processes under a range of conditions—low-order/high-order, self-regulating/integrating, different controller settings and measurement noise, different stiction levels—show that the proposed optimisation can provide stiction-model parameter estimates accurately and reliably. The stiction quantification technique has been successfully demonstrated on two simulation case studies and on many data sets from different industrial control loops. Whenever possible, it is helpful to have a good estimate of the time delay because of its effect on the accuracy of the estimated models.

The relatively high CPU time required for the identification process (particularly when the time delay is simultaneously estimated) is not critical, as the analysis is performed offline. Also, this work is inexpensive compared to the savings in experimentation with the process or in down time costs when invasive methods for stiction quantification would be applied. A faster algorithm for time delay estimation and a more efficient implementation of the algorithms, e.g. as C-code, should significantly accelerate the computation.

The stiction estimation method has also extended to a complete oscillation diagnosis approach, which allows the determination of the cause(s) of oscillating loop behaviour. This could be stiction in the control valve, poorly tuned controller or the presence of an external perturbation. The unique feature of the technique is that it can also detect multiple loop faults when present. The diagnosis approach can be extended easily to other nonlinear valve problems or to nonlinear final elements to detect if they work properly.

Part III
Performance Improvement

Chapter 13
Performance Monitoring and Improvement Strategies

The final and most challenging objective of applying performance assessment methods and indices should always be to suggest measures to improve the control or process/plant performance. Even in the case where the loop is found to work at acceptable performance level, it is useful to know how the performance can be improved to attain top level. In this chapter, possible performance improvement measures are briefly discussed (Sect. 13.1). Some paradigms and strategies for monitoring the performance of complex process-control systems are introduced in Sect. 13.2. Section 13.3 presents a comprehensive control-performance assessment procedure combining different methods described throughout the previous chapters of the book.

13.1 Performance Improvement Measures

Three categories of methods for performance improvement can be distinguished:

- *Controller Re-tuning*. One of the direct results of a CPM procedure is to decide whether the running controller should be re-tuned to achieve improved loop performance, compared to that performance of the selected benchmark. Re-tuning is usually the easiest and cheapest way to improve the performance of control loops. Traditional controller tuning is usually undertaken based on active experiments, e.g. step responses, with the plant. In Chap. 14, non-invasive methods for controller re-tuning will be presented.
- *Control System Re-design*. When it is predicted that controller re-tuning does not help achieve the desired performance, a complete re-design of the controller including its structure may be required. As controller re-design is a complicated and thus time-consuming task, the economical benefits of this measure should always be quantified in advance. Beyond the commonly implemented PID controllers, the introduction of specialised or advanced strategies is then needed. These include *anti-windup* schemes (Åström and Wittenmark 1997; Glattfelder and Schaufelberger 2003), *feedforward control* (Seborg et al. 2004; Åström and Hägglund 2006), *cascade control* (Shinskey 1996; Visioli 2006), *multi-loop and*

M. Jelali, *Control Performance Management in Industrial Automation*,
Advances in Industrial Control, DOI 10.1007/978-1-4471-4546-2_13,

multivariable control (Skogestad and Postlethwaite 1996; Goodwin et al. 2001; Lunze 2008), *time delay compensation* techniques (SPC, IMC, MPC) (Smith 1957; Morari and Zafiriou 1989; Camacho and Bordons 1999; Maciejowski 2001) and *gain scheduling* or *adaptive control* (Rugh 1991; Isermann et al. 1992; Åström and Wittenmark 1995).

- *Maintenance and Process/Plant Modifications*. Abnormal process operation owing to equipment problems (wear, fouling, etc.) or instrumentation malfunctions (stiction in control valves, faulty sensors, etc.) should always be handled within the framework of plant inspection and maintenance. Also, re-selection or replacement of sensors or actuators should be checked.

  In some cases, performance improvement will only be attained by changing the process flow, e.g. adding a bypass, or changing the sensor location. In other cases, some part of the random shocks, $\varepsilon(k)$, is due to measurement error and can be reduced by a more precise sensor (Stanfelj et al. 1993). We found many processes in the metal industry, where sensors are installed up to 7 m or more away from the process itself.

  In cases where control input saturation is too frequently detected, it may be useful to check actuator sizing. In other cases, some quality deviations can only be eliminated by introducing additional actuators. An example from the metal processing industry is strip flatness control, which requires certain actuator for each type of control error to fight with; see Sect. 16.3.2. When the installed actuator system does not provide suitable components for controlling any flatness error type, the flatness controller, no matter how it is tuned, will not have the chance to correct for these errors.

  The disturbance structure and variance could be modified by introducing a well-mixed inventory upstream of the process considered. This would attenuate the variance of disturbance variables like feed composition and temperature. In metal processing, the variance of key quality variables in every production stage has to be minimised to ensure final product quality. For cold rolling, it is desirable to have uniform strip entry thickness and entry profile from hot rolling mill. For hot-dip galvanising, it is essential to get minimised strip flatness errors from cold rolling mill. Good transfer properties of the aforementioned quality features are necessary because it is then difficult to affect them in the down-stream stage. Specifically, thickness profile cannot be changed much in cold rolling, and flatness errors produced in cold rolling cannot be removed in hot-dip galvanising, and thus non-uniform zinc-layer thickness may result. This would affect the corrosion-resistance of the steel required by the automotive industry.

  It is worth noting that process changes can be costly and that some approaches requiring plant or sensor changes can be implemented only during infrequent plant shutdowns. Therefore, every effort should be made to achieve the best performance from the existing system, minimising the need for process modifications.

These performance-improvement measures are ingredients of the integrated framework illustrated in Fig. 13.1.

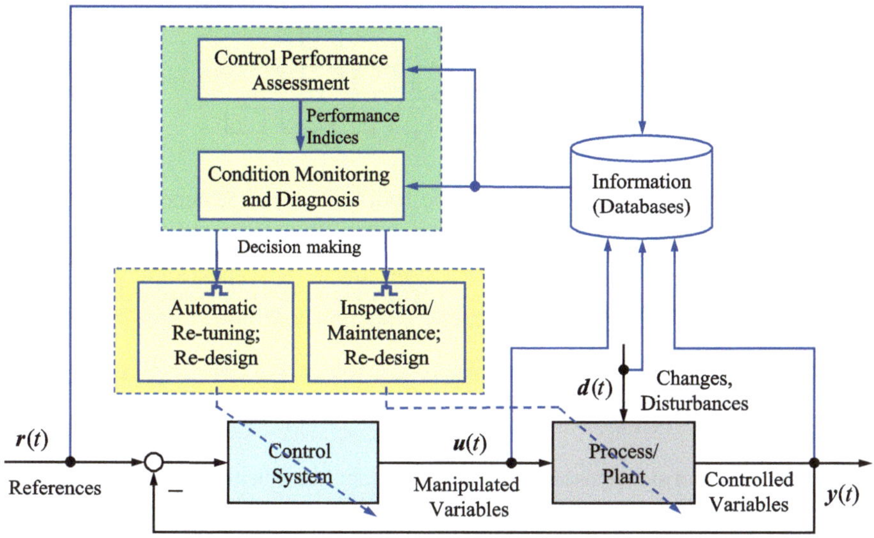

Fig. 13.1 Proposed framework for control-performance-monitoring-based optimisation of control loops

13.2 Loop Monitoring Paradigms

Any modern manufacturing facility, such as an oil refinery, a chemical plant, a paper mill, a power plant or a rolling mill, consists of many hundreds and sometimes thousands of control loops. Most manufacturing plants are automated, and the system components can be organised into the following hierarchy:

- Main process.
- Device level (sensors, actuators and communication devices).
- Basic (secondary) control level (mostly PID).
- Main (primary) control level (multivariable control, MPC, etc.).
- Real-time optimisation level (plant-wide and individual plant optimisation, parameter estimation, supervisory control, data reconciliation, etc.).
- Planning and scheduling level (demand forecasting, raw materials and product planning/scheduling, etc.).

Therefore, it is nearly impossible to monitor the performance of more than a few of the most critical control loops without some systematic/formalised procedures and automatic assessment tools.

13.2.1 Bottom-Up and Top-Down Approaches

The control loops are usually located in different but clearly defined levels of hierarchy. Therefore, the first task of a monitoring strategy is to decide whether a

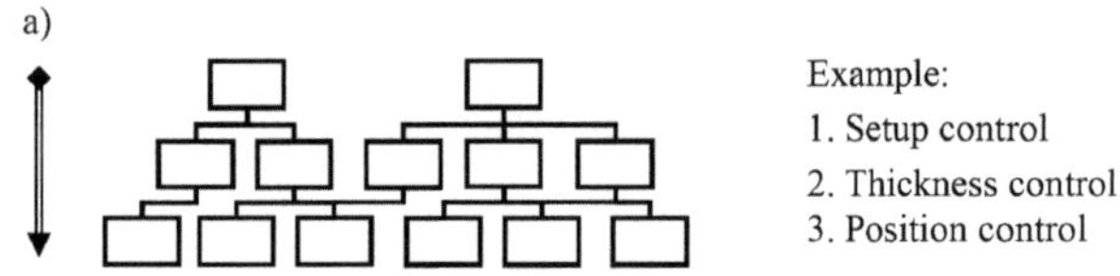

Fig. 13.2 Principle of top-down performance assessment strategy

bottom-up or top-down strategy should be followed. It is not necessary to further diagnose a plant component and controller when its performance is entirely satisfactory with respect to safety, process-equipment service factor and plant profit. Only those control loops which are not adequately performing and offer potential benefit are considered in the subsequent diagnostic steps.

13.2.1.1 Top-down Approach

Control-loop performance assessment as important ingredient of maintenance work should be tied more closely to economic aspects. Key loops that are economically critical require top priority. These loops usually lie at higher level of hierarchy (i.e. supervisory or primary loops). An effective strategy for CPM could thus start with assessing the performance of higher-level control loops and then move to lower-level loops (Fig. 13.2). This means that monitoring and maintenance effort is first focused on loops exhibiting performance problems in higher levels, which have direct impact on the economical performance of the plant. The number of control loops that need to be investigated can be significantly reduced in a first run. The benchmarking problem becomes not only manageable but also meaningful.

Deterioration in a performance index for a higher level could indicate a more severe problem in a lower hierarchical level. In fact, the performance indices for the higher levels could help identify problematic loops, the repair of which provides a better return on maintenance effort. This means, for instance, introducing time-delay compensation, adding appropriate strategies for disturbance rejection and multivariable control to remedy loop interaction.

13.2.1.2 Bottom-Up Approach

Many loop problems appear in the lowest level of the automation hierarchy, such as excessive sensor noise or the very common problem of valve hysteresis and stiction. The problems first manifest themselves in drift, performance deterioration and the indices for the corresponding levels. As problems diffuse upward, they affect performance indices at higher levels. Therefore, there is also enough incentive to follow a bottom-up strategy for the assessment of complex control systems (Fig. 13.3). This means starting with isolating low-level loop problems and then moving towards resolving performance faults at higher levels. However, it should be recognised that the number of control loops remarkably increases at lower levels.

When applying a bottom-up CPM strategy, it is important to determine the influence of each lower loop on the performance of the upper loops and thus on the final

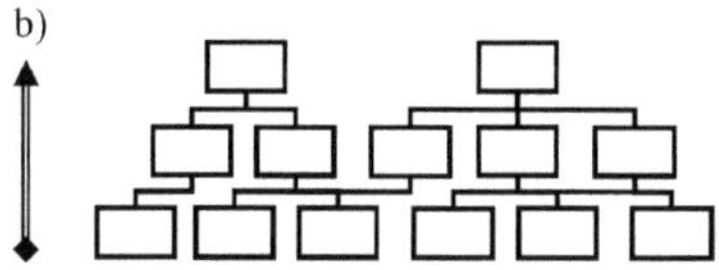

Fig. 13.3 Principle of bottom-up performance assessment strategy

product. For instance, a translation of the variance reduction of basic loops into the primary loops is desirable. For this purpose, the variability matrix introduced by Farenzena and Trierweiler (2006) may be considered. It is a matrix (similar to a static gain matrix) where the elements quantify how the change in the variance of the control loop produces a change in the variance of the main loop. Note, however, that it is not easy to identify the variability matrix in practice, since it requires some experimentation with the loops, or expensive modelling work.

13.2.2 Loop Prioritisation and Ranking

Within a hierarchical level, the question is now how to determine which loops are more important and thus should be assessed first. This problem of so-called loop *prioritisation* involves the use of control performance-assessment algorithms to determine the *critical* loops in large multivariable interacting processes. By focusing on the benchmarking of loops with the lowest performance indices the effort needed can be significantly reduced. This also promises to provide the biggest payback, particularly if a higher economic priority is assigned to the considered loops. The economic priority indicates the relative importance of each loop to others in the plant. If two loops have the same performance index but different economic priorities, the one with the higher priority will bubble to the top of the ranking.

13.2.3 Relationship to Economical Benefits

CPM techniques showed widespread use in the last years because improved performance of control loops has been recognised to have positive economic effects. However, it is usually difficult to quantify these benefits.

The analysis of variance, for instance, can help quantify how much the performance of the control loop can be improved, which can be translated in terms of increased product quality and/or material/energy consumption. Principally, the variance can be transformed into an economic measure by multiplying it by a weighting factor w:

$$b_{\text{eco}}[€] = w\sigma^2. \qquad (13.1)$$

However, substantial effort is required to obtain weighting factors for specific control loops: a deep understanding of the process and information about the interactions between the control loops are needed. This considers the fact that each loop usually contributes in a complicated way to the overall process performance.

Moreover, the economical benefits must be derived from finding and fixing problem loops throughout a plant on the basis of data gathered over a long time period. In addition to these hard benefits, the soft benefits resulting from better maintenance will emerge. This includes reduction of unnecessary preventative maintenance actions, improved facility stability and process operability and increased equipment life cycle (Vatanski et al. 2005).

13.3 Comprehensive Procedure for Performance Monitoring

After selecting the suitable assessment strategy, work is focused on evaluating the performance each loop belonging to the group considered. It is not sufficient and sometimes dangerous to rely on a single statistic, or statistical analysis, by itself for performance monitoring and diagnosis, as each criterion has its merits and limitations (Kozub 1996; Ogawa 1998). The best results are often obtained by the collective application of several methods that reflect control performance measures from different aspects. Based on our experiences using different performance monitoring methods in steel processing automation, the following systematic procedure for automatic and continuous control performance monitoring and optimisation is strongly advisable.

Procedure 13.1 Comprehensive control performance monitoring and optimisation.

1. *Select Assessment Objectives*. Gather as much information as possible about the control loops to be evaluated. Of particular importance is to decide which performance benchmark(s) should be considered; see Chap. 1.
2. *Pre-process the Data*. Use raw data collected at a proper sampling frequency. Strictly avoid filtering/smoothing or compression of the data. It is always important to spent time previewing and processing the data before proceeding to the analysis. Certain treatment of the data, such as the removal of outlines or bad data, mean entering and scaling, is recommended; see Chap. 7. For nonlinearity detection, varying drifts, abrupt changes and non-stationary trends should be removed from the data by appropriate band-pass filtering (Sect. 9.8).
3. *Find out Interactions between Control Loops when Dealing with MIMO Systems*. Multivariate control performance assessment is only required when the loops are strongly coupled. This can be found out by applying standard interaction measures, such as relative gain array (RGA). Cross-correlation analysis is also useful to assess the interaction between the control loops. Even in the case of significant interactions, one can apply a performance assessment approach, which does not require the interactor matrix of the process; see Chap. 6.
4. *Determine or Estimate the Time Delay(s)*. In metal processing applications, usually the strip processing speed is either measured or can be indirectly estimated. However, the speed is often varying. Therefore, the time delay should be automatically estimated or continuously updated based on input/output measurements, when sufficient excitation can be ensured. When the time delay is un-

known, varying, or its determination is costly or even not possible, the extended prediction horizon approach can be applied. See Chaps. 3 and 7.

5. *First-pass Analysis*. The correlation (covariance) analysis of the control error is simple and should be always considered as a "first-pass" test before carrying out further performance analysis. The cross-correlation between measured disturbances and the control error can be used to qualitatively assess feedforward controls. Also spectral analysis of the closed-loop response, which allows one to detect oscillations, offsets, nonlinearities and measurement noises present in the process data, should be performed; see Chap. 2.
6. *Detect Oscillating Loops*. As the presence of oscillation affect the performance index calculation, oscillations, due to aggressive tuning, the presence of nonlinearities, loop interactions, etc., need to be screened out using techniques presented described in Chap. 9.
7. *Apply Nonlinearity Detection Tests*. It is important to detect loop nonlinearities by applying one or several nonlinearity tests described in Chap. 10. When hard nonlinearities, such as hysteresis or stiction, are detected, immediate correction or compensation is needed (Chap. 11). If changes in operating point or time-varying disturbances are present, the data must be properly segmented prior to the performance analysis, i.e. non-overlapping sliding data windows have to be used.
8. *Detect Sluggish Control, Evaluate Set-point Response or Load Disturbance Performance*. Many undesired and easy to detect characteristics of the closed-loop behaviour can be detected using specialised methods and indices, provided in Chap. 5. These techniques are particularly important when deterministic performance is of the key point.
9. *Apply the Minimum-variance-control-based Assessment*. This should be the standard benchmark to be applied. When the Harris index signals that the loop is performing well, then further assessment is neither useful nor necessary. In the case where a poor performance relative to MVC is detected, there is a potential to improve the control loop performance, but no guarantee that this will be attained by means of re-tuning the existing controller. Further analysis is then warranted. See Chaps. 2 and 6.
10. *Apply User-specified or Advanced Control Performance Benchmarking*. Baselines and thresholds (historical benchmark values) using data with "perfect" controller performance or other user design specifications can be applied; see Chap. 3. Moreover, the use of more advanced (LQG/GMV/MPC) benchmarking can be an option, particularly in the cases where performance improvement cannot be achieved by re-tuning the running controller and/or for supervisory control loops (usually model-based); see Chap. 4.
11. *Re-tune the Control Loop*. Adjust some parameters of the control loop(s) detected to be poorly performing. Techniques are presented in Chap. 14. When re-tuning is not necessary or does not improve the control performance, modifications of the instrumentation, control system structure or the process itself will be required if the current product quality variation is deemed unacceptable by plant personal.

12. *Modify the Control Structure*. When re-tuning does not improve the control performance, introduce some structural components such as anti-windup, feedforward control, cascade control or time-delay compensating techniques. This may also mean to employ a nonlinear controller.
13. *Repair/re-design System Components*. For instance, repair valves or sensors. When loops are identified to have an oscillation problem probably caused by the valve, additional (stiction and hysteresis) tests on the valve should be carried out to pinpoint and verify the root cause. Another modification might be to alter the feedback dynamics, reducing the time delay by changing the process flow, e.g. adding a bypass, or changing the sensor location. Also, disturbance sources might be eliminated, or supplementary sensors be installed, to make FFC possible.

In this procedure, many parameters have to be carefully selected by the user, as shown in Chap. 7. In our experiences, it is necessary and well spent time to carefully test, inspect and compare performance assessment results using different parameter choices. Usually, each control loop (category) will have its individually suitable parameters.

A principal decision procedure, partly adopted from Hugo (1999), for carrying out the right measures for improving the performance of a control system is illustrated in Fig. 13.4. A continuous CPM system indicates whether the control performance is acceptable, i.e. meets the required specification in terms of standard deviation or other measures, product quality, energy consumption or even safety. This level is thus more related to the direct economical performance of the plant. When proceeding with the performance analysis, the current performance is compared to that of the selected benchmark, to find out whether the installed controller is performing well under the current process conditions, using the techniques of Part I of this book.

When the assessment reveals significant potential to improve the control performance either by re-tuning the controller or maintaining plant components, the diagnosis methods and tests presented in Part II have to be applied. This helps identify the source of non-acceptable performance and suggests the right corrective actions.

In general, the major difference between a PID controller and a single-loop MPC controller is that the latter has some form of time-delay compensation and should perform much better on loops with significant time delays. Other proven strengths of MPC is its straightforward applicability to multivariable systems with constraints. If the PID performance index indicates good tuning, but the process show dominant time delays, strong interactions, or constraints, then the MPC controller performance index may be computed (Sect. 4.2.6) to determine whether there would be any improvement if a model predictive controller were applied to the process.

When a model-based, e.g. MPC, controller is installed at the plant, the reason for non-acceptable control can be improper tuning or poor modelling, i.e. mismatch between the plant and the (internal) model. Consequently, either the controller has to be re-tuned, or the model should be re-identified. If the final assessment indicated

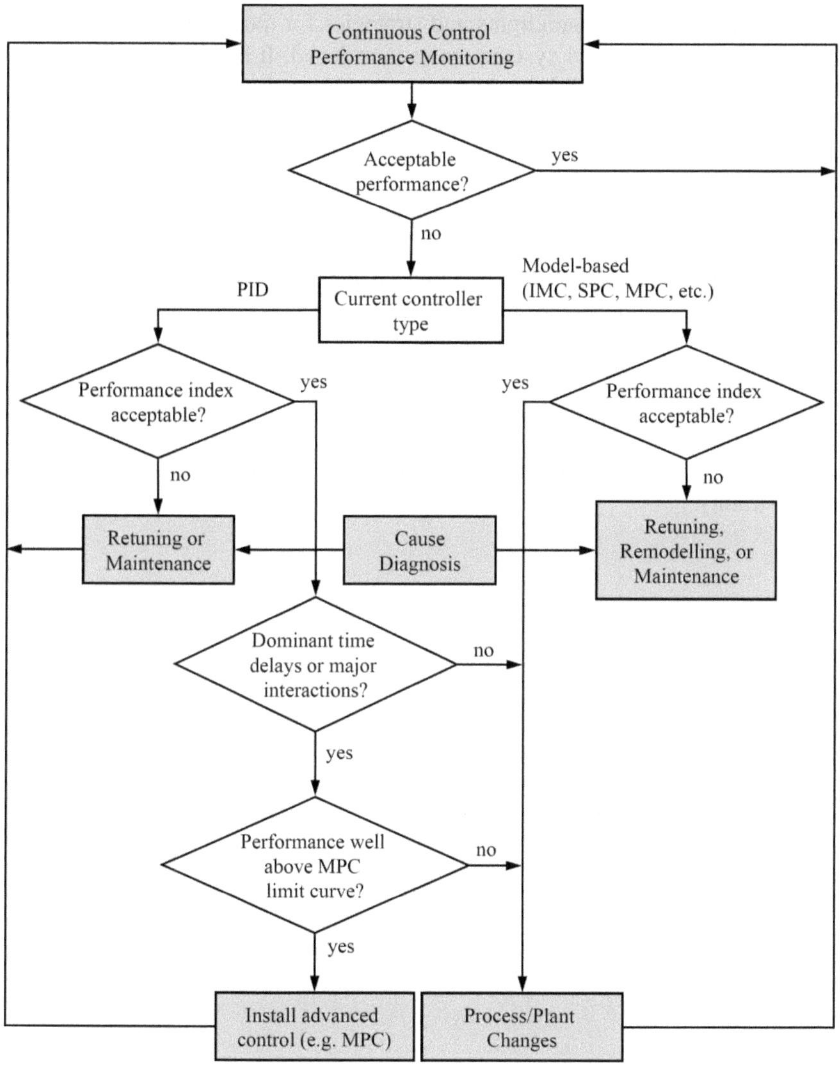

Fig. 13.4 Decision procedure for performance improvement with differentiation between PID and MPC

that the desired level of control performance cannot be achieved by re-tuning or re-designing the controller, the only chance is to introduce substantial modifications to the process.

13.4 Summary and Conclusions

In this chapter, the measures for improving performance of control loops have been mentioned, and an integrated framework for performance monitoring and optimisation proposed. Important paradigms and strategies for monitoring the performance of complex process-control systems were introduced. It has been pointed out that a top-down strategy for CPM should be preferred because of the direct relationship between the performance of upper loops and economical factors. However, under certain circumstances, the bottom-up strategy has its strengths, particularly when basic control loops are believed to perform poorly. The impact of improved loop performance on financial performance has been briefly addressed. We also proposed a comprehensive controller performance assessment procedure combining different methods described throughout the previous chapters. The main aim is to have a systematic and thus efficient way for loop performance assessment avoiding relying on a single performance measure or a single assessment method, which may be misleading. It is also fundamental to recognise the key differences between performance assessment of PID controllers and MPC controllers. Although the assessment of MPC systems is much more difficult, it can give valuable hints on how to attain top performance, particularly in the case of multivariable systems with time delay and constraints.

Chapter 14
Controller Auto-Tuning Based on Control Performance Monitoring

In this chapter, we assume that the first assessment stage has indicated that the control performance can be improved by re-tuning the controller. In other words, all other control-loop components are found to be healthy, i.e. the process is well designed, and actuators and sensors have no major faults.

In practice, it is the norm to perform controller tuning only at the commissioning stage and never again. A control loop that worked well at one time is prone to degradation over time unless regular maintenance is undertaken. Typically, 30 % of industrial loops have poor tuning, and 85 % of loops have sub-optimal tuning. There are many reasons for the degradation of control loop performance, including changes in disturbance characteristics, interaction with other loops, changes in production characteristics (e.g. plant throughput, product grade), etc. Also, many loops are still "tuned by feel" without considering appropriate tuning methods—a practice often leading to very strange controller behaviour. The effects of poor tuning are then (Buckbee 2008):

- Sluggish loops do not respond to upsets, causing disturbances to propagate and deteriorate the performance of other interacting loops.
- Overly-aggressive loops oscillate, creating new disturbances and increasing the risk of plant shut-down.
- Operators put the loops in manual. The loops then are unable to respond properly, leading to degradation of product quality, higher material and energy consumption and decreased productivity.

Continuous performance monitoring is therefore recommended to detect performance degradation and re-tune the controller and sustain top performance; see Fig. 14.1.

Controller tuning is a traditional topic in standard control texts, such as Åström and Hägglund (1988, 2006) and Seborg et al. (2004). A large number of tuning methods and rules are found. A comprehensive collection of more than 400 PID-controller tuning rules is given by O'Dwyer (2003). It is therefore not within the scope of this chapter to consider the design or commissioning of any controllers using such methods which normally require extensive experimental testing on the

M. Jelali, *Control Performance Management in Industrial Automation*,
Advances in Industrial Control, DOI 10.1007/978-1-4471-4546-2_14,

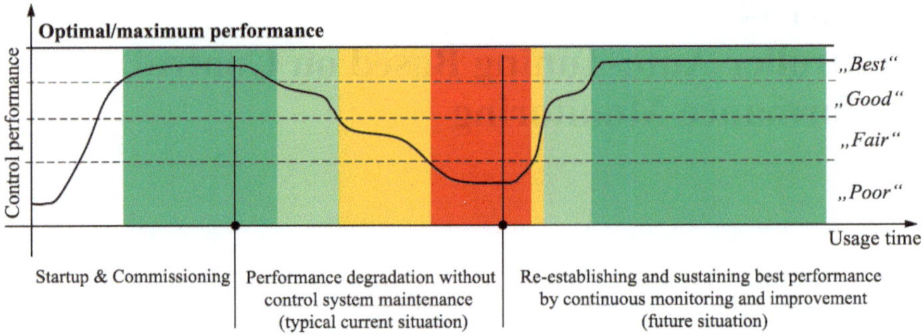

Fig. 14.1 Development of control performance without and with continuous monitoring

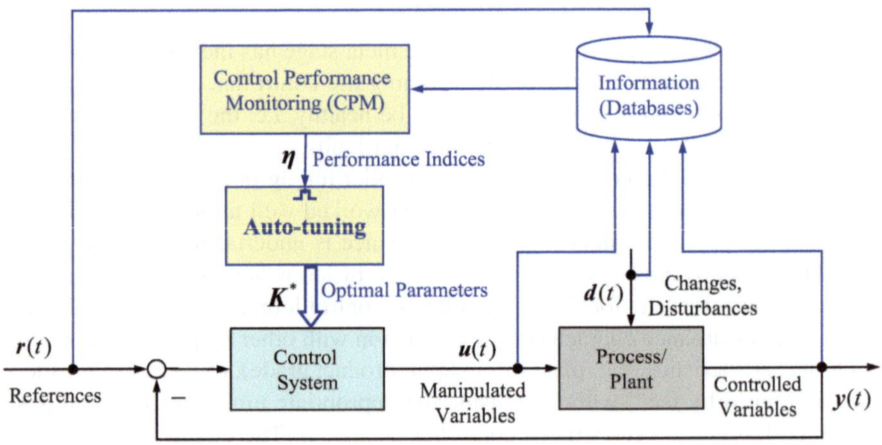

Fig. 14.2 Basic principle of CPM-based controller re-tuning

plant. The main innovation of the tuning methods presented in this work is to treat controller tuning in the context of control performance monitoring and thus substantially extend the traditional field of controller auto-tuning. This means that control performance measures are *continuously monitored* on a regular basis, i.e. during normal operation, and performance statistics used to schedule loop re-tuning and automatically determine the optimal controller parameters; see Fig. 14.2. In other words, the re-tuning operation is initiated automatically whenever the performance is below a user-specified threshold. The overall aim is to find controller settings that maximise the control performance index, i.e. guarantee best achievable controller performance, despite changes in the process operation conditions.

This chapter presents innovative techniques for *automatic and non-invasive generation of optimal controller settings from normal operating data*. It starts with recalling the basic concepts of PID auto-tuning and adaptation in Sect. 14.1 and a classification of CPM-based controller re-tuning methods in Sect. 14.2. Techniques, which deliver optimal controller parameters by solving an optimisation problem, are described in Sect. 14.3. Section 14.4 presents new re-tuning methods, which simultaneously provide the assessment of the controller performance and finding the optimal controller settings in an iterative way on the closed loop. Particularly, a new performance index based on the damping factor of the disturbance impulse response is introduced in Sect. 14.4.2.5. Section 14.5 discusses some strategies for variation of controller parameters during the optimisation process. In Sect. 14.6, simulation studies are presented to compare the different techniques and make suggestions for using them.

14.1 Basic Concepts of Controller Auto-Tuning and Adaptation

There are many definitions of auto-tuning in the literature. According to Leva et al. (2001), an auto-tuner is something capable of computing the parameters of a controller connected to a plant *automatically* and, possibly, without any user interaction apart from initiating the operation. The auto-tuner is not part of the regulator: when no auto-tuning is in progress, the computation of the control signal in no sense depends on the auto-tuner's presence.

It is also useful at this point to distinguish between *auto-tuning* and *adaptation* (or adaptive control). In the latter case, the controller parameters are computed without user intervention, while in the auto-tuning context the system may at best suggest the user to re-tune but does not initiate a tuning operation. Therefore, we shall distinguish four cases of controller tuning (Leva et al. 2001):

1. Tuning is initiated by the user as a deliberate decision, either explicitly or by making some manoeuvre to initiate a tuning, e.g. modifying the set point.
2. Tuning is initiated by the user as a deliberate decision, but the regulator can suggest re-tuning. In this case, the suggestion logic should be clearly documented and configurable.
3. Tuning occurs automatically when some condition occurs, e.g. the error becomes "too big" for "a certain time". In this case, the logic should also be precisely documented and configurable. Moreover, it must be possible to disable this functionality in the regulator configuration and to inhibit it temporarily from outside the regulator.
4. Tuning occurs continuously.

Cases (1) and (2) are to be classified as *auto-tuning*, case (4) is clearly *continuous adaptation*, and case (3) is somehow *hybrid*, but if the logic is properly configured, it is much more similar to auto-tuning than to continuous adaptation. It is important, when selecting an auto-tuner, to understand in which category it falls so as to forecast how it will possibly interact with the rest of the control system (Leva et al. 2001). The tuning methods presented in this chapter can be classified as CPM-based auto-tuning, where the decision for re-tuning is automatically taken based on the performance indices determined in the controller assessment stage. For safety reasons, it is recommended that the user should always confirm the need for the re-tuning action.

14.2 Overview and Classification of CPM-Based Tuning Methods

The first approaches for simultaneous controller performance assessment and tuning were proposed by Eriksson and Isaksson (1994) and Ko and Edgar (1998). These techniques calculate a lower bound of the variance by restricting the controller type to PID only (*optimal PID benchmarking*) and allow for more general disturbance models. The optimal controller parameters are found by solving an optimisation problem with respect to the controller parameters. The *PID-achievable* lower bound determined is generally larger than that calculated from MVC but is possibly achievable by a PID controller. That is, one is interested in determining how far the control performance is from the "best" achievable performance for the pre-specified controller.

Optimal (IMC-based) PID benchmarking has been implemented and studied by Bender (2003) using an iterative solution of the optimisation problem. An explicit "one-shot" solution for the closed-loop output was derived by Ko and Edgar (2004) as a function of PID settings. Recent developments in this (pragmatic) direction have been worked out in Horton et al. (2003) and Huang (2003). Note that these approaches require the process/disturbance model to be known or identified from measured input/output data and the use of (usually constrained) optimisation algorithms to calculate the optimal controller settings. Although the optimisation is often a hard task, optimal parameters of the controller are a nice by-product. In the author's experience, an IMC-based parameterisation of PID controllers is highly recommended to simplify the optimisation problem and improve its conditioning (only one parameter λ has to be selected).

Grimble (2002a, 2002b) provides criteria by which restricted structure controllers (such as the PID controller) can be assessed and tuned. Though this approach takes control activity into account while defining the "cost of control", it is a model-based procedure (as opposed to a data-based procedure) and places rather heavy emphasis on the *not so easily available* process knowledge.

Recently, Ingimundarson and Hägglund (2005) discuss an approach where they used the extended horizon performance index (EHPI) curve to detect problematic control loops. Two parameters of their monitoring scheme (the horizon length and the alarm limit) are set based on the loop tuning itself. They consider loops in a pulp and paper mill, where λ-tuning is predominantly employed. While useful for detecting problems, their method does not consider the follow-up problem which is to recover the performance by controller re-tuning.

In the light of above description, CPM-based controller tuning methods can be classified into *direct* and *indirect* groups (Jelali 2007a):

- *Direct Methods* (Sect. 14.3). The optimal controller settings are determined by solving a parameter-optimisation problem based on available or identified process models and suitable measured data. These methods require more or less experimentation with the process or specific set-point or load-disturbance changes.
- *Indirect Methods* (Sect. 14.4). The controller settings are determined from iterative tuning on the closed loop, following the basic procedure:

(a) Collection and pre-processing of normal operating data;
(b) Computation of the actual performance indices;
(c) Comparison with benchmark/desired values;
(d) Decision whether the performance is optimal/sufficient; if so, then break the procedure; else, change the controller parameters, apply them on the process, then go to Step 1 and repeat the procedure.

The methods will be presented for PI(D) controller but can be extended to more complicated controller types. It is important to note that the indirect methods are different from what is known as "iterative feedback tuning" (IFT) proposed by Hjalmarsson et al. (1998). Although IFT also iteratively determines the controller settings on the closed loop using special calculation of the gradient of the control error, it requires several guided, so-called "recycling experiments" on the system. Moreover, IFT was introduced within the traditional field of controller auto-tuning rather than in the framework of control performance assessment.

14.3 Optimisation-Based Assessment and Tuning

As learned from Chap. 2, the minimum variance is only exactly achievable when a minimum-variance controller is used with perfectly known system and disturbance model, which requires at least an SPC structure for systems with (dominant) time delays. In practice, however, more than 90 % of industrial control loops are of PID type without time-delay compensation. Therefore, no matter how the PID parameters are tuned, the MVC-based variance is not exactly achievable for PID controllers when time delay is significant or the disturbance is non-stationary. Some experiments performed by Qin (1998) showed that the minimum variance can be achievable for a PID controller when the time delay is very small or very large, but it is not achievable for a PID controller when the time delay is medium. Practical experience shows that about 20 % loops in refinery can achieve minimum variance using PID controllers (Kozub 1996).

Eriksson and Isaksson (1994) and Ko and Edgar (1998) addressed this point and proposed more realistic benchmarks for control performance monitoring and assessment by introducing *PID-achievable performance indices*. These approaches calculate a lower bound of the variance by restricting the controller type to PID only (optimal PID benchmarking) and allow for more general disturbance models. The PID-achievable lower bound is generally larger than that calculated from MVC but is possibly achievable by a PID controller. That is, one is interested in determining how far the control performance is from the "best" achievable performance for the pre-specified controller.

14.3.1 Methods Based on Complete Knowledge of System Model

When accurate plant models are available or can be estimated from gathered data, it is a straightforward task to determine the optimal controller parameters by using an optimisation algorithm. However, the generation of the models usually requires experimentation with the process, such as introducing extra input signal sequences. This is usually allowed only in the commissioning stage, but not when the system is in normal operation.

14.3.1.1 PID-Achievable Performance Assessment

Eriksson and Isaksson (1994) introduced an index that makes a comparison with the optimal PID controller instead of the MVC, i.e.

$$\eta = \frac{\sigma^2_{\mathrm{PID,opt}}}{\sigma^2_y}, \tag{14.1}$$

where $\sigma^2_{\mathrm{opt,PI}}$ denotes the minimum value of the integral (Ko and Edgar 2004)

$$\sigma^2_y = \frac{1}{2\pi}\int_{-\pi}^{\pi} \Phi_y(w)\mathrm{d}w = \left(\frac{1}{2\pi j}\oint H_\varepsilon(z)H_\varepsilon\left(z^{-1}\right)\frac{\mathrm{d}z}{z}\right)\sigma^2_e, \tag{14.2}$$

when the controller structure is restricted to PID. $\Phi_y(w)$ represents the spectrum of the output signal y, and $\oint$ denotes the counter-clockwise integral along the unit circle in the complex plane. A procedure for the evaluation of the integral in Eq. 14.2 can be found in Åström (1979) when H_ε has all its zeros inside the unit circle. With this method, however, it is difficult to obtain the output variance as an explicit function of the PID-controller parameters.

Instead, the PID-achievable performance can be numerically determined by solving the following the optimisation problem:

$$\boldsymbol{K}^*_{\mathrm{PID}} = \min_{\boldsymbol{K}_{\mathrm{PID}}} \sigma^2_y(G_\mathrm{p}, G_\varepsilon), \tag{14.3}$$

once the process and disturbance models, i.e. G_p and G_ε, are given. Recall the relationship between the controlled variable and the external signals, i.e. set point and noise, under closed loop

$$y(k) = \frac{G_\mathrm{c}G_\mathrm{p}}{1+G_\mathrm{c}G_\mathrm{p}}r(k) + \frac{G_\varepsilon}{1+G_\mathrm{c}G_\mathrm{p}}\varepsilon(k) = G_r r(k) + H_\varepsilon \varepsilon(k). \tag{14.4}$$

The complete procedure for calculating the PID-achievable performance index is given in the following algorithm, called *approximate stochastic disturbance realisation* (*ASDR*) method (Ko and Edgar 1998).

Procedure 14.1 PID-achievable performance-assessment algorithm

1. Preparation. Select the time-series-model types and orders.
2. Determine/estimate the system time delay τ.

3. Identify the closed-loop (noise) disturbance model from collected output samples based on the installed PID controller.
4. Identify the open-loop system model from collected input–output samples.
5. Estimate and calculate the series expansion (impulse response) for the closed-loop transfer function (Eq. 2.36).
6. Derive the PID-achievable variance by numerically solving the optimisation problem in Eq. 14.3.
7. Estimate the actual output variance from Eq. 2.39 or directly from measured data (Eq. 1.1).
8. Compute the performance index (Eq. 14.1) to see how far the actual performance from the optimal PID performance.

It is apparent that this procedure is much more complex and difficult to implement than that based on MVC. The disadvantage is that to evaluate Eq. 14.2, we first need to calculate the closed-loop transfer function H_ε. This requires knowledge of the current controller parameters and an explicit model G_p of the plant. Hence it is not possible to use only the output for identification as done for the calculation of the Harris index. Furthermore, to be able to estimate the plant, it may be necessary to perturb the process with extraneous test signals. The closed-loop transfer function H_ε can be approximated by a high-order AR model. Alternatively a low-order ARIMA$(p, 1, 1)$ model with $2 \leq p \leq 5$ can be used, as recommended by Ko and Edgar (1998).

Note that there are some special situations where it is not required to know the process model:

- If the process time delay is "large" enough relative to the settling time of the disturbance, the process model can be estimated from routine operating data; refer to Sect. 7.2.7.
- If the process model is assumed of the first-order type and the controller of the PI-type, the optimal PID performance index (but not the optimal PID parameters) can be estimated only from normal operating data and knowledge of the time delay; see Hugo (2006).

The numerical solution of the optimisation problem in Eq. 14.3 can be obtained using, for example, the MATLAB Optimization Toolbox (function fmincon or fminsearch) or the Genetic Algorithms and Direct Search (GADS) Toolbox (function ga or patternsearch). An implementation of the method using the fmincon function was carried out by Bender (2003) for control loops with PID controllers and IMC-tuned PID controllers. Later, a similar solution using Newton's iterative method has been applied by Ko and Edgar (2004), to give the best-achievable performance in an existing PID loop with the process output data and the nominal process model (assumed in step response form).

The author implemented a solution of the optimisation problem using the pattern search algorithm. The results of this approach are illustrated in the following examples. We use the following discrete representation for PI controllers:

$$G_{\mathrm{PID}}(q) = \frac{K_1 + K_2 q^{-1}}{1 - q^{-1}}. \tag{14.5}$$

The corresponding parameters K_c and T_I of the continuous counterpart

$$G_{\mathrm{PI}}(s) = K_c \left(1 + \frac{1}{T_I s}\right) \tag{14.6}$$

can be calculated as (based on the backward difference approximation $s \approx (1 - q^{-1})/(T_s q^{-1})$)

$$K_c = K_1, \qquad T_I = \frac{T_s}{1 + \frac{K_2}{K_1}}. \tag{14.7}$$

Other digital controller descriptions with corresponding parameter sets can be considered as well.

Example 14.1 Consider the following system ($T_s = 1$ s; $\sigma_\varepsilon^2 = 0.01$):

$$y(k) = \frac{0.1}{1 - 0.8q^{-1}} q^{-6} u(k) + \frac{1}{(1 - 0.7q^{-1})(1 - q^{-1})} \varepsilon(k). \tag{14.8}$$

This example has been used by Ko and Edgar (2004) to illustrate their optimisation solution. We here demonstrate the use of pattern search to achieve similar results. However, pattern search is more robust against falling in local minima. Initially, the PI controller

$$G_{\mathrm{PI}}(q) = \frac{1.93 - 1.71q^{-1}}{1 - q^{-1}}$$

is adopted, resulting in a Harris index of $\eta = 0.47$ ($\sigma_y^2 = 0.6066$), indicating poor performance compared with MVC. Running the optimisation leads to the optimal PI controller

$$G_{\mathrm{PI,opt}}(q) = \frac{2.16 - 1.94q^{-1}}{1 - q^{-1}}.$$

The Harris index is now $\eta = 0.48$ ($\sigma_y^2 = 0.5996$), which is very near to the initial performance. This is however the maximally achievable performance when using a PI controller. Next, an optimal PID controller is sought to give

$$G_{\mathrm{PID,opt}}(q) = \frac{7.79 - 12.90q^{-1} + 5.50q^{-2}}{1 - q^{-1}},$$

giving a Harris index of $\eta = 0.74$ ($\sigma_y^2 = 0.3947$). This implies a clear performance improvement compared with the PI controller, but the variance is still not in the neighbourhood of the MV. However, when relating the performance of the actual

Table 14.1 Minimum variance, PID-best-achievable variance and Harris index for the FOPTD process with different disturbance models

Disturbance model	Parameters of optimal PI/PID $[K_1; K_2; K_3]$	Best-achievable variance	Minimum variance	Harris index
$\frac{1-0.2q^{-1}}{(1-0.3q^{-1})(1+0.4q^{-1})(1-0.5q^{-1})}$	[0.142; −0.143]	1.19	1.07	0.90
	[0.135; −0.137; 0.0]	1.18	1.07	0.91
$\frac{1+0.6q^{-1}}{(1-0.6q^{-1})(1-0.5q^{-1})(1+0.7q^{-1})}$	[0.142; −0.143]	3.73	3.11	0.83
	[0.135; −0.137; 0.0]	3.67	3.10	0.85
$\frac{1-0.2q^{-1}}{(1-0.3q^{-1})(1+0.4q^{-1})(1-0.5q^{-1})(1-q^{-1})}$	[0.208; −0.188]	17.71	11.05	0.62
	[0.757; −1.279; 0.56]	12.71	10.98	0.86
$\frac{1+0.6q^{-1}}{(1-0.6q^{-1})(1-0.5q^{-1})(1+0.7q^{-1})(1-q^{-1})}$	[0.234; −0.214]	123.67	53.98	0.44
	[0.853; −1.449; 0.64]	80.36	54.12	0.67

PI controller to that of the optimal PI/PID controller, we have $\eta_{\mathrm{PI,opt}} = 0.99$ and $\eta_{\mathrm{PID,opt}} = 0.65$, respectively. Therefore, using the PID-achievable performance index is more sensible than using the Harris index, because the former is related to what is achievable in practice.

Example 14.2 Consider again the FOPTD process in Eq. 14.8, but with unity gain and unity variance. Inspired by Qin (1998), the purpose of this example is to study the PI/PID-achievable performance as a function of some fundamental performance limitations in control systems, namely time delay and the minimum phase behaviour and non-stationarity of disturbances affecting the process.

The minimum variance and the PID-best-achievable variances have been computed for four different disturbance models, as given in Table 14.1. It can be deduced that the PID-best-achievable variance is always higher than the minimum variance, so the Harris index for the best possible PID controller is always smaller than unity. The gap between both variance values increases when the disturbance becomes non-stationary, i.e. ARIMA instead of ARMA. We should also learn that PID control is able significantly improve the performance for non-stationary disturbances, compared with PI control. This feature of PID control can be explained by the "prediction capability" of the derivative action.

Next, we consider the system model

$$y(k) = \frac{0.1}{1-0.8q^{-1}} q^{-\tau} u(k) + \frac{1-0.2q^{-1}}{(1+0.4q^{-1})(1-0.5q^{-1})(1-cq^{-1})} \varepsilon(k) \quad (14.9)$$

with the free parameters τ and c to test the effect of the time delay and disturbance stationarity on the PI/PID-best-achievable performance; see Figs. 14.3, 14.4, 14.5 and 14.6. On the one hand, it is observed that the minimum variance can be achievable for a PI controller when the time delay is very small or very large, but it is not achievable for a PI controller when the time delay is medium. This observation cannot be confirmed for PID controllers. On the other hand, it can be concluded that

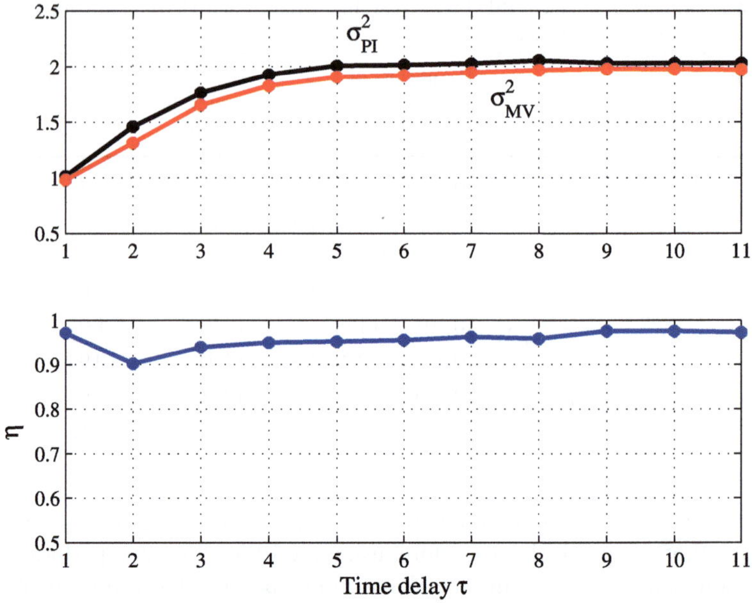

Fig. 14.3 Effect of time delay on variances and Harris index for a FOPTD process ($c = 0.7$) with optimal PI controller

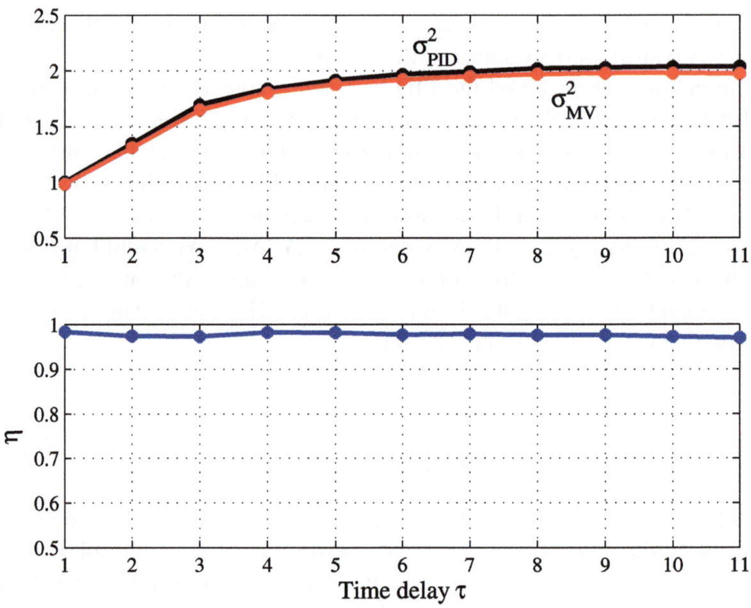

Fig. 14.4 Effect of time delay on variances and Harris index for a FOPTD process ($c = 0.7$) with optimal PID controller

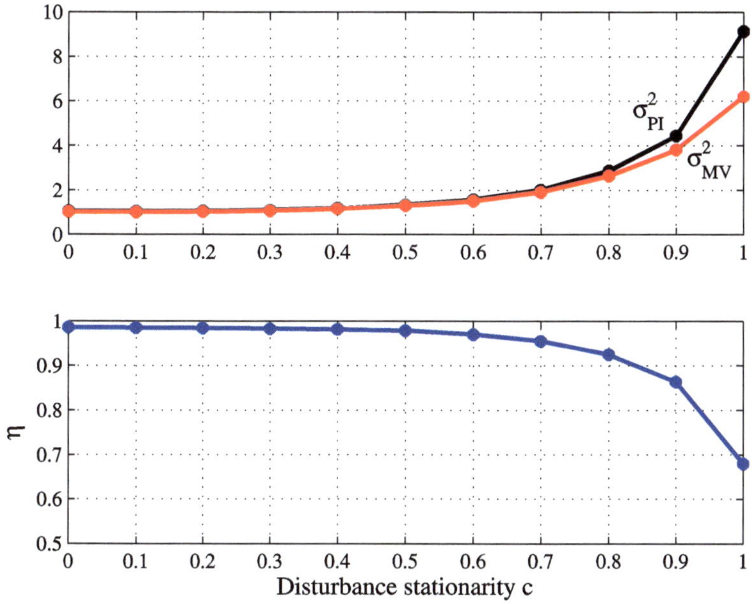

Fig. 14.5 Effect of disturbance stationarity on variances and Harris index for a FOPTD process ($\tau = 6$) with optimal PI controller

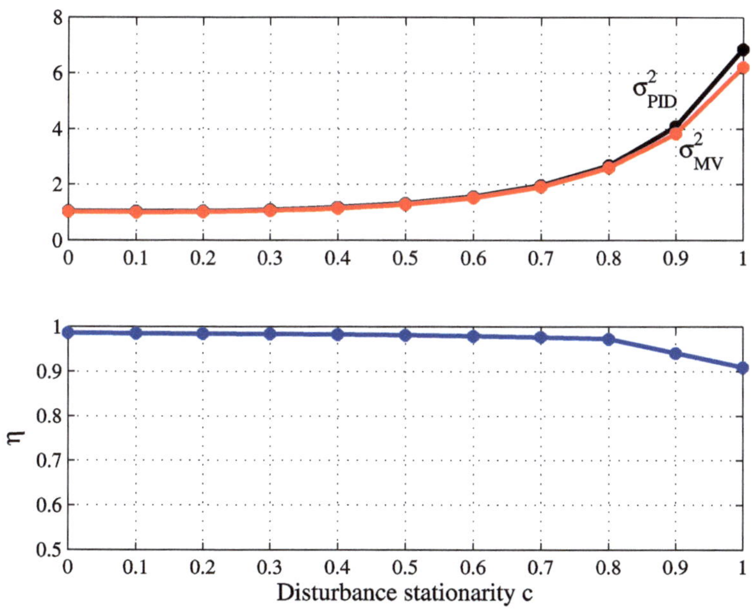

Fig. 14.6 Effect of disturbance stationarity on variances and Harris index for a FOPTD process ($\tau = 6$) with optimal PID controller

non-stationary disturbances generally make the MVC performance more difficult to achieve by PID than PI controllers.

14.3.1.2 Maximising Deterministic Performance

The PID-achievable assessment presented above aims to maximise the stochastic disturbance rejection performance but may be not so appropriate if the objective is to change the output from one set point to another. However, the optimisation task can accordingly be re-formulated and solved if set-point tracking is the main control objective. Set-point changes are injected into the closed-loop model. The PID controller parameters are determined so that the mean square error is minimised:

$$\boldsymbol{K}_{\mathrm{PID}}^{*} = \min_{\boldsymbol{K}_{\mathrm{PID}}} \frac{1}{N} \sum_{k=1}^{N} \bigl(r(k) - y(k)\bigr)^2. \tag{14.10}$$

Therefore the same methods (and functions) can be used to solve the tracking optimisation task. At this point it is important to recall that disturbance rejection performance is more important in the process industries.

Example 14.3 The process from Example 14.2 with the second disturbance model, i.e.

$$y(k) = \frac{1}{1-0.8q^{-1}} q^{-6} u(k) + \frac{1+0.6q^{-1}}{(1-0.6q^{-1})(1-0.5q^{-1})(1+0.7q^{-1})} \varepsilon(k), \tag{14.11}$$

is considered again. The results for maximal stochastic disturbance rejection are given in Table 14.1 (second row). However, looking at the resulting tracking behaviour shown in Fig. 14.7 (left) reveals unacceptable performance. If we now optimise the PI controller to yield best tracking performance, we get the response shown in Fig. 14.7 (right). The optimal PI controller settings are found to be $K_1 = 0.199$ and $K_2 = -0.178$. The high performance is confirmed by applying the assessment indices based on set-point response: $T_{\mathrm{set}}^{*} = 4.3$, $IAE_{\mathrm{d}} = 2.6$, and $\alpha = 18\ \%$ (Sect. 5.2). The side-effect is, however, that the Harris index is significantly lower, $\eta = 0.75$, but still indicating a satisfactory stochastic performance.

The same analysis undertaken for the same process but with the fourth disturbance model, i.e.

$$y(k) = \frac{1}{1-0.8q^{-1}} q^{-6} u(k) + \frac{1+0.6q^{-1}}{(1-0.6q^{-1})(1-0.5q^{-1})(1+0.7q^{-1})(1-q^{-1})} \varepsilon(k), \tag{14.12}$$

yields the step responses shown in Fig. 14.8. In this case, the PI controller was only slightly detuned ($K_1 = 0.199$; $K_2 = -0.178$) to give optimal tracking performance, without significant change in the Harris index. Moreover, these controller

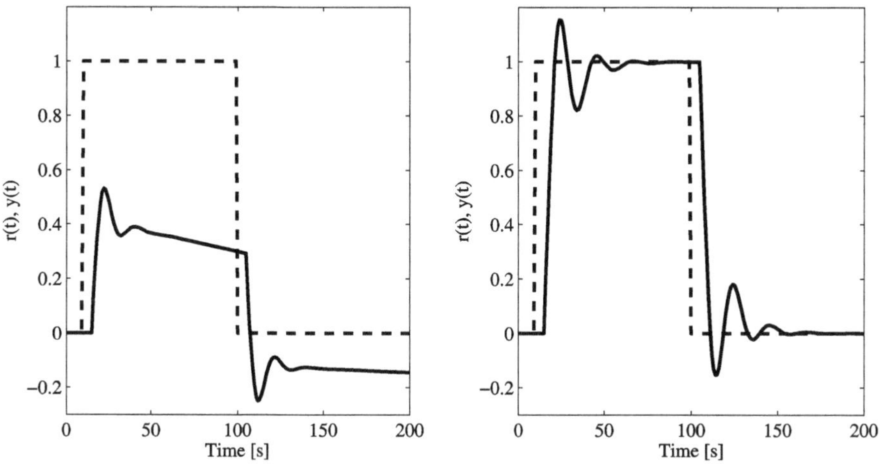

Fig. 14.7 Sept-point responses with PI controller optimised for best stochastic disturbance rejection (*left*) and tracking (*right*) (system Eq. 14.11)

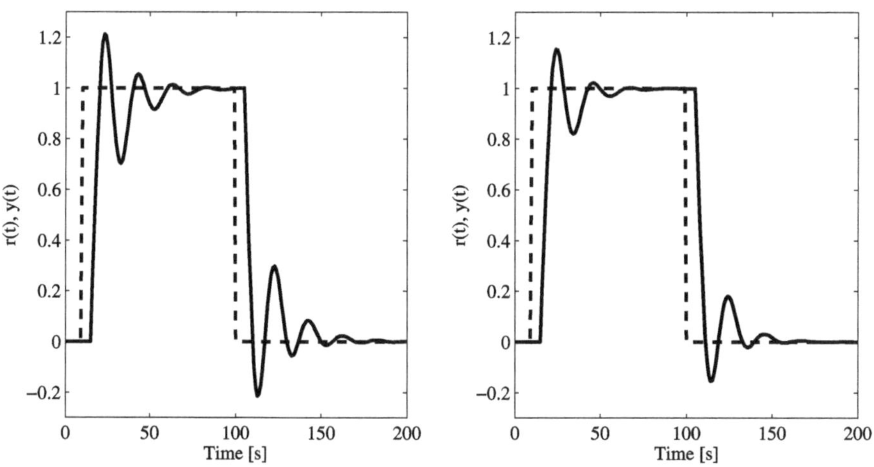

Fig. 14.8 Sept-point responses with PI controller optimised for best stochastic disturbance rejection (*left*) and tracking (*right*) (system Eq. 14.12)

settings seem to provide a good trade-off between stochastic disturbance rejection ($\eta = 0.55$–0.75) and tracking performance ($T^*_{\text{set}} = 4.3$), irrespective of the disturbance model used. This is very useful, since in practice one is always interested in using one set of controller parameters that can handle the whole range of disturbances acting on the process.

14.3.1.3 Restricted Structure Optimal Control Benchmarking

A method of restricted structure (RS) optimal control benchmarking has been introduced by Grimble (2000), in which the controller structure may be specified. If, for example, a PID structure is selected, the algorithm computes the best PID parameters to minimise an objective function, an LQG cost function, where the dynamic weightings are chosen to reflect the desired economic and performance requirements, as in the case of GMV benchmarking. The benchmarking solution is obtained by solving an optimal control problem directly leading optimal controller parameters. The actual optimisation involves a transformation into the frequency domain and numerical optimisation of an integral cost term. Detailed descriptions of the restricted structure benchmarking method can be found by Grimble (2000, 2002b) or Ordys et al. (2007: Chap. 4).

It is important to note that the RS-LQG algorithm does not use plant data to compute the performance index. Instead, the process transfer function is required. Moreover, the user has to specify the type of restricted structure controller (P, PI or PID) against which the existing controller should be benchmarked. The user must also define the models of the system disturbance and reference as transfer functions and specify the error and control weightings. The choice of these weightings must be consistent with the choice of optimal RS controller and the objectives of the control problem. The weighting selection for the RS-LQG design plays a decisive role on its success. The accuracy of the results achieved ultimately depends on the accuracy of the model used for benchmarking Uduehi et al. (2007b). All in all, it can be concluded that RS-LQG benchmarking is much more involved than other performance assessment and controller tuning methods presented in this chapter.

14.3.2 Techniques Based on Routine and Set-Point Response Data

If the open-loop process model is not known, then an obvious procedure to calculate the PI-achievable performance consists of the following steps:

1. Obtain the open-loop process model using closed-loop experimental data. This experiment can involve a sequence of acceptable set-point changes. Suitable identification methods can be used to obtain the open-loop process model.
2. Use the optimisation-based methods in Sect. 14.3.1 with the identified process model to calculate the PID-achievable performance.

This approach is, however, somewhat invasive and may be undesirable in practice. Agrawal and Lakshminarayanan (2003) proposed an alternate way of determining the PID-achievable performance from closed-loop experimental data without the need for identifying the open-loop process and noise models. This method is described next.

14.3.2.1 Optimisation Based on Set-Point Responses

For this purpose, system identification can be employed to determine the closed-loop servo transfer function, which takes the form of an ARMAX model. Equation 14.4 gives

$$G_{\mathrm{p}} = \frac{G_r}{(1-G_r)G_{\mathrm{c}}}, \qquad G_\varepsilon = \frac{H_\varepsilon}{(1-G_r)}. \tag{14.13}$$

Assuming the time-invariant process (G_{p}) and noise dynamics (G_ε), the optimal closed-loop disturbance impulse response $\tilde{G}_\varepsilon^*$ can be given as

$$\tilde{G}_\varepsilon^* = \frac{G_\varepsilon}{1+G_{\mathrm{c}}^* G_{\mathrm{p}}} = \frac{H_\varepsilon}{1+G_r(\frac{G_{\mathrm{c}}^*}{G_{\mathrm{c}}}-1)}, \tag{14.14}$$

where G_{c}^* is the optimal controller to be determined. Equation 14.14 implies that, with the knowledge of the current closed-loop disturbance impulse response (H_ε), the closed-loop servo transfer function (G_r) and the controller G_{c}^*, it is possible to estimate the closed-loop disturbance impulse response $\tilde{G}_\varepsilon^*$ for any given controller G_{c}^*. Specifically, to determine the optimal PI controller G_{c}^* (parameters K_{c}^* and T_{I}^*), the objective function to be minimised is

$$\boldsymbol{K}_{\mathrm{PID}}^* = \min_{\boldsymbol{K}_{\mathrm{PID}}} (1-\eta)^2, \qquad \eta = \frac{\sigma_{\mathrm{PID}}^2}{\sigma_y^2} = \frac{\sum_{i=0}^{\infty} h_{\tilde{G}_\varepsilon}^2 \sigma_\varepsilon^2}{\sigma_y^2}. \tag{14.15}$$

This equivalently maximises the Harris index value η. Again, for instance, the fminsearch/fmincon function from the MATLAB Optimization Toolbox or patternsearch function from the MATLAB GADS Toolbox can be employed to obtain the optimal controller parameters. Recall that noise variance σ_ε^2 is estimated as the prediction error from ARMAX fitting to the data with the set-point change.

To summarise, the optimal PID controller settings can be computed—at least theoretically—using only one set of closed-loop experimental data, without the need of estimating the open-loop process or noise models. The obtained controller parameters will, however, be applied on the data set used to determine the closed-loop transfer functions. Therefore, it is important to use plant data that contain typical disturbances expected to affect the process. Our experience showed that it is sometimes necessary to repeat the estimation step once again to ensure convergence to the optimal controller settings. Also, for solving the optimisation task involved in this assessment and tuning technique, we have good experience with employing the pattern search algorithm of the MATLAB GADS Toolbox. Note that two separate sets of routine operating data are needed to calculate values of the Harris index before and after controller re-tuning.

14.3.2.2 Filter-Based Approach

The set-point-response-based optimisation method can be reformulated to avoid usage of the closed-loop disturbance impulse response H_ε. Consider again Eq. 14.4 with zero set point, to write

$$y(k) = \frac{G_\varepsilon}{1 + G_c G_p}\varepsilon(k). \tag{14.16}$$

This equation expressed with the optimal controller G_c^* gives the "new" output

$$y^*(k) = \frac{G_\varepsilon}{1 + G_c^* G_p}\varepsilon(k). \tag{14.17}$$

Assuming a time-invariant process and noise dynamics, we can write

$$\frac{y^*(k)}{y(k)} = \frac{1 + G_c G_p}{1 + G_c^* G_p}. \tag{14.18}$$

Inserting the first relationship into Eq. 14.13 yields

$$\frac{y^*(k)}{y(k)} = \frac{G_c}{(1 - G_r)G_c + G_r G_c^*}. \tag{14.19}$$

The right-hand side of this equation represents the filter that gives the "new routine closed-loop data series" $y^*(k)$ when the current output (routine data) data series $y(k)$ passes through it. Thus, the method is termed *filter-based approach* (Jain and Lakshminarayanan 2005). For any new controller G_c^*, the filter is specified using the closed-loop servo model G_r and the installed controller G_c. The original routine data $y(k)$ can then be used to "generate" the routine closed-loop data $y^*(k)$ that would be obtained with the controller G_c^*. Using this $y^*(k)$, it is possible to calculate the Harris index

$$\eta^* = \frac{\sigma_{y^*}}{\sigma_y} \tag{14.20}$$

corresponding to the new controller G_c^*. Incorporating this methodology into an optimisation task, it is possible to determine the optimal controller parameters that maximise the control loop performance

$$\boldsymbol{K}_{\mathrm{PID}}^* = \min_{\boldsymbol{K}_{\mathrm{PID}}}\left(1 - \eta^*\right)^2. \tag{14.21}$$

This requires the knowledge of the installed controller, a corresponding routine data set and the closed-loop servo model G_r, which has, as before, to be identified from a data set with a set-point change. The noise model G_ε is not needed any more but is implicitly included in the measured data $y(k)$.

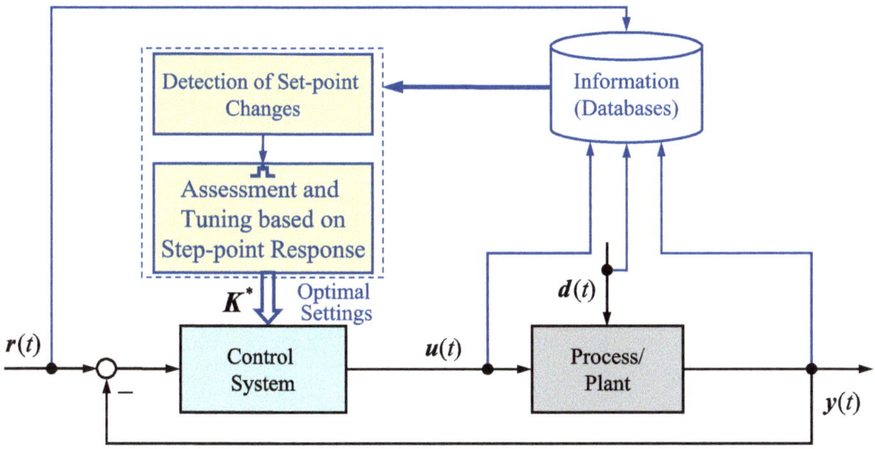

Fig. 14.9 Controller assessment and tuning based on load and set-point change detection

14.3.2.3 Detection of Set-Point Changes

For the controller assessment and tuning method presented in this section, it is essential to extract data windows with distinctive load changes occurring during normal process operation. Techniques are thus needed for automatic detection of these changes to trigger the assessment and tuning task. In other words, the assessment and tuning algorithms must be provided with a supervisory shell that takes care of those operating conditions, in which the algorithm would give wrong performance indications. Only set-point changes that are significantly larger than the noise level of the process variable should be used. The automatic detection of naturally occurred set-point changes (if any) is introduced here to make the method non-invasive.

A convenient approach for excitation detection was suggested by Hägglund and Åström (2000) within adaptive control. The basic idea is to make a high-pass filtering of the measurement signals u and y:

$$G_{\mathrm{hp}}(s) = \frac{s}{s + \omega_{\mathrm{hp}}} \tag{14.22}$$

to give the corresponding high-pass filtered signals u_{hp} and y_{hp}. ω_{hp} is chosen to be inversely proportional to the process time scale T_{p}. When the magnitude of the filtered variable exceeds a certain threshold, it is concluded that the excitation is high enough to trigger the performance assessment and tuning block; see Fig. 14.9. Assuming that the process has a positive static gain, i.e., $G_{\mathrm{p}}(0) > 0$, and that all zeros are in the left half-plane, both u_{hp} and y_{hp} then go in the same direction after a set-point change; both variables go in opposite directions when a load disturbance occurs.

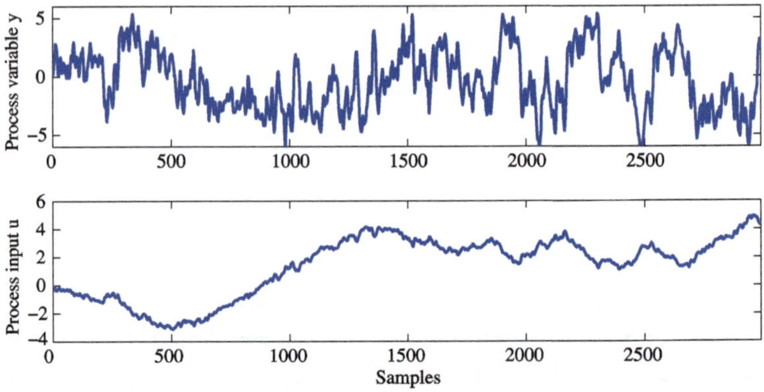

Fig. 14.10 Normal operating data set under the initial controller

14.3.2.4 Considering Stochastic and Deterministic Performance

The aim of the optimisation-based controller assessment and tuning presented so far is to obtain a set of optimal controller settings that will provide good disturbance rejection and a high value of the Harris index. However, since the method requires a set-point change in closed loop, set-point tracking performance can be simultaneously evaluated. This is done by computing the deterministic performance indices, the normalised settling time T^*_{set} and $\mathrm{IAE_d}$, and related robustness margins, presented in Sect. 5.2, based on the set-point response data recorded. Moreover, it is possible to modify the objective function used for optimisation to

$$\boldsymbol{K}^*_{\mathrm{PID}} = \min_{\boldsymbol{K}_{\mathrm{PID}}} \big[(1-w)(1-\eta) + w J_{\mathrm{det}}\big]^2. \tag{14.23}$$

This objective function provides a trade-off between the stochastic and deterministic performance measures. w $(0 \le w \le 1)$ represents the weight given to the deterministic performance measure J_{det}, e.g. $\mathrm{IAE_d}$ or related measures such as gain and phase margins.

Example 14.4 We consider again the system in Eq. 14.12 and apply the technique based on set-point response data on the closed-loop with an initial PI controller with $K_{\mathrm{c}} = 0.1$ and $T_{\mathrm{I}} = 25$ (i.e., $K_1 = 0.1$; $K_2 = -0.096$). The closed loop is simulated without any set-point change, and the recorded "normal operating data" (3000 samples; $\sigma_y^2 = 0.02$) are analysed to give the Harris index $\eta = 0.17$; see Fig. 14.10.

A set-point change experiment is performed on the closed loop. The gathered data for the process input and output are shown in Fig. 14.11. From these data, an ARMAX(5, 3, 1, 6) model has been identified to give the closed-loop transfer function G_r. Pattern search is then run based on the identified model to yield the optimal controller settings as $K_{\mathrm{c}} = 0.22$ and $T_{\mathrm{I}} = 10.6$ (i.e., $K_1 = 0.226$; $K_2 = -0.205$).

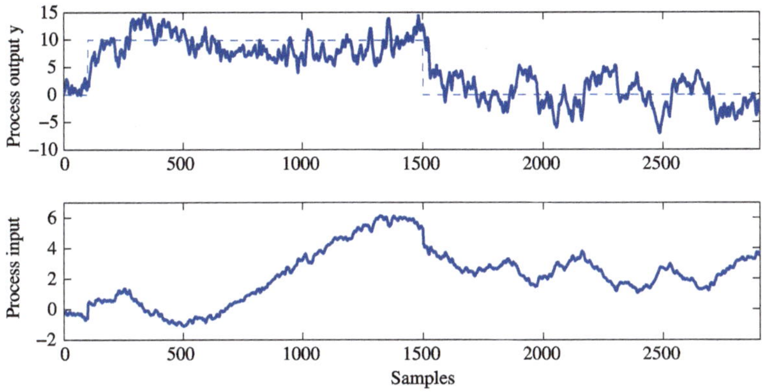

Fig. 14.11 Set-point response data used for identification

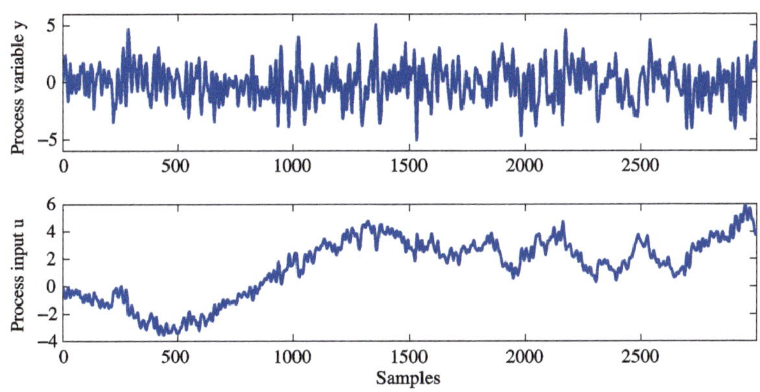

Fig. 14.12 Normal operating data set under the optimal controller

When the obtained controller is applied on the process, we get a Harris index $\eta = 0.47$ from the data set in Fig. 14.12. The step response obtained when simulating the closed loop under the optimal controller is shown in Fig. 14.13. It is observed how the variability of the input increases while the output variability decreases when the initial controller is replaced by the optimal one. The substantial transfer of variability also symbolises the improvement in the stochastic control performance.

When compared with the results in Example 14.3, the Harris index obtained here is slightly lower. Also the corresponding normalised settling time $T^*_{\mathrm{set}} = 4.8$ is slightly higher. This deviation can be mainly attributed due to modelling errors which are unavoidable in practice, irrespective of the method used. Recall that the technique applied here does not require the knowledge of the process model, as is the case for the method of Sect. 14.3.1.

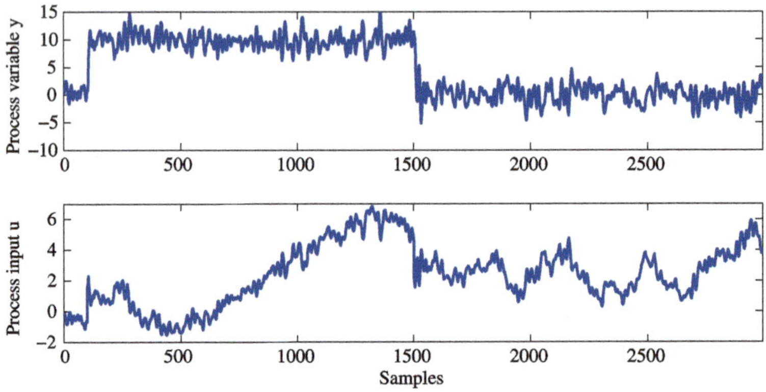

Fig. 14.13 Set-point response with the optimal controller

14.4 Iterative Controller Assessment and Tuning

The methods presented in this section do not require any effort to model the process dynamics. Only the time delay is assumed to be known. This is especially advantageous having in mind that process models are not often available in industry and that their development is expensive. Desborough and Miller (2002) estimated that process models are available for only *ca.* 1 % of chemical processes. The novel techniques of this section simultaneously provide the assessment of the controller performance and finding the optimal controller settings in an iterative way on the closed loop, following the procedure described in Sect. 14.1. Appropriate performance criteria or empirical characteristics of the impulse response will be used to control the progress of the iteration towards finding the optimal controller parameters.

14.4.1 Techniques Based on Load Disturbance Changes

The aim of the proposed methodology based on Visioli's area index (Sect. 5.4) is to verify, by evaluating an abrupt load disturbance response whether the tuning of the adopted PI controller is satisfactory, in the sense that it guarantees a good IAE. Based on this assessment, possible modifications of the controller parameters are suggested when the controller performance can be improved.

14.4.1.1 Basic Approach

In this section, a new method is proposed to generate a rule base for the tuning of PI controllers. It is based on the combination of the *area index* I_a, *idle index* I_i and the *output index* I_o, already defined in Sects. 5.3 and 5.4. The suggested tuning rules

Table 14.2 PI-controller tuning rules generated from Visioli's assessment rules in Table 5.7

	$I_i < -0.6$ (low)	$I_i \in [-0.6,\ 0]$ (medium)	$I_i > 0$ (high)
$I_a > 0.7$ (high)	Increase K_c	Increase K_c, Increase T_I	Increase K_c, Decrease T_I
$I_a \in [0.35, 0.7]$ (medium)	**K_c ok, T_I ok**	Increase K_c, Increase T_I	Increase K_c, Decrease T_I
$I_a < 0.35$ (low)	Decrease K_c; $I_o < 0.35$: decrease T_I	Decrease T_I	Decrease T_I

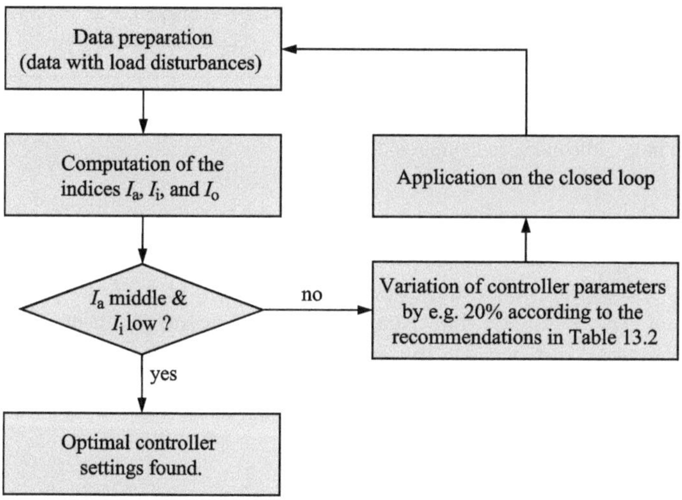

Fig. 14.14 Flow chart of the iterative controller tuning based on the combination of the area index, the idle index and the output index

are given in Table 14.2. The values 0.35 and 0.7 for I_a as well as -0.6 and 0 for I_i are just default values derived from many simulation studies and may be slightly modified depending on the application and design specifications at hand.

The tuning rules in Table 14.2 provide the basis for the iterative assessment and tuning procedure illustrated in Fig. 14.14. This new method assumes that data windows with step-wise or abrupt load changes exist, or additional OP steps $l(t)$ can be applied to the closed loop (Fig. 5.11). Model identification or the knowledge of the time delay is not needed. Since the indices used are sensitive to noise, an appropriate filtering is required. Possible filtering techniques have been described in Sect. 5.3.3.

The procedure in Fig. 14.14 starts with using a recorded data set containing load disturbances to compute the three indices, the area index I_a, the idle index I_i and the output index I_o. If the target region ($0.35 \leq I_a \leq 0.7$ & $I_i \leq -0.6$) is attained, the procedure terminates, and the current controller settings are the optimal ones. Otherwise, the controller parameters are changed according to Table 14.2, the current

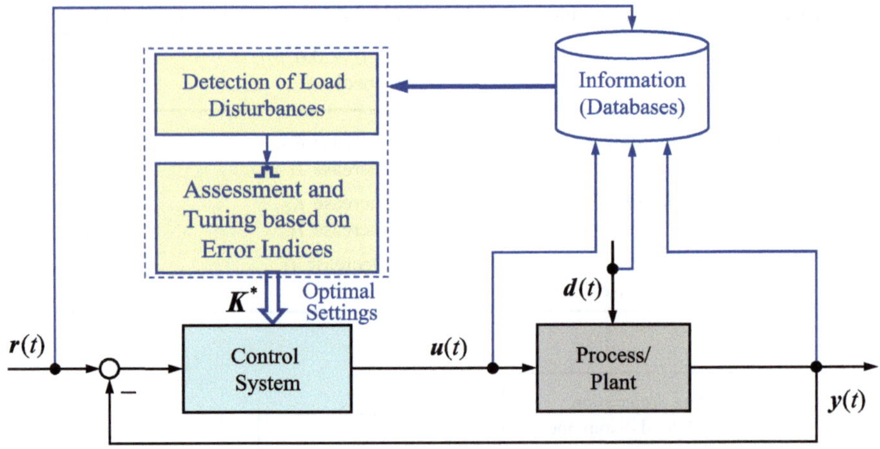

Fig. 14.15 Controller assessment and tuning based on load change detection and control error indices (area index, idle index and output index)

controller settings (ensuring a stable closed loop) are applied on the process, and a new operating data set is recorded. The same aforementioned steps are repeated. Of course, one can specify a target performance point (e.g., $I_a = 0.6$ & $I_i = 0.7$) rather than a "fuzzy" region, but usually this is not necessary in practice. We also do not recommend this because the number of iterations required would be much larger.

14.4.1.2 Detection of Load Disturbances

For this assessment and tuning method, it is essential to extract data windows with distinctive load changes occurring during normal process operation. Techniques are thus needed for automatic detection of these changes to be applied before activating the assessment and tuning task; see Fig. 14.15. In other words, the assessment and tuning algorithms must be provided with a supervisory shell that takes care of those operating conditions, in which the algorithm would give wrong performance indications.

The automatic detection of naturally occurred load disturbances (if any) is introduced here to make the method non-invasive. In addition to the method mentioned in Sect. 14.3.2.3, the technique proposed by Hägglund (1995) based on computing the IAE between zero-crossings of the control error (Sect. 9.4.3) can be used.

Example 14.5 To illustrate this new iterative tuning procedure based on the combination of the area index, the idle index and the output index, consider the following FOPTD process:

$$G_p(s) = \frac{1}{10s+1}e^{-5s}. \tag{14.24}$$

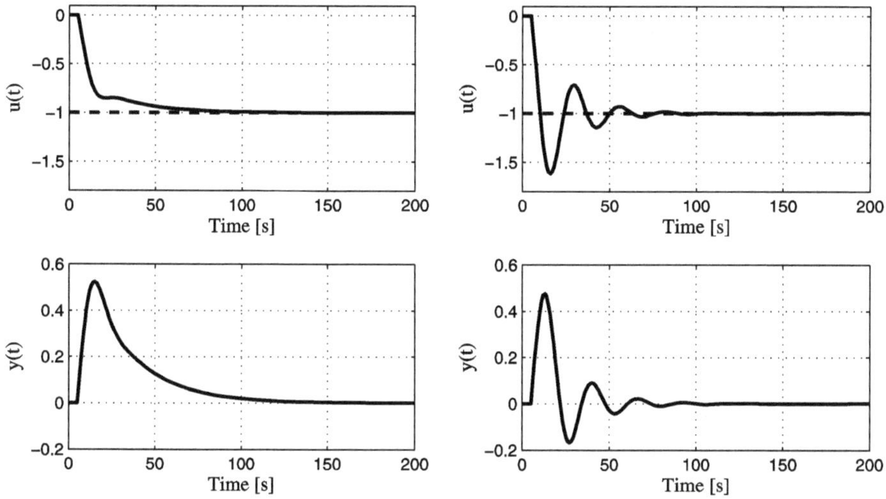

Fig. 14.16 Load change responses for initial controller (*top*) and optimal controller (*bottom*)

Table 14.3 Details of the iterative tuning process for Example 14.5

Iteration no.	K_c	T_I	I_a	I_a	*IAE*
0	0.90	5.00	0.68	−0.31	10.31
1	1.08	5.50	0.68	−0.40	9.21
2	1.30	6.05	0.66	−0.52	8.30
3	1.56	6.66	0.60	−0.57	7.70
4	**1.87**	**7.32**	**0.50**	−0.66	**7.68**

The initial PI controller was set to $K_c = 0.90$ and $T_I = 5.0$. Running the procedure in Fig. 14.14 with the rates of change $\Delta K_c = 20$ % and $\Delta T_I = 10$ % and applying a unit step in load disturbance on the process in each iteration leads to the final controller settings $K_c = 1.87$ and $T_I = 7.32$. The history of the iterative tuning process is shown in Table 14.3. The found settings are close to the optimal controller parameters $K_c^* = 1.81$ and $T_I^* = 10.36$ (corresponding to $IAE = 6.11$) given by Visioli (2005), minimising the IAE index. The responses to a unit load disturbance change before and after controller re-tuning are shown Fig. 14.16. It is observed how the proposed method correctly recognised the sluggish control and adjusted the controller setting to attain optimal behaviour from deterministic load disturbance performance view point.

14.4.2 Methods Based on Routine Data and Impulse Response Assessment

Usually, when the control system commissioning is completed and the plant is in normal operation, it is undesirable to perform even closed-loop experiments, required for the determination of the PID-achievable performance indices. This is particularly the case when the process is operated under regulatory control with only noise dynamics affecting the process. Obviously, this is not true when set-point changes naturally occur, but this situation is not the rule in process industries.

A methodology that can be very useful in the situations where experimentation with the process is not possible or undesirable at all (neither in the open- nor the closed-loop) has been proposed by Goradia et al. (2005). In this method, optimisation is carried out directly on the control loop by carefully and systematically changing controller parameters, thereby eliminating the identification step altogether. The objective is to iteratively improve the present controller settings until the PID-achievable performance is attained. To control the progress of the iteration, the Harris index is used as a measure of control loop performance improvement and the closed-loop disturbance IR curve as a diagnostic tool. This heuristic method seems to be appealing, easy to use and effective. Note again that this method intends to find the PID-achievable performance using routine data only, a highly desirable property in the industrial practice.

In the following, we first present the technique by Goradia et al. (2005) in detail and then provide many improvements to it.

14.4.2.1 Classification of Control Performance

Three classes of control behaviour have been defined by Goradia et al. (2005) to serve as a basis for the assessment and tuning of the controllers; each class contains three categories, as illustrated in Fig. 14.17.

These patterns were obtained by Goradia et al. from analysing more than 20 simulated case studies involving a wide range of process and noise dynamics. The process tried range from first order to higher orders, open loop stable to open loop unstable and noise dynamics from integrating noise to noise affecting the process at more than one place. The three classes of control behaviour defined above are used to assess and tune the controllers are characterised as follows:

1. *Under-Tuned Controllers*. The first class of under-tuned controllers shows similar impulse responses, which can be divided into three categories. The first category is *very sluggish* with or without offset (Fig. 14.17a) and has no undershoot and no oscillations. The other category shows no offset, no oscillation and no undershoot (Fig. 14.17b). These responses characterise slightly aggressive tuning compared to that of the previous category. The third category in this classification (Fig. 14.17c) shows impulse responses with slight undershoot and mild (one or two) oscillations obtained by keeping the controller settings slightly less aggressive than that of optimally tuned controller, termed *slightly detuned controller*.

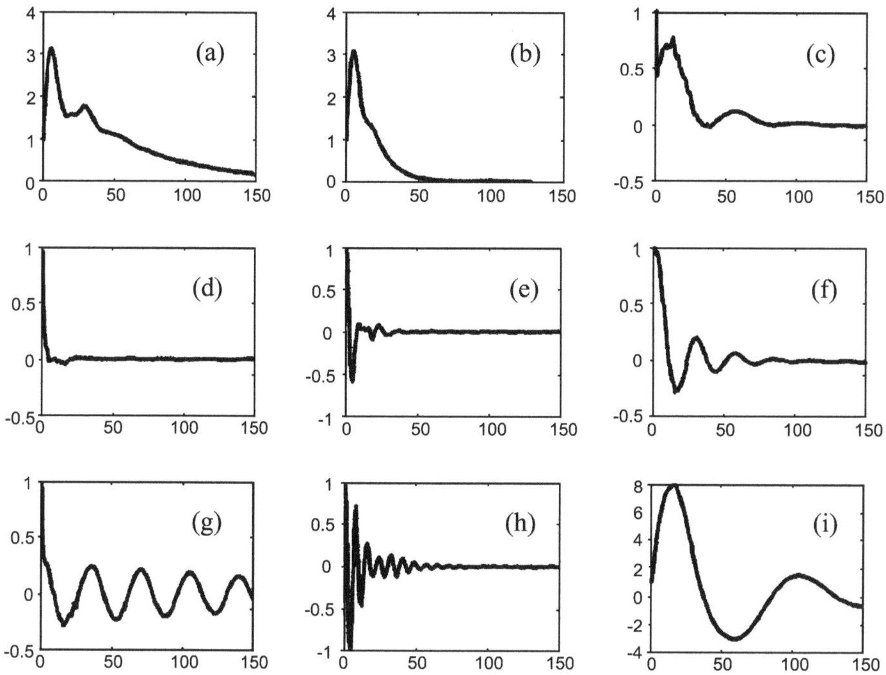

Fig. 14.17 Standard nine signature patterns of the disturbance impulse response for controllers: (**a**) extremely detuned; (**b**) detuned; (**c**) slightly detuned; (**d**) optimally tuned; (**e**) optimally tuned; (**f**) optimally tuned; (**g**) extremely aggressive; (**h**) aggressive/very oscillatory; (**i**) mildly aggressive (Goradia et al. 2005)

2. *Optimally-Tuned Controllers*. The second class was obtained by keeping the controller parameters at optimal settings obtained via optimisation with known process and noise models for various processes and noise dynamics. All the impulse responses for this class of tunings are divided into three categories. The first category shows under-shoot of -0.05 with a few oscillations (Fig. 14.17d). If the IR is similar to Fig. 14.17d, the PI-achievable performance is often close to minimum variance performance, i.e. $\eta \approx 1$, and one may not wish to tune the controller any further. This type of response is characteristic for systems that are less sensitive to controller parameters and when the disturbance affecting the loop is not severe. The impulse responses in Fig. 14.17e and Fig. 14.17f were from optimally tuned loops of all the other types, where PI-achievable performance is far from the minimum variance performance. As we move from Fig. 14.17d to Fig. 14.17f, the PI-achievable performance will further drift from unity. One of the reasons for decreasing the performance index is increase in settling time of IR, when moving from Fig. 14.17d to Fig. 14.17i. Since we do not make any effort in modelling either the noise or process dynamics, it is better to check whether one can still obtain better performance index by retuning the controller (even though the closed-loop disturbance impulse response suggests that the con-

troller is in proximity of optimally tuned controller, i.e. IR similar to Fig. 14.17e and Fig. 14.17f).
3. *Aggressive Controllers*. The third class of impulse response was obtained by keeping the controller parameters aggressive compared to that of optimally tuned controller to different degrees. The first category shows oscillations that do not decrease in amplitude (Fig. 14.17g), i.e. limit cycle. This very oscillatory IR plot with undershoot of −0.2 or more is obtained when controller is extremely aggressive and the loop is on the verge of stability. The second and third categories are of undershoot greater than −0.55 and more than four oscillations (Fig. 14.17h). IR plot of a mildly aggressive controller (Fig. 14.17i) has a few oscillations and undershoot of −1.0 or more.

14.4.2.2 Basic Methodology

The step-by-step iterative procedure to attain the PI-achievable performance for linear processes with time delay is summarised following the original work by Goradia et al. (2005); see also the flow chart in Fig. 14.18.

Procedure 14.2 Iterative controller tuning based on routine data and assessment of impulse response (IR).

1. The first step is to obtain routine operating data of PV, calculate the Harris index (η) of the loop with a priori knowledge of process delay and to plot the estimated IR coefficients. The IRs are estimated using time series analysis, i.e. as by-product of the Harris index (Sect. 2.4). For a reliable estimate of η, several sets of routine data collected over different periods should be considered, and the average Harris index subsequently used.
2. The IR plot is compared with standard nine signature patterns in Fig. 14.17. The IR plot computed from routine data will fit in with one of the following possibilities:

   - *Case A*. If the plot is similar to the pattern of detuned controller (Fig. 14.17a to Fig. 14.17c), then the existing controller is under-tuned and needs to be made aggressive to attain PI achievable performance;
   - *Case B*. If the plot resembles the pattern of an optimally tuned controller (Fig. 14.17e to Fig. 14.17f), then the existing controller may be performing near the PI achievable performance. If the IR pattern is similar to that of Fig. 14.17d, the Harris index will be very close to 1, and one may not wish to tune the controller any further. To confirm this, one should make the controller aggressive and check for the improvement in η as suggested in Fig. 14.16. However, it depends on the application at hand and on the desired performance, i.e. which IR pattern is specified as optimal.
   - *Case C*. If the plot is similar to the pattern of aggressively tuned controller (Fig. 14.17g to Fig. 14.17i), then the existing controller is aggressively tuned and needs to be detuned to attain PI achievable performance.

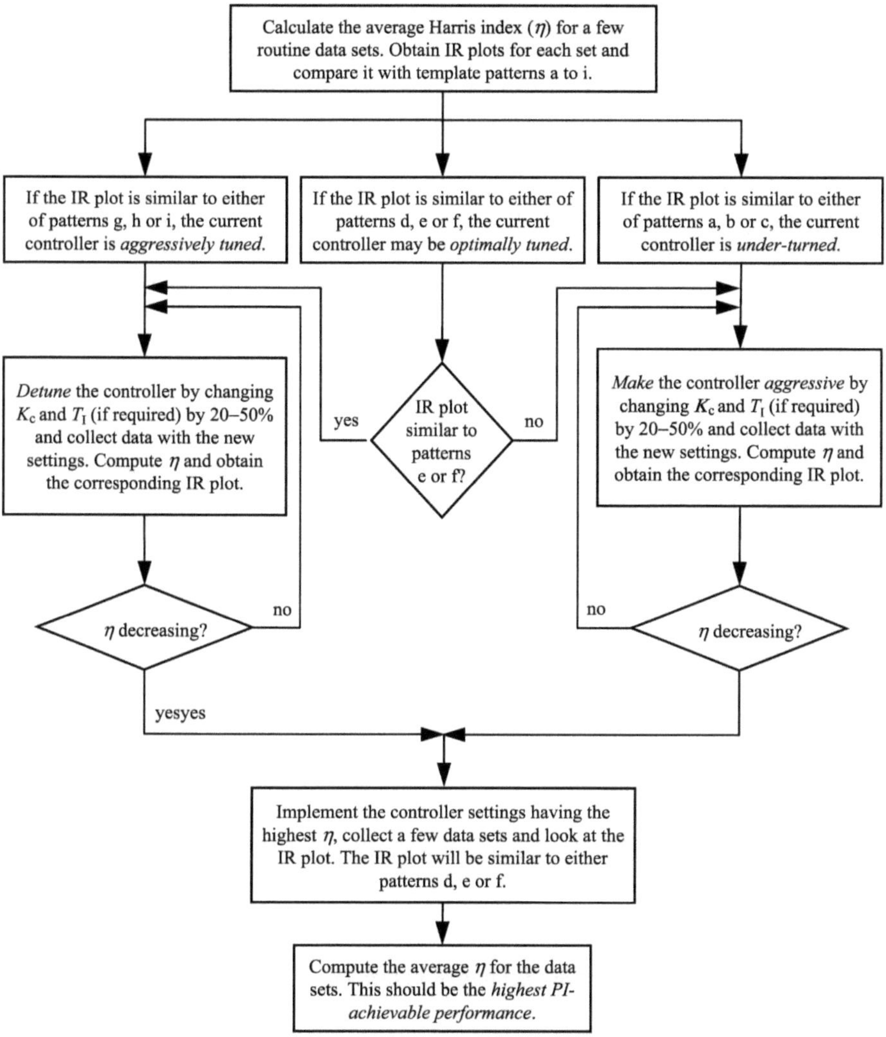

Fig. 14.18 Flow chart of the methodology proposed by Goradia et al. (2005) for attaining PI-achievable performance

In either Case A or C, the controller needs to be re-tuned to attain PI-achievable performance.

3. This is described assuming a detuned control (Case A); for the aggressively tuned controller (Case C), the action indicated in the bracket should be implemented. In Case B (optimally tuned controller), one needs to confirm that the controller is indeed performing close to the PI-achievable performance. If the IR plot is similar to Fig. 14.17e or 14.17f, consider the controller as aggressively tuned (Case C) and follow this step correspondingly. The reason for this is that one

should try to get the IR pattern similar to Fig. 14.17d, as this IR pattern is very close to IR pattern of minimum-variance control.

Increase (decrease) the controller gain or integral time depending on the obtained IR plot by, e.g. 20 %. Implement the new controller settings and collect the routine operating data. In simulations, these data are obtained with new controller settings along with a new random number seed. Using the newly obtained routine data, calculate a new η value and generate the IR plot. Compare the new η value with that in the previous iteration. If the change is greater than the inherent variation in the Harris index, then one can accept it as an increase in performance and proceed in that direction of making controller more aggressive (detuned). If the change in η is less than the inherent variation in η compared to the previous iteration, then one can increase the rate of change in controller gain up to 50 % for subsequent iterations.

Repeat Step 3 to make controller aggressive (detuned) until IR plot matches that of optimal PI settings for PI-achievable performance (any one of Fig. 14.17d to Fig. 14.17f). Once the optimal settings are reached, further increase (decrease) in controller parameters will decrease the Harris index, and the IR plot will resemble that of aggressive (detuned) controller. This indicates that the maximum Harris index has just been crossed. The optimal controller parameters are those at which η is the highest and IR plot resembles the signature of optimally tuned controller (Fig. 14.17d to Fig. 14.17f).

4. Implement the optimal controller parameters and collect some sets of routine operating data. Calculate η for each data set and then the average of the Harris index values. This average Harris index is expected to be close to the theoretical PI-achievable performance.

The reader should consult Goradia et al. (2005) for different examples, which show how this procedure works.

14.4.2.3 Introducing Impulse Response Features

One shortcoming of the method by Goradia et al. (2005) is that it may take up to a few days to arrive at PI-achievable performance if one is dealing with slowly sampled loops. But this is the price one must pay for not having a process model or avoiding any experimentation and still wanting to reach top performance. Nevertheless, the technique by Goradia et al. (2005) will now be modified to be suitable for the integration into an automated CPM and controller tuning software tool.

For this purpose, the following modifications and extensions are introduced here:

- The method is simplified by only considering the three main categories of the IR plot for sluggish, optimally tuned and aggressive controllers, shown in Fig. 14.19.
- Suitable measures are used to characterize the IR response and thus avoid the visual inspection to classify the controller behaviour.
- Pattern recognition techniques, such self-organising maps, are investigated for the automatic classification of the controller behaviour.

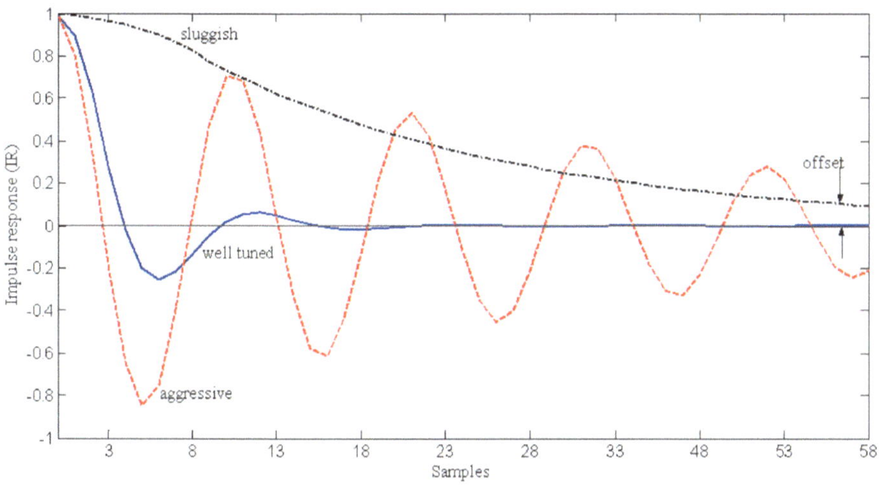

Fig. 14.19 Impulse-response plots for sluggish, optimally tuned and aggressive controllers

Characteristics are introduced for automatically determining the signature pattern (out of the three signature patterns) that best matches the estimated IR trajectory. One possibility is to calculate the IR characteristics *offset*, *undershoot* and *number of oscillations*, as described in Procedure 14.1, and use them for pattern matching.

An alternative way is to apply the area index (Sect. 5) to the IR trajectory rather than the control signal, i.e.

$$I_{\mathrm{ai}} := \begin{cases} 1 & \text{for } n < 3, \\ \frac{\max(A_1,\ldots,A_{n-2})}{\sum_{i=1}^{n-1} A_i} & \text{elsewhere,} \end{cases} \tag{14.25}$$

termed as *impulse-response area index* (*IRAI*). A_i stands for the area over or under the zero line of the IR curve:

$$A_i = \int_{t_i}^{t_{i+1}} \left|g(t)\right| \mathrm{d}t, \quad i = 0, 1, \ldots, n-1. \tag{14.26}$$

Note that load disturbance changes are not required here.

The procedure in Fig. 14.20 starts with using a routine operating data set to determine the impulse response (IR), the Harris index (η) and the impulse-response area index (I_{ai}). If the I_{ai} value lies in the target region, the controller parameters are slightly changed. It is then checked whether η is decreasing: if yes, the tuning is terminated, and the optimal controller settings are found. In the cases where the target region is not attained or the Harris index is not decreasing, the controller parameters are changed according to the selected variation strategy, the current controller settings (ensuring a stable closed loop) are applied on the process, and a new operating data set is recorded. The same aforementioned steps are repeated. The controller settings are systematically changed and applied to the closed loop until the IR area

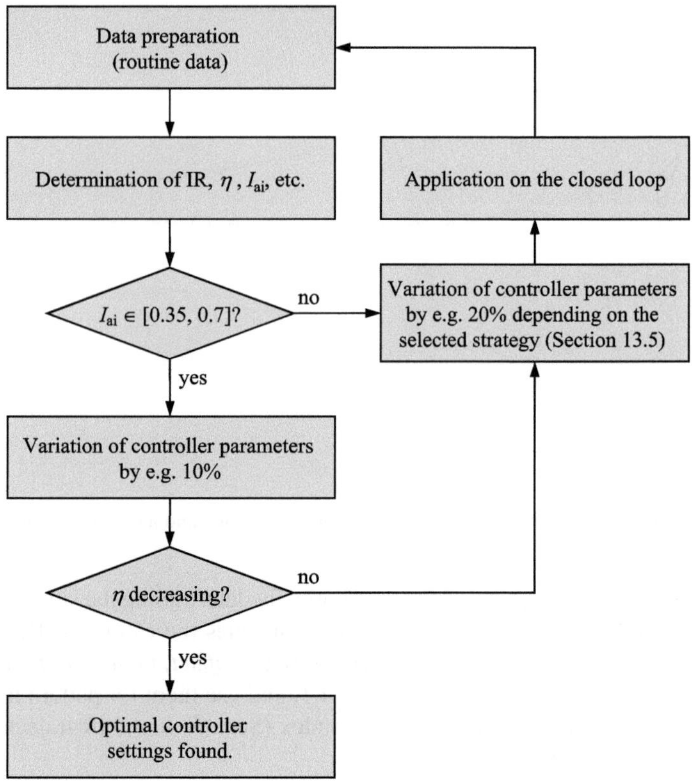

Fig. 14.20 Flow chart of iterative controller assessment and tuning based on routine data and the impulse-response area index

index I_{ai} is within the range $[0.3, 0.7]$ *and* the Harris index value η attains a satisfactory value, which should be close to the optimal (PID-achievable) performance. Figure 14.20 illustrates the flow chart of the new controller-tuning procedure.

Example 14.6 The iterative assessment and tuning procedure based on routine data and impulse response area index are illustrated on the process model

$$y(k) = \frac{q^{-1}}{1-0.8q^{-1}}u(k) + \frac{1-0.2q^{-1}}{1-q^{-1}}\varepsilon(k) \tag{14.27}$$

with $T_{\mathrm{s}} = 1$ s and $\sigma_\varepsilon^2 = 0.01$. For the initial PI controller settings $K_{\mathrm{c}} = 0.14$ and $T_{\mathrm{I}} = 7.0$, we get a Harris index value of $\eta = 0.34$, indicating poor control performance. Applying the proposed assessment and tuning procedure leads to the optimal controller settings $K_{\mathrm{c}} = 0.78$ and $T_{\mathrm{I}} = 4.1$, giving a Harris index value of $\eta = 0.97$. This means that approximately minimum variance can be achieved when re-tuning the PI controller. The result confirms the fact that PI controllers achieve the minimum-variance control performance for output disturbances represented by

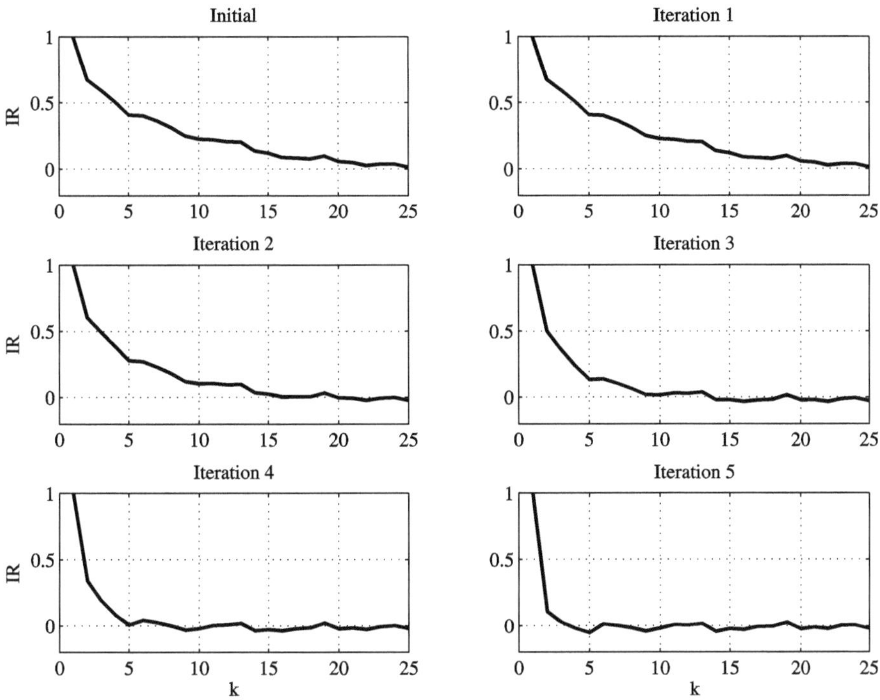

Fig. 14.21 Impulse response sequence obtained during the iterative tuning process for Example 14.6

Table 14.4 Details of the iterative tuning process for Example 14.6

Iteration no.	K_c	T_I	I_{ai}	η
0	0.14	7.0	1.00	0.34
1	0.21	5.6	0.90	0.50
2	0.32	4.5	0.72	0.69
3	0.47	3.6	0.70	0.85
4	0.71	2.9	0.66	0.92
5	**0.78**	**4.1**	**0.61**	**0.97**

an integrated moving average (MA(0, 1, 1)) process (Ko and Edgar 2004). The procedure has been terminated by achieving an IR area index value of $I_{ai} = 0.66$, which lies near the upper bound of the target region [0.3, 0.7]. The results obtained here are in excellent agreement with those achieved by Jain and Lakshminarayanan (2005), who analysed this example using their filter-based assessment method. The theoretically determined optimal controller has the parameters $K_c^* = 0.79$ and $T_I^* = 5.0$. The details of the iterative tuning process are given in Table 14.4 and Fig. 14.21.

Example 14.7 An open-loop stable, time delay process with integrating noise is considered in this example (Ko and Edgar 1998):

$$y(k) = \frac{0.1q^{-5}}{1-0.8q^{-1}}u(k) + \frac{1}{(1-0.6q^{-1})(1-0.3q^{-1})(1-q^{-1})}\varepsilon(k). \qquad (14.28)$$

The process is regulated by a (discrete) PI controller with the initial settings $K_c = 1.1$ and $T_I = 18.0$. These give a Harris index value of $\eta = 0.24$, indicating poor performance. Recall that these model details are only used for simulation, but only the time delay information is needed for the assessment and tuning using the presented method. This was run on a data set of 2000 samples gathered by simulation ($T_s = 1$ s and $\sigma_\varepsilon^2 = 1$), to yield the optimal PI parameters $K_c = 2.28$ and $T_I = 11.8$, using the step sizes $\Delta K_c = 20\ \%$ and $\Delta T_I = 10\ \%$. Four iterations were necessary to get these settings, which are close to the theoretical values $K_c^* = 2.32$ and $T_I^* = 11.2$ (Ko and Edgar 1998) and to those found by Goradia et al. (2005), $K_c^* = 2.72$ and $T_I^* = 11.0$.

14.4.2.4 Use of Pattern Recognition Techniques

As the shapes of the impulse response are clearly categorised in Fig. 14.19, they can also be detected using pattern recognition techniques. The basic principle is to match between the template shapes and those generated from data for the impulse response. A distance measure is defined to capture the similarity between the template and the test data. To achieve the best minimum possible, time alignment and normalisation are essential. For instance, dynamic time warping or artificial neural networks, e.g. self-organising maps (SOMs), can be used for pattern recognition.

Neural networks have been shown to be good at pattern-recognition problems. Their utility for pattern classification and fault diagnosis has been shown before, e.g. by Venkatasubramanian et al. (1990) and Kavuri and Venkatasubramanian (1994). Neural networks have been selected because of two main advantages. The number of classes to be delineated is quite small, and thus one can expect good performance from the network. Further, the networks are not sensitive to the discretisation size. Since the identification of primitives is solved purely as a pattern-recognition problem, treating windows with different numbers of data points is relatively easy (Rengaswany and Venkatasubramanian 1995).

Two key issues in training are that the training set should cover the space of interest as wide as possible and that a proper normalisation scheme for the training set is selected. In the case of this work, there are only a small number of output classes, i.e. three (or nine), to be trained, and hence exhaustive training is possible. Kohonen feature maps[1] were used in this work. Note that the training does not require measured data, i.e. simulation data to generate the reference pattern suffice.

---

[1]The used function for Kohonen feature maps has been implemented by Norbert Link.

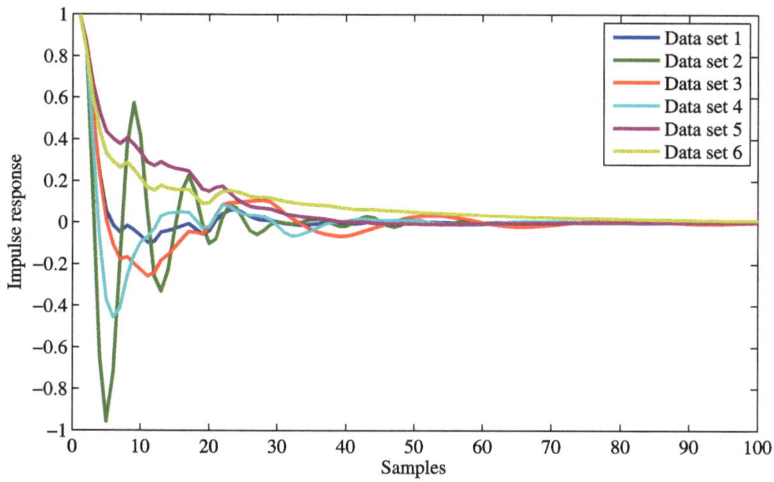

Fig. 14.22 Test data sets for the Kohonen feature maps

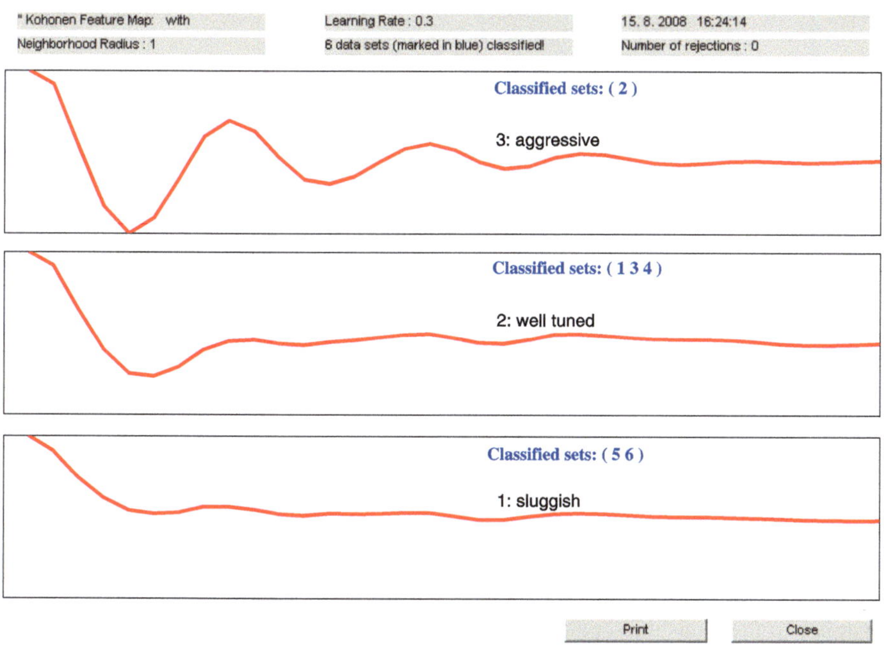

Fig. 14.23 Results of pattern detection using Kohonen feature maps for the test data in Fig. 14.19

As an illustrative example, we consider the IR test patterns shown in Fig. 14.22. The results of the pattern detection procedure are illustrated in Fig. 14.23. It is clearly observed that the algorithm correctly detects the pattern template (Fig. 14.19) for each test data set. The data set no. 2 has been classified as ag-

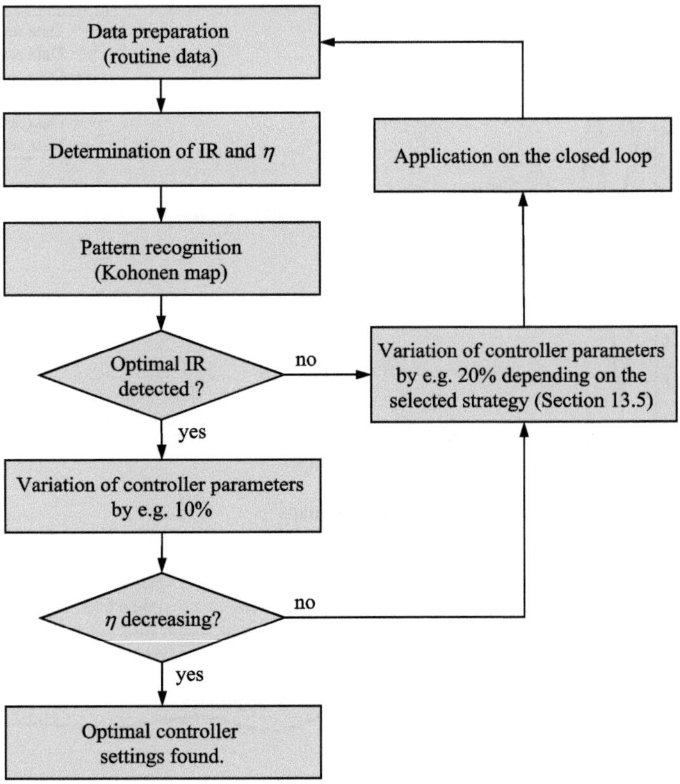

Fig. 14.24 Flow chart of iterative controller assessment and tuning based on routine data and pattern recognition

gressive, the data sets 1, 3 and 4 as well tuned, and the data sets 5 and 6 as sluggish. Recall that SOMs involve minimising a (Euclidean) distance measure to the prototypes.

The procedure in Fig. 14.24 starts with using a routine operating data set to determine the impulse response (IR) and the Harris index (η). Pattern recognition is run to detect the template IR shape giving the minimal dissimilarity measure to the test IR shape. If the IR shape for well-tuned control is detected, i.e. a specified minimum distance to the well-tuned pattern is achieved, the controller parameters are slightly changed. It is then checked whether η is decreasing: if yes, the tuning is terminated, and the optimal controller settings are found. In the cases where target region is not attained or the Harris index is not decreasing, the controller parameters are changed according to the selected variation strategy, the current controller settings (ensuring a stable closed loop) are applied on the process, and a new operating data set is recorded. The same aforementioned steps are repeated until a minimal distance to the IR pattern for well-tuned controller (to be specified) is achieved.

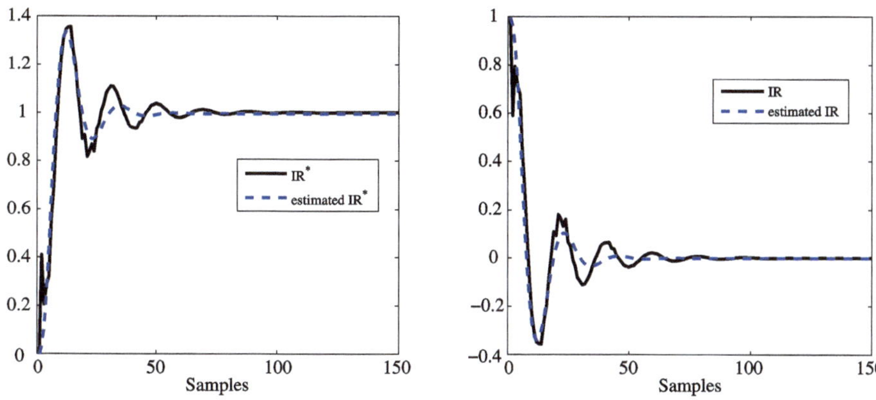

Fig. 14.25 Fitting of a SOPTD model to transformed IR (*left*); back-transformed IR and its approximation (*right*)

14.4.2.5 A New Control Performance Index: Relative Damping Index

An approach to automate the IR analysis is to fit a second-order-plus-time-delay (SOPTD) continuous model

$$G_{\mathrm{IR}}(s) = \frac{K_{\mathrm{IR}} e^{-T_{\mathrm{d,IR}} s}}{T_{0,\mathrm{IR}}^2 s^2 + 2T_{0,\mathrm{IR}} D_{\mathrm{IR}} s + 1} \tag{14.29}$$

to the impulse coefficients. The model estimation can be easily carried out, e.g. using the fminsearch function of the MATLAB Optimization Toolbox; see Sect. 5.2.3.3. The estimated parameters, the time delay $T_{\mathrm{d,IR}}$ and the damping factor D_{IR} provide measures of the disturbance rejection performance. $T_{0,\mathrm{IR}}$ gives an indication of how fast the disturbance is rejected by the controller. D_{IR} is related to its aggressiveness: if D_{IR} is greater than unity, the controller behaviour is over-damped; a value smaller than unity indicates that controller behaviour is over-damped with tendency to oscillate. Note that the IR sequence has be transformed is such a way that it shows a behaviour comparable with the step response of a SOPTD system. This can be achieved by $\mathrm{IR}^* = \max(\mathrm{IR}) - \mathrm{IR}$; see Fig. 14.25. For this example, a SOPTD model with the parameters $K_{\mathrm{IR}} = 1.0$, $T_{\mathrm{d,IR}} = 0.0$, $T_{0,\mathrm{IR}} = 3.35$ and $D_{\mathrm{IR}} = 0.35$ has been estimated.

To show how the IR damping factor provides an assessment of the IR responses and thus indicates in which direction the controller has to be re-tuned, a process described by

$$y(k) = \frac{0.2q^{-5}}{1-0.8q^{-1}} u(k) + \frac{1}{(1+0.4q^{-1})(1-q^{-1})} \varepsilon(k), \tag{14.30}$$

controlled by a PI controller is considered. Figure 14.26 illustrates the tuning map, i.e. the IR pattern and the values of D_{IR} as a function of the PI controller settings

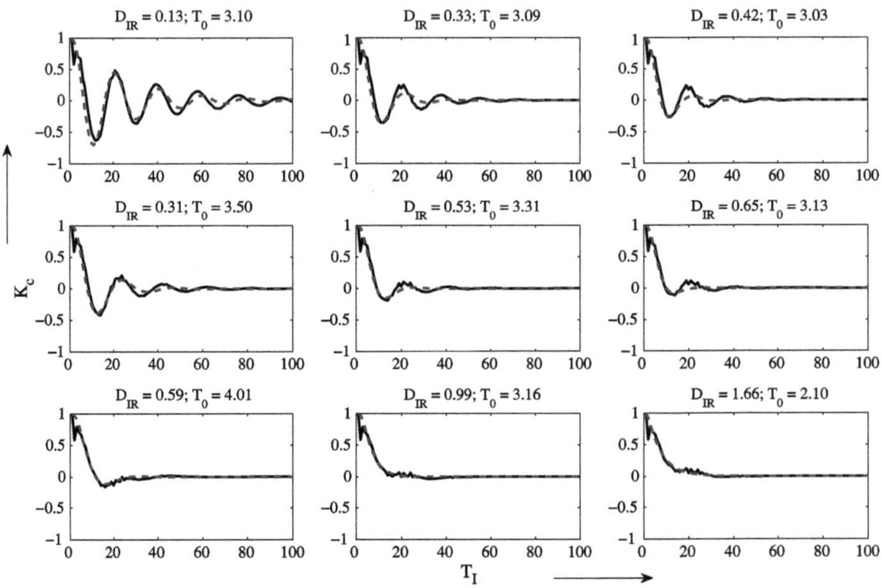

Fig. 14.26 A tuning map based on the IR damping factor

for this example. The central IR plot was generated using $K_c = 1.0$ and $T_I = 8.0$, which can be considered as close to the optimal PI parameters. For the surrounding plots, K_c and T_I have been decreased and increased by 40 % and 30 %, respectively.

To get a relative measure of performance, the (IR-based) relative damping index (RDI) is defined as

$$RDI = \frac{D_{IR} - D_{IR,aggressive}}{D_{IR,sluggish} - D_{IR}}, \tag{14.31}$$

where D_{IR} is the damping factor of the fitted model, $D_{IR,aggressive}$ the limit of aggressive controller behaviour, and $D_{IR,sluggish}$ the limit of aggressive controller behaviour. These performance limits should be selected according to the desired performance specification, typically $D_{IR,aggressive} = 0.3$ and $D_{IR,sluggish} = 0.6$. Note that a similar performance index has been recently introduced by Howard and Cooper (2008, 2010), but in relation with the auto-correlation function.

The RDI can be interpreted as follows:

- If $RDI \geq 0$, i.e. $D_{IR,aggressive} \leq D_{IR} \leq D_{IR,sluggish}$, the control performance is good.
- If $-1 \leq RDI < 0$, i.e. $D_{IR} < D_{IR,aggressive}$, the control behaviour is aggressive.
- If $RDI < -1$, i.e. $D_{IR} > D_{IR,sluggish}$ the control behaviour is sluggish.

Based on the RDI, a new straightforward strategy for optimal controller re-tuning is proposed, as shown in Fig. 14.27. The procedure starts with using a routine operating data set to determine the impulse response (IR) and the Harris index (η). The IR pattern is fitted to a SOPTD model to compute the damping factor D_{IR} and

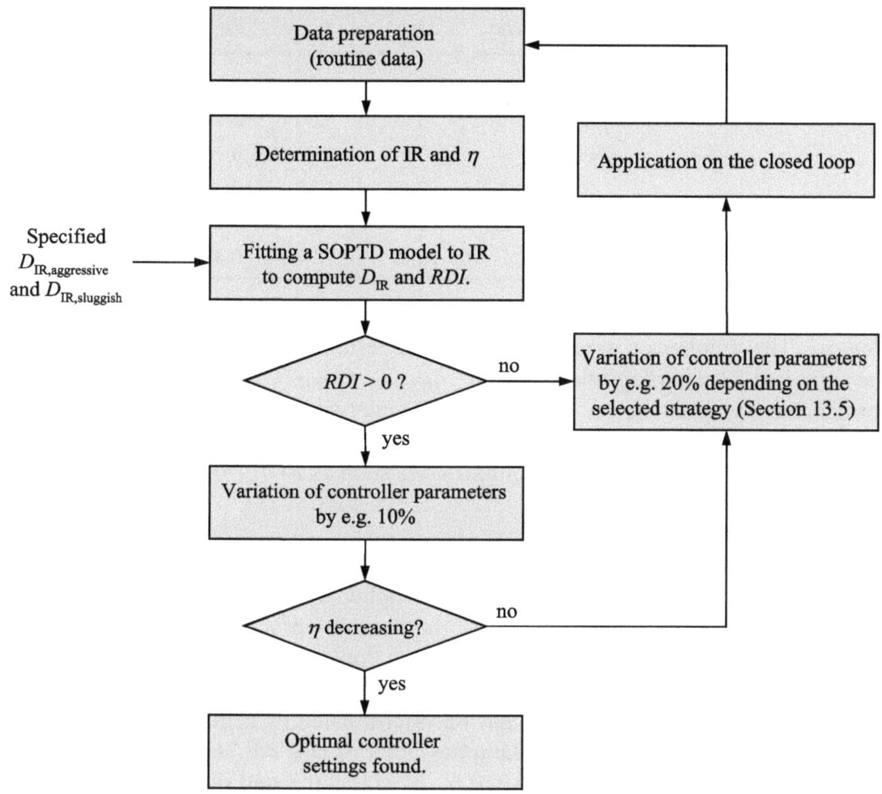

Fig. 14.27 Flow chart of iterative controller assessment and tuning based on the relative damping index

the RDI. As long as $RDI < -1$ or the Harris index is not decreasing, the controller parameters are changed according to the selected variation strategy, the current controller settings (ensuring a stable closed loop) are applied on the process, and a new operating data set is recorded.

This method is intuitive and can be completely automated. Simultaneously, the user has the possibility to specify the target performance region by selecting the corresponding limits $D_{\mathrm{IR,aggressive}}$ and $D_{\mathrm{IR,sluggish}}$. For stricter performance requirements, $D_{\mathrm{IR,aggressive}}$ has to be increased, and $D_{\mathrm{IR,sluggish}}$ decreased.

Remark As the iterative techniques presented above mimic the work of an optimisation routine, there is a risk of trapping in local minima. This can happen when the objective function has such minima and bad starting parameters are selected. Therefore, it may be sometimes necessary to repeat the re-tuning task for different starting points, which means that the optimisation work takes longer time.

Table 14.5 Details of the iterative tuning process for Example 14.8

Iteration no.	K_c	T_I	D_{IR}	RDI	η
0	0.14	7.00	1.24	−1.47	0.34
1	0.21	5.60	0.57	9.54	0.49
2	0.27	6.72	0.70	−3.95	0.55
3	0.40	5.38	0.44	0.86	0.68
4	0.53	6.45	0.42	0.67	0.74
5	**0.69**	**7.74**	**0.38**	**0.36**	**0.77**

Remark The number of iterations needed for re-tuning the controller depends on the specified IR damping interval $[D_{IR,aggressive}, D_{IR,sluggish}]$. The stricter the performance requirement is, i.e., the narrower the interval limits are, the more iterations will be required. Moreover, it is recommended to start with a large step size if the initial controller behaviour is too sluggish. As soon as $RDI \geq 0$ is reached, the step size should be reduced.

Example 14.8 Example 14.6 is reconsidered here. The iterative RDI-based assessment and tuning procedure in Fig. 14.27 was applied to the loop with $D_{IR,aggressive} = 0.3$ and $D_{IR,sluggish} = 0.6$. The results of the re-tuning process are illustrated in Table 14.5. The proposed technique leads to the "optimal" controller settings $K_c = 0.69$ and $T_I = 7.74$, giving a Harris index value of $\eta = 0.77$. This tuning is close to the optimal one. The procedure has been terminated by achieving an RDI value of 0.36. This corresponds to an IR damping factor of $D_{IR} = 0.36$, which lies near the lower bound of the target region [0.3, 0.6]. Note that the step size has been reduced after reaching $RDI \geq 0$ to the first time.

14.5 Strategies for Variation of Controller Parameters

Iterative controller tuning requires the specification of a proper step size for each controller parameter, usually given as percentage. Cautious adjustments to the controller parameters are necessary to guarantee closed-loop stability and performance improvement. There are many strategies for varying the controller settings in each iteration. Some of them are described in the following, including their strengths and weaknesses. Basically, a large step size helps reduce the number of iterations required but may increase the risk to converge to controller parameters that are far from optimum.

14.5.1 *Variation of Proportional Gain Alone and Fine Tuning of Integral Time*

The simplest approach is to vary only the proportional gain (K_c) unless the existing controller is either too sluggish or too aggressive. In such cases, the integral time

(T_I) can also be changed (but only one parameter at a time). Otherwise, the variation of K_c only should lead one to a value near the optimum, and then a "fine tuning" could be done by slight variation of T_I. Results from many simulations showed that an initial change of 20 % in K_c or T_I in each iteration is reasonable to improve the controller without destabilising the loop. If the resulting change in the performance index η is not significant, then the controller gain or integral time can be gradually increased up to 50 % in the subsequent iterations (Goradia et al. 2005). However, this assumes that the current T_I value is not far from the optimum. It should be clear that the number of iterations required for finding the optimum depends directly from the step size.

Practically, some integral action is always desired in industrial environment for offset free set-point tracking and rejection of step-type disturbances. Hence, there should be at least moderate integral action in the final controller suggested, though this could come with a marginal drop in η.

Moreover, Goradia et al. (2005) pointed out that the performance index takes a unimodal locus. Once the proper direction to improve the controller performance is determined, i.e. to make the controller aggressive or detuned, one can proceed iteratively in that direction as long as η continues increasing. After reaching the peak, η starts to decrease even though we are moving in the same direction. The peak value of η is the PI-achievable performance. The controller parameters are to be changed according to the guidelines mentioned in Sect. 14.4.2.2. The change in CPI in any iteration should be accepted only if it is greater than or less than the inherent variation in the CPI (determined in Step 1 in the procedure given in Sect. 14.4.2.2).

14.5.2 Simultaneous Variation

The simultaneous adjustment of the controller settings, i.e. decreasing K_c and increasing T_I (and possibly decreasing T_D) when the controller is aggressive and vice versa in the case of a sluggish controller, is the fastest method to find the optimum tuning. However, this approach is not transparent in practice and should be only considered by well-qualified users.

14.5.3 Successive Variation

In this approach, the proportional term is tuned first until the highest performance index value is reached. This is followed by tuning the integral time and possibly derivative time, which may lead to further improvement in η. The three cases of controller tuning in this strategy are as follows:

1. If the estimated IR is similar to the sluggish IR profile, increase the proportional gain, K_c, until the highest η has been reached. Then, check the IR against the signature IR plots. If the controller is still sluggish (aggressive), decrease (increase) T_I until the highest η is obtained.

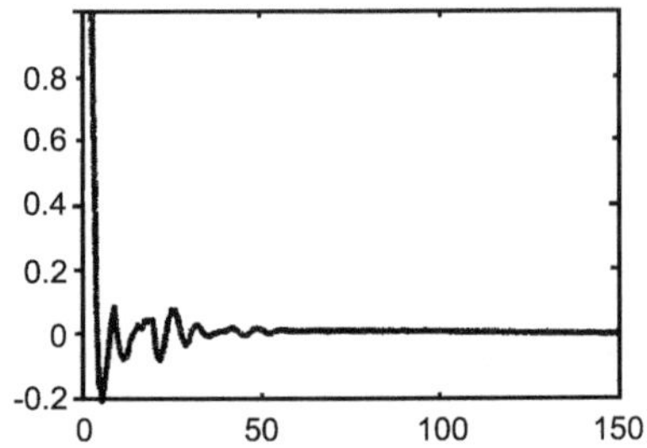

Fig. 14.28 Closed-loop disturbance impulse response with optimally-tuned PI controller for integrating process undershoot of −0.2 and a few oscillations (Goradia et al. 2005)

2. If the estimated IR is similar to the optimal IR profile, the controller is approaching optimal performance and does not need to be tuned too much. Slight tuning of the parameters by about 10 % is sufficient to obtain the maximum performance index, i.e. to make sure that the maximum η has been crossed.
3. If the estimated IR is similar to the aggressive IR profile, decrease K_c until the highest η has been reached. Then, check the IR used to determine the optimal controller settings against the signature IR plots. If the IR plot is sluggish (aggressive), decrease (increase) T_I until the highest performance index is obtained.

This approach is highly recommended in practice owing to its transparency, although it usually takes more iterations than the simultaneous strategy. In this context, one should be careful when the controller has low proportional gain and high integral action, resulting in a slow oscillatory closed-loop disturbance IR. This might be misunderstood as aggressive controller tuning unless other careful observations are made. Besides undershoot and number of oscillations, one should also look at the time when first "zero crossing" of the IR occurs. If it is very long compared to the process delay (which is assumed to be known), it suggests that the oscillations may be due to a less aggressive controller regulating an integrating or lag dominant process. Hence, one should try to decrease the controller integral action to the minimum required (to reject the low-frequency disturbance) as a first step and then compare its IR to the standard patterns (Fig. 14.17). This will eliminate the possibility of confusing the oscillatory IR due to higher integral action with that of real aggressive tuning. One should replace Fig. 14.17f with Fig. 14.28 in the standard templates as the only change in Goradia's procedure when dealing with such processes.

14.5.4 Constraints and Loop Stability

In the context of parameter optimisation considered here for controller tuning, it is decisive to carefully formulate the optimisation task. This is because even not correctly formulated problems can be "solved". Therefore,when using optimisation, it is essential:

- To carefully formulate the criterion to be minimised;
- To include all relevant constraints (in terms of stability and robustness);

Table 14.6 Comparison of the CPM-based controller tuning methods

No.	Method	Required data and models/parameters
1	Parameter optimisation based on complete knowledge of process models (Sect. 14.3.1)	• Knowledge of the process model (incl. time delay) and routine operating data for the estimation the disturbance model. • Data from active experiments on the open system for the identification of the process model and routine operating data for the estimation the disturbance models.
2	Parameter optimisation based on routine and set-point response data (Sect. 14.3.2)	• Time delay and • Data for the identification of ARMAX model of the closed loop.
3	Iterative tuning based on load disturbance changes (Sect. 14.4.1)	Data for the calculation of the area index, idle index and output index.
4	Iterative tuning based on impulse response assessment (Sect. 14.4.2)	• Time delay and • Routine operating data for the estimation the disturbance model (incl. impulse response).

- To be aware of the several pitfalls of optimisation, such as the existence of local minima;
- To realise that the computation burden may be excessive.

If a system model is available, e.g. from the commissioning stage, or can be estimated from system identification, it is always advised to simulate the closed loop and ensure that each controller parameter combination produces a stable system. For this purpose, the poles of the controlled system have to be checked: if they are found to be unstable, i.e. outside the unit circle (for time-discrete systems), the corresponding parameter combination must be removed. Note that this also applies for the iterative tuning methods introduced in Sect. 14.4. As there is no need for having a model for these methods, knowledge of the ranges of the controller parameters ensuring a stable closed loop helps avoid trial-and-error and the related risk of plant shut-down.

14.6 Comparative Studies

Table 14.6 provides a summary of the required parameters for the CPM-based controller tuning methods presented in this chapter.

Inspecting this table reveals the following points:

- Iterative tuning based on impulse response assessment is the most appealing strategy in practice, as it is completely non-invasive and necessitates a minimum of process knowledge.

Table 14.7 Tuning results for Example 14.9 (Initial parameters: $K_c = 0.14$; $T_I = 7.0 \Rightarrow \eta_0 = 0.33$)

Method No.	$\boldsymbol{K}^*_{PI} = [K^*_c; T^*_I]$	η^*	N_{iter} (Variation steps)
1	[0.79; 5.00]	0.98	–
2	[0.80; 5.30]	0.99	2
3	[0.61; 5.70]	0.93	12 ($\Delta K_c = 30$ %, $\Delta T_I = 20$ %)
4	[0.78; 4.1]	0.97	5 ($\Delta K_R = 20$ %, $\Delta T_I = 10$ %)

- If set-point changes occur during normal process operation, parameter optimisation based on routine and set-point response data is simple and effective.
- If step-wise/abrupt changing load disturbances act on the process and these changes can be detected properly, iterative tuning based on load disturbance changes may be useful.
- Parameter optimisation based on complete knowledge of process models is the most involved tuning technique and will not be the first choice in practice, unless accurate models are available or can be estimated from routine operating data or with a minimum of experimentation on the closed loop.

The four strategies have been implemented by the author in MATLAB and tested on numerous and different simulated processes under a range of conditions, i.e. self-regulating/integrating, delay-free/delayed, stable/unstable, different initial controller settings, measurement noise, etc. In the following the results are presented for three examples which have often been used in literature as benchmarks for controller tuning methods. The tables below contain the optimal controller settings $\boldsymbol{K}^*_{PI}$ achieved, the corresponding value of the performance index η^* and the number of iterations N_{iter} required for the iterative techniques.

Example 14.9 A delay-free process described by

$$y(k) = \frac{q^{-1}}{1-0.8q^{-1}}u(k) + \frac{1-0.2q^{-1}}{1-q^{-1}}\varepsilon(k) \tag{14.32}$$

controlled by a PI controller is considered. The disturbance noise has the variance $\sigma_\varepsilon^2 = 0.01$. In this case, a PI controller may achieve the minimum variance. Indeed, the results given in Example 14.9 show that MVC performance is attained. All methods yield (nearly) optimal (stochastic) performance and parameter settings in the neighbourhood of the optimal ones (given by Method no. 1); see Table 14.7. We found that Method no. 2 may be sensitive to the choice of the ARMAX model parameters. The results in Table 14.7 were obtained based on the identification of an ARMAX(3, 2, 2, 1) model, and two iterations were needed. However, this is not generally the case.

Table 14.8 Tuning results for Example 14.10 (Initial parameters: $K_c = 1.6$; $T_I = 15.0 \Rightarrow \eta_0 = 0.37$)

Method No.	$\boldsymbol{K}^*_{PI} = [K^*_c; T^*_I]$	η^*	N_{iter} (Variation steps)
1	[2.57; 10.25]	0.50	–
2	[2.61; 9.77]	0.50	–
4	[2.30; 9.60]	0.50	3 ($\Delta K_c = 20$ %, $\Delta T_I = 20$ %)

Example 14.10 Now we consider a process with time delay and affected by non-stationary disturbances ($\sigma^2_\varepsilon = 0.001$):

$$y(k) = \frac{0.1q^{-5}}{1-0.8q^{-1}}u(k) + \frac{1}{(1-0.6q^{-1})(1-0.3q^{-1})(1-q^{-1})}\varepsilon(k). \tag{14.33}$$

In this case, a PI controller has no chance to attain MVC performance. The results in Table 14.8 confirm that 50 % of the minimum variance can be maximally achieved, $\eta = 0.5$. Method no. 3 is not suitable for such scenarios. This case typically show that MVC-based benchmarking may not be the right (i.e. realistic) benchmarking option for PID-controlled loops. Rather the optimal PID controller itself should be taken as benchmark. Considering this, even the initial controller used does not have poor performance. Note that a predictive controller (compensating the time delay) is able to further reduce the variance, so that a re-design of the controller would pay off.

Example 14.11 Consider a second-order process with time delay, represented by the time-continuous model

$$y(s) = \frac{1}{s^2+2s+1}u(s). \tag{14.34}$$

The same disturbances as in Example 14.9 act on the process. The preferred Methods no. 2 and 4 yield optimal controller tunings, which are very similar to those found by other researchers by applying traditional tuning methods; see Table 14.9. It is worth stressing that Method no. 4 only needs solely routine data and the knowledge of the time delay, i.e., no invasive experimentation with the plant is necessary neither in open- nor closed-loop.

14.7 Summary and Conclusions

This chapter has provided new techniques for automatic generation of controller settings based on the continuous assessment of the control loops using normal operating data. Four categories of CPM-based re-tuning methods have been presented,

Table 14.9 Tuning results for Example 14.11 (Initial parameters: $K_c = 1.0$; $T_I = 3.0$; $T_D = 2.0 \Rightarrow \eta_0 = 0.51$)

Method No.	$\boldsymbol{K}^*_{PI} = [K^*_c; T^*_I; T^*_D]$	η^*	N_{iter} (Variation steps)
2	[1.41; 1.62; 1.03]	0.64	–
4	[1.76; 1.47; 0.59]	0.57	6 ($\Delta K_c = 20\,\%$, $\Delta T_I = 10\,\%$, $\Delta T_D = 5\,\%$)
Krishnaswamy et al. (1987)	[1.89; 2.13; 0.53]	0.60	–
Yuwana and Seborg (1982)	[1.53; 2.42; 0.61]	0.64	–

and their properties discussed. Numerous illustrative examples showed the relative efficiency of the techniques.

It can be concluded that parameter optimisation based on complete knowledge of process models is the most involved tuning technique and should not be the first choice in practice, unless accurate models are available. If set-point changes occur during normal process operation, parameter optimisation based on routine and set-point response data can be effective. If step-wise/abrupt changing load disturbances act on the process, iterative tuning based on load disturbance changes may be useful, provided that the changes can be detected properly.

Therefore, iterative tuning based on impulse response assessment is the best suited strategy in practice, as it is completely non-invasive and necessitates a minimum of process knowledge. The approach mimics the way of solving model-based optimisation problems. Starting from the Harris index value computed from routine data under the installed controller, the controller settings are cautiously updated, and applied on the process, new data are used to recalculate the Harris index, until the optimal controller settings, which maximise the performance index, are attained. Impulse-response features and pattern-recognition have been introduced, to automate the iterative controller assessment and tuning. There is no need for the injection of any input or reference dither signals, as is typically the case for closed-loop identification or for assessment methods based on such experiments. There is also no need for any performing recycling experiments as is the case in iterative feedback control. Some guidelines have been given how to select the step size for updating the controller settings.

Although the methods presented attempt to re-tune PI(D) controllers, the strategies included can be adopted for other controller types.

Part IV
Applications and Tools

Chapter 15
Industrial CPM Technology and Applications

The growing acceptance of the CPM technology in some industries is due to the awareness that control software is recognised to be a capital asset that should be maintained, monitored and revised routinely. Control systems permanently showing top performance significantly reduce or even avoid product-quality degradation, loss of energy resources, waste of production, lost production time and shortened lifetimes for plant components. The CPM field has now matured to the point where a large number of industrial applications and some commercial algorithms and/or vendor services are available for control performance auditing or monitoring.

The primary purpose of this chapter is to present an overview of CPM industrial applications and (available) software products. The requirements for CPM algorithms and packages are given in Sect. 15.1. Section 15.2 contains a comprehensive overview of published CPM applications to industrial processes. Commercially available CPM products are given in Sect. 15.3.

15.1 Demands on Performance Monitoring Algorithms

CPM algorithms aim to calculate performance indices repeatedly over time and comparing them to alert limits. Each alert limit can be decided from statistical characteristics of the index or by some other criteria. Desired properties of CPM algorithms and tools have been stated by many authors, e.g. Horch (2000), Vaught and Tippet (2001) and Ingimundarson (2003). The most important of these properties are:

- *Non-invasiveness.* CPM procedures should run without disturbing the normal operation of the control loops. Data needed for assessment shall be acquired under normal plant production conditions without any additional excitations. However, careful inspection of the collected (routine operating) data is recommended, as they may be not as informative as they would be if a substantial external excitation were introduced in the system.

M. Jelali, *Control Performance Management in Industrial Automation*, Advances in Industrial Control, DOI 10.1007/978-1-4471-4546-2_15,

- *Ability to Run Automatically.* Ideally, a CPM system does need only little or no manual intervention of the operators or engineers.
- *Use of Raw Data.* The use of archived (usually modified) data is not advisable at all. For instance, data smoothing, data compression and data quantisation affect the calculated performance indices (the loop performance will be over-estimated) and thus should be avoided.
- *Detection of Bad or Under-performing Control Loops.* This is the core aim of the monitoring and assessment of the control loops. Usually, a host of control loops, usually embedded in different levels, will be evaluated. Different methods for performance assessment should be applied; see Chaps. 2–7.
- *Low Error Rate.* False alerts occur when the algorithm signals bad performance even though the performance is actually good. Missed detections are those situations where the algorithm should give alert but does not. Too many false alerts or missed detections result in a reduced trust and use of the CPM system. In practice, it is then very likely that the system is ignored or even switched off.
- *Diagnosis of Under-performance Causes.* The determination of the reason(s) of poor performance is a much harder task than detecting poorly performing loops, as there are only a few systematic ways of detecting the underlying causes. Candidate reasons for poor control performance are (i) limitations on achievable performance arising due to a combination of system and controller design, (ii) changes in system dynamics, (iii) varying disturbances, (iv) sensor faults, (v) system nonlinearity and (vi) unknown sources; see Chaps. 9–12.
- *Suggestion of Suitable Measures to Remove the Root-cause(s) of the Performance Deterioration.* Ideally, the measures should indicate what should be done to improve the control, whether the problems may be overcome by re-tuning the controller(s), introducing a new (advanced) controller structure (e.g. to compensate for time delays), or re-designing some system components (such as valves due sticking), etc. (Chaps. 13 and 14).
- *Appropriate Presentation of the Results to the User (Human–Machine Interface).* The interface is often the key to the user acceptance and therefore must be intuitive and easy to use. The interface provides a summary of problem areas that may exist in the plant, as well as a detailed presentation of the data collected and the analysis done. The results will serve for plant staff and for maintenance purposes. Thereby, one should avoid providing too much information content, as this leads to an increase of complexity, requiring more knowledge for the interpretation of the results and suggested measures; see Chap. 15.

The desirable properties mentioned cannot be simultaneously attained. Therefore, only a compromise between these conditions has to be achieved in actual implementations of CPM and diagnosis systems, depending on the preferences of the specific customers/users. Also, the performance of control loops is always subject to a number of practical limitations. These arise from plant dynamics, such as time delay, non-minimum phase behaviour, saturations and dynamics of actuators, noise characteristics, the effect of model uncertainty (particularly when the controller is model-based) and nonlinearities; see Patwardhan and Shah (2002) for a nice discussion of this topic. All this is important information when monitoring/assessing the performance of control loops. Methods have been presented in Chap. 14 to monitor performance to the best achievable subjected to the known limitations, such as controller structure limitations.

One of the main advantages of this approach is that only information about these limitations is required. One of the disadvantages is that there are many types of limitations and it is difficult to know them at all operating levels. Furthermore, performance far away from the optimal one might be perfectly acceptable in some situations (Horch 2000). In other situations, performance may be not acceptable even if it is close to what is optimally achievable considering the limitations. Sometimes, the control objective is not to keep the process at a set point, e.g. in level control of surge vessels. The purpose is to dampen the changes in controlled flow while keeping the liquid level in the vessel between limits. This is indeed contrary to the implicit design of conventional (PID) controllers (Hugo 2001). A variety of techniques have been development by Horton et al. (2003) for the specific assessment of industrial level controllers.

15.2 Review of Control Performance Monitoring Applications

Control performance monitoring applications have been found in about 112 publications (including a few Ph.D. theses and technical reports) appeared during 1989–2010 (without claiming completeness). Note that only those applications have been included which are published and where a minimum of statements (description of plant, control objectives and assessment results) is given. Pilot studies, simulations and advertisement-oriented "success stories" of CPM product vendors were not considered. The survey revealed a remarkable number of application-case studies to date.

Besides, other published industrial case studies can be found in Stanfelj et al. (1993), Hägglund (1999), Haarsma and Nikolaou (2000), Desborough and Miller (2002), Bode et al. (2004) and Li et al. (2003). Pilot plant studies are found in Harris (1989), Harris et al. (1996a), Huang and Shah (1999), Jämsa-Jounela et al. (2002), Horton et al. (2003) and Thornhill et al. (2003a).

The used abbreviations in the following tables are:

- BA: Bayesian approach
- CB: covariance-based
- CC: cascade control
- CLPI: closed-loop potential index
- CMV: constrained MV
- COR: correlation method
- DCB: design-case benchmarking
- DDS: dynamic disturbance sensitivity
- EHPI: extended horizon performance index
- FD: frequency domain method
- FBC: feedback control

- FCOR: filtering and correlation-based MV
- FFC: feedforward control
- GMV: generalised minimum variance
- GPC: generalised predictive control
- HIS: historical benchmarking
- HVAC: heating, ventilating and air-conditioning
- ICA: independent component analysis
- Ii: idle index
- IMC: internal model control
- ISE: integral of squared error
- LCD: load-change detection
- LL: likelihood method
- LQG: linear-quadratic Gaussian
- LTV: linear time-variant
- MB-PI: model-based performance index
- MCPD: multivariate-contribution-based performance diagnosis
- MIMO: multi-input multi-output (multivariable)
- MPC: model predictive control
- MPC-PI: MPC-based PI control
- MSPC: multivariate statistical process control
- MV: minimum variance benchmarking
- NGPA: performance assessment based on normalised gradient
- NLI: nonlinearity index
- NLD: nonlinearity detection
- NNM: nearest neighbour method
- OCI: output covariance-based index
- OD: oscillation detection
- OI: oscillation index
- OPI(D): optimal PI(D) control
- OPPI: overall plant performance index
- PDA: plant-wide disturbance analysis
- PfI: profit index
- PrI: predictability index
- QI: quality index
- PCA: principle component analysis
- PLS: partial least squares
- RCA: root-cause analysis
- RDI: relative damping index
- RPI: relative performance index
- RSI: relative sensitivity index
- RS-LQG: restricted structure LQG
- RVI: relative variance index
- SCVT: selective constraint and variability tuning
- SET: settling-time benchmarking
- SgI: sigma index

- SPA: spectrum analysis
- TMP: thermo-mechanical pulp

15.2.1 Analysis of Fields of Application

The greatest number of applications has been registered in the fields of refining, petrochemicals and chemicals (Table 15.2). Likewise, a wide range of application from pulp & paper mills is found, as given in Table 15.3. However, applications appeared in the mining, mineral and metal processing sectors are scarce; see Table 15.4. The author and his group have recently completed many successful applications in the metal processing field; some of them will be presented in Chap. 15.

15.2.2 Analysis of Type of Implemented Methods

Analysis of the implementations according to the kind of assessment (benchmarking) methods shows that MV benchmarking has found wide application (about 15 % of the case studies). This fact indicates that this technique is mature and is the standard method in everyday use at most plants. In about 15 % of the case studies, oscillation detection methods are applied, confirming the frequent occurrence of oscillating loops in industrial processes. The use of advanced (model-based) benchmarking methods is found in about 15 % of the case studies, concentrated over the last decade. Compared with the situation in 2006 (Jelali 2006c), the number of advanced assessment applications has increased significantly during the last few years. This is due to the increasing application of advanced control methods and to the increasing interest in monitoring the performance of supervisory control loops in the process industries.

15.3 Review of Control Performance Monitoring Systems

15.3.1 CPM Tools and Prototypes

Many authors have developed systems/tools/packages for CPM based on one or more of the methods described in the previous chapters of the book.

- A CPM system called QCLiP (Queen's/QUNO Control Loop Performance Monitoring) making use of an MVC-based performance index and other analyses of closed-loop process data was reported by Jofriet et al. (1995); see also Harris et al. (1996b). This system requires the time delay of each loop to be specified by the user. An open-loop test and analysis for each controller was suggested to determine this parameter.

- Owen et al. (1996) have set up a prototype online system for automatic detection and location of malfunctioning control loops. Particular emphasis was placed on the description of features, which allow this system to perform reliably in nonlinear, highly interactive dynamic environments in paper mills.
- A data analysis and graphical representation system for control loop performance assessment has been developed by Ogawa (1998) and was installed for an integrated paper mill with three paper machines.
- Miller et al. (1998) described a comprehensive system for CPM developed by Honeywell Hi-Spec Solutions. This system is now offered to the process industries as an Internet service called Loop Scout™.
- A CPM-software tool implemented in MATLAB has been mentioned by Horch (2000), containing the algorithms related to his PhD thesis. This tool is not supposed to work in an autonomous manner.
- Paulonis and Cox (2003) have presented a large-scale CPM system (over a huge number of controllers) developed by the Eastman Chemical Company. Emphasis is placed on the description of the web-based system architecture (software, hardware, interfaces) and features/capabilities (performance and diagnostic reports) from practical viewpoint.
- Also, DuPont developed its own control performance monitoring package, called Performance Surveyor™ (Hoo et al. 2003). It monitors large numbers of process variables/control loops and generates valuable performance metrics and reports used in detecting degradation in process conditions, process equipment, instrumentation or control equipment.
- In 2003, the ACT (Advanced Control Technology) Club launched its own (offline) control-loop-benchmarking tool, called PROBE. This tool allows the performance of control loops to be compared against a number of benchmarks, including MV, GMC and LQG benchmarks. This tool is only available for the member companies of the ACT Club.
- Very recently, Huang and co-workers provided free academic, trial versions of different CPA tools, including univariate performance assessment (UVPA), multivariate performance assessment (MVPA), stiction detection, subspace performance assessment, LQG benchmarking and LMI-based economic performance assessment of MPC applications (LMIPA). These tools are called process assessment technologies and solutions (PATS) (Lee et al. 2010).
  Look at: www.ualberta.ca/~bhuang/Research/Research.htm.

15.3.2 Commercial Products

Some process equipment suppliers have already realised the importance of tools to assess the control performance of systems. Commercially available products are listed in Table 15.1. In our opinion, only the first three products do meet most of the desired features discussed in Sect. 15.1.

Table 15.1 Commercially available control-performance assessment and tuning products

Company (web address)	Product Name (Acronym)
Matrikon (www.matrikon.com)	ProcessDoctor
ExperTune (www.expertune.com)	PlantTriage
ProControl Technology (www.pctworld.com)	PCT Loop Optimizer Suite (PCT LOS) (Older versions were known as the ABB Loop Optimizer Suite and the ABB AdvaControl Loop Tuner.)
ABB (www.abb.com)	Optimize^{IT} Loop Performance Manager (LPM)
Honeywell (www.acs.honeywell.com)	Loop Scout
Emerson Process Management (www.emersonprocess.com)	EnTech Toolkit, DeltaV Inspect
ControlSoft (www.controlsoftinc.com)	INTUNE
KCL (www.kcl.fi)	KCL-Control Performance Analysis (KCL-CoPA)
OSIsoft (www.osisoft.com)	PI ControlMonitor
AspenTech (www.aspentech.com)	Aspen Watch
Control Arts Inc. (www.controlartsinc.com)	Control Monitor
Invensys (www.invensys.com)	Loop Analyst
PAS (www.pas.com)	ControlWizard
Metso Automation (www.metsoautomation.com)	LoopBrowser
PAPRICAN (www.paprican.ca)	LoopMD
Shell-Yokogawa (www.yokogawa.com)	MD^{Pro}

- Perhaps, the most complete CPM and diagnosis package is Matrikon's *ProcessDoctor*^{TM}, as it is the only product that provides assessment and monitoring of both regulatory (PID) controls and supervisory (MPC) controls.
- *PlantTriage*^{TM} from ExperTune is also very recommended for industrial use. It provides components for process modelling, basic statistics, controller performance assessment, oscillation detection and diagnosis, and PID analysis and tuning. A demo version of this CPM package is available.
- The *PCT Optimizer Suite*^{TM} is a powerful package, which includes components for effective control-loop performance monitoring (*PCT Loop Audit Evaluator*), PID-tuning and redesign (*PCT Loop Optimizer*). This package developed by ProControl Technology (PCT) is used/licensed by many other (consulting) companies.
- Honeywell's *Loop Scout*^{TM} seems to be more an audit tool rather than a continuous real-time monitoring tool. Loop Scout requires transmitting process data over the Internet to Honeywell for processing and provision of reports. Since this is not acceptable to many companies, these (e.g., Eastman Chemical and DuPont) prefer to develop in-house products rather than use a commercial tool.
- Emerson's *DeltaV Inspect*^{TM} provides (graphical) tools for identification of under-performing loops and quantification/statistics of different (loop) operating

conditions. Controllers (PID and fuzzy) can be tuned by means of Emerson's *DeltaV Tune*. Emerson (Process Management) also offers the *EnTech Toolkit* for signal conditioning, data collection, controller monitoring (*Analyse Module*) and tuning (*Tuner Module*).
- ControlSoft's *INTUNE*^{TM} software tools automatically generate PID parameters, re-tune control loops for optimal performance and monitor multiple PID loops to determine how complete systems are being controlled.
- The *PI ControlMonitor*^{TM} offered by OSIsoft oversees plant control systems and keeps historical system record in terms of different statistical figures.
- KCL's *KCL-CoPA* is a system for analysing the performance of control loops, providing a ranking list and performance parameter histories (different indices).
- MD^{Pro} is Shell's software suite for monitoring base level and commercially available multivariable control systems. It also features off-line applications that can be used to perform advanced diagnostics on single-output and multivariable control loops, and another for analysing PID loops only.

There is scarce information about other CPM products, so no evaluation is possible to date. Also, only a small portion of the methods presented in this book can be found in these packages.

15.4 Summary and Conclusions

The CPM technology has progressed steadily in the last decades since the key research step taken by Harris (1989). This chapter has provided a comprehensive review of the current status in industrial applications of this emerging field. The survey reveals a remarkable number of application case studies to date, with a solid foundation in refining, petrochemical, chemical sectors and pulp & paper plants. However, only a few applications appeared in other industrial sectors.

Analysis of the implementations according to the kind of assessment methods used shows that, whereas MV benchmarking and oscillation detection has found wide application, advanced benchmarking methods are still relatively seldom applied but found increasing interest in the last few years.

The field of CPM has matured to the point where several commercial algorithms and/or vendor services/products are available for control performance auditing or monitoring. However, only a small portion of the methods presented in this book can be found in these packages.

15.5 Summary of Industrial Case Studies

Table 15.2 Control performance assessment applications in refining, petrochemical and chemical sectors

References	Applications	Methods
Horch and Isaksson (1999), Horch (2000)	2 distillation columns: tray end temperature controls	MV, EHPI
Miao and Seborg (1999)	Distillation column: 4 control loops (2 flow, level, pressure)	OD
Harris (1989), Desborough and Harris (1992, 1993), Harris et al. (1996a), Harris and Seppala (2001)	• Polymer production data • Chemical process: cascade level control • Distillation column: duty and temperature controls • Distillation column: flow and level loops	MV CC MV MIMO MV MIMO MV
Kozub and Garcia (1993), Kozub (1996)	Different distillation columns: analyses of several control loops	MV
Stanfelj et al. (1993)	Distillation column: inferential temperature control	FFC + FBC COR
Harris et al. (1996a)	Fractionation column	MIMO MV
Tyler and Morari (1996)	Distillation column: overhead temperature control	LL
Huang et al. (1997a), Badmus et al. (1998), Huang and Shah (1999), Huang et al. (1999, 2000)	• Absorption process: 2 level control loops • Nitrid acid production facility: ammonia flow rate/gauze temperature cascade loop • Composition (SO_2/H_2S) control loop	MIMO MV (FCOR) MIMO MV (FCOR), LQG MV (FCOR)
Kendra and Çinar (1997)	Autothermal tubular packed-bed reactor: regulation of exit concentration and bed temperature	FD
Thornhill and Hägglund (1997)	Analysis of 10 oil refinery loops (pressure, flow, temperature, level)	OD
Vishnubhotla et al. (1997)	Distillation columns: tray end temperature control	FFC + FBC MV
Patwardhan et al. (1998), Patwardhan (1999), Gao et al. (2003)	• Para-Xylene distillation unit: data from 6 controlled variables (Xylene feed, temperature, internal reflux ratio, OX hold up, OX reflux drum level, OX reflux ratio) • Propylene splitter (C_3F) column: top and bottom impurity controls • Hydro-cracker unit: recycle surge drum level control	MIMO MV (FCOR), SET, HIS MIMO SET, HIS (MPC) LQG, DCB (DMC)
Huang (1999)	Distillation column: tray end temperature control	LTV MV

Table 15.2 (Continued)

References	Applications	Methods
Thornhill et al. (1999, 2001)	• Analysis of 12 refinery loops (pressure, flow, temperature, level) • Oscillation detection examples (sticking valves)	EHPI OD
Swanda and Seborg (1999)	Distillation column: • Temperature control loop • Reflux drum level control loop	SET
Huang et al. (2000)	Combined gas oil (CGO) coker: MPC system (level, temperature, flow)	COR, SPA
Huang et al. (2000)	Cracking furnace + distillation column: 2 tension control loops	MIMO FBC + FFC MV (FCOR)
Ko and Edgar (2000)	Distillation column: level-to-flow cascade control system	CC MV
Huang (2002)	Cascade (flow/temperature) reactor control loop	LTV MV
Kinney (2003)	Catalytic cracking unit: several regulatory loops (flow, pressure, temperature, level)	MV, OD
Shah et al. (2001)	• Capacitance drum control loops (pressure, level) • Ethane cracking furnace: regulatory layer + advanced (MPC) layer	MIMO MV (FCOR) MIMO LQG (MPC)
Thornhill et al. (2002), Thornhill et al. (2003a, 2003b, 2003c)	• Plant consisting of 3 distillation columns, two decanters and several recycling streams: 15 control loops (flow, level, temperature, etc.) • 5 level and flow loops: oscillation detection examples (sticking valves)	OD MV, OD
Hoo et al. (2003)	• Polymer reactor: composition control data • Chemical process: temperature control loop	MV –
Horton et al. (2003)	Level control loops from • Gasoline splitter reflux drum • Stripper cold feed surge drum • Fractionator reflux drum • Gas oil tower reflux drum	MPC-PI, OPI, LQG
Paulonis and Cox (2003)	Analysis of 14.000 control loops in 40 plants at 9 sites worldwide (flow, pressure, level, temperature)	–
Xia and Howell (2003)	Chemical plant: 9 control loops (flow, pressure, level, temperature)	OLPI
Choudhury et al. (2004)	Chemical complex: 2 flow control loops	NLD (OD)
Olaleye et al. (2004a, 2004c)	Sulphur recovery unit: tail gas ratio control	LTV MV
Kano et al. (2004)	Data from 4 chemical plants: 2 flow and 2 level control loops	OD

Table 15.2 (Continued)

References	Applications	Methods
Ko et al. (2004)	Hydrobon unit • HP separator level control • Stripper ov'hds receiver level and reflux flow controls • Rich Amine Regen feed flow control	RPI, OD
Bonavita et al. (2004)	Polymer production plant: flow control loop	OD
Yamashita (2006)	Chemical plant: 4 control loops (level, flow)	OD
PAM (2001)	Divided-wall distillation column: 2 control loops (temperature, pressure)	MV, GMV, RS-LQG, MIMO LQGPC
Xia and Howell (2005)	Chemical plant: 9 control loops (flow, pressure, level, temperature)	OD, ICA
Xu et al. (2006a, 2006b)	Sulphur recovery unit	MV, OCI
Agarwal et al. (2007a, 2007b)	Coker plant	MV, BA, MPC
Bauer et al. (2007a)	Distillation unit	PDA, NNM
Bauer et al. (2007b)	Refining unit: acid reactor	PfI
Ghraizi et al. (2007)	Petrochemical plant	PrI
	MV, OD, FBC + FFC	
Konda and Rangaiah (2007)	Hydrodealkylation of toluene to produce benzene	PDA, DDS
Ramachandran et al. (2007)	Fluid catalytic cracking unit	PDA, RCA
Hanna et al. (2008)	Falconbridge smelter: • Steam flow control • Feed temperature control	CMV
Lee et al. (2008)	Hydrogen unit	MV, SCVT
Xu et al. (2008)	Gas oil hydrotreating unit	LTV
AlGhazzawi and Lennox (2009)	Condensate fractionation process	MSPC, PCA, PLS
Yuan and Lennox (2009)	Gas processing plant	RVI
Zhao et al. (2009)	Distillation column	LQG
Badwe et al. (2010)	Kerosene hydrofiner unit	RSI
Lee et al. (2010)	Bitumen distillation column: 19 temperature loops, 14 level loops, 26 flow loops, 8 pressure loops	MV, MPC
Veronesi and Visioli (2010)	Pharmaceutical plant: flow control	CLPI, SgI

Table 15.3 Control performance assessment applications in pulp and paper mills

References	Applications	Methods
Perrier and Roche (1992)	Kamyr digester: consistency, level and level-flow cascade control	MV
Eriksson and Isaksson (1994)	• Wood chip refiner: dilution water control • Paper mill: consistency loop	MV, OPI MV, OPI
Hägglund (1995, 2005)	• Pulp concentration control • Pulp flow control	OD
Jofriet et al. (1995), Harris et al. (1996a, 1996b)	Thick stock system + dry end of a paper machine: several control loops	MV
Lynch and Dumont (1996)	• Reject refiner: motor load control • Kamyr digester: chip level control	MV MV
Owen et al. (1996)	TMP mill + paper machine: • Dryer section pressure control loops • Broke consistency and flow control loops • Overview of complete analysis of 124 loops of TPM and 52 loops pf PM	MV
Huang et al. (1997b)	Headbox control system: • Total head (pressure + level) control • Pond level control	MV (FCOR) MIMO MV (FCOR)
Forsman (1998)	Paper mill stock prep area: 11 flow loops, 3 level loops, 6 concentration loops, 1 pH loop	OD, Ii, NLD
Ogawa (1998)	Paper mill: • Clay flow loop • Two consistency loops	MV
Forsman and Stattin (1999)	Stock preparation process: 30 control loops (flow, consistency, level, etc.)	OD
Horch (1999, 2000)	TMP refiner: motor load control several consistency, flow and level control loops from different pulp & paper mills	MV, EHPI OD
Ingimundarson (2003), Ingimundarson and Hägglund (2005)	Analysis of 19 pulp & paper loops (flow, temperature, level)	EHPI (λ-monitoring)
Stenman et al. (2003)	Paper mill: steam pressure control	OD
Hugo (2006)	Paper machine: steam temperature control	MV, MB-PI
Ingimundarson (2006)	Pulp & paper mill: • Pressure control loop • Flow control loop	NGPA
Hölttä and Koivo (2009)	Paper mill	PDA, QI, OPPI

Table 15.4 Control performance assessment applications in other industrial sectors

References	Applications	Methods
Qin (1998)	Wastewater treatment pH reactor: pH control	MV
Horch and Isaksson (1999)	Lime kiln: front end temperature control	MV, EHPI
Foley et al. (1999), Huang et al. (1999)	Acid leaching process: pH control	MV
Ettaleb (1999)	Lime kiln: control of feed-end temperature and excess of oxygen	MIMO MV
Hägglund (1999)	Heat exchanger: water temperature control	Ii
Haarsma and Nikolaou (2000)	Snack-food frying process: moisture and oil content control (MPC)	MIMO (FCOR) MV
Bender (2003), Gorgels et al. (2003)	Cold tandem mill: strip thickness control	MV
Jämsa-Jounela et al. (2003)	Zinc plant: flotation cells: 3 level control loops	MV, OD, ISE
Li et al. (2003)	Water flow control loop	RPI
McNabb and Qin (2003, 2005)	Wood waste burning power boiler: 5 flow and pressure control loops	MIMO MV (CB)
Bode et al. (2004)	Semiconductor manufacturing: run-to-run control (MPC) of overlay in lithography	MV
Salsbury (2005)	HVAC: air-handling units: temperature control loops	LCD
Singhal and Salsbury (2005)	Commercial building: room temperature control loop	OD
PAM (2001)	Coal-fired power plant: steam turbine: speed and load control	RS-LQG
Jelali (2007b)	Rolling mill: • Strip thickness control • Strip flatness control	MV, OD, FBC + FFC
Ooi and Weyer (2008)	Irrigation channel control loops	PDA, OD, COR
Yu and Qin (2008a, 2008b)	Wood waste burning power boiler: 10 control loops	MIMO MV (CB), MCPD
Yuan et al. (2009)	Sludge combustor unit: furnace process control	MV, HIS, LQG
Howard and Cooper (2010)	• Boiler drum level control • Import/export control loop regulating the CTG operation • Attemperator control loop	COR, RDI
Zhao et al. (2010)	Coking unit: fractionation column	CLPI

Chapter 16
Performance Monitoring of Metal Processing Control Systems

The number of control loops used in process industry is growing continuously, whilst manpower is being reduced. Consequently even companies that have embraced new control technologies struggle to maintain satisfactory performance over the long term. Numerous investigations have shown that the performance of control systems in the process industries is not satisfactory, as mentioned in Chap. 1. This particularly applies for the steel industry, where it is the norm to perform controller tuning only at the commissioning stage and then never again. A loop that worked well at one time is prone to degradation over time unless regular check and maintenance is undertaken.

The field of metal processing continues to provide challenges in the application of process control and supervision at every level of the automation hierarchy, enterprise optimisation and system integration. To re-instate good performance for the large number of control loops in the different automation and production systems of a steel processing line requires highly skilled personnel, making it time consuming, costly and error-prone. The use of automatic and online performance monitoring and re-tuning systems is therefore highly desirable to detect control performance degradation and rapidly restore and sustain peak performance level. This is not intended to eliminate the role of maintenance/control engineers, but to reduce their workload and allow them to focus on higher operational issues and other production-related "fire-fighting" duties.

Techniques successfully used in other process industries have to be adapted to the specific properties and conditions of steel processing, particularly rolling mills, showing high sample rates, varying time delays and semi-continuous operation. This chapter provides a contribution in this direction. In Sect. 16.1, a brief introduction to the metal processing technology and automation is given. Then, in Sect. 16.2, many practical aspects of CPM in metal rolling are discussed, including the special effect of oscillations, the batch-wise performance evaluation, the time-based vs. length-based setting and the use of specific performance indices. Successfully completed industrial case studies and tailored CPM tools are presented in Sect. 16.3. The studies involve the application of different CPM methods to different plants in the rolling area.

M. Jelali, *Control Performance Management in Industrial Automation*, Advances in Industrial Control, DOI 10.1007/978-1-4471-4546-2_16,

16.1 Introduction to the Metal Processing Technology

This section is devoted to a brief introduction to metal processing and automation technology, so that readers that are not familiar with rolling processes will understand the problems addressed and the methods applied. Steel processes are a wide class of industrially important processes, which mainly include iron-making, steelmaking, casting, hot and cold rolling and coating. Improving the control performance of steel processes is of substantial industrial interest. Enhanced control of the strip-quality attributes leads to significant reductions in material consumption, greater production rates for existing equipment, improvement in product quality, elimination of product rejects and reduced energy consumption. Steel processes have many characteristics that challenge the development and application of advanced control and monitoring methods:

- Metal processing plants are very complex, consisting of mechanical, electrical and hydraulic components, sensors, software and hardware, and control systems, which are nonlinear and multivariable in nature.
- Metal processing plants are usually subject to a wide range of dynamic disturbances, parameter variations and constraints, i.e. actuator saturations, inequality constraints, couplings, etc.
- Most of metal processes (particularly in rolling mills) have fast dynamics (with small time constants) and dominant time delays, as many quality parameters can only be measured some distance away from the plant. Moreover, the time delay is usually varying (as a function of the strip velocity) during the dynamic phases (acceleration and deceleration).

16.1.1 Steel Processing Route and Control Objectives

Steel plates, coils and sheets destined for end customer (e.g. automotive industry) undergo many processes before final delivery; see Fig. 16.1. A typical route starts from raw materials to slab and includes cooking, sintering, ironmaking in blast furnaces, steelmaking and casting. From a slab to a coil, there is then pre-heating, multiple passes on a roughing mill, processing through a hot strip (tandem) mill, coiling, cooling and a sequence of repeated passes of uncoiling, cold reduction, recoiling and cooling, followed by hot-dip galvanising (or coating). At some points between the cold mill passes, the coil may be annealed and then cooled again. In all cases, the basic principles of the rolling operation are similar. There may be also finishing processes such as slitting or tension levelling, anodising or painting, before the coil is prepared for shipment to the customer.

In a typical single-stand reversing strip mill (Fig. 16.2), the strip is paid off a coiler at one side of the stand, reduced in thickness as it passes between the work rolls (WRs), supported between a pair of larger diameter backup rolls (BRs) and re-coiled by a coiler at the other side of the stand. A main-drive train provides rotation of the work rolls with desired speed and rolling torque. Roll gap adjustment

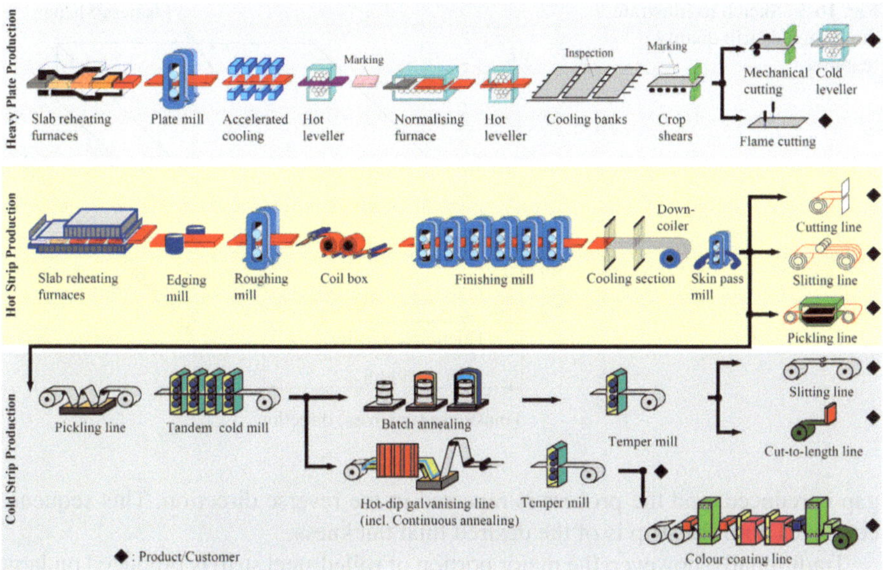

Fig. 16.1 Typical metal (steel) processing routes for production of heavy plates, hot strips and cold sheets

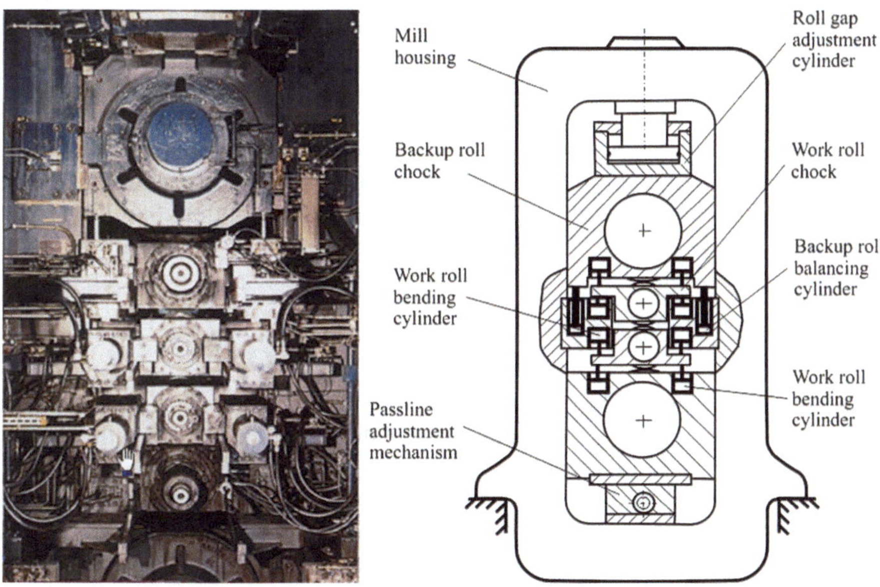

Fig. 16.2 Four-high mill stand: inside view (*left*) and schematic arrangement (*right*)

mechanisms provide setting of the required gap between WRs and may also allow the elevation of the pass line to be adjusted. The mill housing is designed to contain the mill stand components and to withstand the rolling load. In the next pass, the roll

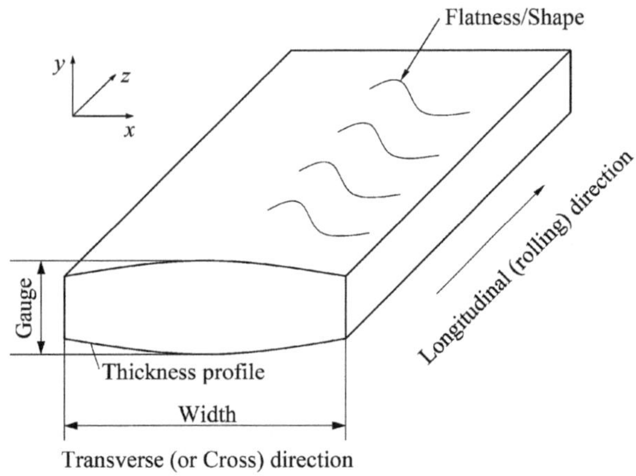

Fig. 16.3 Sketch to illustrate dimensional strip quality features

gap is reduced, and the process is repeated in the reverse direction. This sequence continues until the strip is of the desired final thickness.

Traditionally, however, the major portion of rolled steel strip is produced on large tandem rolling mills, where the strip is put through a series of rolling stands, typically 5–7 stands in hot rolling and 4 or 5 stands in cold rolling. In cold tandem mills, steel strip is reduced from 2–6 mm thickness to 0.4–3 mm in a single pass. In each of the stands, the steel is exposed to a roll force of 10–20 MN. Because steel undergoes a high degree of work hardening during cold rolling, most of the total reduction takes place at the first stands. The maximum speed of the strip is 250–300 m/min at the first stand and, owing to reduction, 1200–1250 m/min at the last stand. Besides the conventional 4-high mill configuration, modern 6-high mills, Z-high mills and 20-high mills (for production of stainless steel) are available; see Roberts (1978) and Ginzburg (1989) for more detailed descriptions of rolling mills.

16.1.2 Control Objectives

From the point view of the customer, there are two major issues associated with the rolling of flat metal, these being the metallurgical properties and the dimensional characteristics of the strip. The *metallurgical* properties are principally concerned with strength and ductility of the material and are influenced by the heat and deformation caused by the rolling process and at a later stage by annealing in galvanising plants (Sect. 16.3.3). The most relevant *dimensional* requirements (i.e. strip gauge/thickness, profile, shape/flatness and surface finish) will be introduced in this section; see Fig. 16.3.

The main dimensional requirement of most rolled products is the final thickness. As the strip is essentially a two-dimensional product: thickness variations may occur in the longitudinal direction (or machine direction, at the centreline), which is commonly referred to as the strip thickness (or *strip gauge*); thickness variation occurring in the transverse direction (or cross-direction) is known as the *thickness profile*. This is usually expressed in terms of its "*crown*", i.e. the difference in thickness between the centre and positions in the region of strip edges, commonly at some arbitrary distance, typically 40 mm, from the actual strip edge. In practice, it is known that a strip-thickness profile is substantially created during the hot rolling operation, i.e. in the earlier stands of a hot-strip mill. It is then believed that the relative profile will remain about the same as the overall thickness of the cold-rolled strip is reduced, except for narrow regions near the strip edges.

Strip shape (or online *flatness*) is the next, most important quality issue in the rolling of flat products. The term *strip shape* is rather ambiguous in the sense that it may refer to the cross-sectional geometry of the strip or to the ability of the strip to remain flat on a horizontal planar surface for subsequent processing. Usually, the emphasis is upon the second meaning of shape. This is important as the strip, which buckles, is non-uniform and difficult to process and thus may be responsible for equipment damage or the need for additional (expensive) processing. In addiction, it may not be appealing—from an aesthetic viewpoint—when such product reaches the market place. As a consequence, over the last decades, the measurement, control and investigation of the problem of the strip shape have become a crucial area of research in rolling. Though considerable progress has been made in improving the shape of the rolled strip, there remain areas of uncertainty in the modelling and mechanisms, which generate it (Jelali et al. 2001a, 2001b, 2001c). The exceedingly complex interactions, which determine the resultant strip shape, have made significant progress in this area difficult.

The cause of strip shape will be explained and defined with the aid of Fig. 16.4. When the strip is reduced in thickness, a corresponding length increase results, provided that the width remains constant (volume conservation), which is almost the case for cold rolling. Shape problems occur when the reduction of the strip is not uniform across the width of the strip. Now, the strip is slit into numerous longitudinal ribbons, each of length l_i. The difference between the stress value for the ith ribbon and a basic stress value (which may be the lowest stress value, the mean stress value or the stress value in the strip centre), $\Delta\sigma_i$, corresponds to a difference in elastic strain $\Delta l_i / l_0$, according to Hook's law:

$$\frac{\Delta l_i(x)}{l_0} = -\frac{\Delta\sigma_i(x)}{E_s}. \tag{16.1}$$

It is important to note that Eq. 16.1 is only valid as long as no major (manifest) shape defects occur. Generally, shape is expressed in terms of the so-called *flatness index I-units* (IU), i.e.

$$\Omega := \frac{\Delta l}{l_0} \times 10^5 \text{ [IU]} \quad \text{or} \quad \Omega = -\frac{\Delta\sigma}{E_s} \times 10^5 \text{ [IU]}. \tag{16.2}$$

For a steel strip with an elastic modulus $E_s = 210000$ N/mm^2, 1 IU corresponds to a stress difference of 2.1 N/mm^2 or a length variation of 10 μm/m.

In the rolling of strip, poor shape of the rolled strip principally results from a *mismatch* between the loaded roll-gap geometry and the cross-sectional profile of

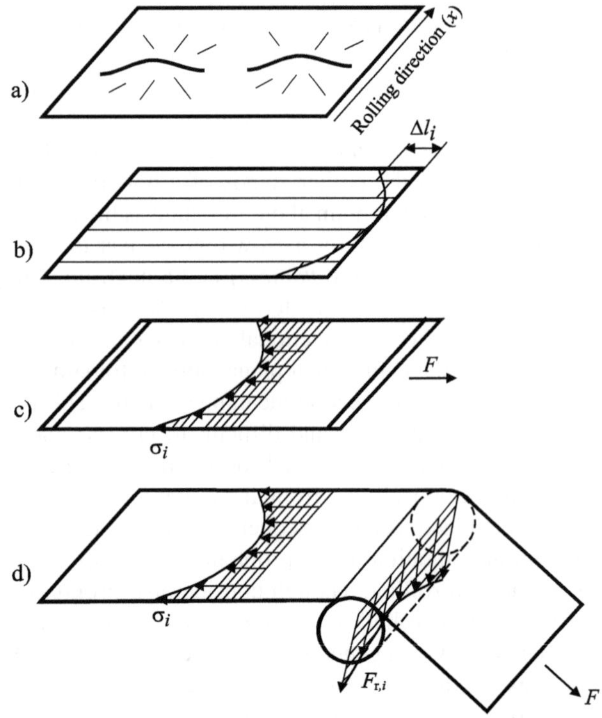

Fig. 16.4 Correlation between strip-length differences and longitudinal stresses for an uneven sheet sample (**a**) with centre buckles; (**b**) length differences after division into ribbons; (**c**) longitudinal stresses on tension; (**d**) principal of radial force measurement based on deflector roll (Keck and Neuschütz 1980)

the strip. Major causes of such incompatibility may be a non-uniform incoming strip profile (e.g. with ridges), non-uniform crowning of the rolls, non-uniform lubrication across the width of the strip in the roll bite, or non-uniform metallurgical properties, such as striations of coarse-grained material, which exhibits a different resistance to deformation than the bulk of the work piece. A number of non-uniform stress patterns may appear in the longitudinal and transverse directions and give rise to shape defects, such as centre buckles, edge waves, quarter buckles, etc.; refer to Roberts (1978) for description of shape defects.

The term *strip flatness* (or offline flatness) is known as a measure of the ability of the strip to lay flat when placed on a level surface with *no* externally applied loads (tension). Flatness is related to shape in that the transverse variation in stress may result in a buckled strip, when the tension applied during rolling is removed. From the definitions given above, flatness seems to be the quality parameter that is of real interest to the end customer. However, flatness cannot be directly measured online during the rolling operation as tension is always applied. Hence, it is common practice to measure the shape (or online flatness) using a *shapemeter* (usually a flatness measuring roll) as an indicator for flatness and to use the shape signal for online flatness control systems. Thus, the terms shape and flatness are very often used synonymously (particularly in the German-speaking region). A shapemeter measures the radial forces (locally) exerted on the strip portions by using sensors placed on the perimeter of the flatness roll. The radial forces are then transformed into tension-

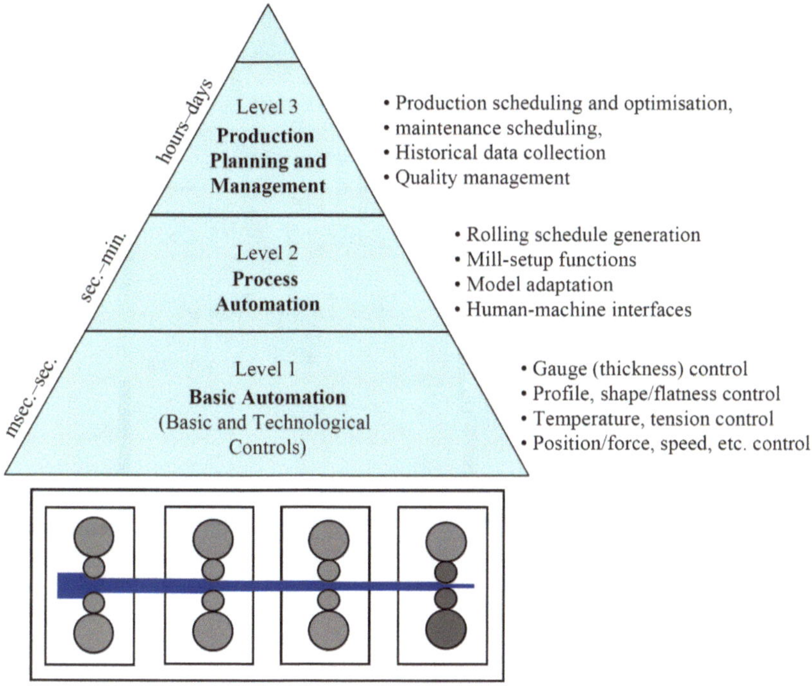

Fig. 16.5 Typical hierarchy of rolling mill automation

stress values (Fig. 16.4d). Details on flatness measurement can be found, e.g. in Keck and Neuschütz (1980) and Mücke and Gorgels (2007).

16.1.3 Mill Automation

The field of metal processing provides challenges in the application of process control, enterprise optimisation and system integration. Figure 16.5 shows the main functional levels (including relative time scales) in the metal processing automation hierarchy, where various monitoring, control and optimisation activities are employed. The relative position of each block is intended to be conceptual because there can be overlap in the functions carried out, and often several levels may utilise the same computing platform. In Fig. 16.6, a typical automation configuration, partly adopted from Lackinger et al. (2002), for cold tandem mills is illustrated:

- *Level 1. Basic Automation.* This level includes all equipment and systems for measurement and dynamic control. The control systems are usually divided into *basic (actuator) controls* (such as position/force controls, drive controls, cooling spray controls) and *technological controls* (such as gauge controls, temperature

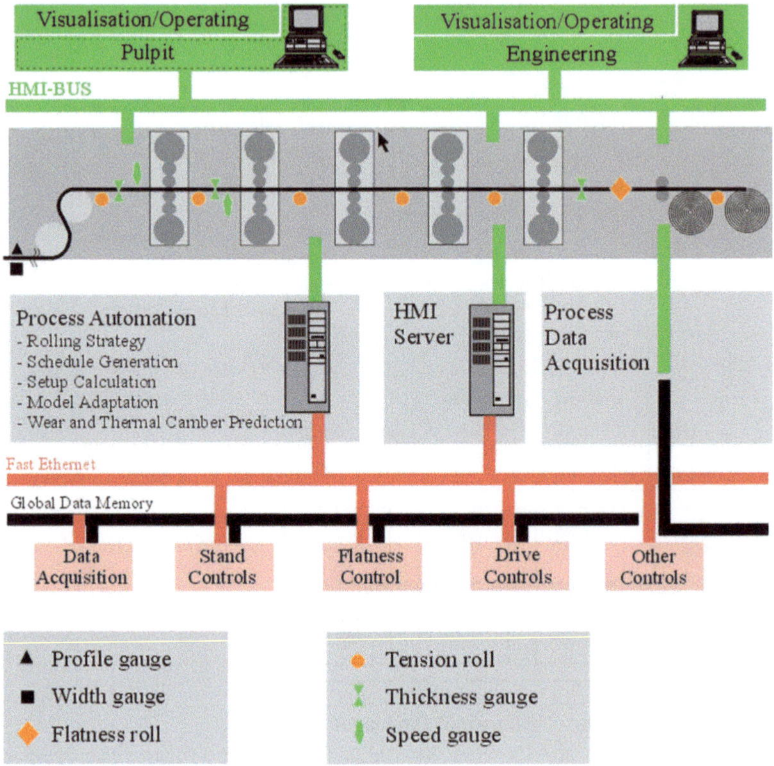

Fig. 16.6 Typical automation structure for a tandem cold mill (Lackinger et al. 2002)

controls and profile and shape/flatness controls). Sometimes, the drive system controls and dedicated hardware systems and measuring devices are isolated as Level 0 (not shown in the figure).

- *Level 2. Process Automation.* This level consists of the rolling scheduling and setup systems (including mathematical models, rolling strategies, optimisation and adaptation), diagnostic functions as well as the human–machine interfaces (HMI), which are usually located in an enclosed pulpit. Level 2 functions play a critical role by ensuring the process is operating safely and satisfies mill constraints and throughput targets.
- *Level 3. Production Planning and Management.* It includes order processing, production scheduling, sequencing and optimisation, maintenance scheduling, as well as historical data collection and quality management. Production rates of all intermediate and final products are planned and coordinated, based on equipment constraints, storage capacity, sales projections and the operation of other plants, sometimes on an enterprise-wide basis.

In metal rolling, the (nominal) *schedule calculation* produces a set of nominal values (called rolling schedule) for gauges and tensions, bending forces and

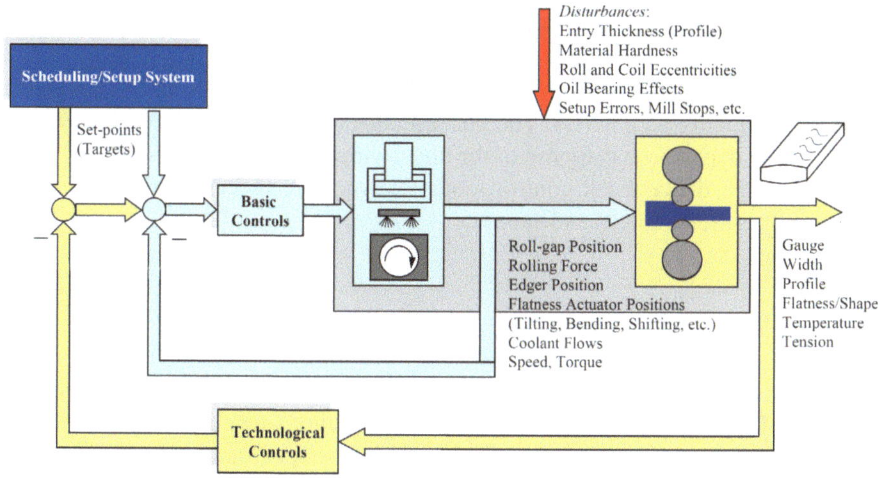

Fig. 16.7 Main controllers in rolling mills (Jelali 2006b)

mill speed (for each pass or stand). The *setup calculation* of the references (set-points) for stand speeds, roll-gap positions and shape control actuators (bending forces, shifting positions, cooling sprays) ensures achieving the specified schedule(s). These set-point calculations are usually performed *before* the strip enters the mill; hence, the term *pre-setting* is also widely used in steel industry. An accurate pre-setting firstly reduces the initial work required by the (online) control systems and secondly results in a higher percentage of the strip meeting the quality targets. Gauge, profile and shape controls are activated as soon as corresponding reliable measurements are available.

16.1.4 Overview of Metal Processing Control Systems

This section presents a brief review of the state-of-the-art in the process automation and control of metal processing, especially in rolling mills; see Fig. 16.7. Control loops in metal processing plants are usually classified into the two following categories:

1. *Main/Primary/Technological Loops*. Loops that directly control the product-quality attributes. Their performance improvement causes the reduction in product variability, which can be directly translated into profitability.
2. *Auxiliary/Secondary/Basic Loops*: Subordinated loops that do not directly control the product quality, but can indirectly affect the product variability.

In this chapter, emphasis is placed on technological control systems. The role of basic control systems should, however, not be underrated, as they build a prerequisite for optimal operation of the higher technological control systems.

16.1.4.1 Basic (Actuator) Controls

Actuators are the second vital element in that they provide the capability to influence the product quality parameters. The performance of technological control systems substantially depends on response of the underlying actuator-control systems. These usually consist of feedback control loops that regulate process variables such as pressures, positions, forces and speeds.

16.1.4.2 Drive Control Systems

There are two principal types of main-drive motors, which are used in rolling mill stands: (i) DC (direct current) motors and (ii) AC (alternating current) motors. Traditionally, DC drives are used, which provide reverse rotation of the rolls in a wide range of the roll-speed control. However, due to remarkable advances in the progress of semiconductor, microprocessor and control technology, the adoption of AC drive to main motor of strip mills has been accepted recently.

16.1.4.3 Hydraulic Roll-Gap Adjustment Systems

The performance of gauge control substantially depends on the response of the roll-gap-control device. Older rolling mills have mechanical screw-down systems. Since the 1970s, hydraulic roll-adjustment systems became standard, as they provide much faster and more accurate operation than screw systems. Position and force control (typical sampling time = 2 ms) of the hydraulic systems is commonly implemented at every modern mill.

16.1.4.4 Profile and Shape Actuators

It should be remembered that poor shape basically results from a mismatch of the cross-sectional profile of the strip and the loaded roll gap. Thus, automatic shape-control systems attempt to modify the loaded roll-gap geometry by activating the actuator systems installed on the mill to influence the *degree of compatibility* between the strip profile and the roll-bite geometry. The most common actuators are (see Figs. 16.8 and 16.9, where the actuators are indicated by numbers between parentheses):

- *Roll Tilting* (*4-high*, *6-high*). Differential adjustment of the hydraulic roll-adjustment systems can serve to correct the mainly linear asymmetric components of profile and shape errors. This operation is known as *tilting* (or skewing) and means that the actuator in each side of the mill is moved by the same amount but in opposite direction.

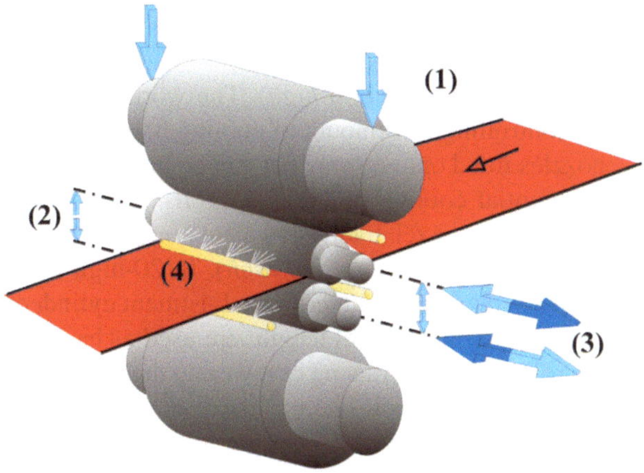

Fig. 16.8 Arrangement of typical profile and flatness actuators for 4-high rolling mills (*1*: tilting; *2*: work roll bending; *3*: work roll shifting; *4*: roll cooling)

- *Roll Bending* (*4-high*, *6-high*, *Z-high*). The application of positive or negative bending between the work-roll chocks enables the roll-gap profile to be modified dynamically. Work-roll bending is *the* fundamental and—without any doubt—the most available actuator for controlling profile and flatness in rolling mills. In modern 6-high mills and Z-high mills, intermediate-roll bending is also commonly used.
- *Roll Side Shifting* (*4-high*, *6-high*, *Z-high*, *20-high*). By moving the WRs sideways, the foundation between the WR and BR can be changed leading to reducing the natural crown due to roll deflection, and thus to an enhancement of the work-roll bending effect. The resultant crown achieved is dependent on the form of the ground camber on the rolls. The most known form is the continuously variable crown (abbreviated as CVC), which is not strictly dynamic. On cluster mills,

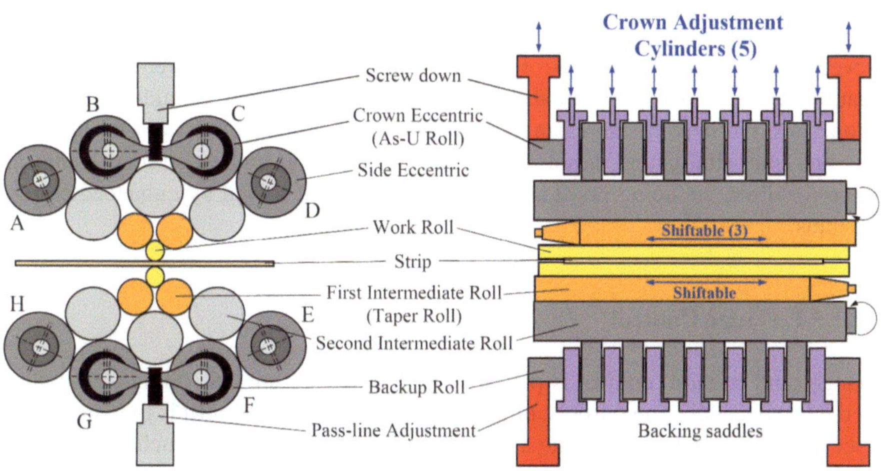

Fig. 16.9 Typical arrangement of 20-high mills from side (*left*) and front view (*right*)

shifting is realised by (relatively slow) axial displacement of the first intermediate (taper) rolls; see Fig. 16.9.

- *Roll Cooling* (*4-high*, *6-high*). The work-roll thermal camber can be influenced by controlling the amount of cooling being sprayed onto them. A series of spray nozzles fitted onto a spray bar can be individually controlled with valves to allow differential cooling of the work rolls. Spray cooling can control local flatness defects but is a relatively slow actuator.
- *Backing-Shaft Bending* (*20-high*). In 20-high rolling mills, so-called crown eccentrics, equipped with crown adjustment cylinders (Fig. 16.9) in equidistant positions over their barrel length, are used as backup rolls. They specifically adjust the roll contour by raising or lowering the position of the saddles. These in turn affect (locally) the roll-bite geometry so that strip of closest flatness tolerances can be produced. Note that there are many other configurations of cluster mills; see Sendzimir (1993). Usually, 20-high mills are used for rolling hard metal, such as stainless steel, to produce extremely thin sheet metal with small-diameter WRs.

Many other actuator systems, such as roll-crossing mechanisms, hydraulically inflated backup rolls and the dynamic shape roll (DSR), have been developed and installed on dispersed mills, mainly in Japan. Note that most of actuators mentioned are hydraulic servo-systems.

16.1.5 Technological Control Systems

Since the 1960s, the computer control systems have been installed in rolling mills. Nowadays, the control area covers all stages of metal processing. Control systems are widely applied in the whole steel-processing route, ranging from treating the raw materials to producing the final products. In the rolling area, the most important technological controls and their subordinated basic controls are summarised in Fig. 16.7. The automation technology of steel processes has matured; however, the development of advanced control systems in this area is still active. The reason is that the demand to improve quality (dimensional accuracy, mechanical and surface properties) has become increasingly severe. Moreover, the market need for high tensile and ultra-thin gauge in hot rolling and for high and highest strength materials in cold rolling, inducing problems, which can only be solved by enhanced control systems.

16.1.5.1 Gauge Control

Gauge (and tension) control is perhaps the most active area in rolling-mill control in the last decades. The purpose of gauge control for a rolling mill is to maintain the specified strip thickness despite a large number of disturbances, such as variations of the strip hardness, the entry strip thickness and tension (typical sampling time = 10 ms).

Since the strip gauge cannot be measured directly at the exit of the roll gap, it has to be estimated using any kind of observer. Common gauge control strategies are feedback (or monitor) control, gaugemeter control, feedforward control and mass-flow control, or combinations of these. Smith-predictor control has also been proposed for gauge control (Jelali et al. 1998). Overviews of current practice in gauge (and tension) control and related problems for rolling mills are provided by Ginzburg (1989), Rigler et al. (1996), Rath (2000) and Bilkhu (2001).

16.1.5.2 Profile and Flatness Control

Thickness profile and flatness are two of the key quality attributes in strip rolling. Ever since the first on-line measurement of shape was recorded using a shapemeter in the early 1960s, shape control systems have been developed and refined to meet the high standards of product flatness required by the end users (typical sampling time = 100 ms). With market demands driving the rolled products to thinner gauges, which are more prone to flatness defects, the understanding and control of the profile and flatness become any more crucial. A great deal of effort has gone into improving the shape of rolled strips for the various rolling mill types. In addition to feedback control (known as monitor control), it is common to use a feedforward controller to compensate for variations in the rolling force by symmetric adjustment to the roll bending force. The state-of-the-art and trends of flatness control are described by Jelali et al. (2001a, 2001b, 2001c) and Gorgels et al. (2003).

16.2 Practical Aspects of Performance Assessment in Metal Processing

There are some special issues that have to be considered when assessing the performance of control systems in metal processing. Some of these are discussed in this section.

16.2.1 Online vs. Batch-Wise Evaluation

Metal rolling is a batch process where time between two coils or passes is in the range of minutes. Therefore, control performance evaluation is usually carried out offline after completing the batches. In our experience, there are, however, some situations in which the analysis should be done online by moving-window or recursive calculation of the performance index (Sect. 2.4.2). In this way, abrupt performance degradation, for instance, induced by specific oscillations could be detected, and an alert can be given to the operator who may initiate countermeasures, such as speed changes during the rolling of the same coil. This situation has been observed at

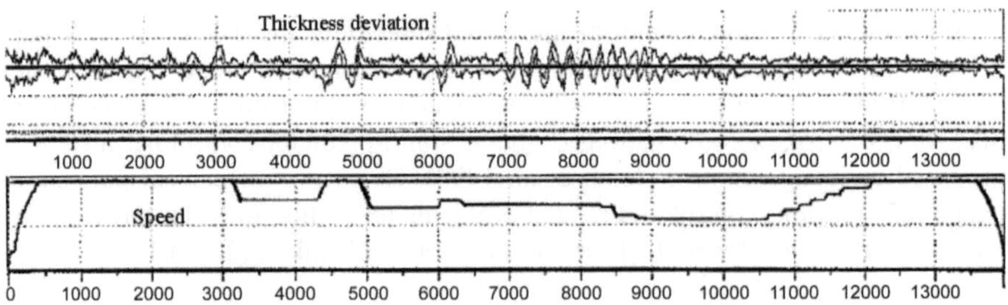

Fig. 16.10 Data from a temper mill indicating the need for online control performance monitoring

temper rolling mills, where the rolling of just one coil takes up to 10 minutes, corresponding to a strip length of up to 15 km and more. Therefore, a lot of things can happen, and there is enough time to take action during a batch in such cases. Figure 16.10 illustrates an example of such a situation, where the control error (thickness deviation) shows intermittent oscillation and how the mill operator changes the speed to avoid the oscillation. This clearly indicates the need for CPM, i.e. oscillation detection and performance-index calculation, to be implemented online. Since a sampling time of 10 ms is usual for thickness control, an online implementation is challenging.

16.2.2 Oscillation Diagnosis

Oscillations may be a very drastic form of plant performance degradation in rolling mills. Oscillation-related problems in this area are similar to those in other industries in that oscillations are caused either by aggressive controller tuning, the presence of nonlinearities, e.g. static friction, dead-zone, hysteresis, or disturbances. Some aspects, however, do differ: oscillation-free signals, e.g. strip thickness, do not exist due to the large number of mechanical components present in a rolling mill and due to many defects induced by the strip surface and geometry variations, mainly determined by the upstream processing stages.

Thus, the primary aim of vibration/condition monitoring and diagnosis in rolling mills is to detect the source of oscillation, such as deformed rolls, roll eccentricity, faults in incoming strip, bearing defects, etc. and then compensate for some of them or keep their amplitudes as small as possible rather than fully avoiding them. A major task when diagnosing the causes of periodic strip-quality faults is the identification of components, process signals and incoming strip defects that can be characterised by "defect" frequencies, proportional to the rolling speed. Chatter marks on the strip are perhaps the most known defects produced in rolling. For thorough discussions of this topic, the reader should consult Markworth et al. (2003) and Polzer et al. (2003).

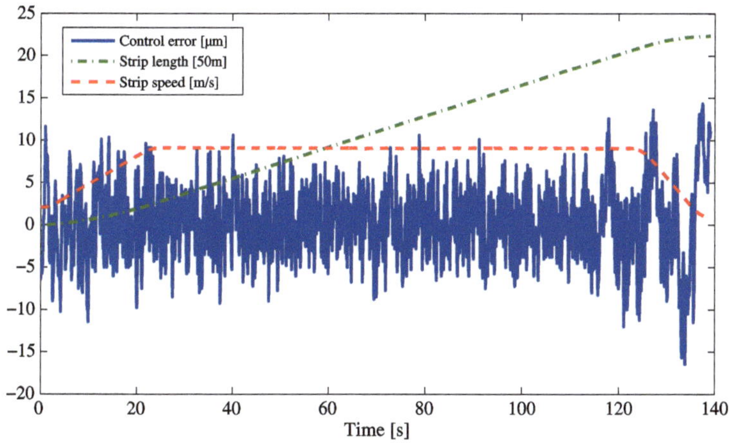

Fig. 16.11 Typical traces of coil data considered

Table 16.1 Harris index values for different evaluation scenarios

	Stationary phase	Whole rolling phase
Time-based setting	0.74	–
Length-based setting	0.74	0.77

16.2.3 Time-Based vs. Length-Based Assessment

In rolling mills, the speed is not constant (acceleration, deceleration); see Fig. 16.11. This means that strip-length samples are not equally distant in time. This could motivate to carry out performance evaluation in a length-based setting rather than in a time-based setting. This would also avoid the need to estimate any time delay, as the distance from actuator to sensor is constant in the length-based scenario. Indeed, this is a simple and straightforward approach to calculate the performance indices for the whole rolling phase, including acceleration and deceleration. Note, however, that data are usually gathered in a time-based setting and transformed into a length-based setting using the strip speed.

For the data shown in Fig. 16.11, the Harris index values have been calculated in both scenarios, i.e. time-based and length-based settings, for the stationary phase and for the whole rolling phase but excluding head and tail ends. The results are given in Table 16.1. As expected, when the speed is constant, both evaluation modes give the same value for the Harris index. When the data of dynamic phases are included in the length-based assessment, the index value indicates consistent performance for the whole coil.

When the strip speed, and thus the time delay, is not accurately known, an estimate of the time delay is necessary and may be determined by applying one of the many time-delay estimation methods (Björklund 2003), usually requiring special experimentation with the process. Otherwise, the prediction horizon method suggested by Thornhill et al. (1999) is recommended; see Sect. 3.3.

16.2.4 User-Specified Indices

Although the minimum-variance benchmark should be the standard one against which the performance of other controllers may be compared, it is well known that it has some drawbacks, such as the assumption of unlimited control action and controller order. This often results in a pessimistic benchmark: in the case where the minimum-variance index signals poor performance, further investigations are warranted. The job can be carried out by considering more realistic performance methods in terms of *user-specified benchmarks*, such as specifications in terms of close-loop dynamics, or assessment values extracted from historical data during a time period when the control system was satisfactory running from the viewpoint of control/maintenance engineers. Such criteria are called baselines, historical data benchmarks or reference data set benchmarks (Chap. 3).

The performance of gauge control is usually measured by the cumulative strip length percentage, e.g. 95.4 % or two standard deviations, lying within a prescribed thickness tolerance, typically around 1 % of the target thickness.

For evaluating the flatness control performance, the average of all maximum values of flatness error is often used as a performance criterion:

$$K_{\text{flatness}} = \frac{1}{N}\sum_{i=1}^{N} \max(\Omega_{i,1}, \ldots, \Omega_{i,n_w}), \tag{16.3}$$

where N is the number of data samples over the strip length, and n_w the number of discrete points over the strip width.

Moreover, it is useful to determine the percentage of time the controller is in AUTO mode, also known as time-on-control, and the ratio of time in saturation to the total time period considered for each coil. The time-in-saturation may give hints about poor performance due to inadequate actuator sizing rather than poor controller tuning.

16.3 Industrial Cases Studies and Developed Monitoring Tools

The majority of applications of CPM found are related to regulatory (basic) control loops. No big attention is paid to the *performance assessment of the more important technological control systems*. In our experience, companies are only interested in the monitoring and tuning of those control loops, which have a significant impact on economic factors such as production rate, product quality, energy and material consumption and plant availability, or maintenance costs. Suitable methods are

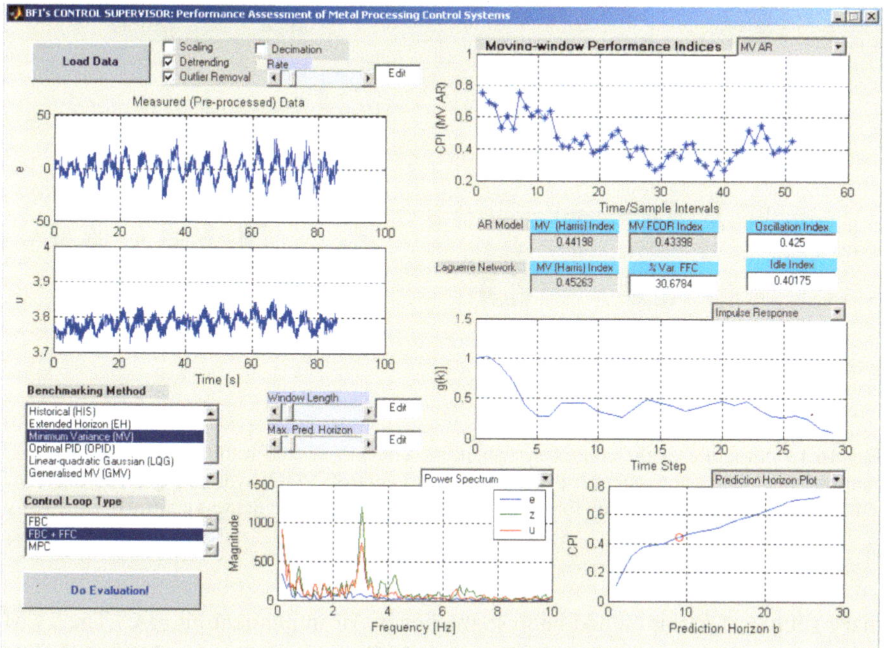

Fig. 16.12 MATLAB offline environment for performance assessment

therefore required to assess the more important technological control loops, whose performance can directly be related to the economical factors mentioned. This book demonstrates the effect of control performance assessment and control tuning on technological performance and revenue optimisation in metal processing.

Some of the control performance monitoring methods presented in the previous chapters have been successfully applied to the following processes:

- Thickness and flatness control of a tandem mill.
- Temperature control of an annealing plant in a hot-dip galvanising line.
- Zinc-layer-thickness control in a hot-dip galvanising line.
- Temperature control in the run-out table in a hot strip mill.
- Thickness and flatness control at a two-stand temper rolling mill.
- Thickness control at a Sendzimir mill.

A MATLAB offline tool for testing and comparing different CPM methods has been developed, as shown in Fig. 16.12. In this section, we present and discuss the results of the first three industrial cases studies.

16.3.1 Gauge Control in Cold Tandem Mills

The main objective of this study is to assess the current performance of the strip thickness control system in a tandem cold rolling mill (TCM), identify their pri-

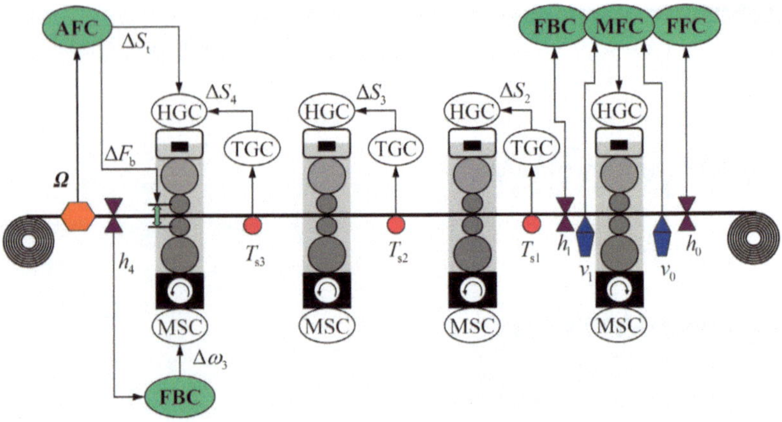

Fig. 16.13 Structure of the control system in the considered cold rolling mill (FBC: feedback control; MFC: mass flow control; FFC: feedforward control; HGC: hydraulic gap control; TGC: tension by gap control; MSC: motor speed control; AFC: automatic flatness control)

mary source of variation and analyse the benefit of implementing FFC. The TCM, where the strip is reduced in thickness typically from 3 mm (at the entry of the first stand) to 0.8 mm (at the exit of the fourth stand), consists of four rolling stands. The mill is equipped with different measurements systems as depicted in Fig. 16.13.

The control objective is to keep the thickness deviation (Δh_1) at the exit of the first stand as small as possible despite a large number of disturbances, such as variations of the strip hardness, the entry strip thickness and tension. The control strategy implemented consists of the combination of feedback control (known as monitor control), mass flow control and feedforward control. The manipulating variable is the position (S) of the (controlled) hydraulic capsule. The thickness feedback control uses the actual thickness deviation measured by the thickness gauge at the exit of the first stand. Based on the two strip-speed-measuring devices (for v_0 and v_1) and the thickness gauge (h_0) at the entry side of the first stand, the mass flow equation (law of volume conservation) allows for the calculation of an estimation of the strip exit thickness at the moment when the considered strip element is under deformation in the roll gap. The thickness estimate is the core of the mass flow control, which allows faster reaction to thickness deviations with bypassing of the varying time delay between the roll gap and the exit gauge. The thickness feedforward control is used to compensate for the strip thickness deviation (Δh_0) measured at the entry of the first stand. The purpose of the thickness feedback control at the last stand is to correct for remaining thickness deviations (Δh_4) mainly resulting from disturbances at the stands 2 to 4. In the following, we concentrate on the evaluation of the key thickness control at the first stand of the mill.

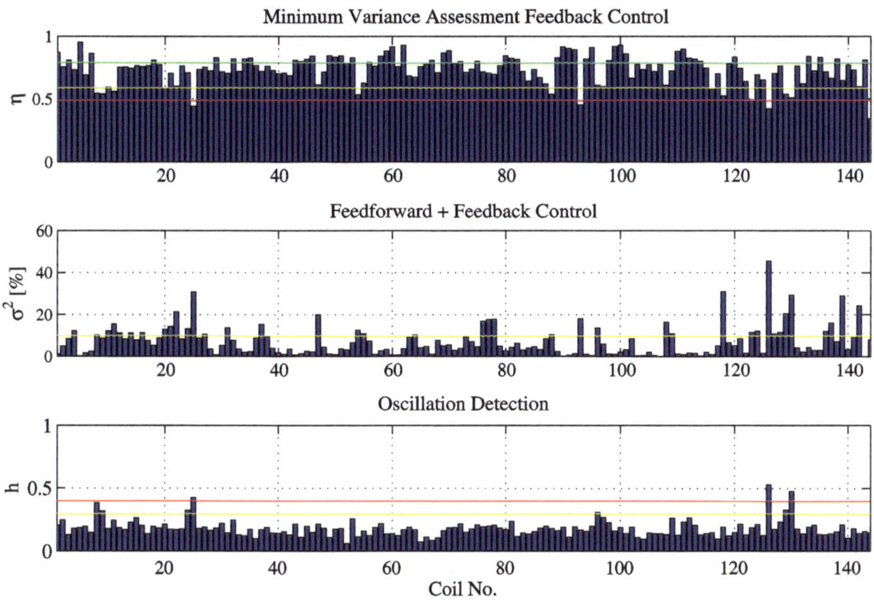

Fig. 16.14 Results of performance assessment of the thickness control at the first stand

16.3.1.1 Data Pre-processing

After acquisition of suitable data sets, pre-processing is needed to verify the appropriate sampling time, eliminate bad data and outliers, to mean-centre the data, etc. The closed-loop data used for thickness control monitoring were available at 0.050 s sampling rate, with a length N varying between 1200 and 7300 (typically $N = 3000$) depending on the strip length. The discrete time delay varies between 5 and 16. Only the steady-state operation phases, where the rolling speed, and thus the time delay, is constant, are considered for the first assessment.

16.3.1.2 Performance Evaluation in Terms of Minimum-Variance Benchmark

The performance analysis has been performed coil-wise for the steady-state phase (with constant rolling speed), and bar charts for the performance indices have been generated. Figure 16.14 illustrates the individual minimum-variance indices and the oscillation indices (Forsman and Stattin 1999) for a representative product mix of 140 coils (corresponding to approximately one-day production).

It can be concluded that the thickness control delivers good ($\eta \geq 0.6$) to optimal ($\eta \geq 0.8$) performance, and significant improvement is possible for only a few coils. Since the delay in disturbance path is longer than the output delay, the contribution of FF to the total variance vanishes for all coils, indicating that an FFC tuning alone cannot affect the control performance. In this case study, AR(MA) modelling was

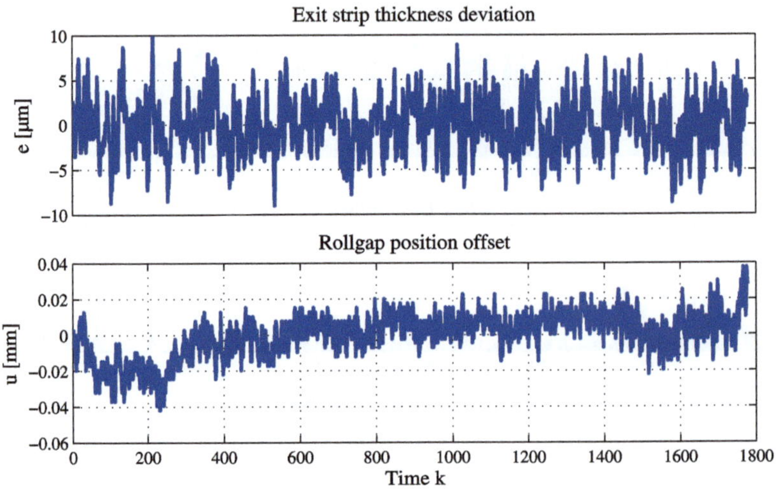

Fig. 16.15 Measured variables for coil no. 88 showing good thickness control performance

Table 16.2 Variance analysis table for coil no. 88

Disturbance	MV	FB	FF	FB/FF	Total
$\varepsilon(k)$	39.7 (93 %)	1.7 (4 %)	–	–	41.4 (97 %)
$w = \Delta h_0$	0	–	0	1.3 (3 %)	1.3 (3 %)
Total	39.7 (93 %)		3.0 (7 %)		42.7 (100 %)

sufficient; we did not found any advantage in applying subspace model identification for calculating (MVC) control performance indices.

The data of a typical coil (no. 88) with very good control performance is shown in Fig. 16.15. The minimum variance index is close to unity. The analysis of variance in Table 16.2 reveals that the disturbance is only responsible for 3 % of the total variance. The control performance is thus close to optimal (i.e. MVC), and no actions are suggested.

The values of the oscillation index for coils no. 4, 21, 122 and 126 indicate distinct oscillatory patterns in the output signals. For instance, based on the data in Fig. 16.16 and the variance analysis results in Table 16.3, the measured thickness of coil no. 21 clearly contains an oscillation due to the presence of a disturbance coming from the entry strip thickness. This also explains the low minimum-variance index value. Since only the variance contribution for FB + FF (31 %) is large, one can expect an FBC tuning (for disturbance rejection) to be able to sufficiently reduce the variance (i.e. $\sigma^2_{\mathrm{FB+FF},w}$) for these coils.

As the material gets harder and harder, from stand 1 to stand 4, the controller will have more and more problems to remove e.g. incoming disturbances. This can best

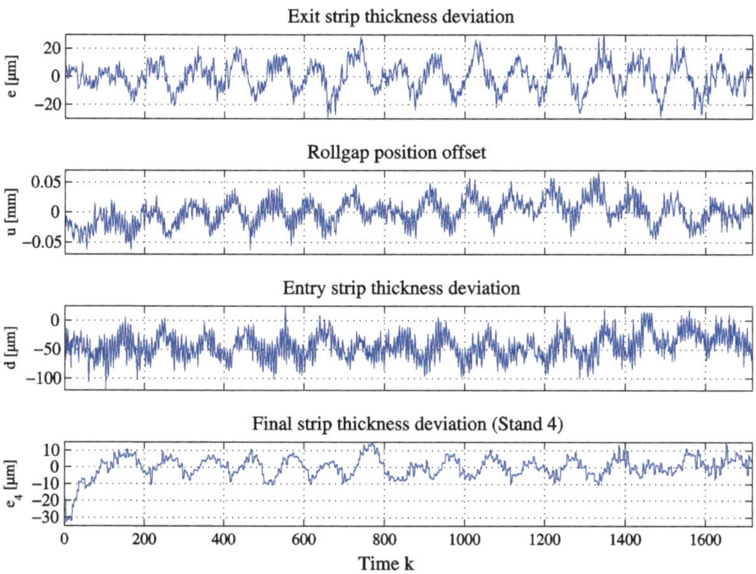

Fig. 16.16 Measured variables for coil no. 21 showing bad thickness control performance due to the presence of an oscillation coming from the entry strip thickness ($d \equiv \Delta h_0$)

Table 16.3 Variance analysis table for coil no. 21

Disturbance	MV	FB	FF	FB/FF	Total
$\varepsilon(k)$	41.7 (44 %)	24.0 (25 %)	–	–	65.7 (69 %)
$w = \Delta h_0$	0	–	0	29.1 (31 %)	29.1 (31 %)
Total	41.7 (44 %)		53.1 (56 %)		94.8 (100 %)

be seen in Fig. 16.16, i.e. the lowest subplot, showing that the incoming thickness deviations propagate till the final thickness at the exit of stand 4. This behaviour is reflected in the Harris index, as shown in Fig. 16.17: the index values are significantly lower than those given in Fig. 16.14. This observation applies also for the case of rolling in single-stand reversing mills, where the material becomes harder from pass to pass. This also implies that controller tuning should take into account this behaviour, i.e. using gain scheduling as function of hardness.

After revamping of the automation system on the considered mill, the performance of the gauge control was evaluated again. Figure 16.18 illustrates the Harris index for 80 coils, showing that the control performance is now good to optimal for all coils. There are also no significant oscillations in the loop. This assessment study was carried out after commissioning of the controller, and it reveals that the tuning is very good, as expected from a "fresh" revamping measure.

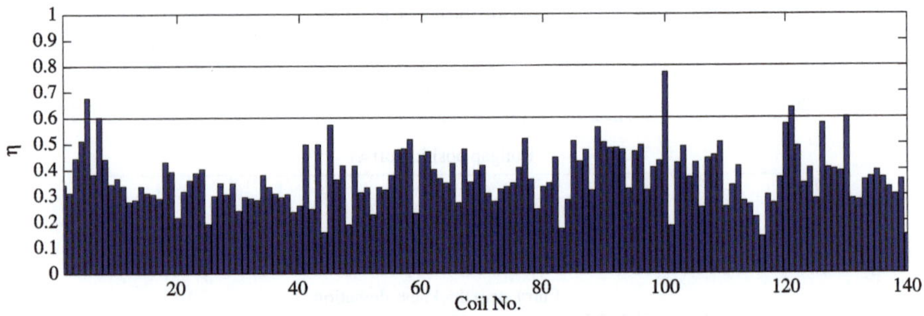

Fig. 16.17 Results of performance assessment of the thickness control at the last stand

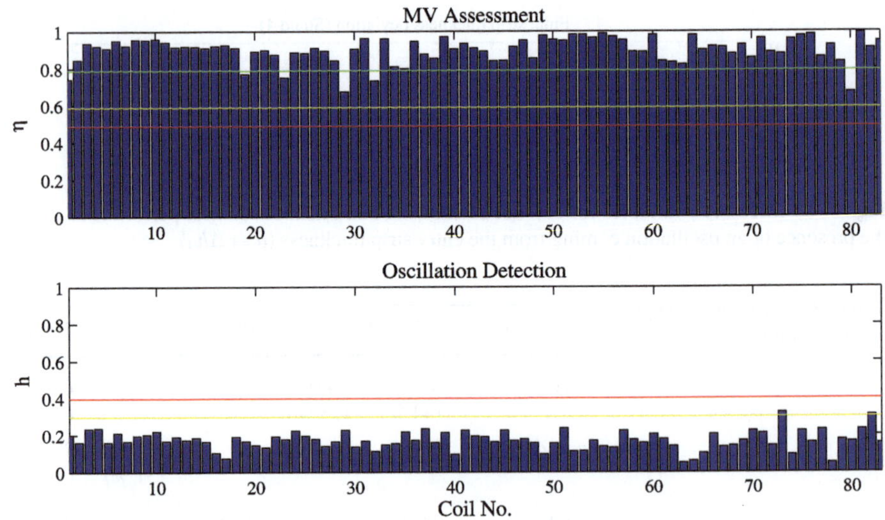

Fig. 16.18 Results of performance assessment of the thickness control at the first stand after re-vamping

16.3.2 Flatness Control in Cold Tandem Mills

The control objective in this study is to maintain the strip flatness deviation ($\Delta\boldsymbol{\Omega}$), at the exit of the mill at a minimum, despite a large number of disturbances, such as non-uniform incoming strip profile, non-uniform crowning of the rolls, non-uniform lubrication across the width of the strip in the roll bite, or non-uniform metallurgical properties.

Shape measurement is performed by measuring a differential tension (or equivalently length) profile across the strip width at some (n_s) discrete points (in this case $n_\mathrm{s} = 32$) using a shapemeter (or flatness roll). The output of the system is therefore a profile represented in vector form, whose dimension depends on the strip width, and thus flatness control is a multi-input multi-output problem. In order to reduce

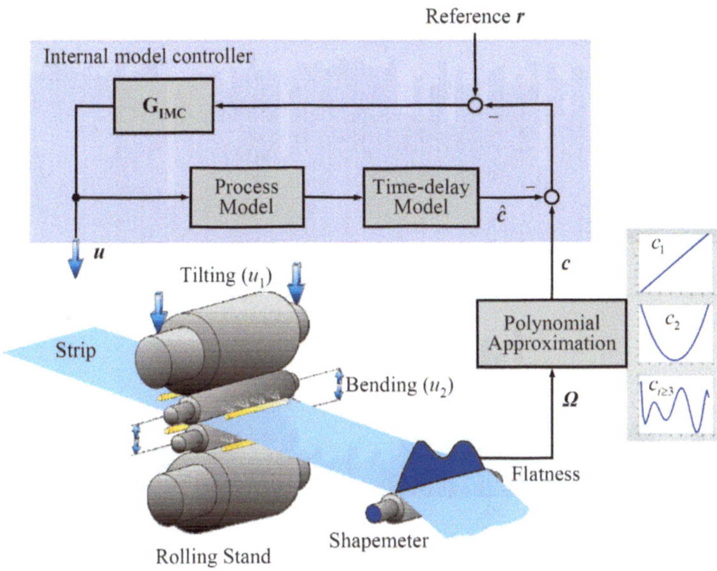

Fig. 16.19 Structure of the strip-flatness controller

the dimension of the MIMO system (and thus reduce its condition number), it is transformed into a parameterised form using basis-function expansions. Among a large number of different, but related, basis-function expansions proposed to analyse cross-directional control systems in the metal industries (Duncan et al. 1998; Ringwood 2000), the Gram polynomials are best suited to approximate the flatness (distribution), as the shapemeter installed on the exit of the mill under consideration is equipped with equidistant sensors. The polynomial coefficients c_i represent then the actually controlled variables. The reference flatness (distribution) is usually non-zero (i.e., the target flatness is *not* necessarily a perfectly flat strip). Its form is determined by the alloy and the requirements of the subsequent processes of the strip.

The available mechanical actuators are tilting (or skewing) ($S_t \equiv u_2$) and work-roll bending ($F_b \equiv u_1$). Tilting is used to control the linear part of the polynomial approximation, i.e. $c_1 \equiv y_2$. Bending serves as the manipulating variable for the quadratic part, i.e. $c_2 \equiv u_1$. The segmented cooling sprays are used to correct residual errors comprising all higher-order defects ($u_{i\geq 3}$). A three-loop internal model control scheme (Fig. 16.19) is implemented for flatness control since it provides an excellent trade-off between control system performance and robustness to modelling errors and process changes. Since the influence of the mechanical actuators on the strip flatness varies with the product properties, the internal model gains are schedule dependent, i.e. functions of the strip width and thickness. More details about the design, implementation and industrial results of this flatness control system can be found by Jelali et al. (2009).

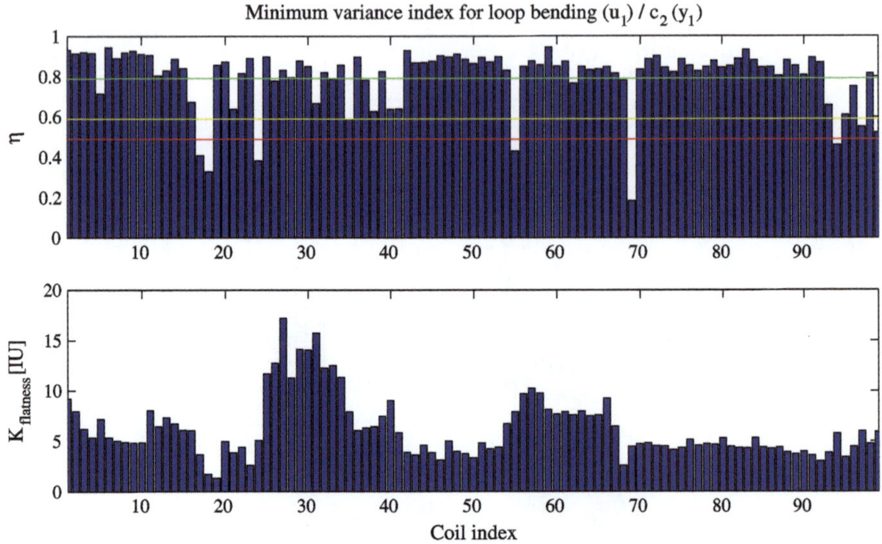

Fig. 16.20 Results of performance assessment of the flatness control for the u_2/y_2 (bending/c_2) loop

16.3.2.1 Data Provision

The closed-loop data used for flatness control monitoring were available at 0.1 s sampling rate, with a length varying between 900 and 4400 (typically $N = 2000$). The discrete time delay varies between 1 and 3. Due to the limited space, only the results for the u_2/y_2 (bending/c_2) loop are described. The results presented are based on data taken after the revamping of the automation system, including installation and commissioning of the new flatness controller by Jelali et al. (2009); see also Wolff et al. (2006). The old flatness control has seldom been used in automatic mode due to its poor performance; this was easy to see without systematic control performance analysis. Moreover, only compressed data were available, which are not suitable for performance assessment.

16.3.2.2 Assessment results

Figure 16.20 shows the individual minimum-variance indices and the user-specified indices for 45 coils. It can be concluded that the flatness control delivers good to optimal performance. The data of a typical coil with optimal control performance, i.e., $\eta = 0.92$, is shown in Fig. 16.21. Even for the coils no. 17, 18 and 43–45, the control performance can be considered to be satisfactory, as the values of the user-specified criterion K_{flatness} (Eq. 16.3) signal very good strip flatness. These coils are relatively thick ($h_4 > 1.4$ mm), so that they are non-critical with respect to flatness. The values of the oscillation index lie in between 0.04 and 0.2, indicating no oscillation problems. Therefore, no actions are suggested.

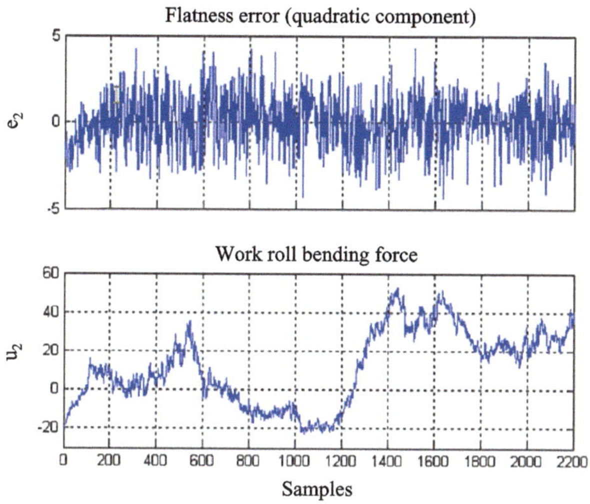

Fig. 16.21 Measured variables for coil no. 4 showing good flatness control performance

16.3.3 Temperature Control in Annealing Lines

Continuous annealing processes are highly efficient heat treatment processes after cold rolling in steel works. It aims to produce steel strips of high tensile strength and high formability. Continuous annealing is also required in continuous hot-dip galvanising before zinc coating. Such lines produce high-quality galvanised sheet for the automotive industry. Figure 16.22 shows the galvanising line considered. The material for annealing is a cold strip coil, which is on pay-off reel at the entry side of the line. The head of the coil is then pulled out and welded with the tail of preceding coil. Then the strip runs through the galvanising process (including annealing). On the delivery side, the strip is cut into a product by a shear machine and coiled again by a tension reel.

16.3.3.1 Control System Description and Analysis Strategy

The challenges of the performance assessment of this controller are:

- The annealing process considered consists of the heating, the soaking (annealing and retaining) and the fast cooling. Strip temperature is measured with a radiation pyrometer and controlled at the exit of each furnace. Temperature must be controlled within defined ranges from heat pattern; see Fig. 16.23. The heat patterns are determined according to the composition and product grade of the strip.
- The overall control structure of the temperature controller of the annealing furnace is very complex, as shown in Fig. 16.24. It consists of several temperature controllers for each furnace and a coordination level, which provides the reference temperature for each temperature controller. Each temperature controller itself consists of several sub-controllers. Sub-controllers are used to manipulate

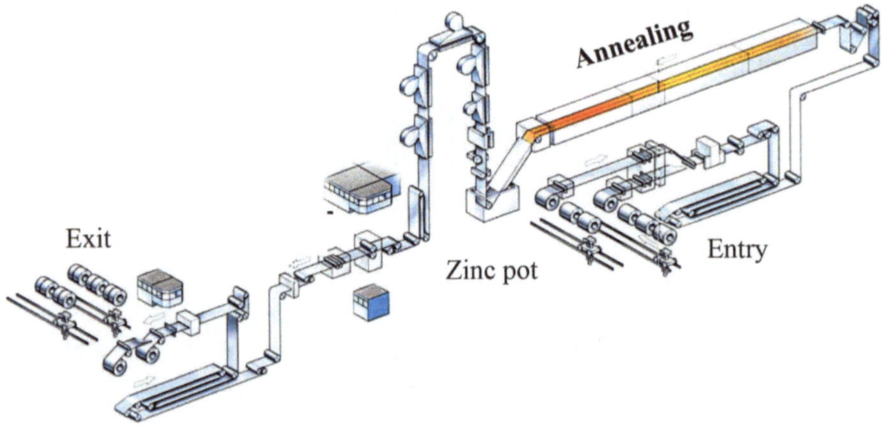

Fig. 16.22 Layout of the hot-dip galvanising line, where the temperature control system is installed

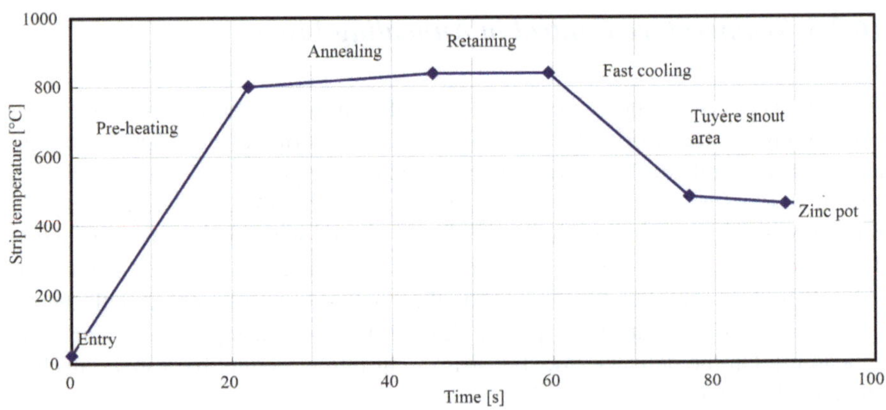

Fig. 16.23 Reference temperature trajectory to be ensured by the control system

the air flow and the natural gas flow within each furnace. Strip temperature shows some complicated characteristics because of slow dynamics, time delay, thermal interactions between the strip and hearth rolls which support the strip and setup changes, i.e. changes of strip thickness, strip width or reference temperature.

- The performance of this control system has large effect on the production rate, the strip quality in terms of mechanical properties (tensile strength, yield strength, elongation to fracture) and the stability of operation in the whole galvanising line.

Taking a look at the temperature control system for the annealing furnace in Fig. 16.25, we can see that it mainly consists of the upper (strip) temperature controller (TC3001) subordinated by split-range control that adjusts the reference (room) temperatures in the furnace zones. In the lower level, combustion controllers control the zone temperatures by adjusting air/gas flows. The other temperature con-

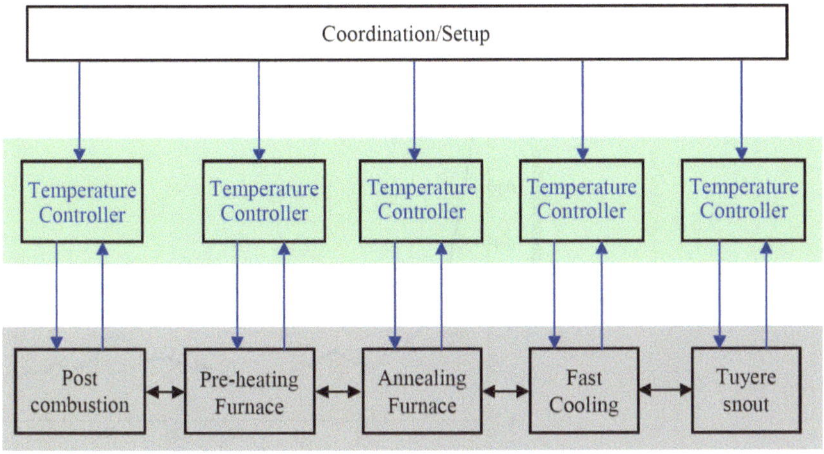

Fig. 16.24 Structure of the temperature control system in the annealing line

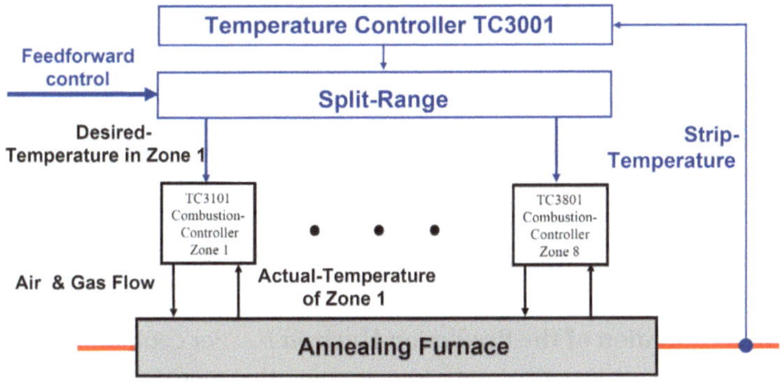

Fig. 16.25 Structure of the temperature control system in the annealing furnace

trol systems for the different furnaces have similar structures. A top-down assessment strategy was chosen to analyse the performance of the control systems. In this study, the results of the evaluation of the upper temperature controllers are presented. The lower level controller were not analysed because the upper controllers (after re-tuning some components) showed satisfactory performance, so that there was not enough incentive to continue the performance analysis.

16.3.3.2 Data Processing and Assessment Procedure

The evaluation procedure applied for the assessment of the performance of the temperature control system consists of following steps.

Step 1. Determination of the Appropriate Sampling Time The measured data were collected with a sample time of 2 ms. To achieve an impulse response, which

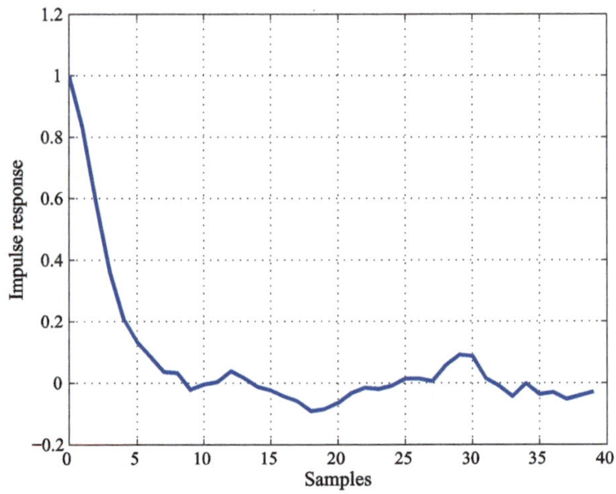

Fig. 16.26 Typical impulse response indicating good control performance

vanishes after 30 to 40 samples, as shown in Fig. 16.26, the data has been down-sampled to a sample time of about 40 s.

Step 2. Merging Data Files For computing the performance index, data files containing at least 1200 samples are necessary. Due to the down sampling, the data files of each strip are reduced to about 90 samples. Therefore, data files of a whole production day have been merged together to a single data file as shown in Fig. 16.27.

Step 3. Determination of the Prediction Horizon b For computing the extended performance index, it was necessary to determine the appropriate prediction horizon. Values for the time delay were not always available for this application. Therefore, an approach based on historical data was applied here: a data file of one-day production is selected, where the performance index is high in comparison to other production days. The trend of the extended performance index vs. the prediction horizon b is computed, as typically shown in Fig. 16.28 for the pre-heating furnace.

An estimate of the prediction horizon can be read from the intersection of the two tangents drawn at the initial point and the settling point of the curve. In this case, the estimated prediction horizon equals 3 and was kept constant for computing the extended performance index of the pre-heating furnace. This procedure has been applied for all other controllers analysed. Another choice could fall on the region where the control-performance index does not vary, i.e. $b = 7$ in this example. This value may be, however, too optimistic.

Step 4. Computation of the Extended Performance Index The extended horizon index is calculated based on AR modelling with higher order, $n = 30$. The results are discussed in what follows.

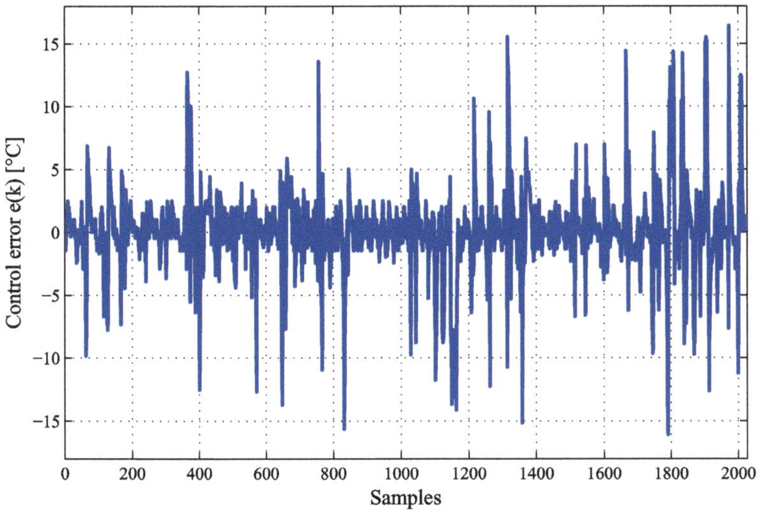

Fig. 16.27 Temperature control error for one-day production

Fig. 16.28 Extended horizon performance index of the strip temperature controller of the pre-heating furnace as function of the prediction horizon

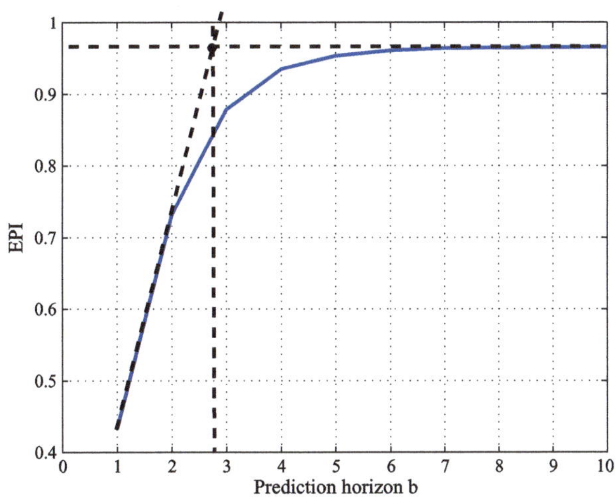

16.3.3.3 Assessment Results

The aforementioned procedure has been applied[1] to the temperature controller for pre-heating, annealing, fast cooling and tuyère snout. The assessment results are shown in Fig. 16.29. It can be clearly seen that the performance of the pre-heating controller varies too much from day to day. The same is observed for the fast cooling controller performance. The performance of the tuyère snout controller is quite

[1]The main analysis work was undertaken by Andreas Wolff.

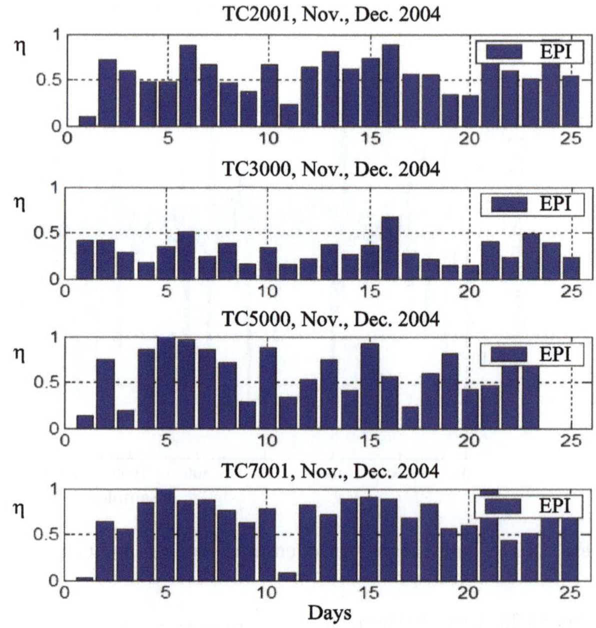

Fig. 16.29 Extended performance index (EPI) values for the different temperature controllers for pre-heating (TC2001), annealing (TC3000), fast cooling (TC5000) and tuyère snout (TC7001)

good, with a few exceptions. By far worst performance is achieved by the annealing controller. Because of these findings, an emphasis was then placed on the further study of the temperature control for the annealing furnace.

16.3.3.4 Diagnosis and Re-tuning of the Annealing Temperature Controller

A closer examination of the temperature traces revealed that the strip temperature reference is not always reached and the maximum furnace temperature of 950 °C is often exceeded in the last zones; see Fig. 16.30. This figure also shows that the heating power produced by the feedforward controller (y_R) in the first zones is very low. The controller is not fast enough to adjust the needed heating power. In the last zones, however, enough heating power is generated. These findings were considered as the main weakness of the control system.

Therefore, a re-distribution of the heating power produced by the feedforward controller has been performed. The feedforward controller consists of the following equation:

$$u_{R,i} = (\theta_{r,\mathrm{TC3000}} - \theta_{r,\mathrm{TC2001}})T_A K_1 K_2 + K_3 + R\theta_{\mathrm{TC3000}}. \qquad (16.4)$$

Details are given by Müller (2005). Relevant for variations of the feedforward controller is the difference between the reference temperature of the annealing furnace ($\theta_{r,\mathrm{TC3000}}$) and the reference temperature of the preheating furnace ($\theta_{r,\mathrm{TC2001}}$). For small values of this difference, the part of the feedforward controller reduces a lot.

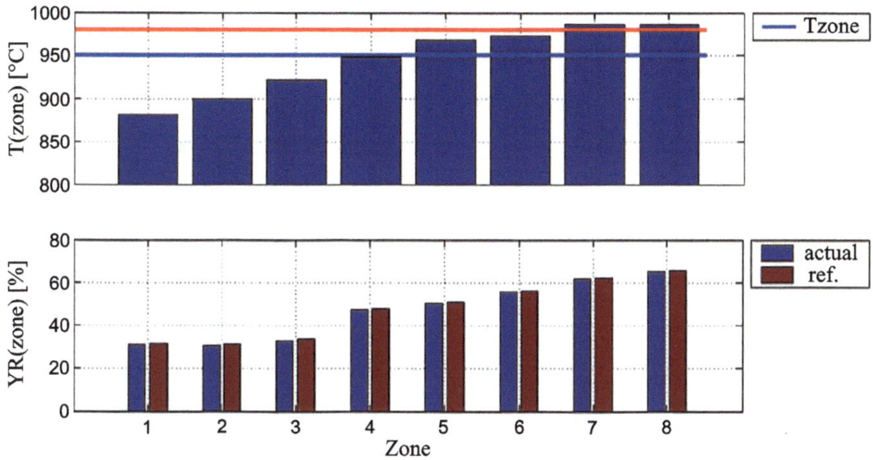

Fig. 16.30 Temperature overrun and heating power of the feedforward controller

When an offset in the difference between the reference temperatures is integrated, we get the modified feedforward control law

$$u_{R,i} = \left[15 + 0.85(\theta_{r,\mathrm{TC3000}} - \theta_{r,\mathrm{TC2001}})\right] T_A K_1 K_2 + K_3 + R\theta_{\mathrm{TC3000}}. \qquad (16.5)$$

This should generate a higher heating power in the first zones without modifying the heating power in the last zones.

16.3.3.5 Improvement Analysis and Benefits

The modified feedforward controller has been implemented and tested on-site. The control performance was then analysed again. The performance results before and after re-tuning are shown in Fig. 16.31. It is clearly observed that the re-adjustment of the heating power generated not only better control performance, but also significant reduction of the frequency of maximum furnace temperature violations (Fig. 16.32). More specifically, the mean performance index is increased from 0.26 to 0.70, and the frequency of maximum temperature violations is decreased from 7.0 to 0.5 %. To typically compare controller errors before and after re-tuning, Fig. 16.33 is provided. The figure indicates the substantial variance decrease after re-tuning.

Note that the exact date of introducing the modification has not been communicated to the analysis team, to test the capability of the CPM tool to flag performance changes. Therefore, the performance analysis was also used to recover the time point of the release of modified control system. This predicted time point was in good agreement with the real one.

The changes resulting from the outcomes of the performance analysis led to improved uniformity of thermal profile and thus to decreased energy consumption and

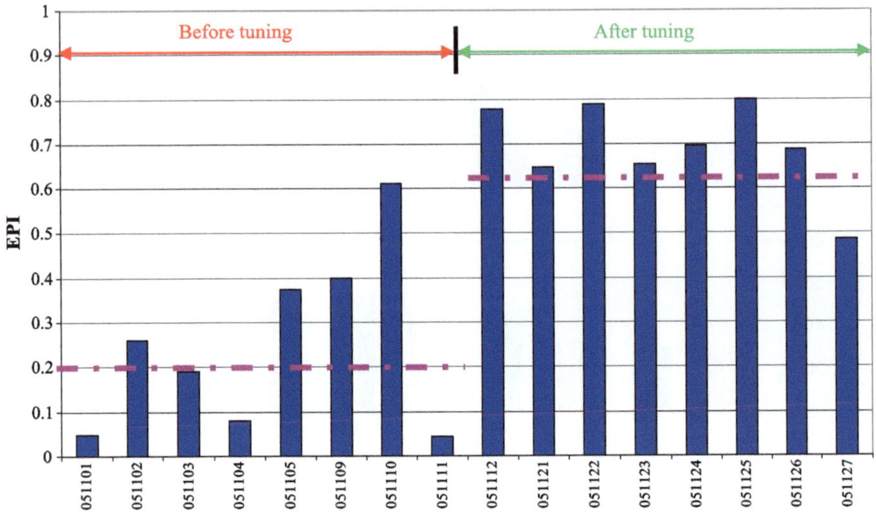

Fig. 16.31 Control performance index values before and after the modification

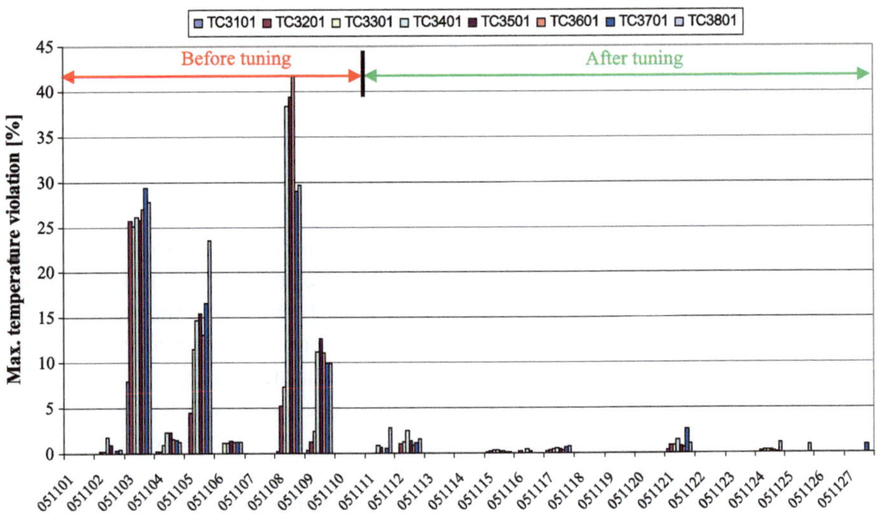

Fig. 16.32 Violations of maximum temperature before and after the modification

material processing rate. Analysis of the natural gas consumption during the year before the controller modification and the year after that showed that a reduction of around 90 k/a has been achieved.

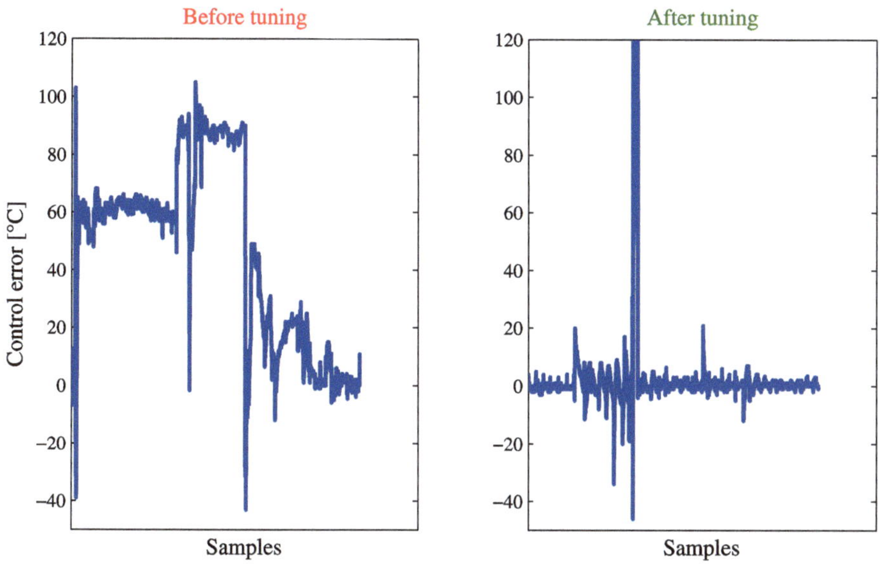

Fig. 16.33 Typical temperature deviations from set point for periods before and after re-tuning the annealing controller

16.3.3.6 Control Performance Monitoring Tools

Control performance monitoring tools have been developed and tailored for all these applications. For instance, the temperature-control monitoring tool is illustrated in Fig. 16.34. After selecting the data sets, the user starts the evaluation to get the overall performance index of the complete temperature control.

Additionally, the tools provide the user with the performance index, the standard deviation and the mean control error for each part of the annealing line, i.e. the pre-heating furnace, the annealing furnace, the fast cooling and the tuyère snout. Moreover, an assessment comment is shown, which explains the results of the performance index. In-depth analysis of data and control performance in terms of impulse response, control error (Fig. 16.35), temperature limit violations (Fig. 16.36), etc. can also be carried out by the control engineers by selecting the corresponding popup menus.

The control performance monitoring tools[2] have been implemented in MATLAB. A special MATLAB compiler translates these source codes in a C++-Code, which is integrated in LabVIEW. LabVIEW offers options for fine and practical user-interface and for embedding C-routines. Also, Oracle data can be imported from a central database (ZQDB), where all measured data are stored (see Fig. 16.37).

[2]The tools described in this section were mainly created by Martina Thormann.

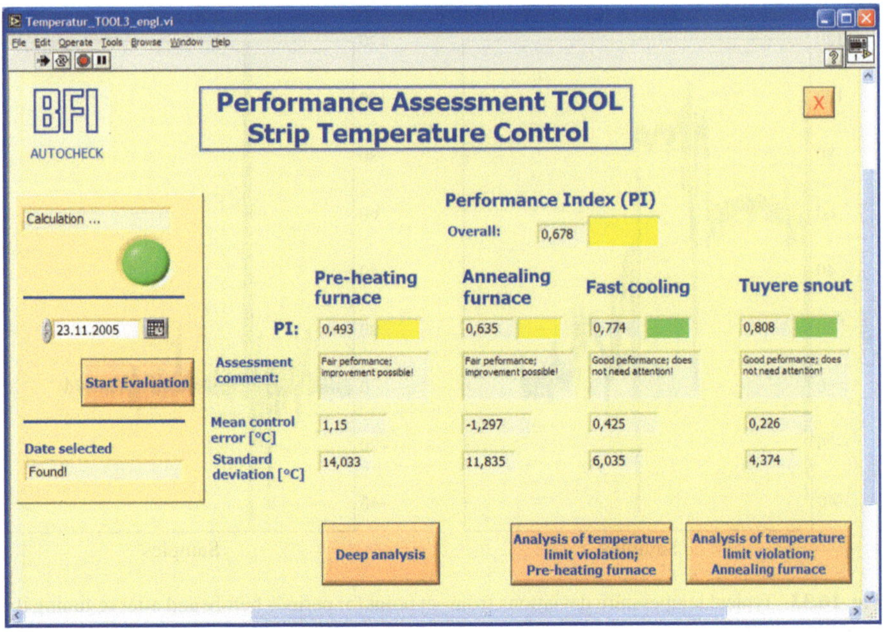

Fig. 16.34 Control-performance-monitoring tool for temperature control

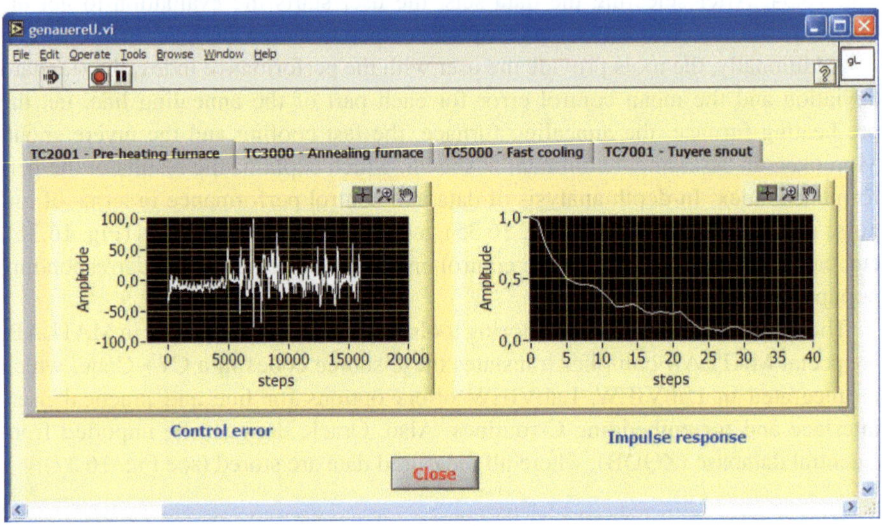

Fig. 16.35 Further analysis of the temperature control performance

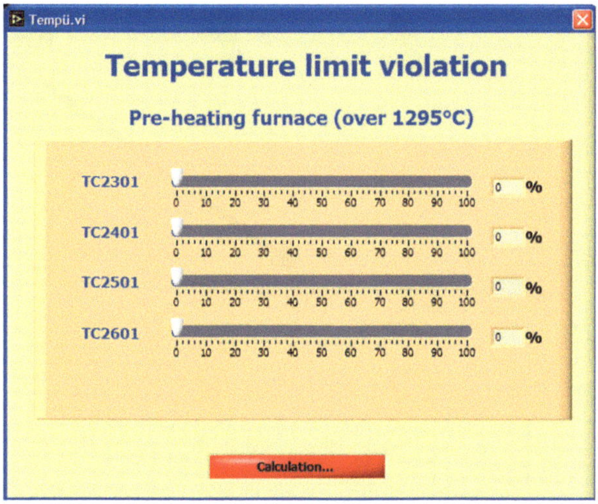

Fig. 16.36 Statistics of the temperature-limit violations of the temperature control

The CPM tools are implemented in a stand-alone PC but integrated into the automation infrastructure of the customer, and analysis can be carried out on demand. Work is running to get a system analysing automatically or in batch mode.

16.3.3.7 Comprehensive CPM Realisation and Integration Concept

The basic concept for the realisation of the framework proposed above is shown in Fig. 16.38. It is based on the exploitation of the information transfer via intranet/internet technology and data exchange media available at the plants. The following information paths are intended for implementation (Jelali et al. 2008):

- Full excess to the databases for selecting data required for control-performance monitoring and setup supervision.
- *Alerts* are sent to the operator when serious performance problems are detected, with plausible action suggestions. The algorithms implemented should calculate performance indices repeatedly over time and comparing them to alert limits. Each alert limit can be decided from statistical characteristics of the index or by some other criteria.
- An *email with detailed evaluation results*. Reports include the individual performance indices of all related control loops, current controller settings, a summary of loops whose performance significantly changed since the last observations, a ranking of the controllers regarding performance, information about model prediction quality behind the benchmark, etc. can be provided. When the automatic communication of alerts by e-mail is established, the plant engineers will not need to access periodically the system to see if some alert has been created, because the system will automatically warn them. Of course, engineers can still access to the server from their personal computers placed at their offices and visualise the

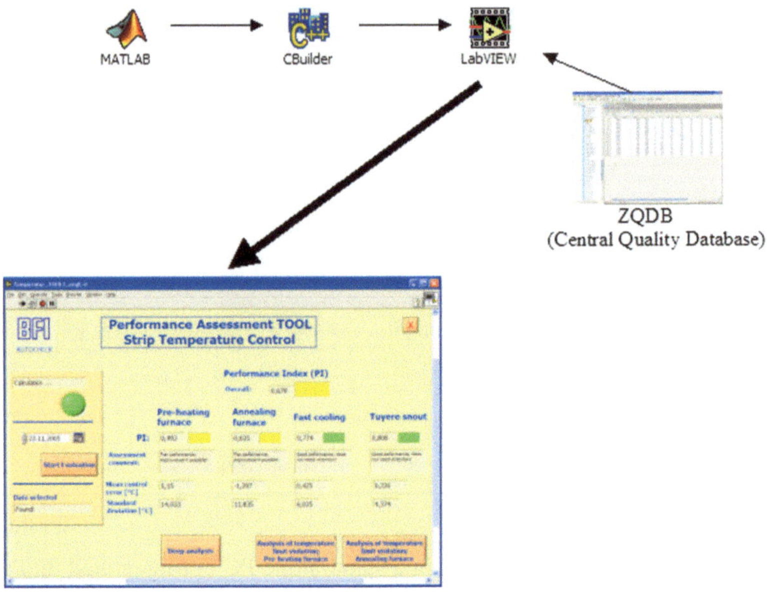

Fig. 16.37 Work flow of the tools implementation

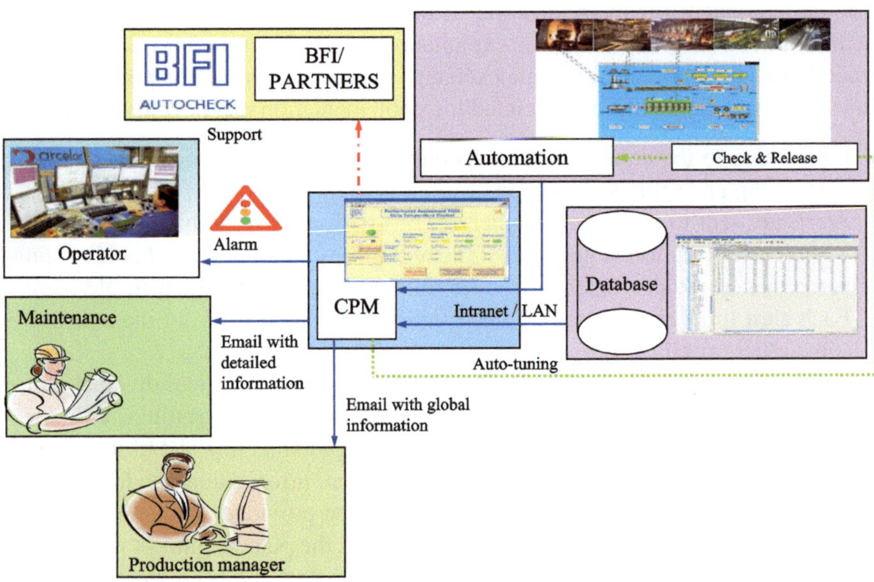

Fig. 16.38 Concept of control system performance supervision based on Intranet/Internet technology

Table 16.4 Overview of possible actions to be implemented

Possible actions	**Advantages**	**Drawbacks**
Alert to mill operator (online adjustments)	Process continuity. Controllability.	Control system still to be optimised.
Sending online email to process control technician (offline adjustments)	Process continuity. Control system (quasi-) optimised. Controllability.	Delayed parameter tuning.
Direct intervention on process control system (online adjustments)	Prompt reaction and adaptation.	Dangerous / need of complicated rules to avoid process instabilities.

current status of performances (on demand) without the need for going personally to the controllers.

- The production manager is provided with only *global information/indication* about the control system including setup status. Information can be given in form of performance indices of the loops, the percentage of time they are working on or frequency of constraint attaining. Also quality indicators related to the performance of the control system should be provided.
- *Internet-information transfer* could be possibly sent to an *external partner* such as BFI for the purpose of fast troubleshooting.
- Based on the results and indices of the control-performance monitoring procedures, an *automatic tuning* of some controllers could be implemented. This topic is however still open and thus planned within a forthcoming project.

An overview about the advantages and drawbacks of possible actions within the above framework is shown in Table 16.4. Each of these measures should be extensively discussed with the plant personnel prior to its implementation. In practice, only a few parameters will be accepted to automatically changed by the tuning and supervision systems. At least, a mechanism has to be implemented to enable the operator or the technician to check the new settings and decide whether they are released, i.e. down-loaded to the control system.

16.4 Summary and Conclusions

It has been shown how control performance monitoring (CPM) algorithms perform well in the environment of metal processing. Special aspects when applying the techniques in this industrial area have been presented. This includes online vs. batch-wise performance evaluation, time-based vs. length-based assessment and oscillation diagnosis.

Control performance monitoring tool for calculating performance indices has been developed and successfully applied to the performance evaluation of the main technological controllers in a tandem cold rolling mill: a strip thickness controller consisting of feedback plus feedforward control components and a flatness controller designed according to the internal model control approach. The results indicate that tuning the feedback thickness controller is suggested to better handle entry thickness disturbances for specific coils. After revamping, the controller shows best performance. The performance of the flatness controller after revamping is found to be optimal, and thus no actions are needed.

Moreover, a complex temperature control system in an annealing plant has been evaluated and re-tuned, leading to an increase in performance of the controller and to a significant reduction of energy consumption for the customer. CPM tools have been tailored to this application and implemented in the infrastructure of the mill, to work in a semi-automatic mode at time.

Appendix A
Basic Signal Processing and Statistics

This section contains a brief introduction to basic statistics. Comprehensive descriptions can be found in many standard books, e.g. Papoulis (1984) and Oppenheim and Schafer (1989). All the results presented here are for discrete-time signals, because we are interested in digital signal processing applications and the data we work with are assumed to be sampled. Note that the statistical characteristics given in this section can be expressed in terms of ω_k, f_k or simply k, all related by $\omega_k = 2pf_k = 2pk/N$, where N is the length (period) of the considered signal.

A.1 Ergodicity

Real signals are almost noisy, i.e. the source signal of interest is superposed by a stochastic (random) phenomenon, called noise. The estimation of the statistical properties of stochastic signals given below depends upon the *ergodic* assumption: (we hope that) the expectation over all realisations (or observations) of a stochastic process can be calculated as the time average of *one* realisation of the process. Thus, ergodicity implies that one realisation of the stochastic process contains all information about the statistical properties as $k \to \infty$. An ergodic process is stationary in the strict sense, i.e. all its statistical properties are time-independent. A signal is said to be stationary in the wide sense if its first two moments (i.e. the mean and the variance) are time-independent.

A.2 Expectation and Variance

The *expected value* or *mean* of a discrete random variable or (stationary) stochastic process (in practice a set of data) can be estimated by [mean]

$$\mathrm{E}\{x(k)\} \approx \bar{x} = \lim_{N\to\infty} \frac{1}{N}\sum_{k=1}^{N} x(k), \tag{A.1}$$

M. Jelali, *Control Performance Management in Industrial Automation*,
Advances in Industrial Control, DOI 10.1007/978-1-4471-4546-2,

where E{·} denotes the ensemble expectation operator. The estimation of the expected value is often written as $\bar{x}$ for shortness and in order to explicitly express the experimental determination of the mean from a data set of finite length N.

The *variance* of a random variable is the *mean squared deviation* from its arithmetic mean [var]:

$$\sigma_x^2 = \mathrm{E}\big\{\big[x(k) - \bar{x}\big]^2\big\} = \lim_{N\to\infty} \frac{1}{N} \sum_{k=1}^{N} \big[x(k) - \bar{x}\big]^2. \tag{A.2}$$

It is often easier to think in terms of the *standard deviation* σ_x [std], i.e. the square root of the variance.

A.3 Correlation and Covariance

Both correlation and covariance measure the similarity between two random variables or stochastic processes. The *auto-correlation function* describes the internal coherence of the variable (i.e. between the variable itself and its time-shifted version). It can be estimated as

$$\Phi_{xx}(\tau) = \mathrm{E}\big\{x(k)x(k+\tau)\big\} = \lim_{N\to\infty} \frac{1}{N} \sum_{k=1}^{N} x(k)x(k+\tau). \tag{A.3}$$

$\Phi_{xx}(\tau)$ is a symmetric function about $\tau = 0$, i.e. $\Phi_{xx}(\mathrm{t}\tau)\mathrm{g} = \Phi_{xx}(\tau)$; hence, $\Phi_{xx}(\tau)$ is a zero-phase function, which means that all phase information about $x(k)$ is lost.

The dependence of two random variables $x(k)$ and $y(k)$ are expressed by the *cross-correlation function* [xcorr]

$$\begin{aligned}\Phi_{xy}(\tau) &= \mathrm{E}\big\{x(k)y(k+\tau)\big\} = \lim_{N\to\infty} \frac{1}{N} \sum_{k=1}^{N} x(k)y(k+\tau) \\ &= \mathrm{E}\big\{x(k-\tau)y(k)\big\} = \lim_{N\to\infty} \frac{1}{N} \sum_{k=1}^{N} x(k-\tau)y(k) = \Phi_{yx}(-\tau).\end{aligned} \tag{A.4}$$

Note that $\Phi_{xy}(\tau)$ is not a symmetric function about $\tau = 0$, but that $\Phi_{yx}(\mathrm{t}\tau) = \Phi_{xy}(\tau)$. Correlation between two variables is high (low) if the variables are closely (weakly) related. Two variables are called *uncorrelated* if

$$\Phi_{xy}(\tau) = \mathrm{E}\{x\}\mathrm{E}\{y\} = \bar{x}\,\bar{y}. \tag{A.5}$$

The *covariance functions* are defined by [xcov]

$$\begin{aligned}C_{xx}(\tau) &= \mathrm{E}\big\{\big(x(k) - \bar{x}\big)\big(x(k+\tau) - \bar{x}\big)\big\} \\ &= \lim_{N\to\infty} \frac{1}{N} \sum_{k=1}^{N} \big[x(k) - \bar{x}\big]\big[x(k+\tau) - \bar{x}\big],\end{aligned} \tag{A.6}$$

$$C_{xy}(\tau) = \mathrm{E}\{(x(k) - \bar{x})(y(k+\tau) - \bar{y})\}$$
$$= \lim_{N\to\infty} \frac{1}{N} \sum_{k=1}^{N} [x(k) - \bar{x}][y(k+\tau) - \bar{y}]. \tag{A.7}$$

Obviously, if the means are equal to zero, correlation and covariance are identical. The special case follows from comparing Eqs. A.2 and A.6:

$$C_{xx}(0) = \sigma_x^2. \tag{A.8}$$

The ratio (bounded between 0 and 1)

$$\rho = \frac{\mu_{xy}}{\sqrt{\mu_{xx}\mu_{yy}}} = \frac{\mathrm{E}\{(x - m_x)(y - m_y)\}}{\mathrm{E}\{(x - m_x)^2\}\mathrm{E}\{(y - m_y)^2\}} \tag{A.9}$$

is called the correlation coefficient [corrcoef].

The so-called *covariance matrix* frequently used in the context of parameter estimation is defined for a vector $\boldsymbol{x}$ of n random variables as [cov]

$$\mathrm{cov}\{\boldsymbol{x}\} = \mathrm{cov}\{\boldsymbol{x}, \boldsymbol{x}\} = \begin{bmatrix} \mathrm{cov}\{x_1, x_1\} & \mathrm{cov}\{x_1, x_2\} & \dots & \mathrm{cov}\{x_1, x_n\} \\ \mathrm{cov}\{x_2, x_1\} & \mathrm{cov}\{x_2, x_2\} & \dots & \mathrm{cov}\{x_2, x_n\} \\ \vdots & \vdots & \ddots & \vdots \\ \mathrm{cov}\{x_n, x_1\} & \mathrm{cov}\{x_n, x_2\} & \dots & \mathrm{cov}\{x_n, x_n\} \end{bmatrix}. \tag{A.10}$$

Strictly speaking, the main diagonal entries of this matrix contain variances. Therefore, the matrix is sometimes referred to as the variance-covariance matrix.

A covariance vector is obtained as the covariance of a vector $\boldsymbol{x}$ of n random variables and a scalar random variable y:

$$\mathrm{cov}\{\boldsymbol{x}, y\} = \begin{bmatrix} \mathrm{cov}\{x_1, y\} \\ \mathrm{cov}\{x_2, y\} \\ \vdots \\ \mathrm{cov}\{x_n, y\} \end{bmatrix}. \tag{A.11}$$

Correspondingly to these covariance matrices and vectors, correlation matrices and vectors can also be defined in a straightforward manner.

A.4 Discrete Fourier Transform

The relation between time domain and frequency domain measures forms the foundations of much of modern signal processing The discrete Fourier transform (DFT) provides a means for transforming from time to frequency domain and vice versa. This is useful because signal properties do not always manifest themselves in the signal waveform and transforming to the frequency domain can expose periodicities in the measured signal and can aid understanding the processes which produced the signal.

DFT is used for periodic, discrete-time or digital signals. The DFT for a signal $x(k)$ with period N is

$$X(k) = \sum_{i=0}^{N-1} x(i) e^{-j\frac{2\pi k}{N} i}. \tag{A.12}$$

The inverse transform (IDFT) is

$$x(k) = \frac{1}{N} \sum_{i=0}^{N-1} X(i) e^{j\frac{2\pi i}{N} k}. \tag{A.13}$$

These equations say that the (complex) coefficients $X(k)$ represent the periodic discrete-time $x(k)$. Note that $X(-k) = X^*(k)$, where '*' denotes the complex conjugate.

To represent the time signal, $X(k)$ needs to be computed for N values of k, and for each of these values, N multiplications and (N t 1) additions must be performed. Therefore, computing the DFT requires $N(2N$ t 1) arithmetic operations, i.e. it has a *computational complexity* of $O(N^2)$ (order of N^2). However, especially if N is a power of 2, many of the DFT calculations are redundant. By carefully re-arranging the order of multiplications and additions, the computational complexity can be reduced to $O(N \log_2 N)$. The resulting algorithm is called the *fast Fourier transform* (FFT) [fft, ifft]. The difference between $O(N^2)$ and $O(N \log_2 N)$ can be substantial: If $N = 1024$ and $\log_2 1024 = 10$, the computational complexity of the FFT is only 1 % of that of a DFT.

As noted above, the DFT and IDFT are defined for periodic time signals. We may, however, be interested in a digital representation of the spectrum of non-periodic discrete-time signals. For time-limited signals, i.e. for signals that differ from zero only for $0 \le k < N$, we found that the DFT results in samples of the DTFT. For non-periodic, infinite (or long) time signals, we have to restrict the calculation of the DFT to a number of N samples, meaning that we may only use a number of $L \le N$ signal samples. The procedure to pick only a limited number of samples from a possibly infinitely long signal is called *windowing* [window]. Various windowing methods, e.g. rectangular window, triangular (Barlett) window, Hamming window, Kaiser window and Chebyshev window, and their effects are well documented in the literature; see Oppenheim and Schafer (1989).

A.5 Power Spectrum and Coherence Function

The *power spectrum* (or power spectral density: PSD) is formally defined as the FT of the auto-correlation sequence (known as the *Wiener*–Khintchine theorem) [psd]

$$S_{xx}(f) = \sum_{\tau=-\infty}^{\tau=+\infty} \Phi_{xx}(\tau) e^{-j2\pi f \tau}, \tag{A.14}$$

where f denotes the frequency. An equivalent definition is given by

$$S_{xx}(f) = \mathrm{E}\{X(f)X^*(f)\}. \tag{A.15}$$

A sufficient, but not necessary, condition for the existence of the PSD is that the auto-correlation be absolutely summable. The PSD is real-valued and non-negative, i.e. $S_{xx} \geq 0$; if $x(k)$ is real-valued, then the PSD is also symmetric, i.e. $S_{xx}(f) = S_{xx}(-f)$.

Similarly, the cross-power spectrum is defined by [csd]

$$S_{xy}(f) = \sum_{\tau=-\infty}^{\tau=+\infty} \Phi_{xy}(\tau)e^{-j2\pi f\tau}, \tag{A.16}$$

$$S_{xy}(f) = \mathrm{E}\{X(f)Y^*(f)\}. \tag{A.17}$$

An indication about the power and phase coherence between two signals $x(k)$ and $y(k)$ at a given frequency can be gained by computing the coherence function (or index or coefficient)

$$C_{xy}(f) = \frac{S_{xy}(f)}{\sqrt{S_{2x}(f)S_{2y}(f)}}. \tag{A.18}$$

From the Schwarz inequality it follows that $0 \leq C_{xy}(f) \leq 1$. Usually, the squared coherence is used instead [cohere]:

$$C_{xy}^2(f) = \frac{|S_{xy}(f)|^2}{S_{2x}(f)S_{2y}(f)}. \tag{A.19}$$

Thorough discussion of the properties of PSD can be found in Marple (1987).

Appendix B
Higher-Order Statistics

The first- and second-order statistics (e.g. mean, variance, autocorrelation, power spectrum) introduced above are only sufficient for describing linear and Gaussian processes. In practice, many situations occur where linearity and Gaussianity do not hold, e.g. when the process exhibits a nonlinear behaviour. To conveniently study these systems, higher-order statistics (HOS) are needed. Particularly, HOS enable us to extract information due to deviations of signals from Gaussianity to recover their true phase character and to detect and quantify nonlinearities in time series.

In this section, the definitions, properties and computation of HOS, i.e. moments, cumulants and their corresponding higher-order spectra (polyspectra) are introduced. Emphasis of the discussion is placed on 2nd- and 3rd-order statistics and their respective Fourier transforms, power spectrum and bispectrum. An application of the bispectrum for nonlinearity detection is introduced in Sect. 10.2. The presentation is mainly adopted from Nikias and Mendel (1993), Nikias and Petropulu (1993) and Fackrell (1996).

B.1 Moments and Cumulants

Given a set of n real random variables $\{x_1, x_2, \dots, x_n\}$, their joints *moments* of order $r = k_1 + k_2 + \cdots + k_n$ are expressed as

$$\begin{aligned}\mathrm{Mom}\{x_1^{k_1}, x_2^{k_2}, \dots, x_n^{k_n}\} &:= \mathrm{E}\{x_1^{k_1}, x_2^{k_2}, \dots, x_n^{k_n}\} \\ &= (-j)^r \frac{\partial^r \Phi(\omega_1, \omega_2, \dots, \omega_n)}{\partial \omega_1^{k_1} \partial \omega_2^{k_2} \cdots \partial \omega_n^{k_n}}\bigg|_{\omega_1 = \omega_2 = \cdots = \omega_n = 0}, \end{aligned} \tag{B.1}$$

where

$$\Phi(\omega_1, \omega_2, \dots, \omega_n) = \mathrm{E}\{\exp(j(\omega_1 x_1, \omega_2 x_2, \dots, \omega_n x_n))\} \tag{B.2}$$

is their joint characteristic function (the moment-generating function).

M. Jelali, *Control Performance Management in Industrial Automation*,
Advances in Industrial Control, DOI 10.1007/978-1-4471-4546-2,

Similarly, the coefficients in the Taylor expansion of the cumulant-generating function, also known as the second characteristic function,

$$\Psi(\omega_1, \omega_2, \ldots, \omega_n) := \ln \mathrm{E}\big\{\Phi(\omega_1, \omega_2, \ldots, \omega_n)\big\} \tag{B.3}$$

are the *cumulants* of order r:

$$\mathrm{Cum}\big\{x_1^{k_1}, x_2^{k_2}, \ldots, x_n^{k_n}\big\} := (-j)^r \frac{\partial^r \Psi(\omega_1, \omega_2, \ldots, \omega_n)}{\partial\omega_1^{k_1}\partial\omega_2^{k_2}\cdots\partial\omega_n^{k_n}}\bigg|_{\omega_1=\omega_2=\cdots=\omega_n=0}. \tag{B.4}$$

Thus, moments and cumulants are related to each other. The general relationship is given by

$$\mathrm{Cum}(x_1, x_2, \ldots, x_n) = \sum (-1)^{p-1}(p-1)!\mathrm{E}\bigg\{\prod_{i\in s_1} x_i\bigg\}\mathrm{E}\bigg\{\prod_{i\in s_2} x_i\bigg\}\cdots\mathrm{E}\bigg\{\prod_{i\in s_p} x_i\bigg\}, \tag{B.5}$$

where the summation extends over all partitions $(s_1, s_2, \ldots, s_p)$, $p = 1, 2, \ldots, n$, of the set of integers $(1, 2, \ldots, n)$. See Nikias and Petropulu (1993) for more details and examples. As cumulants possess certain properties, which lend themselves well to the development of new HOS techniques, most HOS methods are developed in terms of cumulants and not moments.

Perhaps the most important of these properties are those concerning Gaussian processes: a Gaussian process is completely characterised by its mean and variance only. Moreover, the first-order cumulant of a Gaussian process is equal to the mean, the second-order cumulant to the variance, and all higher-order cumulants are identically zero. This property suggests that measurement noise, which is often assumed to be Gaussian, disappears at third and higher orders. This raises the possibility that if the process of interest is non-Gaussian, then its properties will "shine through" the noise in the higher-order domains (Fackrell 1996). This remains one of the key motivations for research in HOS methods. The reader is referred to Nikias and Petropulu (1993) for a detailed description of the properties of cumulants.

Consider now a real *stationary* random process $x(k)$ whose moments up to order n exist. Then,

$$\mathrm{Mom}\big\{x(k), x(k+\tau_1), \ldots, x(k+\tau_{n-1})\big\} = \mathrm{E}\big\{x(k)x(k+\tau_1)\cdots x(k+\tau_{n-1})\big\} \tag{B.6}$$

will depend only on the time differences τ_i, and the nth-order moments m_{nx} of $x(k)$ are written as

$$m_{1x} = \mathrm{E}\big\{x(k)\big\}, \tag{B.7}$$

$$m_{2x}(\tau) = \mathrm{E}\big\{x(k)x(k+\tau)\big\}, \tag{B.8}$$

$$m_{3x}(\tau_1, \tau_2) = \mathrm{E}\big\{x(k)x(k+\tau_1)x(k+\tau_2)\big\}, \tag{B.9}$$

$$m_{4x}(\tau_1, \tau_2, \tau_3) = \mathrm{E}\big\{x(k)x(k+\tau_1)x(k+\tau_2)x(k+\tau_3)\big\}, \tag{B.10}$$

$$\vdots \tag{B.11}$$

$$m_{nx}(\tau_1, \tau_2, \ldots, \tau_{n-1}) = \mathrm{E}\big\{x(k)x(k+\tau_1)x(k+\tau_2)\cdots x(k+\tau_{n-1})\big\}. \tag{B.12}$$

Similarly, the nth-order cumulants of a real stationary random process are expressed in the form [cumest]

$$c_{nx}(\tau_1, \tau_2, \ldots, \tau_{n-1}) := \mathrm{Cum}\{x(k), x(k+\tau_1), \ldots, x(k+\tau_{n-1})\}. \tag{B.13}$$

Combining Eqs. B.5, B.12 and B.13, the relationships between moment and cumulant sequences can be obtained.

The second-, third- and fourth-order cumulants of $x(k)$ are

$$c_{1x} = m_{1x}, \tag{B.14}$$

$$c_{2x}(\tau) = m_{2x}(\tau) - m_{1x}^2, \tag{B.15}$$

$$\begin{aligned} c_{3x}(\tau_1, \tau_2) = m_{3x}(\tau_1, \tau_2) - m_{1x}\big[m_{2x}(\tau_1) + m_{2x}(\tau_2) + m_{2x}(\tau_2 - \tau_1)\big] \\ + 2m_{1x}^3, \end{aligned} \tag{B.16}$$

$$\begin{aligned} c_{4x}(\tau_1, \tau_2, \tau_3) = m_{4x}(\tau_1, \tau_2, \tau_3) - m_{2x}(\tau_1)m_{2x}(\tau_3 - \tau_2) - m_{2x}(\tau_2)m_{2x}(\tau_3 - \tau_1) \\ - m_{2x}(\tau_3)m_{2x}(\tau_2 - \tau_1) - m_{1x}\big[m_{3x}(\tau_2 - \tau_1, \tau_3 - \tau_1) \\ + m_{3x}(\tau_2, \tau_3) + m_{3x}(\tau_2, \tau_4) + m_{3x}(\tau_1, \tau_2)\big] \\ + m_{1x}^2\big[m_{2x}(\tau_1) + m_{2x}(\tau_2) + m_{2x}(\tau_3) + m_{2x}(\tau_3 - \tau_1) \\ + m_{2x}(\tau_3 - \tau_2) + m_{2x}(\tau_2 - \tau_1)\big] - 6m_{1x}^4. \end{aligned} \tag{B.17}$$

The zero-lag cumulants have special names:

$$\gamma_{2x} = \sigma_x^2 = \mathrm{E}\{x^2(k)\} = c_{2x}(0) \quad \text{(variance)}, \tag{B.18}$$

$$\gamma_{3x} = \mathrm{E}\{x^3(k)\} = c_{3x}(0, 0) \quad \text{(skewness)}, \tag{B.19}$$

$$\gamma_{4x} = \mathrm{E}\{x^4(k)\} = c_{4x}(0, 0, 0) \quad \text{(kurtosis)}. \tag{B.20}$$

These equations give the variance (as a measure of the spread of the PDF), the skewness (as a measure of asymmetry) and kurtosis (as a measure of sharpness of peak or "flatness") measures in terms of cumulant lags. The normalised quantities $\gamma_{3x}/\sigma_{2x}^3$ and $\gamma_{4x}/\sigma_{2x}^4$ (both shift and scale invariant) are also defined. If $x(k)$ is symmetric distributed, its skewness is necessarily zero (but not vice versa). If $x(k)$ is Gaussian distributed, its kurtosis is necessarily zero (but not vice versa).

With zero-mean assumption, the second- and third-order cumulants are the same as the second- and third-order moments, respectively. Thus, for the simplification of estimates, if the process has non-zero mean, the mean should be subtracted from it first. However, to generate the fourth-order cumulant, knowledge of the fourth-order and second-order moments is needed. In practice, because of unique linear property of the second characteristic function working with cumulants instead of moments is more common and preferable in the case of stochastic signals.

B.2 Polyspectra and Coherence Functions

The generalisation of the power spectrum to higher orders forms the family of *polyspectra*. They are usually defined in terms of nth-order cumulants as their $(n-1)$-dimensional Fourier transforms

$$S_{nx}(f_1, f_2, \dots, f_{n-1}) = \sum_{\tau_1=-\infty}^{+\infty} \cdots \sum_{\tau_{n-1}=-\infty}^{+\infty} c_{nx}(\tau_1, \tau_2, \dots, \tau_{n-1}) \times \exp\big[-j2\pi(f_1\tau_1 + f_2\tau_2 + \dots + f_{n-1}\tau_{n-1})\big]. \tag{B.21}$$

This is simply a generalisation of the Wiener–Khintchine relation. Nevertheless, again in practice they can be equivalently estimated by statistical averaging of the Fourier amplitudes whose sum frequency vanishes.

Also, cross-cumulants and cross-cumulant spectra may be defined:

$$c_{x_1x_2\dots x_n}(\tau_1, \tau_2, \dots, \tau_{n-1}) := \text{Cum}\big(x_1(k), x_2(k+\tau_1), \dots, x_n(k+\tau_{n-1})\big), \tag{B.22}$$

$$S_{x_1x_2\dots x_n}(f_1, f_2, \dots, f_{n-1}) = \sum_{\tau_1=-\infty}^{+\infty} \cdots \sum_{\tau_{n-1}=-\infty}^{+\infty} c_{x_1x_2\dots x_n}(\tau_1, \tau_2, \dots, \tau_{n-1}) \times \exp\big[-j2\pi(f_1\tau_1 + f_2\tau_2 + \dots + f_{n-1}\tau_{n-1})\big]. \tag{B.23}$$

For instance, the 3rd-order cross-cumulant [cum3x] and the corresponding spectrum are given by

$$c_{xyz}(\tau_1, \tau_2) = \text{Cum}\big\{x(k), y(k+\tau_1), z(k+\tau_2)\big\} \tag{B.24}$$

$$= \text{E}\big\{\big(x(k) - m_x\big)\big(y(k+\tau_1) - m_y\big)\big(z(k+\tau_2) - m_z\big)\big\}, \tag{B.25}$$

where $m_x = \text{E}\{x(k)\}$, $m_y = \text{E}\{y(k)\}$ and $m_z = \text{E}\{z(k)\}$;

$$S_{xyz}(f_1, f_2) = \sum_{\tau_1=-\infty}^{+\infty} \sum_{\tau_2=-\infty}^{+\infty} c_{xyz}(\tau_1, \tau_2) \exp\big[-j2\pi(f_1\tau_1 + f_2\tau_2)\big]. \tag{B.26}$$

Special cases of Eq. B.21 are the *power spectrum* ($n = 2$), the *bispectrum* ($n = 3$) and the *trispectrum* ($n = 4$). Only PSD is real, the others are complex. Besides, one of the most useful functions used for the detection and characterization of nonlinearity in time series is the coherence function. The bispectrum (or bispectral density) and the trispectrum (or trispectral density) can be expressed as

$$S_{3x}(k, l) = \text{E}\big\{X(k)X(l)X^*(k+l)\big\} = \text{E}\big\{X(k)X(l)X\big(-(k+l)\big)\big\}, \tag{B.27}$$

$$S_{4x}(k, l, m) = \text{E}\big\{X(k)X(l)X(m)X^*(k+l+m)\big\}, \tag{B.28}$$

respectively, where k, l and m denote discrete frequencies. These equations show that the bispectrum and trispectrum are complex quantities having both magnitude and phase. Focus of this work is on the 3rd-order polyspectrum, i.e. the bispectrum. The interested reader should consult elsewhere for matters concerning trispectral analysis, e.g. Dalle-Molle (1992), Collis (1996) and Collis et al. (1998).

The bispectrum satisfies the following symmetry relations:

$$S_{3x}(k, l) = S_{3x}(l, k) = S_{3x}^*(-k, -l), \tag{B.29}$$

$$S_{3x}(k, l) = S_{3x}(-k-l, l) = S_{3x}(k, -k-l). \tag{B.30}$$

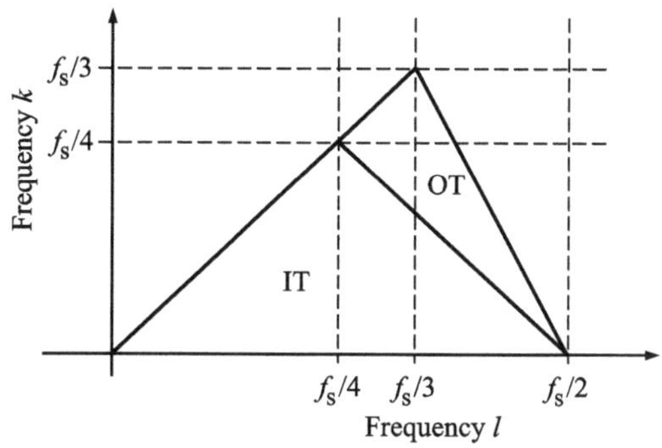

Fig. B.1 Principal domain (PD) of the discrete bispectrum showing the inner triangle (IT) and the outer triangle (OT)

It can be plotted against two independent frequency variables, k and l, in a three-dimensional (3D) plot. Just as the discrete power spectrum has a point of symmetry at the folding frequency, the discrete bispectrum also has 12 regions of symmetries in the (k, l)-plane (Rosenblatt and Van Ness 1965; Nikias and Petropulu 1993).

However, symmetry operations reduce the non-redundant region of the bispectrum (bicoherence and skewness) to a triangular region, called the principal domain: $0 \leq k \leq l, k + l \leq f_s/2$ (f_s is the sampling frequency). The principal domain can be further divided into two regions known as the inner triangle and the outer triangle; see Fig. B.1. Each point in such a plot represents the bispectral content of the signal at the bifrequency (k, l). In fact, the bispectrum at point $(S_{3x}(k, l), k, l)$ measures the interaction between frequencies k and l. This interaction between frequencies can be related to the nonlinearities present in the signal-generating systems (Fackrell 1996), and therein lies the core of its usefulness in the detection and diagnosis of nonlinearities.

B.3 Estimating the Bispectrum from Data

In practice, higher-order spectra have to be estimated from a finite set of (sampled) measurements. Two of the most popular conventional approaches are the *direct* and *indirect* methods, which may be seen as a direct approximation of the definitions of higher-order spectra. Whereas these approximations are straightforward, often the required computations may be expensive despite the use of FFT algorithms. These methods and their extensions for estimating higher-order spectra are well documented in Nikias and Petropulu (1993) [bispeci]. Both have similar computational requirements and give similar results (Elgar and Sebert 1989; Fackrell et al. 1995a). In the following, only the direct method for the estimation of the bispectrum [bispecd] is described.

Let $\{x_i(k), k = 0, 1, 2, \ldots, N-1\}$ be the available set of measured data for bispectrum estimation. The estimation procedure consists of the following steps (Nikias and Mendel 1993):

1. Divide the finite data set of length N is into K (possibly overlapping) segments of M samples each, i.e. $N \geq KM$. Detrend (i.e. substract the mean value of each

segment) and appropriately window[1] the data to provide some control over the effects of spectral leakage. If necessary, delete the last data points or add zeros to obtain a convenient length $M = 2^n$ for FFT.

2. Generate the FFT coefficients for each data segment:

$$X_i(k) = \frac{1}{M} \sum_{l=0}^{M-1} x_i(l) e^{-j\frac{2\pi k}{M}l}. \tag{B.31}$$

3. Form the *raw* spectral estimates based on the FFT coefficients:

$$\hat{S}_{2x,i}(k) = X_i(k) X_i^*(k), \tag{B.32}$$

$$\hat{S}_{3x,i}(k) = X_i(k) X_i(l) X_i^*(k+l). \tag{B.33}$$

4. Compute the segment-averaged estimate of the power spectrum and the bispectrum of the given data set from the average over K pieces,

$$\hat{S}_{2x}(k) = \frac{1}{K} \sum_{i=1}^{K} \hat{S}_{2x,i}(k) \tag{B.34}$$

and

$$\hat{S}_{3x}(k,l) = \frac{1}{K} \sum_{i=1}^{K} \hat{S}_{3x,i}(k,l), \tag{B.35}$$

respectively.

B.4 Skewness and Squared Bicoherence Functions

The properties of the spectral estimate in Eq. B.35 will now be discussed. According to Fackrell (1996), one has to distinguish between the properties for stochastic signals, M1 (Brillinger and Rosenblatt 1967; Huber et al. 1971; Hinich 1982), and those for mixtures of deterministic and stochastic signals, M2 (Elgar and Guza 1988; Chandran and Elgar 1991).

A common measure of bispectral estimate (for M1 signals) is the so-called *skewness* function (Hinich 1982)

$$skew(k,l) := \frac{\mathrm{E}\{S_{3x}(k,l)\}}{\sqrt{\mathrm{E}\{S_{2x}(k)\}\mathrm{E}\{S_{2x}(l)\}\mathrm{E}\{S_{2x}(k+l)\}}} \tag{B.36}$$

$$\Rightarrow \quad skew^2(k,l) := \frac{|\mathrm{E}\{S_{3x}(k,l)\}|^2}{\mathrm{E}\{S_{2x}(k)\}\mathrm{E}\{S_{2x}(l)\}\mathrm{E}\{S_{2x}(k+l)\}}. \tag{B.37}$$

[1] Hamming windows were found to give some of the best results for phase-coupling detection (Fackrell et al. 1995b).

If the denominator of Eq. B.37 was known exactly, then the estimate of the skewness function given by

$$skew^2(k,l) = \frac{|\hat{S}_{3x}(k,l)|^2}{S_{2x}(k)S_{2x}(l)S_{2x}(k+l)} \tag{B.38}$$

would have a flat variance. However, in practice, the power spectral terms are *not* known, and they too have to be estimated from the data using Eq. B.34. Thus, $skew^2(k,l)$ in Eq. B.37 is estimated using

$$skew^2(k,l) = \frac{|\hat{S}_{3x}(k,l)|^2}{\hat{S}_{2x}(k)\hat{S}_{2x}(l)\hat{S}_{2x}(k+l)}. \tag{B.39}$$

Fortunately, the power spectra estimates $\hat{S}_{2x}$ generally have lower variance than the bispectra estimates $\hat{S}_{3x}$, and in practice, in the cases where it can be checked, Eq. B.39 often turns out to be very close to Eq. B.38. The skewness function, which resembles the coherence function (Eq. A.18) not only by name, may be regarded as the normalised third-order polyspectrum . It is also a complex quantity with real and imaginary parts. It is well known in the HOS literature that the bicoherence is a complex normal variable, i.e. both estimates of real and imaginary parts of the bicoherence are normally distributed (Hinich 1982) and asymptotically independent, i.e. the estimate at a particular bifrequency is independent of the estimates of its neighbouring bifrequencies (Fackrell 1996). Note that the skewness generally can assume values *larger* than unity.

The normalised bispectrum (skewness function) can be generalised to the nth-order coherence function (Nikias and Petropulu 1993)

$$P_{nx}(k_1,k_2,\ldots,k_{n-1}) = \frac{S_{nx}(k_1,k_2,\ldots,k_{n-1})}{\sqrt{S_{2x}(k_1)S_{2x}(k_2)\cdots S_{2x}(k_{n-1})S_{nx}(k_1+k_2+\cdots+k_{n-1})}}. \tag{B.40}$$

This becomes useful in studying the phase response of non-Gaussian linear processes, i.e. processes whose spectra are modelled by the same linear filter. The magnitude of the nth-order coherence function $|P_{nx}(k_1,k_2,\ldots,k_n\text{t}1)|$ is called the coherence index (Nikias and Petropulu 1993). This is very useful in the detection and characterisation of nonlinearities in time series and in discriminating linear processes from nonlinear ones. In fact, a signal is said to be a linear non-Gaussian process of order n if the coherence index is constant over all frequencies; otherwise, the signal is said to be a nonlinear process (Nikias and Mendel 1993).

It has been shown that the bispectral estimates are asymptotically unbiased and the variance of the estimator depends on the 2nd-order spectral properties (Hinich 1982; Nikias 1988). This poses a serious problem (which does not occur with power-spectrum estimates) that the variance of the estimate is dependent on the signal energy, i.e. it is higher at a bifrequency where the signal energy is high and lower where the energy is low.

Instead of the skewness function and to avoid the undesired property of the variance of the bispectrum estimate, an alternative normalisation is often used for bispectral characterisation of M2 signals (Kim and Powers 1979; Elgar and Sebert 1989):

$$bic^2(k,l) := \frac{|\mathrm{E}\{S_{3x}(k,l)\}|^2}{\mathrm{E}\{S(k,l)\}\mathrm{E}\{S_{2x}(k+l)\}}; \qquad S(k,l) := \left|X(k)X(l)\right|^2. \tag{B.41}$$

This can be estimated by

$$bic^2(k,l) = \frac{|\hat{S}_{3x}(k,l)|^2}{\hat{S}(k,l)\hat{S}_{2x}(k+l)}; \qquad \hat{S}(k,l) = \frac{1}{K}\sum_{i=1}^{K}\left|X_i(k)X_i(l)\right|^2. \tag{B.42}$$

This normalised bispectrum, which is known as the *squared bicoherence*, *bicoherence spectrum* or *quadratic correlation coefficient*, does not have approximately the same flat variance that the skewness function has. However, it does have the useful property that it is bounded between zero and unity, i.e.

$$0 \le bic^2(k,l) \le 1 \tag{B.43}$$

a property, which $skew^2(k, l)$ *does not share*. This can be shown using the Schwarz inequality, which may be expressed as

$$\left|\mathrm{E}\{z_1 z_2\}\right|^2 \le \mathrm{E}\left\{|z_1|^2\right\}\mathrm{E}\left\{|z_2|^2\right\}, \tag{B.44}$$

where $z_1 = X(k)X(l)$ and $z_2 = X(k+l)$. In fact, in practice $skew^2(k,l)$ and $bic^2(k,l)$ do usually take very similar values, a fact which has led to some confusion between them in the literature. Care should be taken, as the terms skewness and bicoherence have been used interchangeably in the HOS literature. In practice, there is often little to choose between the two measures because both have the same numerator and because of the relative statistical stability of their denominators in comparison with their numerators. Furthermore, Kim and Powers (1979) have shown that the variance of the bispectrum can be expressed as

$$\mathrm{var}\left\{\hat{S}_{3x}(k,l)\right\} \approx \frac{1}{K} S_{2x}(k)S_{2x}(l)S_{2x}(k+l)\left[1 - bic^2(k,l)\right]. \tag{B.45}$$

That is, when the waves at k, l and $k+l$ are nonlinearly coupled ($bic \approx 1$, quadratically coherent), the variance approaches zero, and when the oscillations are statistically independent ($bic \approx 0$, quadratically incoherent), the variance is proportional to the product of power at each spectral component.

When considering a discrete ergodic stationary time series, $x(k)$, which can be represented by

$$x(k) = \sum_{i=0}^{M-1} h(i)e(k-i), \tag{B.46}$$

where $e(k)$ is a sequence of independent identically distributed random variables with $\mathrm{E}\{e(k)\} = 0$, $\sigma_e^2 = \mathrm{E}\{e^2(k)\}$, and $\mu_3 = \mathrm{E}\{e^3(k)\}$, it can be shown that

$$bic^2(k,l) = \frac{\mu_3^2}{\sigma_e^6} = \text{const} \quad \forall k,l. \tag{B.47}$$

Note that the term "bicoherence spectrum", as defined in Eq. B.42, has recently been criticized by Hinich and Wolinsky (2004) to be misleading, since it is really a skewness spectrum and the normalisation depends on the frame length and on the magnitude of the trispectrum. An alternative statistical normalisation has been presented by Hinich and Wolinsky (2004), who provide a measure of quadratic coupling for stationary random nonlinear processes with finite dependence.

Besides the skewness function and the squared bicoherence discussed so far, other methods for normalising the bispectrum have been proposed by Kravtchenko-Berojnoi (1994), Lyons et al. (1995), Thyssen et al. (1995), Collis et al. (1998) and Hinich and Wolinsky (2004). See Fackrell et al. (1995a) and Fackrell (1996) for a comparison. Note that the word normalisation is used in a very liberal sense here, since some of the measures, notably $skew^2(k,l)$, can exceed unity.

Some important properties of normalised bispectra are given as follows (Fackrell 1996):

- The theoretical bispectrum of a Gaussian signal is identically zero.
- The theoretical bispectrum of a linearly filtered Gaussian signal is identically zero.
- The theoretical bicoherence of a Gaussian signal is zero.
- The theoretical bispectrum of a non-Gaussian signal is "blind" to additive Gaussian noise. This theoretical "blindness" to Gaussian noise has been the prime motivation to much of the exploitative HOS research to date.
- The theoretical bicoherence of a signal conforming to either the M1 or M2 models is in general *not* "blind" to Gaussian noise.
- If a signal is filtered by a linear filter, then, provided that the filter has no zeros on the unit circle, the magnitude of the normalised bispectrum is unchanged.
- The theoretical skewness function of linearly filtered non-Gaussian independent identically distributed signals is flat.
- If a signal is filtered by a linear phase filter, then its biphase information is unchanged.
- The theoretical skewness function of a non-Gaussian M1 signal, which has been passed through a nonlinear filter, may not be flat.
- The theoretical bicoherence of a harmonic M2 signal peaks if the signal phases ϕ_1, ϕ_2 and ϕ_3 at frequencies f_1, f_2 and $f_3 = f_1 + f_2$, respectively, have the relation $\phi_3 = \phi_1 + \phi_2$. This sort of phase relation is known as the quadratic phase coupling (QPC) and is an indicator of nonlinear signal generation mechanisms.

Note that applying the bicoherence to real and noisy data results in many spurious bispectral peaks in the (k,l)-plane due to noise. Extracting the "true" peaks, i.e. those reflecting coupled oscillations, is not straightforward. This problem is also encountered when analysing a system using the coherence function. In the latter case, a threshold of 0.5 is typically used to discern peaks, which represent coupled oscillations from spurious peaks caused by noise. This threshold is arbitrary, and its value has been criticized (Toledo 2002). To discriminate real bispectral peaks from spurious ones, the approach suggested by Haubrich (1965) can be used. The principle of this method is to consider the question of a bispectral peak being a real one within the framework of statistical hypothesis testing. The null hypothesis in this case is that the analysed signal is a Gaussian random process. We can reject or accept this hypothesis, depending on the value of the bicoherence. Determining the threshold using this approach provides an objective and quantitative method to differentiate real from spurious peaks. However, the statistical properties of the bicoherence, under the assumption of the null hypothesis, must be known. Therefore, the focus should be on the statistical properties of the bicoherence of Gaussian stochastic processes. See Haubrich (1965) and Toledo (2002) for more details.

Appendix C
Control Loops from Different Industries

The methods presented throughout the book have been demonstrated on different data sets from different process industries, including chemicals, pulp and paper mills, commercial building and metal processing. Some information about these loops are collected in Table C.1.

Table C.1 Information about control loops analysed; some of them are evaluated throughout the book

Loop name	Industrial field	Control loop	T_s [s]	Comments, known Problems or visual observations
BAS 1	Buildings	Temperature control	1	No oscillation
BAS 2	Buildings	Temperature control	1	No oscillation
BAS 3	Buildings	Temperature control	3	Intermittent oscillation
BAS 4	Buildings	Pressure control	3	Intermittent oscillation
BAS 5	Buildings	Pressure control	1	OP not available
BAS 6	Buildings	Temperature control	1	Stiction and tight tuning
BAS 7	Buildings	Temperature control	1	Stiction
BAS 8	Buildings	Temperature control	60	No oscillation
CHEM 1	Chemicals	Flow control	1	Stiction
CHEM 2	Chemicals	Flow control	1	Stiction
CHEM 3	Chemicals	Temperature control	30	Quantisation
CHEM 4	Chemicals	Level control	1	Tuning problem
CHEM 5	Chemicals	Flow control	1	Stiction
CHEM 6	Chemicals	Flow control	1	Stiction
CHEM 7	Chemicals	Pressure control	1	Open-loop data; stiction
CHEM 8	Chemicals	Pressure control	1	Open-loop data; stiction
CHEM 9	Chemicals	Pressure control	1	Stiction
CHEM 10	Chemicals	Pressure control	1	Stiction

M. Jelali, *Control Performance Management in Industrial Automation*, Advances in Industrial Control, DOI 10.1007/978-1-4471-4546-2,

Table C.1 (Continued)

Loop name	Industrial field	Control loop	T_s [s]	Comments, known Problems or visual observations
CHEM 11	Chemicals	Flow control	1	Stiction
CHEM 12	Chemicals	Flow control	1	Stiction
CHEM 13	Chemicals	Analyser control	20	Faulty steam sensor; no stiction
CHEM 14	Chemicals	Flow control	20	Faulty steam sensor; no stiction
CHEM 15	Chemicals	Pressure control	20	Interaction (likely); no stiction
CHEM 16	Chemicals	Pressure control	20	Interaction (likely); no stiction
CHEM 17	Chemicals	Temperature control	20	Faulty steam sensor; no stiction. The OP of CHEM17 is the set point to CHEM14
CHEM 18	Chemicals	Flow control	12	Stiction (likely)
CHEM 19	Chemicals	Flow control	12	Stiction (likely)
CHEM 20	Chemicals	Flow control	12	Stiction (likely)
CHEM 21	Chemicals	Flow control	12	Disturbance (likely)
CHEM 22	Chemicals	Flow control	12	Stiction (likely)
CHEM 23	Chemicals	Flow control	12	Stiction (likely)
CHEM 24	Chemicals	Flow control	12	Stiction (likely)
CHEM 25	Chemicals	Pressure control	12	Possible margin stability
CHEM 26	Chemicals	Level control	12	Stiction (likely)
CHEM 27	Chemicals	Level control	12	Disturbance (likely)
CHEM 28	Chemicals	Temperature control	12	Stiction (likely)
CHEM 29	Chemicals	Flow control	60	
CHEM 30	Chemicals	Flow control	15	No stiction
CHEM 31	Chemicals	Flow control	15	
CHEM 32	Chemicals	Flow control	10	Stiction (likely)
CHEM 33	Chemicals	Flow control	12	Disturbance (likely)
CHEM 34	Chemicals	Flow control	10	Disturbance (likely)
CHEM 35	Chemicals	Flow control	10	Stiction (likely)
CHEM 36	Chemicals	Level control	12	Disturbance (likely)
CHEM 37	Chemicals	Level control	12	Disturbance (likely)
CHEM 38	Chemicals	Pressure control	10	Disturbance (likely)
CHEM 39	Chemicals	Pressure control	60	Disturbance (likely)
CHEM 40	Chemicals	Temperature control	60	No clear oscillation (according to power spectrum)
CHEM 41	Chemicals	Temperature control	60	OP saturation, as assessed by Matsuo et al. (2004).
CHEM 42	Chemicals	Temperature control	60	
CHEM 43	Chemicals	Temperature control	60	
CHEM 44	Chemicals	Temperature control	60	Too few cycles; no clear oscillation; OP saturation

Table C.1 (Continued)

Loop name	Industrial field	Control loop	T_s [s]	Comments, known Problems or visual observations
CHEM 45	Chemicals	Pressure control	60	No clear oscillation (according to power spectrum)
CHEM 46	Chemicals	Pressure control	60	No clear oscillation (according to power spectrum)
CHEM 47	Chemicals	Pressure control	60	No clear oscillation (according to power spectrum)
CHEM 48	Chemicals	Pressure control	60	No clear oscillation (according to power spectrum)
CHEM 49	Chemicals	Pressure control	60	
CHEM 50	Chemicals	Level control	60	
CHEM 51	Chemicals	Level control	60	
CHEM 52	Chemicals	Level control	60	No clear oscillation (according to power spectrum)
CHEM 53	Chemicals	Level control	60	No clear oscillation
CHEM 54	Chemicals	Level control	60	No clear oscillation
CHEM 55	Chemicals	Level control	60	
CHEM 56	Chemicals	Flow control	60	No clear oscillation (according to power spectrum)
CHEM 57	Chemicals	Flow control	60	
CHEM 58	Chemicals	Flow control	60	No clear oscillation (according to power spectrum)
CHEM 59	Chemicals	Flow control	60	No clear oscillation (according to power spectrum)
CHEM 60	Chemicals	Flow control	60	
CHEM 61	Chemicals	Flow control	60	No clear oscillation (according to power spectrum)
CHEM 62	Chemicals	Flow control	60	No clear oscillation (according to power spectrum)
CHEM 63	Chemicals	Flow control	60	
CHEM 64	Chemicals	Gas flow control	60	
PAP 1	Pulp & Papers	Flow control	1	Stiction
PAP 2	Pulp & Papers	Flow control	1	Stiction
PAP 3	Pulp & Papers	Level control	1	Stiction
PAP 4	Pulp & Papers	Concentration control	1	Dead zone and tight tuning; no stiction
PAP 5	Pulp & Papers	Concentration control	0.2	Stiction
PAP 6	Pulp & Papers	Level control	1	No stiction
PAP 7	Pulp & Papers	Flow control	0.2	External disturbance
PAP 8	Pulp & Papers	Level control	5	No stiction
PAP 9	Pulp & Papers	Temperature control	5	No stiction

Table C.1 (Continued)

Loop name	Industrial field	Control loop	T_s [s]	Comments, known Problems or visual observations
PAP 10	Pulp & Papers	Level control	5	
PAP 11	Pulp & Papers	Level control	15	
PAP 12	Pulp & Papers	Level control	15	Stiction
PAP 13	Pulp & Papers	Level control	15	Stiction
POW 1	Power Plants	Level control	5	Stiction
POW 2	Power Plants	Level control	5	Stiction
POW 3	Power Plants	Level control	5	Stiction
POW 4	Power Plants	Level control	5	Stiction
POW 5	Power Plants	Level control	5	Stiction
MIN 1	Mining	Temperature control	60	Stiction
MET 1	Metals	Thickness control	0.05	External disturbance (likely)
MET 2	Metals	Thickness control	0.05	External disturbance (likely)
MET 3	Metals	Thickness control	0.05	No oscillation

References

Agrawal P, Lakshminarayanan S (2003) Tuning proportional-integral-derivative controllers using achievable performance indices. Ind Eng Chem Res 42:5576–5582

Agarwal N, Huang B, Tamayo EC (2007a) Assessing MPC performance. Part 1: probabilistic approach for constraint analysis. Ind Eng Chem Res 46:8101–8111

Agarwal N, Huang B, Tamayo EC (2007b) Assessing MPC performance. Part 2: Bayesian approach for constraint tuning. Ind Eng Chem Res 46:8112–8119

Ahsan Q, Grosvenor RI, Prickett PW (2004) Distributed control loop performance monitoring architecture. In: Proc control 2004, University of Bath, UK, ID-054

Akande S, Huang B, Lee K-H (2009) MPC constraint analysis—Bayesian approach via a continuous-valued profit function. Ind Eng Chem Res 18:3944–3954

Åkesson IN (2003) Plant loop auditing in practice. In: VDI-Berichte 1756, Proc GMA-Kongress: Automation und Information in Wirtschaft und Gesellschaft. Baden-Baden, Germany, pp 927–934

AlGhazzawi A, Lennox B (2008) Monitoring a complex refining process using multivariate statistics. Control Eng Pract 16:294–307

AlGhazzawi A, Lennox B (2009) Model predictive control monitoring using multivariate statistics. J Proc Control 19:314–327

Allgöwer F, Zheng A (eds) (2000) Nonlinear model predictive control. Birkhäuser, Basel

Alvarez J (1995) An internal-model controller for nonlinear discrete-time systems. In: Proc Europ control confer, Rome, pp 301–306

Anderson BDO, Moore JB (1991) Optimal control: linear quadratic methods. Prentice Hall, New York

Armstrong-Hélouvry B, Dupont P, De Wit CC (1994) A survey of models, analysis tools and compensation methods for the control of machine with friction. Automatica 30:1083–1138

Åström KJ (1979) Introduction to stochastic control. Academic Press, San Diego

Åström KJ (1991) Assessment of achievable performance of simple feedback loops. Int J Adapt Control Signal Process 5:3–19

Åström KJ, Hägglund T (1988) Automatic tuning of PID controllers. ISA, Research Triangle Park

Åström KJ, Hägglund T (2006) Advanced PID control. ISA, Research Triangle Park

Åström KJ, Wittenmark P (1995) Adaptive control. Addison-Wesley, Reading

Åström KJ, Wittenmark P (1997) Computer controlled systems: theory and design. Prentice Hall, New York

Badmus O, Banks D, Vishnubhotla A, Huang B, Shah SL (1998) Performance assessment: a requisite for maintaining your APC assets. In: Proc IEEE workshop dynamic modeling and control applications for industry, pp 54–58

Badwe AS, Patwardhan RS, Shah SL, Patwardhan SC, Gudi RD (2010) Quantifying the impact of model-plant mismatch on controller performance. J Process Control 20:408–425

M. Jelali, *Control Performance Management in Industrial Automation*, Advances in Industrial Control, DOI 10.1007/978-1-4471-4546-2,

Bai E-W (2002) Identification of linear systems with hard input nonlinearities of known structures. Automatica 38:853–860

Barnard JP, Aldrich C, Gerber M (2001) Identification of dynamic process systems with surrogate data methods. AIChE J 47:2064–2075

Basseville M (1988) Detecting changes in signals and systems—a survey. Automatica 24:309–326

Bauer M, Cox JW, Caveness MH, Downs JJ, Thornhill NF (2007a) Nearest neighbors method for root cause analysis of plantwide disturbances. Ind Eng Chem Res 46:5977–5984

Bauer M, Cox JW, Caveness MH, Downs JJ, Thornhill NF (2007b) Finding the direction of disturbance propagation in a chemical process using transfer entropy. IEEE Trans Control Syst Technol 15:12–21

Bender M (2003) Auswahl, Implementierung und Test von Algorithmen zur Bewertung der Güte von Reglern. Diploma Thesis, BFI/University of Cologne, Germany

Bergh LG, MacGregor JF (1987) Constrained minimum variance controllers: internal model control structure and robustness properties. Ind Eng Chem Res 26:1558–1564

Bezergianni S, Georgakis C (2003) Evaluation of controller performance-use of models derived by subspace identification. Int J Adapt Control Signal Process 17:527–552

Bialkowski WL (1993) Dreams vs. reality: a view from both sides of the gap. Pulp & Paper Canada 94:19–27

Bilkhu TM (2001) Dynamic control of tension, thickness and flatness for a tandem cold mill. AISE Steel Technol 78:49–54

Bittanti S, Colaneri P, Mongiovi M (1994) The spectral interactor matrix for the singular Riccati equation. In: Proc IEEE confer decision control, Orlando, USA, vol 3, pp 2165–2169

Björklund S (2003) A survey and comparison of time delay estimation methods in linear systems. PhD thesis, Lund Institute of Technology, Sweden

Bode CA, Ko BS, Edgar TF (2004) Run-to-run control and performance monitoring of overlay in semiconductor manufacturing. Control Eng Pract 12:893–900

Bonavita N, Bovero JC, Martini R (2004) Control loops: performance and diagnostics. In: Proc ANIPLA confer, Milan, Italy

Box GEP, Jenkins GM (1970) Time series analysis: forcasting and control. Holden-Day, Oakland

Box GEP, MacGregor J (1974) The analysis of closed-loop dynamic stochastic systems. Technometrics 18:371–380

Box GEP, Jenkins GM, Reinsel GC (1994) Time series analysis: forcasting and control. Prentice Hall, New York

Boyd S, Barratt C (1991) Linear control design. Prentice Hall, New York

Brillinger D, Rosenblatt M (1967) Asymptotic theory of estimates of k-th order spectra. In: Harris B (ed) Spectral analysis of time signals. Wiley, New York, pp 153–188

Brisk ML (2004) Process control: potential benefits and wasted opportunities. In: Proc Asian control confer, Melbourne, Australia, pp 10–16

Buckbee G (2008) The 6 most common PID configuration errors: how to find and fix them. www.expertune.com/articles/WPPIDConfigErrors.pdf

Calvet J, Arkun Y (1988) Feedforward and feedback linearization of nonlinear systems and its implementation using internal model control (IMC). Ind Eng Chem Res 27:1822–1831

Camacho EF, Bordons C (1999) Model predictive control. Springer, Berlin

Cao S, Rhinehart RR (1995) An efficient method for on-line identification of steady state. J Process Control 5:363–374

Casdagli MC, Iasemidis LD, Sackellares JC, Roper SN, Gilmore RL, Savit RS (1996) Characterizing nonlinearity in invasive EEG recordings from temporal lobe epilepsy. Physica D 99:381–399

Chandran V, Elgar SL (1991) Mean and variance of estimates of the bispectrum of a harmonic random process an analysis including leakage effects. IEEE Trans Signal Process 39:2640–2651

Chen J, Wang WY (2010) PCA-ARMA-based control charts for performance monitoring of multivariable feedback control. Ind Eng Chem Res 19:2228–2241

Chen Q, Kruger U, Meronk M, Leung AYT (2004) Synthesis of T^2 and Q statistics for process monitoring. Control Eng Pract 12:745–755

Choudhury MAAS, Shah SL, Thornhill NF (2004) Diagnosis of poor control-loop performance using higher-order statistics. Automatica 40:1719–1728

Choudhury MAAS, Kariwala V, Shah SL, Douke H, Takada H, Thornhill NF (2005a) A simple test to confirm control valve stiction. In: Proc IFAC world congress, Praha

Choudhury MAAS, Thornhill NF, Shah SL (2005b) Modelling valve stiction. Control Eng Pract 13:641–658

Choudhury MAAS, Thornhill NF, Shah SL, Shook DS (2006a) Automatic detection and quantification of stiction in control valves. Control Eng Pract 14:1395–1412

Choudhury MAAS, Shook D, Shah SL (2006b) Linear or nonlinear? A bicoherence based metric of nonlinearity measure. In: Proc IFAC symposium SAFEPROCESS, Beijing, China

Choudhury MAAS, Shah SL, Thornhill NF (2008) Diagnosis of process nonlinearities and valve stiction—data driven approaches. Springer, London

Choudhury MAAS, Thornhill NF, Kano M, Shah SL (2009) Stiction modelling. In: Jelali M, Huang B (eds) Detection and diagnosis of stiction in control loops: state of the art and advanced methods. Springer, Berlin. Chapter 2

Clarke DW, Scattolini R (1991) Constrained receding horizon predictive control. Proc IEE D 138:347–354

Clarke DW, Mohtadi C, Tuffs PS (1987a) Generalized predictive control. Part I: the basic algorithm. Automatica 23:137–148

Clarke DW, Mohtadi C, Tuffs PS (1987b) Generalized predictive control. Part II: extensions and interpretations. Automatica 23:149–160

Clegg A (2002) Benchmarking as an aid to identifying under-performing control loops. www.isc-ltd.com/benchmark/learning_centre/aclegg.html

Collis WB (1996) Higher order spectra and their application to nonlinear mechanical systems. PhD Thesis, University of Southampton

Collis WB, White PR, Hammond JK (1998) Higher-order spectra: the bispectrum and trispectrum. Mech Syst Signal Process 12:375–394

Connell R (1996) Process instrumentation application manual. McGaw-Hill, New York

Cutler CR, Ramaker BL (1980) Dynamic matrix control—a computer control algorithm. In: Proc joint automatic control confer, San Francisco, USA

Dalle-Molle JW (1992) Higher-order spectral analysis and the trispectrum. PhD thesis, The University of Texas at Austin

Davies L, Gather U (1993) The identification of multiple outliers. J Am Stat Assoc 88:782–792

Desborough L, Harris T (1992) Performance assessment measures for univariate feedback control. Can J Chem Eng 70:1186–1197

Desborough L, Harris T (1993) Performance assessment measures for univariate feedforward/feedback control. Can J Chem Eng 71:605–616

Desborough L, Miller R (2002) Increasing customer value of industrial control performance monitoring—Honeywell's experience. AIChE Symp Ser 98(326):153–186

DeVries W, Wu S (1978) Evaluation of process control effectiveness and diagnosis of variation in paper basis weight via multivariate time-series analysis. IEEE Trans Autom Control 23:702–708

Dittmar R, Bebar M, Reinig G (2003) Control Loop Performance Monitoring: Motivation, Methoden, Anwendungswünsche. Automtech Prax 45:94–103

Dumont GA, Kammer L, Allison BJ, Ettaleb L, Roche AA (2002) Control performance monitoring: new developments and practical issues. In: Proc IFAC world congress, Barcelona, Spain

Duncan SR, Allwood JM, Garimella SS (1998) The analysis and design of spatial control systems in strip metal rolling. IEEE Trans Control Syst Technol 6:220–232

Economou CG, Morari M (1986) Internal model control. 6. Multiloop design. Ind Eng Chem Process Des Dev 25:411–419

Economou CG, Morari M, Palsson BO (1986) Internal model control. 5. Extension to nonlinear systems. Ind Eng Chem Process Des Dev 25:403–411

Elgar S, Guza RT (1988) Statistics of bicoherence. IEEE Trans Acoust Speech Signal Process 36:1667–1668
Elgar S, Sebert G (1989) Statistics of bicoherence and biphase. J Geophys Res C 94:10993–10998
Elnaggar A, Dumont GA, Elshafei A-L (1991) Delay estimation using variable regression. In: Proc American control confer, Boston, USA, pp 2812–2817
Ender D (1993) Process control performance: not as good as you think. Control Eng 40:180–190
EnTech (1998) EnTech control valve dynamic specification. Version 3.0
Eriksson P, Isaksson AJ (1994) Some aspects of control loop performance monitoring. In: Proc IEEE confer control applications, Glasgow, Scotland, pp 1029–1034
Ettaleb L (1999) Control loop performance assessment and oscillation detection. PhD thesis, University of British, Columbia, Canada
Fackrell JWA (1996) Bispectral analysis of speech signals. PhD thesis, University of Edinburgh, UK
Fackrell JWA, McLaughlin S, White PR (1995a) Bicoherence estimation using the direct methods. Part 1: theoretical conisderations. Appl Signal Process 3:155–168
Fackrell JWA, McLaughlin S, White PR (1995b) Bicoherence estimation using the direct methods. Part 2: practical conisderations. Appl Signal Process 3:186–199
Farenzena M, Trierweiler JO (2006) Variability matrix: a new tool to improve the plant performance. In: Proc IFAC ADCHEM, Gramado, Brazil, pp 893–898
Favoreel W, Moor BD, van Overschee P (1999) Model-free subspace-based LQG-design. In: Proc Amer control confer, San Diego, pp 3372–3376
Favoreel W, Moor BD, van Overschee P (2000) Subspace state space system identification for industrial processes. J Process Control 10:149–155
Fisher-Rosemount (1999) Control valve handbook. Fisher Controls International Inc., Marshalltown
Foley MW, Buckley PA, Huang B, Vishnubhotla A (1999) Application of control loop performance assessment to an industrial acid leaching process. In: Hodouin D, Bazin C, Desbiens A (eds) Control and optimization in minerals, metals and materials processing, METSOC, pp 3–16
Forsman K (1998) Performance monitoring of sensors, controllers and actuators with applications to paper making. In: Proc control systems, Porvoo, Finland, pp 275–281
Forsman K (2000) On detection and classification of valve stiction. In: Proc TAPPI conf process control, Williamsburg, USA
Forsman K, Stattin A (1999) A new criterion for detecting oscillations in control loops. In: Proc Europ control confer, Karlsruhe, Germany
Frank PM (1974) Entwurf von Regelkreisen mit vorgeschriebenem Verhalten. G Braun Verlag, Karlsruhe
Fu Y, Dumont GA (1993) Optimum Laguerre time scale and its on-line estimation. IEEE Trans Autom Control 38:934–938
Gao J, Patwardhan RS, Akamatsu K, Hashimoto Y, Emoto G, Shah SL, Huang B (2003) Performance evaluation of two industrial MPC controllers. Control Eng Pract 11:1371–1387
García CE, Morari M (1982) Internal model control. 1. A unifying review and some new results. Ind Eng Chem Process Des Dev 21:308–323
García CE, Morari M (1985a) Internal model control. 2. Design procedure for multivariable systems. Ind Eng Chem Process Des Dev 24:472–484
García CE, Morari M (1985b) Internal model control. 3. Multivariable control law computation and tuning guidelines. Ind Eng Chem Process Des Dev 24:484–494
García CE, Morshedi AM (1986) Quadratic programming solution of dynamic matrix control (QDMC). Chem Eng Commun 46:73–87
García CE, Prett DM, Morari M (1989) Model predictive control: theory and practice—a survey. Automatica 25:335–348
Geladi P, Kowalski BR (1986) Partial least squares regression: a tutorial. Anal Chim Acta 185:1–17
Gerry JP (2002) Process monitoring and loop prioritization can reap big payback and benefit process plants. In: Proc ISA, Chicago, USA

Gerry J, Ruel M (2001) How to measure and combat valve stiction online. In: Instrumentation, systems and automation society, Houston, TX, USA. www.expertune.com/articles/isa2001/StictionMR.htm
Ghraizi RA, Ernesto M, de Prada C, Cifuentes F, Martnez JL (2007) Performance monitoring of industrial controllers based on the predictability of controller behavior. Comput Chem Eng 31:477–486
Giannakis G, Tstatsanis M (1994) Time-domain tests for gaussianity and time-reversibility. IEEE Trans Signal Process 42:3460–3472
Ginzburg WB (1989) Steel-rolling technology: theory and practice. Marcel Dekker, New York
Glattfelder AH, Schaufelberger W (2003) Control systems with input and output constraints. Springer, Berlin
Goldberg DE (1989) Genetic algorithms in search, optimization, and machine learning. Addison-Wesley, Reading
Goodwin GC, Sin K (1984) Adaptive filtering, prediction and control. Prentice Hall, New York
Goodwin GC, Graebe SF, Salgado ME (2001) Control system design. Prentice Hall, New York
Gorgels F, Jelali M, Lathe R, Mücke G, Müller U, Ungerer W, Wolff A (2003) State of the art and future trends in metal processing control. In: Proc METEC congress (Europ rolling confer), Düsseldorf, Germany, pp 393–402
Goradia DB, Lakshminarayanan S, Rangaiah GP (2005) Attainment of PI achievable performance for linear SISO process with deadtime by iterative tuning. Can J Chem Eng 83:723–736
Grimble MJ (2000) Restricted-structure LQG optimal control for continuous-time systems. IEE Proc Part D, Control Theory Appl 147:185–195
Grimble MJ (2002a) Controller performance benchmarking and tuning using generalised minimum variance control. Automatica 38:2111–2119
Grimble MJ (2002b) Restricted structure controller tuning and performance assessment. IEE Proc Part D, Control Theory Appl 149:8–16
Grimble MJ (2003) Restricted structure control loop performance assessment for PID controllers and state-space systems. Asian J Control 5:39–57
Grimble MJ (2006a) Robust industrial control systems. Wiley, New York
Grimble MJ (2006b) Design of generalized minimum variance controllers for nonlinear systems. Int J Control Autom Syst 4:281–292
Grimble MJ, Johnson MA (1988) Optimal control and stochastic estimation, volumes 1 and 2. Wiley, New York
Grimble MJ, Majecki P (2004) Weighting selection for controller benchmarking and tuning. Tech report ICC/219/Dec 2004, University of Strathclyde, Glasgow
Grimble MJ, Majecki P (2005) New ideas in performance assessment and benchmarking of non-linear systems. www.isc-ltd.com/benchmark/workshop/NGMV.pdf
Grimble MJ, Uduehi D (2001) Process control loop benchmarking and revenue optimization. In: Proc Amer control confer, Arlington, USA
Gunnarsson S, Wahlberg B (1991) Some asymptotic results in recursive identification using Laguerre models. Int J Adapt Control Signal Process 5:313–333
Haarsma G, Nikolaou M (2000) Multivariate controller performance monitoring: lessons from an application to snack food process. www.chee.uh.edu/faculty/nikolaou/FryerMonitoring.pdf
Haber R, Keviczky L (1999) Nonlinear system identification—input–output modelling approach, vol 2. Kluwer, Dordrecht
Hägglund T (1995) A control-loop performance monitor. Control Eng Pract 3:1543–1551
Hägglund T (1999) Automatic detection of sluggish control loops. Control Eng Pract 7:1505–1511
Hägglund T (2005) Industrial implementation of on-line performance monitoring tools. Control Eng Pract 13:1383–1390
Hägglund T, Åström KJ (2000) Supervision of adaptive control algorithms. Automatica 36:1171–1180
Hanna J, Upreti SR, Lohi A, Ein-Mozaffari F (2008) Constrained minimum variance control using hybrid genetic algorithm—an industrial experience. J Process Control 18:36–44

Harris TJ (1985) A comparative study of model based control strategies. In: Proc Amer control confer, Boston, USA
Harris TJ (1989) Assessment of closed loop performance. Can J Chem Eng 67:856–861
Harris TJ (2004) Statistical properties of quadratic-type performance indices. J Process Control 14:899–914
Harris TJ, Davis JH (1992) An iterative method for matrix spectral factorization. SIAM J Sci Stat Comput 13:531–540
Harris T, Seppala CT (2001) Recent developments in performance monitoring and assessment techniques. In: Proc chemical process control confer, Tuscon, USA
Harris T, Seppala CT, Desborough LD (1999) A review of performance monitoring and assessment techniques for univariate and multivariate control systems. J Process Control 9:1–17
Harris T, Boudreau F, MacGregor JF (1996a) Performance assessment using of multivariable feedback controllers. Automatica 32:1505–1518
Harris T, Seppala CT, Jofriet PJ, Surgenor BW (1996b) Plant-wide feedback control performance assessment using an expert-system framework. Control Eng Pract 4:1297–1303
Haubrich RA (1965) Earth noise, 5 to 500 millicycles per second. J Geophys Res 70:1415–1427
He X, Asada H (1993) A new method for identifying orders of input–output models for nonlinear dynamical systems. In: Proc American control confer, San Francisco, USA, pp 2520–2523
He QP, Wang J, Pottmannn M, Qin SJ (2007) A curve fitting method for detecting valve stiction in oscillating control loops. Ind Eng Chem Res 46:4549–4560
Hegger R, Kantz H, Schreiber T (2004) TISEAN2.1 surrogates manual, periodicity artefacts. www.mpipks-dresden.mpg.de/~tisean/TISEAN_2.1
Henson MA, Seborg E (1991) An internal model control strategy for nonlinear systems. AIChE J 37:1065–1081
Hinich MJ (1982) Testing for Gaussianity and linearity of a stationary time series. J Time Ser Anal 3:169–176
Hinich MJ, Wolinsky M (2004) Normalizing bispectra. www.gov.utexas.edu/hinich/files/Statistics/Normbispec.pdf
Hjalmarsson H, Gevers M, de Bruyne F (1996) For model-based control design, closed-loop identification gives better performance. Automatica 32:1659–1673
Hjalmarsson H, Gevers M, Gunnarsson S, Lequin O (1998) Iterative feedback tuning. IEEE Control Syst 18:26–41
Hölttä V, Koivo H (2009) Quality index framework for plant-wide performance evaluation. J Process Control 19:1143–1148
Holland J (1975) Adaptation in natural and artificial systems. The University of Michigan Press, Ann Arbor
Hoo KA, Piovoso MJ, Schnelle PD, Rowan DA (2003) Process and controller performance monitoring: overview with industrial applications. Int J Adapt Control Signal Process 17:635–662
Horch A (1999) A simple method for detection of stiction in control valves. Control Eng Pract 7:1221–1231
Horch A (2000) Condition monitoring of control loops. PhD thesis, Royal Institute of Technology, Stockholm, Sweden
Horch A (2007) Benchmarking control loops with oscillations and stiction. In: Ordys AW, Uduehi D, Johnson MA (eds) Process control performance assessment. Springer, Berlin, pp 227–257
Horch A, Isaksson AJ (1998) A method for detection of stiction in control valves. In: Proc IFAC workshop on on-line fault detection and supervision in chemical process industry, Lyon, France
Horch A, Isaksson AJ (1999) A modified index for control performance assessment. J Process Control 9:475–483
Horch A, Stattin A (2002) A complete practical implementation of a method for step response performance assessment. In: Proc control systems, Stockholm, Sweden, pp 348–352
Horton EC, Foley MW, Kwok KE (2003) Performance assessment of level controllers. Int J Adapt Control Signal Process 17:663–684
Howard R, Cooper DJ (2008) Performance assessment of non-self-regulating controllers in a cogeneration power plant. Appl Energy 86:212–219

Howard R, Cooper DJ (2010) A novel pattern-based approach for diagnostic controller performance monitoring. Control Eng Pract 18:279–288
Huang B (1999) Performance assessment of processes with abrupt changes of disturbances. Can J Chem Eng 77:1044–1054
Huang B (2002) Minimum variance control and performance assessment of time variant processes. J Process Control 12:707–719
Huang B (2003) A pragmatic approach towards assessment of control loop performance. Int J Adapt Control Signal Process 17:489–608
Huang CT, Chou CJ (1994) Estimation of the underdamped second-order parameters from the system transient. Ind Eng Chem Process Des Dev 33:174–176
Huang CT, Clements WC (1982) Parameter estimation for the second-orderplus-deadtime-model. Ind Eng Chem Process Des Dev 21:601–603
Huang B, Kadali R (2008) Dynamic modelling, predictive control and performance monitoring. Springer, Berlin
Huang B, Shah SL (1997) Practical issues in multivariable feedback control performance assessment. In: Proc IFAC ADCHEM, Banff, Canada, pp 429–434
Huang B, Shah SL (1998) Practical issues in multivariable feedback control performance assessment. J Process Control 8:421–430
Huang B, Shah SL (1999) Performance assessment of control loops. Springer, Berlin
Huang B, Shah SL, Kwok EK (1997a) Good, bad or optimal? Performance assessment of multivariable processes. Automatica 33:1175–1183
Huang B, Shah SL, Kwok EK, Zurcher J (1997b) Performance assessment of multivariate control loops on a paper-machine headbox. Can J Chem Eng 75:134–142
Huang B, Shah SL, Fujii H (1997c) The unitary interactor matrix and its estimation from closed-loop data. J Process Control 7:195–207
Huang B, Shah SL, Badmus L, Vishnubhotla A (1999) Control performance assessment: an enterprise asset management solution. www.matrikon.com/download/products/lit/processdoctor_pa_eam.pdf
Huang B, Shah SL, Miller R (2000) Feedforward plus feedback controller performance assessment of MIMO systems. IEEE Trans Control Syst Technol 8:580–587
Huang B, Ding SX, Qin J (2005a) Closed-loop subspace identification: an orthogonal projection approach. J Process Control 15:53–66
Huang B, Ding SX, Thornhill N (2005b) Practical solutions to multivariable feedback control performance assessment problem: reduced a priori knowledge of interactor matrices. J Process Control 15:573–583
Huang B, Ding SX, Thornhill N (2006) Alternative solutions to multi-variate control performance assessment problems. J Process Control 16:457–471
Huber PJ, Kleiner B, Gasser T, Dummeruth G (1971) Statistical methods for investigating phase relations in stationary stochastic processes. IEEE Trans Audio Electroacoust 19:78–86
Hugo AJ (1999) Process controller performance monitoring and assessment. www.controlartsinc.com/Support/Articles/PerformanceAssessment.PDF
Hugo AJ (2001) Process controller loop performance assessment. Hydrocarbon Process April:85–90
Hugo AJ (2006) Performance assessment of single-loop industrial controllers. J Process Control 16:785–794
Ingimundarson A (2002) Performance monitoring of PI controllers using a synthetic gradient of a quadratic cost function. In: Proc IFAC world congress, Barcelona, Spain
Ingimundarson A (2003) Dead-time compensation and performance monitoring in process control. PhD thesis, Lund Institute of Technology, Sweden
Ingimundarson A (2006) Performance assessment using a normalized gradient. J Process Control 16:1013–1020
Ingimundarson A, Hägglund T (2005) Closed-loop performance monitoring using loop tuning. J Process Control 15:127–133

ISA Subcommittee SP75.05 (1979) Process instrumentation terminology. Technical report ANSI/ISA-S51.1-1979, Instrument Society of America
Isaksson AJ (1997) A comparison of some approaches to time-delay estimation. PhD thesis, Royal Institute of Technology, Stockholm, Sweden
Isaksson AJ, Horch A, Dumont GA (2000) Event-triggered dead-time estimation—comparison of methods. In: Proc confer control systems, Halifax, Canada, pp 171–178
Isermann R (1971) Required accuracy of mathematical models of linear time invariant controlled elements. Automatica 7:333–341
Isermann R (1992) Identifikation dynamischer systeme I + II. Springer, Berlin
Isermann R, Lachman KH, Matko D (1992) Adaptive control systems. Prentice Hall, New York
Jackson J (1991) A user's guide to principal components. Wiley, New York
Jämsa-Jounela S-L, Poikonen R, Halmevaara K (2002) Evaluation of level control performance. In: Proc IFAC world congress, Barcelona, Spain
Jämsa-Jounela S-L, Poikonen R, Vantaski N, Rantala A (2003) Evaluation of control performance: methods, monitoring tool, and applications in a flotation plant. Miner Eng 16:1069–1074
Jain M, Lakshminarayanan S (2005) A filter-based approach for performance assessment and enhancement of SISO control systems. Ind Eng Chem Res 44:8260–8276
Jelali M (2006a) Regelkreisüberwachung in der Metallindustrie Teil 1: Klassifikation und Beschreibung der Methoden. Automatisierungstechnik 54:36–46
Jelali M (2006b) Regelkreisüberwachung in der Metallindustrie Teil 2: Anwendungskonzept und Fallstudie. Automatisierungstechnik 54:93–99
Jelali M (2006c) An overview of control performance assessment technology and industrial applications. Control Eng Pract 14:441–466
Jelali M (2006d) Performance assessment of control systems in rolling mills—application to strip thickness and flatness control. J Process Control 17:805–816
Jelali M (2007a) Automatisches Reglertuning basierend auf Methoden des Control Performance Monitoring. Automatisierungstechnik 55:10–19
Jelali M (2007b) Performance assessment of control systems in rolling mills—application to strip thickness and flatness control. J Process Control 17:805–816
Jelali M (2008) Estimation of valve stiction in control loops using separable least-squares and global search algorithms. J Process Control 18:632–642
Jelali M, Huang B (eds) (2010) Detection and diagnosis of stiction in control loops: state of the art and advanced methods. Springer, Berlin
Jelali M, Kroll A (2003) Hydraulic servo-systems: modelling, identification and control. Springer, Berlin
Jelali M, Totz O, Börgens R (1998) A new dynamic simulator and an open CACSD environment for rolling mills. Preprints IFAC symp automation in mining, mineral and metal processing, Cologne, Germany, pp 205–209
Jelali M, Müller U, Wolff A, Ungerer W (2001a) Einsatz moderner Regelkonzepte zur besseren Nutzung anlagentechnischer Potentiale. Stahl und Eisen 122(8):35–39
Jelali M, Müller U, Wolff A, Ungerer W (2001b) Advanced control strategies in rolling mills. MPT, Metall Plant Technol Int 3:54–58
Jelali M, Müller U, Wolff A, Ungerer W (2001c) Advanced control strategies for rolling mills. MPT Int 3:54–57
Jelali M, Sonnenschein D, Wolff A, Kothe H, Mintken M (2006) New high performance flatness control system for tandem cold mills. Millenn Steel 2007:177–179
Jelali M, Thormann M, Wolff A, Müller T, Loredo LR, Sanfilippo F, Zangari G, Foerster P (2008) Enhancement of product quality and production system reliability by continuous performance assessment of automation systems (AUTOCHECK). Final report to Contract No RFS-CR-03045, EUR 23205. Office for Official Publications of the European Communities, Luxembourg
Jelali M, Wolff A, Sonnenschein D (2009) Internal Model Control zur Regelung der Bandplanheit in Kaltwalzwerken. Automatisierungstechnik 57:208–216

Jofriet P, Seppala C, Harvey M, Surgenor B, Harris T (1995) An expert system for control loop performance analysis. In: Proc annual meeting, technical section, Canadian pulp and paper association, pp B41–B49
Johansson R (1993) System modelling and identification. Prentice Hall, New York
Julien RH, Foley MW, Cluett WR (2004) Performance assessment using a model predictive control benchmark. J Process Control 14:441–456
Jutten C, Herault J (1991) Blind separation of sources, part I: an adaptive algorithm based on neuromimetic architecture. Signal Process 24:1–10
Kadali R, Huang B (2002a) Estimation of the dynamic matrix and noise model for model predictive control using closed-loop data. Ind Eng Chem Res 41:842–852
Kadali R, Huang B (2002b) Controller performance analysis with LQG benchmark obtained under closed loop conditions. ISA Trans 41:521–537
Kadali R, Huang B (2004) Multivariable controller performance assessment without interactor matrix—a subspace approach. In: Proc IFAC ADCHEM, Hong Kong, pp 591–596
Kaiser G (1994) A friendly guide to wavelets. Birkhäuser, Basel
Kammer LC, Bitmead RR, Barlett PL (1996) Signal-based testing of LQ-optimality of controllers. In: Proc IEEE confer decision control, Kobe, Japan, pp 3620–3624
Kano M, Nagao K, Hasebe S, Hashimoto I, Ohno B, Strauss R, Bakshi B (2000) Comparison of statistical process monitoring methods: application to the Eastman challenge problem. Comput Chem Eng 24:175–181
Kano M, Tanaka S, Ohno H, Hasebe S, Hashimoto I (2002) The use of independent component analysis for multivariate statistical process control. In: Proc internat symp advanced contr ind processes (AdCONIP), Kumamoto, Japan
Kano M, Maruta H, Kugemoto H, Shimizu K (2004) Practical model and detection algorithm for valve stiction. In: Proc IFAC symp DYCOPS, Boston, USA
Kantz H, Schreiber T (1997) Nonlinear time series analysis. Cambridge Univ Press, Cambridge
Kaplan DT (1997) Nonlinearity and nonstationarity: the use of surrogate data in interpreting fluctuations. In: Di Rienzo M, Mancia G, Parati G, Pedotti A, Zanchetti A (eds) Frontiers of blood pressure and heart rate analysis. IOS Press, Amsterdam
Kaplan DT, Glass L (1995) Understanding nonlinear dynamics. Springer, Berlin
Karnopp D (1985) Computer simulation of stick-slip friction in mechanical dynamical systems. J Dyn Syst Meas Control 10:100–103
Kavuri SN, Venkatasubramanian V (1994) Neural network decomposition strategies for large-scale fault diagnosis. Int J Control 59:767–792
Keck R, Neuschütz E (1980) German system brings accuracy to flatness measurement. Iron Steel Int 53:215–220
Kendra S, Çinar A (1997) Controller performance assessment by frequency domain techniques. J Process Control 7:181–194
Kim YC, Powers EJ (1979) Digital bispectral analysis and its applications to nonlinear wave interactions. IEEE Trans Plasma Sci PS-7:120–131
Kinney T (2003) Performance monitor raises service factor of MPC. In: Proc ISA, Houston, USA
Ko B-S, Edgar TF (1998) Assessment of achievable PI control performance for linear processes with dead time. In: Proc Amer control confer, Philadelphia, USA
Ko B-S, Edgar TF (2000) Performance assessment of cascade control loops. AIChE J 46:281–291
Ko B-S, Edgar TF (2001a) Performance assessment of constrained model predictive control systems. AIChE J 47:1363–1371
Ko B-S, Edgar TF (2001b) Performance assessment of multivariable feedback control systems. Automatica 37:899–905
Ko B-S, Edgar TF (2004) PID control performance assessment: the single-loop case. AIChE J 50:1211–1218
Ko H-C, Park YS, Vishnubhotla A, Mitchell W (2004) Locating the cause of poor control performance, and obtaining operational excellence at one of the world's largest refiners. In: Proc

NPRA decision support confer, San Antoio, USA
Konda NVSNM, Rangaiah GP (2007) Performance assessment of plantwide control systems of industrial processes. Ind Eng Chem Res 46:1220–1231
Kouvaritakis B, Cannon M (eds) (2001) Nonlinear predictive control: theory and practice. The Institute of Electrical Engineers
Kozub DJ (1996) Controller performance monitoring and diagnosis: experiences and challenges. In: Proc chemical process control confer, Lake Tahoe, USA, pp 83–96
Kozub DJ (2002) Controller performance monitoring and diagnosis. Industrial perspective. In: Proc IFAC world congress, Barcelona, Spain
Kozub DJ, Garcia C (1993) Monitoring and diagnosis of automated controllers in the chemical process industries. In: Proc AIChE, St Louis, USA
Krishnaswamy PR, Mary Chan BE, Rangaiah GP (1987) Closed-loop tuning of process control systems. Chem Eng Sci 42:2173–2182
Kuehl P, Horch A (2005) Detection of sluggish control loops—experiences and improvements. Control Eng Pract 13:1019–1025
Kucera V (1979) Discrete linear control: the polynomial equations approach. Wiley, New York
Kwakernaak H, Sivan R (1972) Linear optimal control systems. Wiley, New York
Kwakernaak H, Sebek R (2000) Ploynomial toolbox. www.polyx.com
Lackinger C, Nettelbeck H-J, Oemkes H (2002) Die neue Tandemstraße von ThyssenKrupp Stahl im Kaltwalzwerk Beeckerwerth. Stahl und Eisen 122(2):25–32
Lee JH, Cooley B (1997) Recent advances in model predictive control and other related areas. In: Proc chemical process control AICHE and CACHE, pp 201—216b
Lee TH, Wang QG, Tan KK (1996) A robust Smith-predictor controller for uncertain delay systems. AIChE J 42:1033–1173
Lee K-H, Huang B, Tamayo EC (2008) Sensitivity analysis for selective constraint and variability tuning in performance assessment of industrial MPC. Control Eng Pract 16:1195–1215
Lee K-H, Tamayo EC, Huang B (2010) Industrial implementation of controller performance analysis technology. Control Eng Pract 18:147–158
Leva A, Cox C, Ruano A (2001) Hands-on PID autotuning: a guide to better utilisation. IFAC Professional Brief. www.ifac-control.org
Lewis RB, Torczon V (1999) Pattern search algorithms for bound constrained minimization. SIAM J Optim 9:1082–1099
Lewis RB, Torczon V (2000) Pattern search methods for linearly constrained minimization. SIAM J Optim 10:917–941
Li Q, Whiteley JR, Rhinehart RR (2003) A relative performance monitor for process controllers. Int J Adapt Control Signal Process 17:685–708
Li Q, Whiteley JR, Rhinehart RR (2004) An automated performance monitor for process controllers. Control Eng Pract 12:537–553
Litrico X, Georges D (1999) Robust continuous-time and discrete-time flow control of a dam–river system. (II) Controller design. Appl Math Model 23:829–846
Liu H, Shah S, Jiang W (2004) On-line outlier detection and data cleaning. Comput Chem Eng 28:1635–1647
Ljung L (1993) Perspectives on the process of identification. In: Proc IFAC world congress, Sydney, Australia, vol 5, pp 197–205
Ljung L (1999) System identification: theory for the user. Prentice Hall, New York
Ljung L, Söderström T (1987) Theory and practice of recursive identification. MIT Press, Cambridge
Loquasto F III, Seborg DE (2003) Model predictive controller monitoring based on pattern classification and PCA. In: Proc Amer control confer, Denver, Colorado, USA
Lunze J (2007) Automatisierungstechnik. Oldenbourg, Munich
Lunze J (2008) Regelungstechnik 2: Mehrgrößensysteme, Digitale Regelung. Springer, Berlin
Lynch C, Dumont GA (1996) Control loop performance monitoring. IEEE Trans Control Syst Technol 18:151–192

Lyons AR, Newton TJ, Goddard NJ, Parsons AT (1995) Can passive sonar signals be classified on the basis of their higher-order statistics? In: Proc IEE colloquium on higher order statistics in signal processing "Are they of any use?", London, pp 6/1–6/6

MacGregor JF (1977) Discrete stochastic control with input constraints. IEE Proc, Control Theory Appl 124:732–734

MacGregor JF, Fogal DT (1995) Closed-loop identification: the role of the noise model and pre-filters. J Process Control 5:163–171

Maciejowski JM (2001) Predictive control with constraints. Prentice Hall, New York

Majecki P, Grimble MJ (2004a) Controller performance design and assessment using nonlinear generalized minimum variance benchmark: scalar case. In: Proc control 2004, University of Bath, UK, ID-232

Majecki P, Grimble MJ (2004b) GMV and restricted-structure GMV controller performance assessment—multivariable case. In: Proc Amer control confer, Boston, USA, vol 1, pp 697–702

Manum H (2006) Analysis of techniques for automatic detection and quantification of stiction in control loops. Diploma thesis, Norwegian University of Science and Technology

Manum H, Scali C (2006) Closed loop performance monitoring: automatic diagnosis of valve stiction by means of a technique based on shape analysis formalism. In: Proc internat congress on methodologies for emerging technologies in automation (ANIPLA), Rome, Italy

Markworth M, Polzer J, Ungerer W (2003) Höhere Qualität von Walzprodukten durch komplexe Überwachung. In: Proc VDI-Schwingungstagung "Schwingungsüberwachung und diagnose von Maschinen und Anlagen", Magdeburg, Germany

Marple SL (1987) Digital spectral analysis. Prentice Hall, New York

Matsuo T, Tadakuma I, Thornhill NF (2004) Diagnosis of a unit-wide disturbance caused by saturation in manipulated variable. In: Proc IEEE advanced process control applications for industry workshop, Vancouver, Canada

Mayne DQ (1997) Nonlinear model predictive control: an assessment. In: Proc chemical process control AICHE and CACHE, pp 217–231

Mayne DQ, Rawlings JB, Rao CV, Scokaert POM (2000) Constrained model predictive control: stability and optimality. Automatica 26:789–814

McIntosh AR, Shah SL, Fisher DG (1991) Analysis and tuning of adaptive generalized predictive control. Can J Chem Eng 69:97–110

McNabb CA, Qin SJ (2003) Projection based MIMO control performance monitoring: I—covariance monitoring in state space. J Process Control 13:739–757

McNabb CA, Qin SJ (2005) Projection based MIMO control performance monitoring: II—measured disturbances and setpoint changes. J Process Control 15:89–102

McMillan GK (1995) Improve control valve response. Chem Eng Prog: Meas Control June:77–84

Miao T, Seborg DE (1999) Automatic detection of excessively oscillatory feedback control loops. In: Proc IEEE confer control applications. Kohala Coast-Island, USA

Michalewicz Z (1994) Genetic algorithms + data structures = evolution programs. Springer, Berlin

Miletic S, Quinn S, Dudzic M, Vaculik M, Champagne M (2004) An industrial perspective on the implementing on-line applications of multivariate statistics. J Process Control 14:821–836

Miller RM, Timmons CF, Desborough LD (1998) CITGO's experience with controller performance monitoring. In: Proc NPRA computer confer, San Antonio, USA

Molle JD, Hinich M (1995) Trispectral analysis of stationary time series. J Acoust Soc Am 97:2963–2978

Montgomery DC, Runger GC (1992) Applied statistics and probability for engineers. Wiley, New York

Morari M, Lee JH (1991) Model predictive control: the good, the bad, and the ugly. In: Proc chemical process control, pp 271–296

Morari M, Zafiriou E (1989) Robust process control. Prentice Hall, New York

Moudgalya KM (2007) Digital control. Wiley, New York

Moudgalya KM, Shah SL (2004) A polynomial based first course in digital control. In: Proc IEEE intern symp computer aided control systems design, Taipei, Taiwan, pp 190–195

Mücke G, Gorgels F (2007) Flatness measurement for high quality cold strip production. MPT Int 1:70–75

Müller T (2005) Regelung Glühofen VZA1. Internal report, EKO Stahl

National Instruments Corporation (2004) LabVIEW^{TM} system identification toolkit user manual

Nelles O (2001) Nonlinear system identification: from classical approaches to neural networks and fuzzy models. Springer, Berlin

Nikias CL (1988) ARMA bispectrum approach to nonminimum phase system identification. IEEE Trans Acoust Speech Signal Process 4:513–525

Nikias CL, Mendel JM (1993) Signal processing with higher-order spectra. IEEE Signal Process Mag 10:10–37

Nikias CL, Petropulu AP (1993) Higher-order spectra analysis: a non-linear signal processing framework. Prentice Hall, New York

O'Dwyer A (1996) The estimation and compensation of processes with time delays. PhD thesis, Dublin City University, Scotland

O'Dwyer A (2003) Handbook of PI and PID controller tuning rules. Imperial College Press, London

Ogawa S (1998) A data analysis and graphical representation system for control loop performance assessment. In: Proc TAPPI process control confer, Vancouver, Canada

Olaleye F, Huang B, Tamayo E (2004a) Performance assessment of control loops with time varying disturbance dynamics. J Process Control 14:867–877

Olaleye F, Huang B, Tamayo E (2004b) Feedforward and feedback controller performance assessment of linear time-variant processes. Ind Eng Chem Res 43:589–596

Olaleye F, Huang B, Tamayo E (2004c) Industrial applications of feedback controller performance assessment of time-variant processes. Ind Eng Chem Res 43:597–607

Olsson H (1996) Control systems with friction. PhD thesis, Lund Institute of Technology, Sweden

Olsson H, Åström KJ, de Wit CC, Gafvert M, Lischinsky P (1998) Friction models and friction compensation. Eur J Control 4:176–195

Ooi SK, Weyer E (2008) Detection of oscillatory control loops in irrigation channels. In: Proc IFAC world congress, Seoul, pp 5501–5505

Oppenheim AV, Schafer RW (1989) Discrete-time signal processing. Prentice Hall, New York

Ordys AW, Uduehi D, Johnson MA (eds) (2007) Process control performance assessment: from theory to implementation. Springer, Berlin

Owen J, Read D, Blekkenhorst H, Roche AA (1996) A mill prototype for automatic monitoring of control loop performance. In: Proc control systems, Halifax, Canada, pp 171–178

PAM (2001) Performance assessment and benchmarking of controls. Research Project No IST-2000-29239, European Community, FP5, 2001–2004. www.isc-ltd.com/benchmark

Panteley E, Ortega R, Gäfvert M (1998) An adaptive friction compensator for global tracking in robot manipulators. Syst Control Lett 33:307–313

Paluš M (1995) Testing for nonlinearity using redundancies: quantitative and qualitative aspects. Physica D 80:186–205

Paplinski A, Rogozinski M (1990) Right nilpotent interactor matrix and its application to multivariable stochastic control. In: Proc Amer control confer, San Diego, USA, vol 1, pp 494–495

Papoulis A (1984) Probability, random variables and stochastic processes. McGraw Hill, New York

Patwardhan RS (1999) Studies in synthesis and analysis of model predictive controllers. PhD thesis, University of Alberta, Canada

Patwardhan RS, Shah SL (2002) Issues in performance diagnostics of model-based controllers. J Process Control 12:413–427

Patwardhan RS, Shah S, Emoto G, Fujii H (1998) Performance analysis of model-based predictive controllers: an industrial study. In: Proc AIChE, Miami, USA

Paulonis MA, Cox JW (2003) A practical approach for large-scale controller performance assessment, diagnosis, and improvement. J Process Control 13:155–168

Peng Y, Kinnaert M (1992) Explicit solution to the singular lq regulation problem. IEEE Trans Autom Control 37:633–636
Perarson RK (2002) Outliers in process modeling and identification. IEEE Trans Control Syst Technol 10:55–63
Perrier M, Roche AA (1992) Towards mill-wide evaluation of control loop performance. In: Proc control systems, Whistler, Canada, pp 205–209
Piipponen J (1996) Controlling processes with nonideal valves: tuning of loops and selection of valves. In: Preprints of control systems, Halifax, Nova Scotia, Canada, pp 179–186
Polzer J, Markworth M, Ungerer W (2003) New developments in monitoring & diagnosis for rolling mills. In: Proc METEC congress (Europ rolling confer), Düsseldorf, Germany, pp 81–90
Press WH, Flannery BP, Teukolsky SA, Vetterling WT (1986) Numerical recipes. Cambridge University Press, Cambridge
Prett DM, García CE (1988) Fundamental process control. Butterworths, Stonehan
Qin SJ (1998) Control performance monitoring—a review and assessment. Comput Chem Eng 23:173–186
Qin SJ (2006) An overview of subspace identification. Comput Chem Eng 30:1502–1513
Qin SJ, Badgwell TA (1997) An overview of industrial model predictive control technology. In: Kantor JC, García CE, Carnaham B (eds) Int confer chemical process control. AIChE symposium series, vol 93, pp 232–256
Qin SJ, Badgwell TA (2003) A survey of industrial model predictive control technology. Control Eng Pract 11:733–764
Qin SJ, Ljung L, Wang J (2002) Subspace identification methods using parsimonious model formulation. In: Proc AIChE, Indianapolis, USA
Rakar A, Zorzut S, Jovan V (2004) Assessment of production performance by means of KPI. In: Proc control 2004, University of Bath, UK, ID-073
Ramachandran R, Rangaiah GP, Lakshminarayanan S (2007) Data analysis, modeling and control performance enhancement of an industrial fluid catalytic cracking unit. Chem Eng Sci 62:1958–1973
Rangaiah GP, Krishnaswamy PR (1994) Estimating second-order plus dead time model parameters. Ind Eng Chem Res 33:1867–1871
Rangaiah GP, Krishnaswamy PR (1996) Estimating second-order dead time parameters from underdamped process transients. Chem Eng Sci 51:1149–1155
Rath G (2000) Model based thickness control of the cold strip process. Diss, University of Leoben, Austria
Ratjen H (2006) Entwicklung und Untersuchung von Verfahren zur Bewertung der Regelgüte bei Regelkreisen für MIMO-Systeme. Internal Tech report, University of Cologne/Germany, Subcontractor of BFI within the EU Project AUTOCHECK
Ratjen H, Jelali M (2006) Performance monitoring for feedback and feedforward control with application to strip thickness control. In: Proc research and education in mechatronics, KTH, Stockholm, Sweden
Rawlings JB, Muske KR (1993) The stability of constrained receding horizon control. IEEE Trans Autom Control 38:1512–1516
Rawlings JB, Meadows ES, Muske K (1994) Nonlinear model predictive control: a tutorial and survey. In: Proc IFAC ADCHEM, Japan
Rengaswany R, Venkatasubramanian V (1995) A syntactic pattern-recognition approach for process monitoring and fault diagnosis. Eng Appl Artif Intell 8:35–51
Richalet J, Rault A, Testud JL, Papon J (1978) Model predictive heuristic control: applications to an industrial process. Automatica 14:413–428
Ricker NL (1991) Model predictive control: state of the art. In: Proc chemical process control, pp 419–444
Rigler GW, Aberl HR, Staufer W, Aistleitner K, Weinberger KH (1996) Improved rolling mill automation by means of advanced control techniques and dynamic simulation. IEEE Trans Ind Appl 32:599–607

Rhinehart R (1995) A watchdog for controller performance monitoring. In: Proc Amer control confer, Seattle, USA, pp 2239–2240

Rhodes C, Morari M (1997) The false nearest neighbors algorithm: an overview. Comput Chem Eng 21:S1149–S1154

Ringwood JV (2000) Shape control systems for Sendzimir steel mills. IEEE Trans Control Syst Technol 8:70–86

Rivera DE, Morari M, Skogestad S (1986) Internal model control. 4. PID controller design. Ind Eng Chem Process Des Dev 25:252–265

Roberts WL (1978) Cold rolling of steel. Marcel Dekker, New York

Rogozinski M, Paplinski A, Gibbard M (1987) An algorithm for calculation of nilpotent interactor matrix for linear multivariable systems. IEEE Trans Autom Control 32:234–237

Rosenblatt M, Van Ness JW (1965) Estimation of the bispectrum. Ann Math Stat 65:420–436

Rossi R, Scali C (2005) A comparison of techniques for automatic detection of stiction: simulation and application to industrial data. J Process Control 15:505–514

Rousseeuw PJ, Leroy AM (1987) Robust regression and outlier detection. Wiley, New York

Ruel M (2000) Stiction: the hidden menace. Control Mag 13:69–75. http://www.expertune.com/articles/RuelNov2000/stiction.html

Ruel M (2002) Learn how to assess and improve control loop performance. In: Proc ISA, Chicago, USA

Ruel M (2003) The conductor directs this orchestra. Intech, November:20–22

Rugh WJ (1991) Analytical framework for gain scheduling. IEEE Control Syst Mag 11:79–84

Ruiz Ordóñez ML (2008) Multivariate statistical process control and case-based reasoning for situation assessment of sequencing batch reactors. Doctoral thesis, University of Girona, Spain

Salsbury TI (2005) A practical method for assessing the performance of control loops subject to random load changes. J Process Control 15:393–405

Salsbury TI (2006) Control performance assessment for building automation systems. In: IFAC workshop on energy saving control in plants and buildings, Bulgaria

Schäfer J, Çinar A (2002) Multivariable MPC performance assessment, monitoring and diagnosis. In: Proc IFAC world congress, Barcelona, Spain

Seborg J, Viberg M (1997) Separable non-linear least-squares minimization—possible improvements for neural fitting. In: Proc IEEE workshop on neural nets for sig pro, vol 7, pp 345–354

Seborg DE, Edgar TF, Mellichamp DA (2004) Process dynamics and control. Wiley, New York

Sendzimir M (1993) Shape control in cluster mills. www.sendzimir.com

Seppala CT, Harris TJ, Bacon DW (2002) Time series methods for dynamic analysis of multiple controlled variables. J Process Control 12:257–276

Shah SL, Mohtadi C, Clarke D (1987) Multivariable adaptive control without a priori knowledge of the delay matrix. Syst Control Lett 9:295–306

Shah SL, Patwardhan R, Huang B (2001) Multivariate controller performance analysis: methods, applications and challenges. In: Proc chemical process control confer, Tucson, USA, pp 187–219

Shah SL, Mitchell W, Shook D (2005) Challenges in the detection, diagnosis and visualization of controller performance data. In: IEE seminar on control loop assessment and diagnosis, University College, London, UK

Shinskey FG (1990) How good are our controllers in absolute performance and robustness? Meas Control 23:114–121

Shinskey FG (1996) Process-control systems: application, design, and tuning. McGraw Hill, New York

Shunta JP (1995) Achieving world class manufacturing through process control. Prentice Hall, New York

Small M, Tse CK (2002) Applying the method of surrogate data to cyclic time series. Physica D 164:187–201

Smith OJM (1957) Closed control of loops with dead time. Chem Eng Prog 53:217–219

Singhal A, Salsbury TI (2005) A simple method for detecting valve stiction in oscillating control loops. J Process Control 15:371–382

Skogestad S, Postlethwaite I (1996) Multivariable feedback control: analysis and design. Wiley, New York

Söderström T, Stoica P (1989) System identification. Prentice Hall, New York

Söderström T, Gustavsson I, Ljung L (1975) Identifiability conditions for linear systems operating in closed-loop. Int J Control 21:243–255

Soeterboeck R (1992) Predictive control: a unified approach. Prentice Hall, New York

Spencer MA, Elliot RM (1997/1998) Improving instrumentation and control systems performance. Pet Technol Q Winter:93–97

Srinivasan R, Rengaswany R, Miller R (2005a) Control loop performance assessment. 1. A qualitative approach for stiction diagnosis. Ind Eng Chem Res 44:6708–6718

Srinivasan R, Rengaswany R, Narasimhan S, Miller R (2005b) Control loop performance assessment. 2. Hammerstein model approach for stiction diagnosis. Ind Eng Chem Res 44:6719–6728

Stam CJ, Pijn JPM, Pritchard WS (1998) Reliable detection of nonlinearity in experimental time series with strong periodic components. Physica D 112:361–380

Stanfelj N, Marlin TE, MacGregor JF (1993) Monitoring and diagnosis of process control performance: the single-loop case. Ind Eng Chem Res 67:856–861

Stenman A, Gustafsson F, Forsman K (2003) A segmentation-based method for detection of stiction in control valves. Int J Adapt Control Signal Process 17:625–634

Strejc V (1959) Approximation aperiodischer Übertragungscharakteristiken. Regelungstechnik 7:124–128

Stribeck R (1902) Die wesentlichen Eigenschaften der Gleit- und Rollenlager. Z Ver Dtsch Ing XXXXVI:1341–1348

Subba Rao TS, Gabr MM (1980) A test for linearity and stationarity of time series. J Time Ser Anal 1:145–158

Sugihara G, May RM (1990) Nonlinear forecasting as a way of distinguishing chaos from measurement error in time series. Nature 344:734–741

Sunan H, Kiong TK, Heng LT (2002) Applied predictive control. Springer, Berlin

Swanda A, Seborg DE (1997) Evaluating the performance of PID-type feedback control loops using normalized settling time. In: Proc IFAC ADCHEM, Banff, Canada, pp 301–306

Swanda A, Seborg DE (1999) Controller performance assessment based on setpoint response data. In: Proc Amer control confer, San Diego, USA, pp 3863–3867

Teo TM, Lakshminarayanan S, Rangaiah GP (2005) Performance assessment of cascade control systems. J Inst Eng, Singapore 45(6):27–38

Theiler J, Eubank S, Longtin A, Galdrikian B, Farmer B, Farmer JD (1992) Testing for nonlinearity in time-series—the method of surrogate data. Physica D 58:77–94

Thornhill NF (2005) Finding the source of nonlinearity in a process with plant-wide oscillation. IEEE Trans Control Syst Technol 13:434–443

Thornhill NF (2007) Locating the source of a disturbance. In: Ordys AW, Uduehi D, Johnson MA (eds) Process control performance assessment. Springer, Berlin, pp 199–225

Thornhill NF, Hägglund T (1997) Detection and diagnosis of oscillation in control loops. Control Eng Pract 5:1343–1354

Thornhill NF, Horch A (2007) Advances and new directions in plant-wide disturbance detection and diagnosis. Control Eng Pract 15:1196–1206

Thornhill NF, Oettinger M, Fedenczuk MS (1999) Refinery-wide control loop performance assessment. J Process Control 9:109–124

Thornhill NF, Shah SL, Huang B (2001) Detection of distributed oscillations and root-cause diagnosis. In: Proc CHEMFAS, Chejudo Island, Korea, pp 167–172

Thornhill NF, Shah SL, Huang B, Vishnubholta A (2002) Spectral principal component analysis of dynamic process data. Control Eng Pract 10:833–846

Thornhill NF, Cox J, Paulonis M (2003a) Diagnosis of plant-wide oscillation through data-driven analysis and process understanding. Control Eng Pract 11:1481–1490

Thornhill NF, Huang B, Shah SL (2003b) Controller performance assessment in set point tracking and regulatory control. Int J Adapt Control Signal Process 17:709–727

Thornhill NF, Huang B, Zhang H (2003c) Detection of multiple oscillations in control loops. J Process Control 13:91–100

Thornhill NF, Choudhury MAAS, Shah SL (2004) The impact of compression on data driven process analyses. J Process Control 14:389–398

Thyssen J, Nielsen H, Hansen SD (1995) Nonlinearities in speech. In: Proc IEEE workshop nonlinear signal image processing, Halkidiki, Greece, pp 662–665

Timmer J, Schwarz U, Voss HU, Wardinski I, Belloni T, Hasinger G, van der Klis M, Kurths J (2000) Linear and nonlinear time series analysis of the black hole candidate Cygnus X-l. Phys Rev E 61:1342–1352

Toledo E (2002) Linear and nonlinear characteristics of the human ECG as markers for cardiovascular functioning. PhD thesis, Tel Aviv University

Torres BS, de Carvalho FB de Oliveira Fonseca M, Filho CS (2006) Performance assessment of control loops—cases studies. In: Proc IFAC ADCHEM, Gramado, Brasil

Tsiligiannis C, Svoronos S (1989) Dynamic interactors in multivariable process control. Chem Eng Sci 44:2041–2047

Tugnait JK (1987) Identification of linear stochastic systems via second- and fourth-order cumulant matching. IEEE Trans Inf Theory 33:393–407

Tyler M, Morari M (1995) Performance assessment for unstable and nonminimum-phase systems. In: Preprints IFAC workshop on-line fault detection supervision chemical process industries, Newcastle upon Tyne, UK

Tyler M, Morari M (1996) Performance monitoring of control systems using likelihood methods. Automatica 32:1145–1162

Uduehi D, Ordys A, Grimble MJ, Majecki P, Xia H (2007a) Controller benchmarking procedures—data-driven methods. In: Ordys AW, Uduehi D, Johnson MA (eds) Process control performance assessment. Springer, Berlin, pp 81–126

Uduehi D, Ordys A, Grimble MJ, Majecki P, Xia H (2007b) Controller benchmarking procedures—model-based methods. In: Ordys AW, Uduehi D, Johnson MA (eds) Process control performance assessment. Springer, Berlin, pp 127–168

Uduehi D, Ordys A, Xia H, Bennauer M, Zimmer G, Corsi S (2007c) Controller benchmarking algorithms: some technical issues. In: Ordys AW, Uduehi D, Johnson MA (eds) Process control performance assessment. Springer, Berlin, pp 259–294

Van Overschee P, De Moor B (1996) Subspace identification of linear systems: theory, implementation, applications. Kluwer, Dordrecht

Van den Hof PMJ, Schrama RJP (1995) Identification and control—closed-loop issues. Automatica 31:1751–1770

Van den Hof PMJ, Heuberger PSC, Bokor J (1995) System identification with generalized orthonormal basis functions. Automatica 31:1821–1834

Vatanski N, Jämsä-Jounela S-L, Rantala A, Harju T (2005) Control loop performance measures in the evaluation of process economics. In: Proc IFCA world congress, Prague

Vaught R, Tippet J (2001) Control performance monitoring: shaman or saviour. Pulp & Paper Canada 102:26–29

Venkatasubramanian V, Vaidyanathan R, Yamamoto Y (1990) Process fault detection and diagnosis using neural networks—I. Steady-state processes. Comput Chem Eng 14:699–712

Venkataramanan G, Shukla V, Saini R, Rhinehart RR (1997) An automated on-line monitor of control system performance. In: Proc Amer control confer, Albuquerque, New Mexico, USA, pp 1355–1359

Veronesi M, Visioli A (2010) Performance assessment and retuning of PID controllers for integral processes. J Process Control 20:261–269

Vishnubhotla A, Shah SL, Huang B (1997) Feedback and feedforward performance analysis of the shell industrial closed-loop data set. In: Proc IFAC ADCHEM, Banff, Canada, pp 295–300

Visioli A (2005) Assessment of tuning of PI controllers for self-regulating processes. In: Proc IFAC world congress, 2005, Prague

Visioli A (2006) Practical PID control. Springer, Berlin

Wahlberg B (1991) System identification using Laguerre models. IEEE Trans Autom Control 36:551–562
Wahlberg B (1994) System identification using Kautz models. IEEE Trans Autom Control 39:1276–1282
Wahlberg B, Hannan EJ (1993) Parametric signal modelling using Laguerre filters. Ann Appl Probab 3:476–496
Wang L, Cluett WR (2000) From plant data to process control. Taylor & Francis, London
Wang J, Qin SJ (2002) A new subspace identification approach based on principal component analysis. J Process Control 12:841–855
Wire B, Gallagher N (1996) The process chemometrics approach to process monitoring and fault detection. J Process Control 6:328–348
Wolff A, Jelali M, Sonnenschein D, Kothe H, Mintken M (2006) New flatness control system at the tandem cold mill of EKO Stahl. In: Internat ATS steelmaking confer, Paris, pp 58–59
Wolovich W, Falb P (1976) Invariants and canonical forms under dynamic compensation. SIAM J Control 14:996–1008
Xia C, Howell J (2003) Loop status monitoring and fault localization. J Process Control 13:679–691
Xia C, Howell J (2005) Isolating multiple sources of plant-wide oscillations via independent component analysis. Control Eng Pract 13:1027–1035
Xia C, Howell J, Zheng J (2005) Commissioning-stage poor performance of control loops in process plants. In: Proc IEEE confer control applications, Toronto, Canada, pp 1461–1466
Xia H, Majecki P, Ordys A, Grimble MJ (2003) Controller benchmarking based on economic benefits. In: Proc Europ control confer, Cambridge, UK, pp 2393–2398
Xia H, Majecki P, Ordys A, Grimble MJ (2006) Performance assessment of MIMO systems based on I/O delay information. J Process Control 16:373–383
Xu F, Huang B, Tamayo EC (2006a) Assessment of economic performance of model predictive control through variance/constraint tuning. In: Proc IFAC ADCHEM, Gramado, Brazil, pp 899–904
Xu F, Lee K-H, Huang B (2006b) Monitoring control performance via structured closed-loop response subject to output variance/covariance upper bound. J Process Control 16:971–984
Xu F, Huang B, Akande S (2007) Performance assessment of model predictive control for variability and constraint tuning. Ind Eng Chem Res 46:1208–1219
Xu F, Huang B, Tamayo EC (2008) Performance assessment of MIMO control systems with time-variant disturbance dynamics. Comput Chem Eng 32:2144–2154
Yamashita Y (2006) An automatic method for detection of valve stiction in process control loops. Control Eng Pract 14:503–510
Youla DC, Bongiorno JJ, Jabr HA (1976a) Modern Wiener–Hopf design of optimal controllers—part I: the single-input-output case. IEEE Trans Autom Control 21:3–13
Youla DC, Bongiorno JJ, Jabr HA (1976b) Modern Wiener–Hopf design of optimal controllers—part II: the multivariable case. IEEE Trans Autom Control 21:319–338
Yu J, Qin SJ (2008a) Statistical MIMO controller performance monitoring. Part I: data-driven covariance benchmark. J Process Control 18:277–296
Yu J, Qin SJ (2008b) Statistical MIMO controller performance monitoring, part II: performance diagnosis. J Process Control 18:297–319
Yuan J (1999) Testing linearity for stationary time series using the sample interquartile range. J Time Ser Anal 21:713–722
Yuan Q, Lennox B (2009) Control performance assessment for multivariable systems based on a modified relative variance technique. J Process Control 19:489–497
Yuan Q, Lennox B, McEwan M (2009) Analysis of multivariable control performance assessment techniques. J Process Control 19:751–760
Yuwana M, Seborg DE (1982) A new method for online controller tuning. AIChE J 28:434–440
Zang X, Howell J (2004) Comparison of methods to identify the root cause of plant-wide oscillations. In: Proc IEEE confer control applications, Taipei, Taiwan, pp 695–700

Zervos CC, Dumont GA (1988) Deterministic adaptive control based on Laguerre series representation. Int J Control 48:2333–2359

Zhang Y, Henson MA (1999) A performance measure for constrained model predictive controllers. In: Proc Europ control confer, Karlsruhe, Germany

Zhao C, Zhao Y, Su H, Huang H (2009) Economic performance assessment of advanced process control with LQG benchmarking. J Process Control 19:557–569

Zhao Y, Chu J, Su H, Huang B (2010) Multi-step prediction error approach for controller performance monitoring. Control Eng Pract 18:1–12

Ziegler JG, Nichols NB (1942) Optimum settings for automatic controllers. Trans Am Soc Mech Eng 64:759–768

Ziegler JG, Nichols NB (1943) Process lags in automatic control circuits. Trans Am Soc Mech Eng 65:433–444

Index

A
Area index (AI), 139, 363
Auto-correlation test, 36

C
Case studies
  annealing temperature control, 427
  strip flatness control, 424
  strip gauge/thickness control, 419
Closed-loop identification, 179, 187
Control performance index, 13, 40, 44, 45, 57, 66, 72, 73, 75, 76, 101, 159, 161, 163, 164
CPM algorithms
  applications, 391
  demands, 389
  FCOR, 44
  packages, 394
  procedure, 125
  procedures, 41, 46, 51, 75, 92, 94, 105, 160, 165, 184, 201
  tools, 435

D
Diophantine equation, 33, 34, 39, 56, 156, 160

E
Extended horizon performance index (EHPI), 73

H
Harris index, 38, 43

I
Identification methods
  AR(MA) modelling, 202
  comparison, 200
  data length, 173
  FCOR, 204
  identifiability, 187, 188
  Laguerre networks, 204
  multivariable, 205
  sampling period, 172
  subspace, 90, 203
Idle index, 133, 363
Impulse response
  classification, 368
  features, 370
Interactor matrix, 147
  computation procedure, 150
  diagonal, 148
  estimation from closed-loop data, 153
  recursive determination, 149
  simple, 148
  unitary, 148, 149

M
Minimum variance
  cascade control, 55
  control, 32
  feedback/feedforward control, 49
  index, 38
Model fitting
  FOPTD, 129, 193
  SOPTD, 129, 193
Model predictive control (MPC)
  infinite horizon, 102
  performance assessment, 96
  performance assessment procedure, 105
  performance index, 101
  potential improvement, 104
  principle, 97
  tuning, 100

M. Jelali, *Control Performance Management in Industrial Automation*, Advances in Industrial Control, DOI 10.1007/978-1-4471-4546-2,

N

Non-Gaussianity index (NGI), 242, 243, 246
Non-linear predictability index (NPI), 250
Non-linearity index (NLI), 243, 246

O

Oscillation detection
  decay ratio of auto-covariance function, 230
  peaks in power spectrum, 222
  procedure, 226, 231, 321
  regularity integral of absolute errors and zero crossings, 228
  regularity of integral of absolute error, 224
  regularity of zero crossings of auto-covariance function, 233
Oscillations
  characterisation, 221
  intermittent, 237
  plant-wide, 261
  sources, 221
Output index, 363

P

Performance assessment/benchmarking
  deterministic, 121
  economic, 110
  extended horizon, 72
  GMVC, 81
  Horch–Isaksson, 75
  IMC-achievable, 66
  LQG, 92, 116
  minimum-variance-based, 29
  MPC, 101, 105
  PID-achievable, 348
  procedure, 348
  RS-LQG, 356
  user-specified, 65
Pre-whitening, 41
Prediction models
  ARMAX, 29
  AR(X), 180, 305
  FIR, 190
  Laguerre networks, 182
  model-order selection, 199
  orthornormal basis functions, 182
Preprocessing
  compression, 177
  detrending, 175
  end-matching, 254
  filtering, 134, 229, 235, 292
  quantisation, 136
  removing of outliers, 175
  scaling, 176
  smoothing, 176

R

Re-tuning
  filter-based, 358
  impulse response, 366
  iterative, 362
  load-disturbance changes, 362
  methods comparison, 384
  optimisation-based, 347, 357
  pattern recognition, 374
  set-point response, 357
  variation of controller parameters, 380
Relative performance index (RPI), 76

S

Statistical process control, 209
  control chart, 209
  cumulative sum, 210
  exponentially-weighted moving-average, 210
  partial least squares, 214
  principal component analysis, 212
Stiction detection
  cross-correlation, 281, 283
  curve fitting, 286
  gain-change test, 295
  multiple faults, 321
  non-linearity and ellipse fitting, 290
  procedure, 297, 321
  relay technique, 288
  shape analysis, 277, 279
  tools, 221
  valve-travel test, 296
Stiction model
  data-based, 272
  describing function, 272
  general, 303
  physically-based, 271
Surrogate data analysis, 248

T

Time-delay estimation (TDE), 196, 198, 311
Total non-linearity index (TNLI), 247

MIX
Papier aus verantwortungsvollen Quellen
Paper from responsible sources
FSC® C105338

If you have any concerns about our products,
you can contact us on
ProductSafety@springernature.com

In case Publisher is established outside the EU,
the EU authorized representative is:
Springer Nature Customer Service Center GmbH
Europaplatz 3, 69115 Heidelberg, Germany

Printed by Libri Plureos GmbH
in Hamburg, Germany